Geology of Coal Fires: Case Studies from Around the World

Edited by

Glenn B. Stracher
Division of Natural Science and Mathematics
East Georgia College
University System of Georgia
131 East College Circle
Swainsboro, Georgia 30401
USA

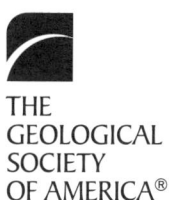

THE
GEOLOGICAL
SOCIETY
OF AMERICA®

Reviews in Engineering Geology XVIII

3300 Penrose Place, P.O. Box 9140 • Boulder, Colorado 80301-9140 USA

2007

Copyright © 2007, The Geological Society of America, Inc. (GSA). All rights reserved. GSA grants permission to individual scientists to make unlimited photocopies of one or more items from this volume for noncommercial purposes advancing science or education, including classroom use. For permission to make photocopies of any item in this volume for other noncommercial, nonprofit purposes, contact the Geological Society of America. Written permission is required from GSA for all other forms of capture or reproduction of any item in the volume including, but not limited to, all types of electronic or digital scanning or other digital or manual transformation of articles or any portion thereof, such as abstracts, into computer-readable and/or transmittable form for personal or corporate use, either noncommercial or commercial, for-profit or otherwise. Send permission requests to GSA Copyright Permissions, 3300 Penrose Place, P.O. Box 9140, Boulder, Colorado 80301-9140, USA.

Copyright is not claimed on any material prepared wholly by government employees within the scope of their employment.

Published by The Geological Society of America, Inc.
3300 Penrose Place, P.O. Box 9140, Boulder, Colorado 80301-9140, USA
www.geosociety.org

Printed in U.S.A.
GSA Books Science Editor: Marion E. Bickford

Library of Congress Cataloging-in-Publication Data

Geology of coal fires : case studies from around the world / edited by Glenn B. Stracher.
    p. cm. (Reviews in engineering geology ; 18)
  Includes bibliographical references and index.
  ISBN 978-0-8137-4118-5 (pbk.)
    1. Coal—Geology. 2. Coal—Combustion. 3. Engineering geology. I. Stracher, Glenn B.

TN800 .G485 2007
553.2'4—dc22

2007032815

Cover, background photo: Gas venting at the surface from an underground coal fire in Hazard, Kentucky. Courtesy of Glenn B. Stracher, 2007. **United States:** Coal and shale on fire in the top split of the Pittsburgh seam (~3 ft thick in the photo), Boyce Park mine fire, Pennsylvania. Courtesy of Steve Jones, 2002. **Russia:** Paralava (0.455–0.328 Ma) formed during an underground coal fire and intruded into clinker, Malinovka Village in the Kuznetsk coal basin, West Siberia. Courtesy of Svetlana N. Zateeva, 2005. **South Africa:** J. Denis N. Pone (left) and Glenn B. Stracher sampling gas vents encrusted with mascagnite and letovicite (white color) in a burning culm bank, Witbank coalfield. Courtesy of Harold J. Annegarn, 2004. **India:** Surface coal fire in the eastern Jharia coalfield. Field of view width is about 3 meters wide. Courtesy of Anupma Prakash, 2006. **Inner Mongolia (back cover):** Gas vent in quartzofeldspathic sandstone, encrusted with salammoniac in the Wuda coalfield. Source: Stracher, G.B., Prakash, A., Schroeder, P., McCormack, J., Zhang, X., Van Dijk, P., and Blake, D., New mineral occurrences and mineralization processes: Wuda coal-fire gas vents of Inner Mongolia: American Mineralogist, v. 90, p. 1729–1739.

10 9 8 7 6 5 4 3 2 1

Contents

Preface v

Acknowledgments vii

Illustrations of coal fires ix

SPONTANEOUS COMBUSTION AND GREENHOUSE GASES

1. Greenhouse gases generated in underground coal-mine fires 1
   Ann G. Kim

2. The spontaneous combustion index and its application: Past, present, and future 15
   Sezer Uludağ

3. Geological models of spontaneous combustion in the Wuda coalfield, ... 23
   Inner Mongolia, China
   Daiyong Cao, Xinjie Fan, Haiyan Guan, Chacha Wua, Xiaolei Shi, and Yuerong Jia

4. Survey of experimental work on the self-heating and spontaneous combustion of coal 31
   Mark Nelson and Xiao Dong Chen

5. A laboratory study of a reactive surface layer for the prevention of spontaneous combustion ... 85
   Rufaro Kaitano, David Glasser, and Diane Hildebrandt

MINERALOGY AND PETROLOGY

6. The origin of gas-vent minerals: Isochemical and mass-transfer processes 91
   Glenn B. Stracher

7. Combustion metamorphic events resulting from natural coal fires 97
   Ellina V. Sokol and Nina I. Volkova

8. Mineralogy and petrography of iron-rich slags and paralavas formed by 117
   spontaneous coal combustion, Rotowaro coalfield, North Island, New Zealand
   M. Naze-Nancy Masalehdani, Philippa M. Black, and Huldrych W. Kobe

9. Paralavas in a combustion metamorphic complex: Hatrurim Basin, Israel 133
   Yevgeny Vapnik, Victor V. Sharygin, Ella V. Sokol, and Reginald Shagam

10. Geochronology of clinker and implications for evolution of the Powder River Basin 155
    landscape, Wyoming and Montana
    Edward L. Heffern, Peter W. Reiners, Charles W. Naeser, and Donald A. Coates

GEOPHYSICS—MODELING

11. Possible sources of magnetic anomalies over thermally metamorphosed 177
carbonate rocks of the Mottled Zone in Israel
Boris Khesin, Shimon Feinstein, and Sophia Itkis

12. Detecting concealed coal fires 199
Hartwig Gielisch

13. Subsurface coal-mine fires: Laboratory simulation, numerical modeling, ...211
and depth estimation
Anupma Prakash and Antony R. Berthelote

GEOPHYSICS—REMOTE SENSING

14. Remotely sensed land-cover changes in the Wuda and Ruqigou-Gulaben coal-mining 219
areas of China
Claudia Kuenzer, Jianzhong Zhang, Stefan Voigt, and Wolfgang Wagner

15. Remote-sensing–based coal-fire detection with low-resolution MODIS data 229
Christoph Hecker, Claudia Kuenzer, and Jianzhong Zhang

16. Application of remote sensing in coal-fire studies and coal-fire–related emissions 239
Prasun K. Gangopadhyay

17. Three-dimensional thermal-imaging methodology for detecting underground coal fires 249
Zhang Jianmin, Huan Zhongdan, Sun Yujing, Tian Yuan, Stefan Voigt, and Zhao Xuejun

COAL FIRES AND PUBLIC POLICY

18. Comparison of Pennsylvania anthracite mine fires: Centralia and Laurel Run 261
Melissa A. Nolter, Daniel H. Vice, and Harold Aurand Jr.

19. Congressional response to coal fires: Illustrating transitions in the policy process 271
Karen M. McCurdy

Index 279

Preface

Dedicated to People the World Over Adversely Affected by Coal Fires

Naturally burning coal fires and those ignited by human activities receive little attention from the media compared to other environmental hazards. This is surprising because coal fires are responsible for human diseases and fatalities; dangerous land subsidence; degradation of floral and faunal habitats; water, air, and soil pollution; and the destruction of an increasingly valuable energy resource. Toxic gases, particulate matter, and the heat from coal fires destroy people's homes and businesses, render roads and rail lines inoperable, and turn communities into ghost towns.

Unfortunately, few university-geoscience curricula if any devote time to the study of coal fires. Coal fires are a part of Earth's history, and their occurrence is preserved globally in the rock record as "burnt-out" coal seams, paralava, and as clinker, rock baked by burning coal. Few people realize that such fires are responsible for topographic features developed in response to subsidence associated with burning. These include ground fissures, slumping, and sinkholes. In the United States, these features occur in the Powder River basin in Wyoming, the Emery coalfield in Utah, the Southern Ute Indian Reservation in Colorado, and the Western Middle anthracite field in Pennsylvania. Other occurrences include the coal basins in northern China and the Jharia coalfield in India, where coal fires are spewing enormous quantities of noxious gases and particulate matter into the atmosphere. The potentially important contribution of these and other coal fires to global warming and their connection to Earth's carbon cycle remain unknown. Likewise, the contribution of prehistoric coal fires to the chemistry of Earth's atmosphere, oceans, and even life on Earth itself has not been evaluated.

The study of coal fires, "coal-fires science," is an exciting and interdisciplinary area of research that is just beginning to gain international attention as is evident from recent Internet, newspaper, and magazine articles, as well as radio broadcasts. This recognition is partly in response to presentations and field trips at international meetings during the past four years including the American Association for the Advancement of Science in Denver, Colorado, in February 2003; the Geological Society of America (GSA) Annual Meeting in Denver, in November 2004; and the International Conference on Coal Fires Research in Beijing, China, from November to December 2005. Since 2004, geoscientists from numerous countries have attended field trips to active coal fires, sponsored by the Coal Geology Division of GSA. These trips were to the South Cañon Number One Mine fire in Colorado (2004, 2007), the Emery coalfield in Utah (2005), and the Centralia mine fire in Pennsylvania (2006). During these trips, the origin and potential environmental effects of these fires were discussed and both gas- and mineral-sample–collecting techniques were demonstrated. Radio talk shows, including those in Japan, the United States, Canada, India, and "A Burning Issue," which the British Broadcasting Corporation has aired multiple times since August 2005 to an estimated 50 million people around the world, are designed to inform the public about human suffering and the environmental dangers posed by modern-day coal fires.

In this Reviews in Engineering Geology volume, the world's leading experts in the developing field of coal-fires science present their latest research findings. Scientists who presented their preliminary research findings during the 2004 GSA Annual Meeting in Denver authored some of the manuscripts contained herein. Their abstracts were presented during GSA Topical Session T62, entitled "Wild Coal Fires: Burning Questions with Global Consequences," convened by Glenn B. Stracher and Edward L. Heffern. In addition, this volume includes papers by numerous authors who did not attend the GSA coal-fires session.

This volume is divided into five sections: (1) Spontaneous Combustion and Greenhouse Gases is devoted to papers about gases generated in underground coal fires, the spontaneous combustion index, models of coal combustion, experimental work with self-heating, and the prevention of spontaneous combustion. (2) Mineralogy and Petrology discusses the origin of gas-vent minerals, combustion metamorphism associated with natural-coal fires, melted sediments and iron-rich slag formed during a spontaneous-coal fire, paralavas in a combustion metamorphic complex, and the geochronology of clinker with implications for landscape evolution. (3) Geophysics—Modeling presents papers about magnetic anomalies and combustion, detection methodologies for concealed coal fires, and numerical modeling and fire-depth estimation. (4) Geophysics—Remote Sensing includes papers about land-cover changes due to coal fires, coal-fire detection with MODIS data, remote-sensing applications in coal-fires studies including emissions, and 3-D thermal imaging for detecting underground coal fires. (5) Coal Fires and Public Policy contains a paper that compares two fires and a paper that explores the political implications of coal fires.

This volume is the first GSA publication devoted exclusively to the topic of coal fires. It is my hope and also that of the contributing authors that readers of this volume will be energized by the information presented herein and that at least some readers will study one or more aspects of the important and intriguing subject of coal-fires science.

Acknowledgments

On behalf of all the authors who contributed to this volume, I would like to thank a number of people who generously donated a great deal of time reviewing the manuscripts presented here. They are:

Uli Barth, *Bergische University of Wuppertal, Germany*
John Carras, *CSIRO Energy Technology, Newcastle, NSW, Australia*
Pete Clapham, *Cleveland State University, Ohio*
Kristan Cockerill, *University of New Mexico, Albuquerque*
Gary Colaizzi, *Goodson and Associates, Inc., Wheat Ridge, Colorado*
Derek Cunnold, *Georgia Institute of Technology, Atlanta*
Steven M. de Jong, *Environmental Hazards and Earth Observation Group, TC Utrecht, The Netherlands*
Rebecca L. Dodge, *University of West Georgia, Carrollton*
Nehar Eroglu, *New South Wales Coal Compensation Board, Maitland, Australia*
Eric Essene, *University of Michigan, Ann Arbor*
Rosemary Falcon, *SABS: Coal and Mineral Technologies (Pty) Ltd, Gauteng, South Africa*
Robert B. Finkelman, Emeritus, *United States Geological Survey, Reston, Virginia*
Mike Fuller, *Honolulu University, Hawaii*
Robert C. Fuller, *North Georgia College and State University, Dahlonega*
John Garver, *Union College, Schenectady, New York*
Arie Gilat, Emeritus, *Geological Survey of Israel, Ramat-HaSharon*
Hal Gluskoter, *United States Geological Survey, Reston, Virginia*
M.J. Hibbard, Emeritus, *University of Nevada, Reno*
Gretchen Hoffman, *New Mexico Bureau of Geology and Mineral Resources, Socorro*
Haiyang Huang, *Beijing Normal University, China*
Clifford Jones, *University of Aberdeen, Scotland*
Allan Kolker, *United States Geological Survey, Reston, Virginia*
Yehoshua Kolodny, *The Hebrew University of Jerusalem, Israel*
Salomon Kroonenberg, *Delft University of Technology, The Netherlands*
Todd T. Kunioka, *California State University, Los Angeles*
Gennadiy Lepezin, *Siberian Branch of the Russian Academy of Sciences, Novosibirsk*
Chun-Zhu Li, *Monash University, Australia*
Nancy Lindsley-Griffin, *University of Nebraska, Lincoln*
B.H.P. (Ben) Maathuis, *International Institute for Geo-Information Science and Earth Observation (ITC), Enschede, The Netherlands*
Ron McDowell, *West Virginia Geological Survey, Morgantown*
Andy C. McIntosh, *University of Leeds, UK*
Uwe Meyer, *Federal Institute for Geosciences and Natural Resources, Hanover, Germany*
Stanley Michalski, *GAI Consultants, Inc., Homestead, Pennsylvania*
Yu Minggao, *Henan Polytechnic University, China*
Deepak R. Mishra, *University of Nebraska, Lincoln*
Thomas R. Moore, *CDX Gas, LLC, Charleston, West Virginia*
Douglas C. Peters, *Peters Geosciences, Golden, Colorado*
David Philbin, *Office of Surface Mining, Reclamation, and Enforcement, Wilkes-Barre, Pennsylvania*

Huw R. Phillips, *University of Witwatersrand, Johannesburg, South Africa*
Steven Renner, *Colorado Division of Minerals and Geology, Grand Junction*
Vladimir V. Reverdatto, *Siberian Branch of the Russian Academy of Sciences, Novosibirsk*
Dallas D. Rhodes, *Georgia Southern University, Statesboro*
Michael Roden, *University of Georgia, Athens*
Timothy Rohrbacher, *United States Geological Survey, Denver, Colorado*
Hagai Ron, *The Hebrew University of Jerusalem, Israel*
Donald Rundquist, *University of Nebraska, Lincoln*
Michael Rybakov, *The Geophysical Institute of Israel, Lod*
Robert R. Seal II, *United States Geological Survey, Reston, Virginia*
R. Jeffrey Swope, *Indiana University–Purdue University, Indianapolis*
David E. Tabet, *Utah Geological Survey, Salt Lake City*
Shuheng Tang, *China University of Geosciences, Beijing*
Tammy P. Taylor, *Los Alamos National Laboratory, New Mexico*
John L. van Genderen, *International Institute for Geoinformation Science and Earth Observation (ITC), Enschede, The Netherlands*
Robert K. Vincent, *Bowling Green State University, Ohio*
Doyle Watts, *Wright State University, Dayton, Ohio*
Roger Wheate, *University of Northern British Columbia, Prince George*
Jens Wiegand, *Universität Würzburg, Germany*
Dong-ke Zhang, *Curtin University of Technology, Perth, Australia*
Xiangmin Zhang, *Utrecht University, The Netherlands*

I am grateful to Stacia A. Spaulding who assisted with the grammatical revisions of several manuscripts authored by non-native English speakers. I also wish to express my gratitude to Abhijit Basu, M.E. (Pat) Bickford, and Charles Welby, Geological Society of America books editors, for their encouragement and guidance throughout this project. In addition, I thank the staff at GSA who assisted me with all technical aspects associated with this publication.

Glenn B. Stracher
East Georgia College
Swainsboro, Georgia

Illustrations of coal fires

Below are illustrations not found in the text that represent the global extent of coal fires. A short description of each fire is also provided.

COAL FIRES	
Location	Figure no.
World Map	1
India	2–3
Inner Mongolia	4–5
Indonesia	6–7
South Africa	8–9
France	10
United States	11–13

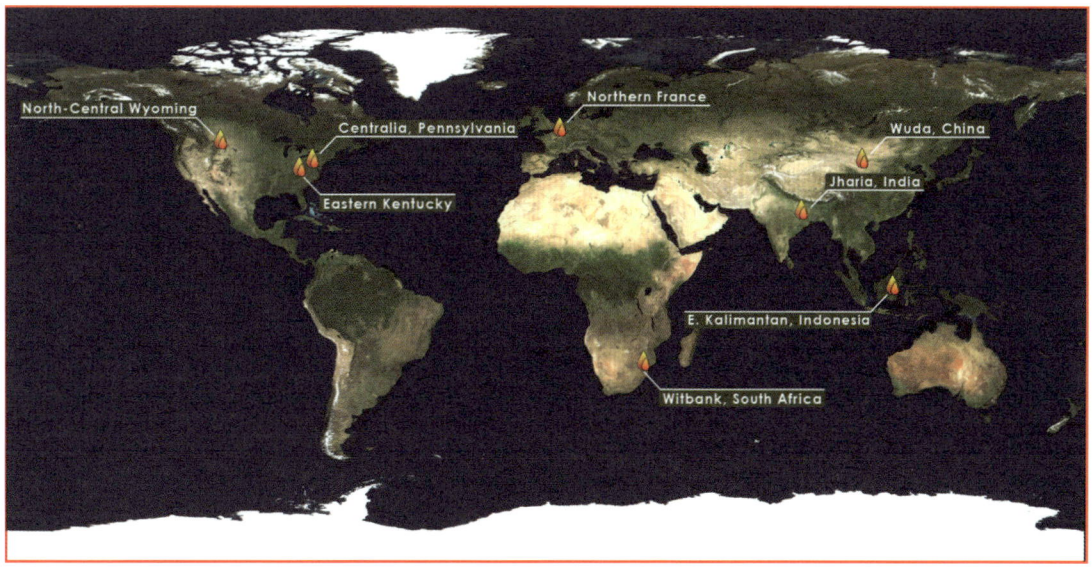

Figure 1. World map highlighting the locations of coal fires illustrated in Figures 2–13. (Source: Anupma Prakash, 2007, University of Alaska, Fairbanks. Background satellite image © the National Aeronautics and Space Administration, Washington, D.C., USA.)

India

Figure 2. Uncontrolled burning in an opencast-coal mine located near Dhanbad, Jharia coalfield, India. Most coal fires in Jharia start by spontaneous combustion subsequent to mining. The central flame is about 7 m (23 ft) high. (Source: Daniel B. Sanger, 1994, GAI Consultants, Inc., Homestead, Pennsylvania, USA.)

Illustrations of coal fires xi

Figure 3. Jharia coalfield, India. (A) Children in front of a Kúli Dhaóra (miners' slum), abandoned by most workers, except the very poorest. Precarious cracks in most buildings are due to subsidence and associated ground fissures. (B) An opencast-coal mine on fire in close proximity to the slum. (Source: Prasun Gangopadhyay, 2006, International Institute for Geo-Information Science and Earth Observation [ITC], The Netherlands.)

China (Inner Mongolia)

Figure 4. Claudia Kuenzer using a thermal radiometer to measure the temperature (200 °C) of bedrock and burning coal at fire number 8 south, Wuda coalfield, Inner Mongolia. The underground fire, located on the east-dipping (< 20°) western limb of the Wuda syncline, occurs in a sulfur-rich coal bed, overlain by 5–10 m of white sandstone and weathered shale. It started spontaneously as a consequence of mining activities in the area. (Source: Christoph Hecker, 2003, International Institute for Geo-Information Science and Earth Observation [ITC], The Netherlands.)

Figure 5. Burning coal at fire number 13, started by spontaneous combustion as a result of mining in the Wuda coalfield, Inner Mongolia. The fire in this high-sulfur coal is overlain by 5 m of white sandstone and other sedimentary rocks on the east-dipping (12–25°) western limb of the Wuda syncline, about 3–5 km southwest of and up section from fire number 8 south (Fig. 4). The hottest temperature recorded here was nearly 830 °C. The burning is currently reduced to smoldering coal. (Source: Claudia Kuenzer, 2003, Vienna University of Technology, Austria.)

Indonesia

Figure 6. Coal fire excavated and extinguished along the only road from Samarinda to Bontang in Lempake, East Kalimantan, Indonesia. The fire burned for nearly two years and is thought to have started in a roadside ditch by either a cigarette or burning roadside trash. (Source: Alfred E. Whitehouse, 1999, U.S. Office of Surface Mining, Washington, D.C., USA.)

Figure 7. A coal fire of unknown origin in East Kalimantan, Indonesia, started in September 1997 in an ephemeral stream valley containing exposed coal beds, adjacent to a home near the Samarinda-Balikpapan road. The fire burned under the house, which was moved across the road and repaired. By November 1999, the fire was excavated and extinguished. (Source: Alfred E. Whitehouse, 1998, U.S. Office of Surface Mining, Washington, D.C., USA.)

South Africa

Figure 8. Coal fire in a bituminous culm bank in an opencast (formerly underground) mine ignited by spontaneous combustion, Witbank coalfield, South Africa. (Source: Glenn B. Stracher, 2004, East Georgia College, Swainsboro, Georgia, USA.)

Illustrations of coal fires xv

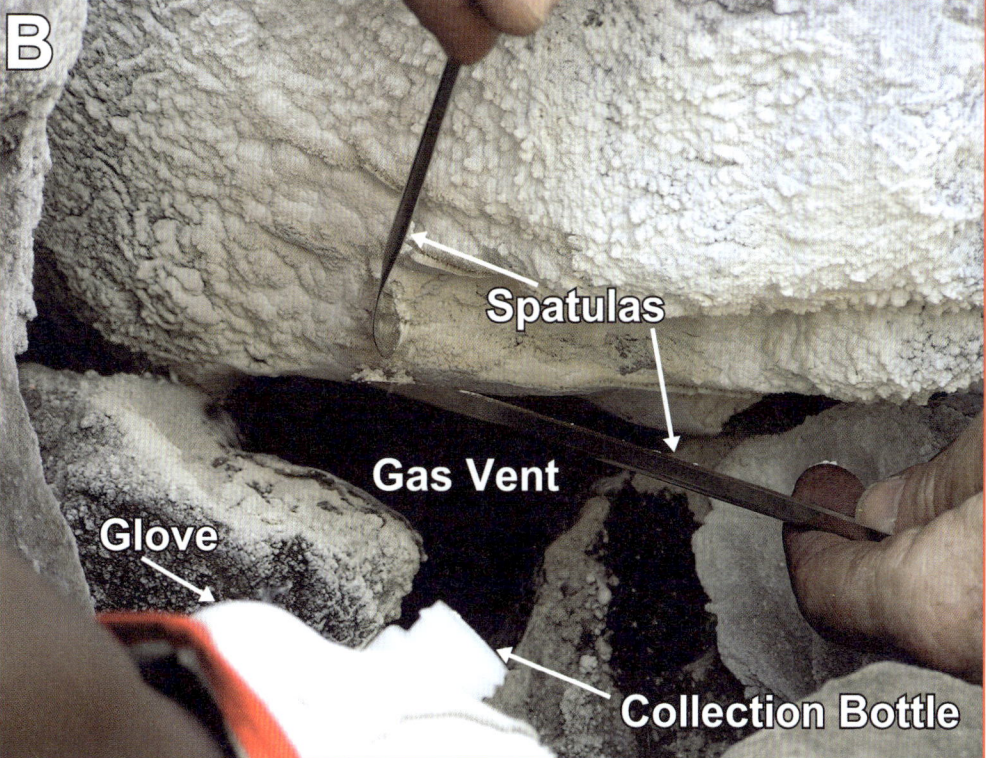

Figure 9. Culm bank consisting of sandstone and shale underlain by coal, Witbank coalfield, South Africa. (A) Glenn B. Stracher (left) and J. Denis N. Pone collecting GPS and temperature (380 °C minimum) data, respectively, from an active gas vent. (B) Close up of the vent and mineralization (white coating on rocks) during sample collecting. X-ray analysis of the coating revealed the presence of mascagnite $(NH_4)_2SO_4$ and letovicite $(NH_4)_3H(SO_4)_2$. (Source: Harold J. Annegarn, 2004, University of Johannesburg, South Africa.)

France

Figure 10. A burning-culm bank with a surface area of nearly 7 ha and a height of ~120 m in the town of Fouquières-les-Lens, Nord-Pas-de-Calais region of northern France. The bituminous coal burning here since 1972 is from the former Courriéres underground mine. Funds are unavailable to extinguish the fire, thought to have started by spontaneous combustion. (Source: M. Naze-Nancy Masalehdani, 2005, Université de Lille 1, Cédex, France.)

United States

Figure 11. The Centralia underground mine fire in eastern Pennsylvania, USA, started in 1962 when a landfill fire ignited the Buck Mountain coal bed. The photo shows Pennsylvania Route 61, now closed, leading to Centralia; replaced by a detour. The road is subsiding into coal-mine tunnels beneath it. Although most homes in Centralia have been demolished, a few occupied dwellings remain. (Source: Janet L. Stracher [small photos], 2003, East Georgia College, Swainsboro, Georgia, USA, and John Barnes [large photo], 2002, Pennsylvania Department of Conservation and Natural Resources, USA.)

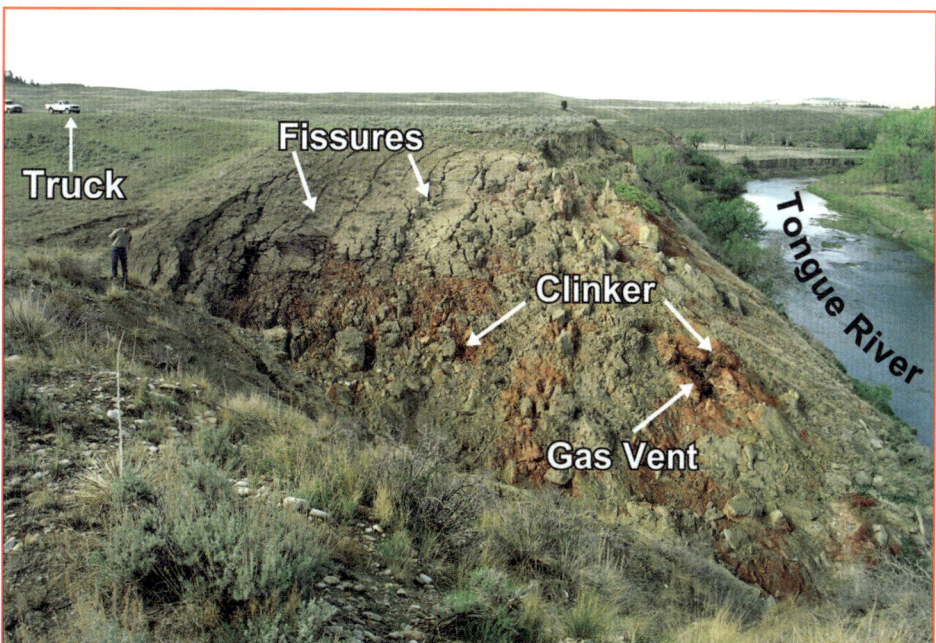

Figure 12. Ranch land slumping into the Tongue River, Powder River basin, ~10 miles north of Sheridan, Wyoming, USA. Coal beds burning underground dip into the base of the cliff. A gas vent, red clinker, and ground fissures are visible upslope. (Source: Glenn B. Stracher, 2004, East Georgia College, Swainsboro, Georgia, USA.)

Figure 13. Underground coal-mine fire in eastern Kentucky, USA. The smoke is coming out of former underground haulage ways, exposed by subsequent strip mining. Explosions during burning blew rock out of the haulage ways, littering the ground outside of and adjacent to the pillars of smoke in the photo. (Source: Stanley Michalski, 1990, GAI Consultants, Inc., Homestead, Pennsylvania, USA.)

Greenhouse gases generated in underground coal-mine fires

Ann G. Kim*
National Energy Technology Laboratory, U.S. Department of Energy, 626 Cochrans Mill Road, P.O. Box 10940, Pittsburgh, Pennsylvania 15236-0940, USA

ABSTRACT

The release of greenhouse gases from underground coal-mine fires is a function of temperature and the concentration of O_2. In a laboratory study on spontaneous combustion, samples of coal, coal refuse, and carbonaceous shale were heated at a controlled rate between ambient temperature and 250 °C. In these experiments, the concentration of O_2 was not limited and the concentration of CO_2 increased with increasing temperature to a maximum of 10%. Carbon monoxide was not detected at temperatures below 100 °C, and the maximum concentration of CO was less than 4%. In field studies, samples of combustion gases were obtained from fires in three abandoned coal mines. These indicated a linear increase in the concentration of CO_2 relative to the decreased concentration of O_2. At an O_2 concentration of 2%, the CO_2 concentration approached 15%, and CO was detected only when the O_2 concentration was less than 8%. At temperatures over 50 °C, the rate of desorption of CH_4 also increased, but the average concentration in the mine atmosphere was 0.20%.

These laboratory experiments and field studies indicate that the rate of gas production is controlled by O_2 concentration and temperature, but physical factors, such as overburden fracturing and differences between surface and subsurface temperature and pressure, control the rate of emission to the atmosphere. In coal-mine fires, both chemical and physical factors control the rate and magnitude of contributions to the atmospheric concentration of greenhouse gases.

Keywords: carbon dioxide, carbon monoxide, methane, coal-combustion gases.

INTRODUCTION

Anthropogenic generation of greenhouse gases is usually attributed to industrial sources, such as coal-fired power production and the use of fossil fuels for transportation. However, coal mining and coal-mine fires also emit significant amounts of CO_2 and CH_4. Of the of the 1% annual increase in the atmospheric concentration of CH_4, 8% is attributed to coal mining (Clayton, 1998; Su and Agnew, 2006), and 2%–3% of annual world emissions of CO_2 are estimated to come from coal fires in China (Stracher and Taylor, 2004). The control of these fires could be a factor in the global reduction of greenhouse gases.

Prevalence of Fires in Abandoned Mines and Waste Banks

Uncontrolled combustion of coal and carbonaceous shales occurs in abandoned underground mines, abandoned surface mines, coal waste banks, and in coal outcrops (Johnson and Miller

---

*Ann.Kim@or.netl.doe.gov

Kim, A.G., 2007, Greenhouse gases generated in underground coal-mine fires, *in* Stracher, G.B., ed., Geology of Coal Fires: Case Studies from Around the World: Geological Society of America Reviews in Engineering Geology, v. XVIII, p. 1–13, doi: 10.1130/2007.4118(01). For permission to copy, contact editing@geosociety.org. ©2007 The Geological Society of America. All rights reserved.

1979; Kim and Chaiken, 1993; McNay 1971). In underground room-and-pillar mines, a relatively large proportion (30%–50%) of the coal may be left in place. Since the roof coals and carbonaceous shales are also left in the mine, the tonnage of combustible material in an abandoned mine could exceed that extracted during mining. Fires can start at the coal outcrop and propagate along the outcrop or through the miles of interconnected underground workings. Heat can move by convection through the mine or by conduction into the overburden. As the overburden becomes warmer, or as the coal pillars fail, the overburden subsides, creating a system of cracks and fractures through which smoke and fumes leave the mine and fresh air enters. Most abandoned mine fires exhibit smoldering combustion with little visible flame and involve relatively small amounts of combustible material at any given time. Such fires can continue to burn in a low O_2 (~2%) atmosphere for 0–80 yr (Chaiken et al., 1983).

According to its Abandoned Mine Land Inventory System (AMLIS), the U.S. Office of Surface Mining Reclamation and Enforcement (OSMRE) lists 141 underground mine fires in 11 states (OSMRE, 2005). These are fires that present an immediate hazard or affect health and safety. Control projects have been completed at 69 of these fires; 11 fire control projects are currently funded and 61 are unfunded. Over $27 million has been spent on the completed projects; the estimated cost for unfunded projects is over $471 million. The high cost and the difficulty of controlling fires in abandoned or inactive underground mines limit the number of fires that have been extinguished.

Gases from Underground Mine Fires

If combustion is considered to be the exothermic reaction of carbon and oxygen to form CO_2, it can be described by the equation:

$$C + O_2 \rightarrow CO_2 + \text{heat}. \quad (1)$$

The amount of heat liberated is 93.7 kcal/mole. However, coal is not composed of elemental carbon; on a dry, mineral matter–free basis, coal contains between 60% and 90% carbon. The rest of the coal "molecule" is composed of hydrogen, oxygen, nitrogen, and sulfur. On this basis, the stoichiometric combustion of coal can be written as (Chaiken, 1977):

$$CH_{1.18}N_{0.15}O_{0.35}S_{0.005} + 1.12 O_2 + 4.15 N_2 \rightarrow CO_2 \\ + 0.58 H_2O + 0.005 SO_2 + 4.157 N_2 + 138.4 \text{kcal} \quad (2)$$

Factors influencing the initiation and propagation of wasted coal fires are related to the three required components of sustained combustion: fuel, oxygen, and energy. The amount of combustible material, its particle size, surface area, and its tendency to spontaneously combust are fuel-related factors. The presence of fractures through which air can be drawn into the fire zone, circulation enhanced by the fire, and changes in barometric pressure control the amount of available O_2. The rate of heat generation versus the rate of heat loss within a mine is controlled by the heat-generating reactions (oxidation of coal, oxidation of pyrite, surface adsorption of water vapor, bacterial activity) and insulation provided by adjacent strata.

While the normal ignition temperature of coal is between 420 °C and 480 °C, under adiabatic conditions, the minimum temperature at which coal will self-heat is between 35 and 140 °C (Smith and Lazarra, 1987).

The former U.S. Bureau of Mines (USBM) conducted research into the causes of outcrop and underground mine fires and into methods to locate, control, or extinguish such fires. Gas chromatographic analyses of the mine atmosphere at field sites and of the gases produced in laboratory studies are the basis for estimating the composition and concentration of greenhouse gases generated in underground coal-mine fires.

GASES GENERATED DURING SELF-HEATING EXPERIMENTS

The spontaneous combustion of coal has been extensively studied (Kim, 1977), but little data are available on the gases generated. In a laboratory study on the self-heating of coal, of carbonaceous shales associated with coals, and of coal refuse (Kim, 1995), the composition of the gas generated from these materials was determined as a function of temperature.

Methods and Materials

To determine the relative self-heating tendencies of coals, coal refuse, and carbonaceous shales, a combination of differential thermal analysis (the sample temperature is measured relative to a standard) and crossing-point temperature (the point at which the sample temperature equals or exceeds a reference temperature) methods was used. Samples were heated in a 20-cm-diameter tube furnace. The sample container was a stainless-steel cylinder with a volume of 3700 cm^3. Movable grids were used to hold the sample in a packed bed. Both ends were closed with threaded caps. The sample container was designed to promote the even distribution of air through the sample and to minimize the effect of discontinuities between the furnace wall and the central thermocouple measuring the sample temperature. The furnace temperature was the reference temperature. An air line, a gas sampling line, and a thermocouple were inserted through the cap into the sample chamber. The air flow rate was 15 cm^3/s. Duplicate tests were performed with wet and dry air.

A large sample (1300–3000 g) was crushed and screened to −0.6 cm by +20 mesh, then dried overnight in the furnace under N_2 at 100 °C to remove surface moisture and reduce the effect of evaporative cooling. Normal air was used to replace the N_2, and the furnace was held to 100 °C for ~24 h (Fig. 1). The temperature of the sample increased until it equaled or exceeded the furnace temperature. The furnace temperature was then raised to 150 °C, and the procedure was repeated. On the third day, the

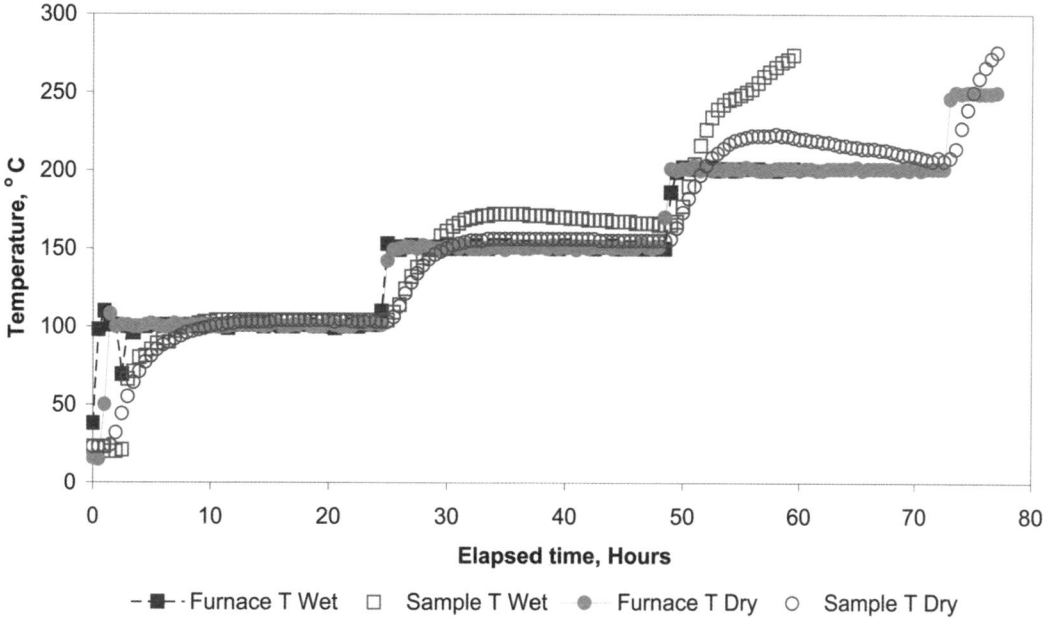

Figure 1. Temperature profile for furnace (reference) and sample temperatures in wet and dry air.

temperature of the furnace was raised to 200 °C and held at that level for 24 h. The rate at which the sample temperature approached the reference temperature and the maximum difference between the sample and the reference temperatures were considered indicators of self-heating behavior.

Analysis of evolved gases was performed by gas chromatography against a standard. Oxygen depletion was calculated as the difference between the O_2 concentration in the effluent from the furnace and that in the influent air.

The samples in this study (Table 1) included three coals that were known to self heat, the Pittsburgh coal which does not generally self heat, two samples of coal refuse, and seven samples of carbonaceous shale. The zone A, B, C, and D shales (Tables 1–4) came from a coal mine that had self-heating problems.

Results

The amount of O_2 consumed and the amount of CO_2 produced were monitored as indicators of heat-producing chemical reactions. Both factors were found to increase with increasing temperature for each sample. The presence of moisture in the air stream had no consistent effect on the change in the concentration of O_2.

When the sample was placed in the furnace, O_2 was adsorbed (Table 2); the amount adsorbed varied from less than 2% to more than 18% of the available O_2. Although less than half of the samples adsorbed more than 5% of the O_2 in the air stream in the first minute in the furnace, one-quarter of the samples adsorbed more than 10%. At temperatures above ambient, the adsorption of O_2 increased with increased temperature.

TABLE 1. PROXIMATE ANALYSIS OF COAL, COAL REFUSE, AND CARBONACEOUS SHALES

Sample	Type	Moisture (%)	Ash (%)	Volatile matter (%)	Fixed carbon (%)
Albright	Coal refuse	4.79	57.50	12.17	25.53
Bailey	Coal refuse	2.18	79.76	12.03	6.03
Blue Creek	Coal	0.77	36.16	19.16	43.91
D Seam	Coal	7.90	5.60	36.20	50.30
F Seam	Coal	9.14	4.04	40.86	45.96
Pittsburgh	Coal	1.40	4.76	37.36	56.48
Pittsburgh roof	Shale	2.14	87.42	8.26	2.18
Zone A	Shale	0.31	79.96	17.66	5.07
Zone B	Shale	0.64	59.98	15.20	27.18
Zone C1	Shale	0.94	75.28	10.51	13.27
Zone C2	Shale	1.28	61.70	14.16	22.83
Zone D1	Shale	1.04	61.58	14.26	23.12
Zone D2	Shale	1.30	21.92	22.85	53.93

TABLE 2. ADSORPTION OF O_2 ($-d[O_2]$) AT AMBIENT TEMPERATURE (T_0), 100 °C (T_1), 150 °C (T_2), AND 200 °C (T_3)

Sample	$-d[O_2] \cong T_0$ (%)	$-d[O_2] \cong T_1$ (%)	$-d[O_2] \cong T_2$ (%)	$-d[O_2] \cong T_3$ (%)
Albright	13.11	3.27	6.42	12.79
Bailey	3.07	1.47	2.48	8.36
Blue Creek	3.73	1.73	3.38	11.56
D Seam	3.77	2.16	6.22	7.83
F Seam	18.98	1.70	7.05	9.29
Pittsburgh	4.31	2.57	5.30	16.68
Pittsburgh roof	1.27	0.57	0.85	7.27
Zone A	2.58	4.35	4.63	9.68
Zone B	1.21	1.81	1.80	7.93
Zone C1	2.96	1.15	4.11	12.77
Zone C2	8.60	0.87	1.94	8.54
Zone D1	1.92	1.17	2.43	8.99
Zone D2	6.51	1.14	5.00	13.17

The production of CO_2 followed the same pattern as the consumption of O_2 (Table 3). If the initial adsorption of O_2 was low, the initial production of CO_2 was also low; if more O_2 was consumed initially, more CO_2 was produced initially. Although the production of CO_2 varied directly with the consumption of O_2 (Fig. 2), the amount of CO_2 produced was, on the average, less than the 25% of O_2 consumed at 200 °C. At lower temperatures, less CO_2 was produced, and the relationship to O_2 depletion was more random. As a function of temperature, CO_2 production can be estimated as:

$$CO_2 = 0.002T - 0.19, R^2 = 0.97, \quad (3)$$

where the CO_2 production rate is in moles d^{-1} kg^{-1} of fixed carbon and temperature is in degrees Celsius.

As a function of O_2 depletion ($-d[O_2]$), the relationship at 200 °C can be described by the equation:

$$[CO_2] = 0.22(-d[O_2]), R^2 = 0.91. \quad (4)$$

TABLE 3. CONCENTRATION OF CO_2 AT 100 °C (T_1), 150 °C (T_2), AND 200 °C (T_3)

Sample	$CO_2 \cong T_1$ (%)	$CO_2 \cong T_2$ (%)	$CO_2 \cong T_3$ (%)
Albright	0.05	0.30	1.60
Bailey	0.06	0.18	1.56
Blue Creek	0.04	0.31	3.30
D Seam	0.25	2.38	3.91
F Seam	0.31	1.67	2.29
Pittsburgh	0.29	1.31	2.89
Pittsburgh roof	<0.01	0.07	1.77
Zone A	0.08	0.28	1.53
Zone B	0.03	0.15	1.21
Zone C1	0.04	0.31	2.48
Zone C2	0.02	0.12	1.60
Zone D1	0.04	0.19	1.48
Zone D2	0.28	0.51	2.35

Both concentrations are in moles d^{-1} kg^{-1} fixed carbon. Correlations are much lower when the CO_2 production and $-d[O_2]$ are calculated per sample weight or as a function of total carbon (volatile matter + fixed carbon).

Less CO was generated than CO_2. At temperatures above 100 °C, the concentration of CO increased as the CO_2 concentration increased (Fig. 3). At temperatures of 100 °C or less, the CO concentration was frequently below the limits of detection (Table 4). On the average, CO accounted for 7% of the reduction in O_2 concentration. At 200 °C, the CO production as a function of O_2 depletion (Fig. 4) can be described by the equation:

$$[CO] = 0.10(-d[O_2]), R^2 = 0.86. \quad (5)$$

Both concentrations are in units of moles d^{-1} kg^{-1} fixed carbon. From this study, it is apparent that when the supply of O_2 is not limited, CO_2 is the primary product of coal oxidation. The concentration of CO_2 and the rate at which it is produced are apparently related to the temperature and the amount of fixed carbon.

GAS COMPOSITION AT UNDERGROUND MINE FIRES

The former U.S. Bureau of Mines performed diagnostic studies based on the composition of the underground atmosphere at several mine-fire sites to determine the location of subsurface fire zones. The method involved obtaining gas samples at a network of cased boreholes and analyzing them by gas chromatography for O_2, CO_2, CO, and low molecular weight hydrocarbons. Baseline samples were taken when the mine atmosphere was at rest, and communication samples were obtained when an exhaust fan attached to a borehole was used to control the direction of air flow in the mine.

Carbondale Mine Fire

At the Carbondale mine-fire site, apartment buildings had been built on the site of a previous mine-fire control project in the

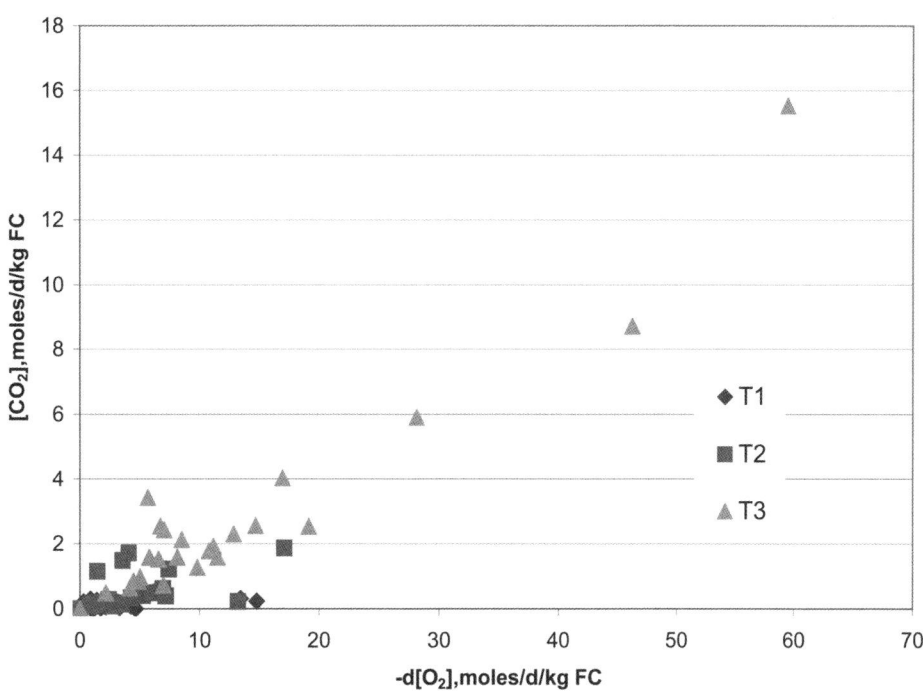

Figure 2. Carbon dioxide (CO_2) concentration as a function of oxygen depletion ($-d[O_2]$) and the concentration of fixed carbon (FC) at T_1 (100 °C), T_2 (150 °C), and T_3 (200 °C).

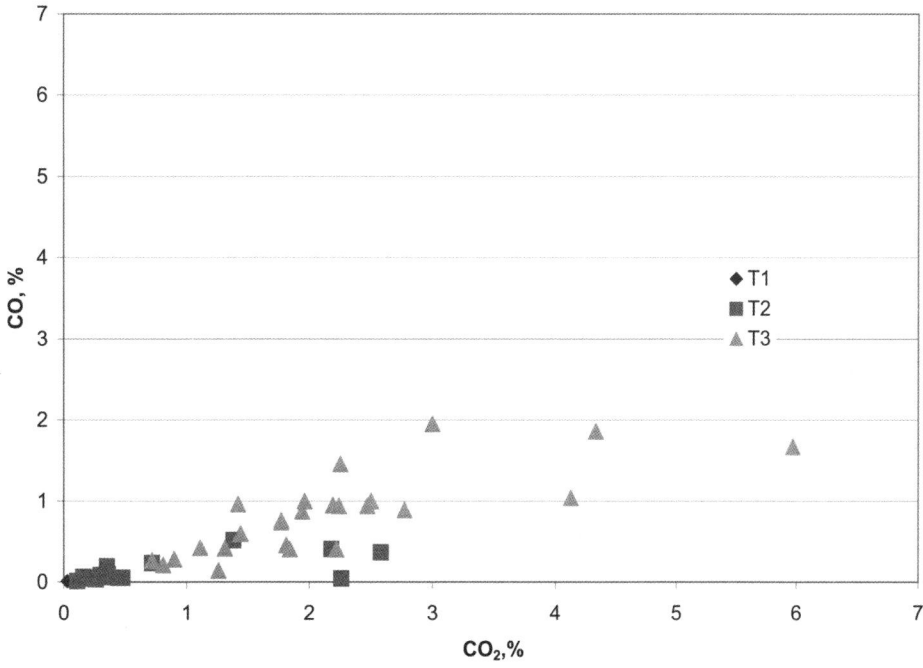

Figure 3. Carbon monoxide (CO) concentration compared to the production of carbon dioxide (CO_2) at T_1 (100 °C), T_2 (150 °C), and T_3 (200 °C).

TABLE 4. CONCENTRATION OF CO AT 100 °C (T_1), 150 °C (T_2), AND 200 °C (T_3)

Sample	CO ≅ T_1 (%)	CO ≅ T_2 (%)	CO ≅ T_3 (%)
Albright	<0.01	0.05	0.86
Bailey	<0.01	0.06	0.34
Blue Creek	0.01	0.19	0.99
D Seam	0.05	0.38	1.04
F Seam	<0.01	0.51	0.67
Pittsburgh	<0.01	0.05	1.23
Pittsburgh roof	<0.01	<0.01	0.73
Zone A	<0.01	0.05	0.84
Zone B	<0.01	0.01	0.42
Zone C1	<0.01	0.06	1.48
Zone C2	<0.01	0.04	0.51
Zone D1	<0.01	0.04	0.60
Zone D2	<0.01	0.16	0.98

anthracite region of eastern Pennsylvania (Kim et al., 1992). Anomalous snow melt indicated that there was a fire in the unreclaimed portion of the mine. Thirty-four boreholes were drilled at the Carbondale mine-fire site within an area of ~3.8 ha (8.5 acres).

To measure the subsurface pressure and to obtain gas samples, a hollow 0.95-cm-inner diameter (ID) plastic (or stainless steel for extremely hot holes) sampling line was permanently connected to a special borehole instrumentation cap and suspended to the bottom of the borehole casing. A portable diaphragm pump was used to withdraw samples of the mine gases via the downhole sample line, and gas samples were collected in completely evacuated 20 cm^3 test tubes. Also suspended along the full length of the casing was a thermocouple probe to monitor temperature near the sample point. Gas samples were analyzed with a HP 550A gas chromatograph.

In both baseline and communication tests, CO_2 production was a linear function of O_2 depletion (Fig. 5), according to the equation

$$[CO_2] = 0.87(-d[O_2]) + 0.17, R^2 = 0.99. \qquad (6)$$

where CO_2 and O_2 concentrations are given in percent.

The concentration of CO was relatively low (0.005% to 0.50%) and showed no correlation to the O_2 depletion (Fig. 6). The CH_4 concentration at the Carbondale site was also low. The average concentration under baseline (static) conditions was less than 1 ppm, but during communication tests, the average concentration was greater than 1 ppm and exceeded 100 ppm when the measured borehole temperature was greater than 50 °C (Fig. 7).

Large Mine-Fire Site

The Large mine fire is located in the bituminous region of western Pennsylvania (Kim and Dalverny, 1994). The fire is in the Pittsburgh coal seam, in what is believed to have been the Walden Pool #2 Mine. At this location, the coal outcrops at the base of a steep hill; the depth of overburden ranges from 6 to 55 m. The surface area affected by the mine fire is ~1 ha (2.5 acres).

Active underground mining probably ceased in the 1920s. Between 1953 and 1964, the Pittsburgh and Redstone coal seams were partially strip mined. The property was restored to

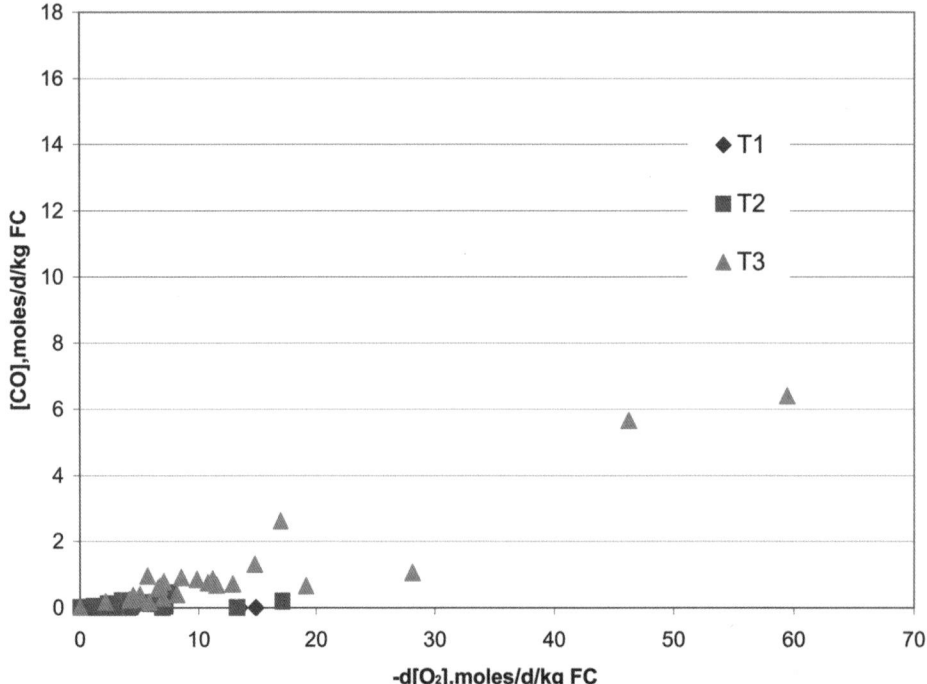

Figure 4. Carbon monoxide (CO) concentration as a function of oxygen depletion (–$d[O_2]$) and the concentration of fixed carbon (FC) at T_1 (100 °C), T_2 (150 °C), and T_3 (200 °C).

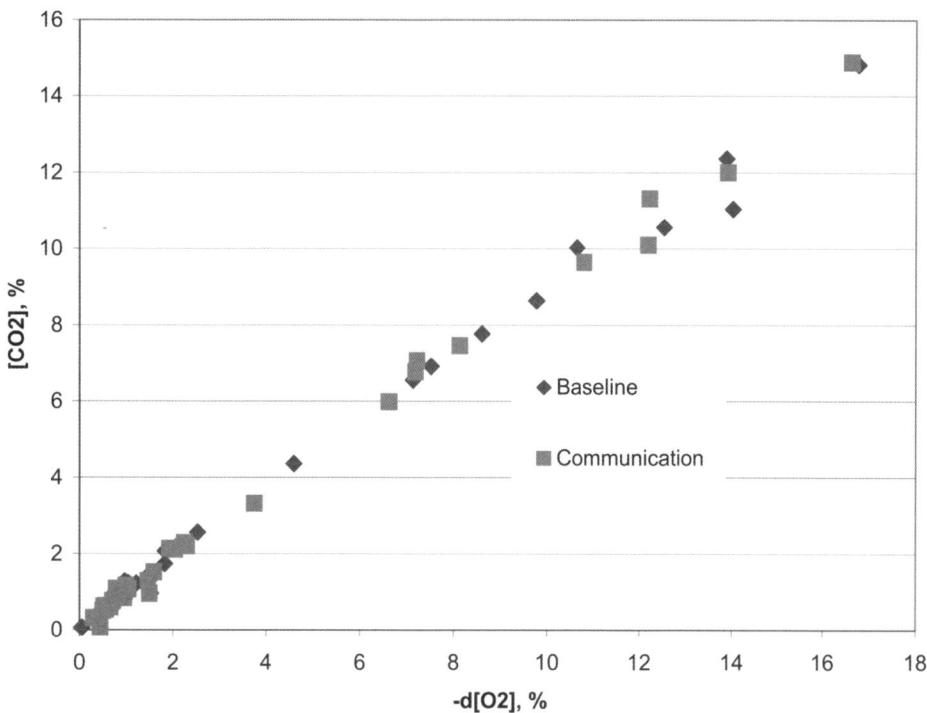

Figure 5. Carbon dioxide (CO_2) concentration at the Carbondale mine-fire site as a function of oxygen depletion ($-d[O_2]$) during baseline and communication tests.

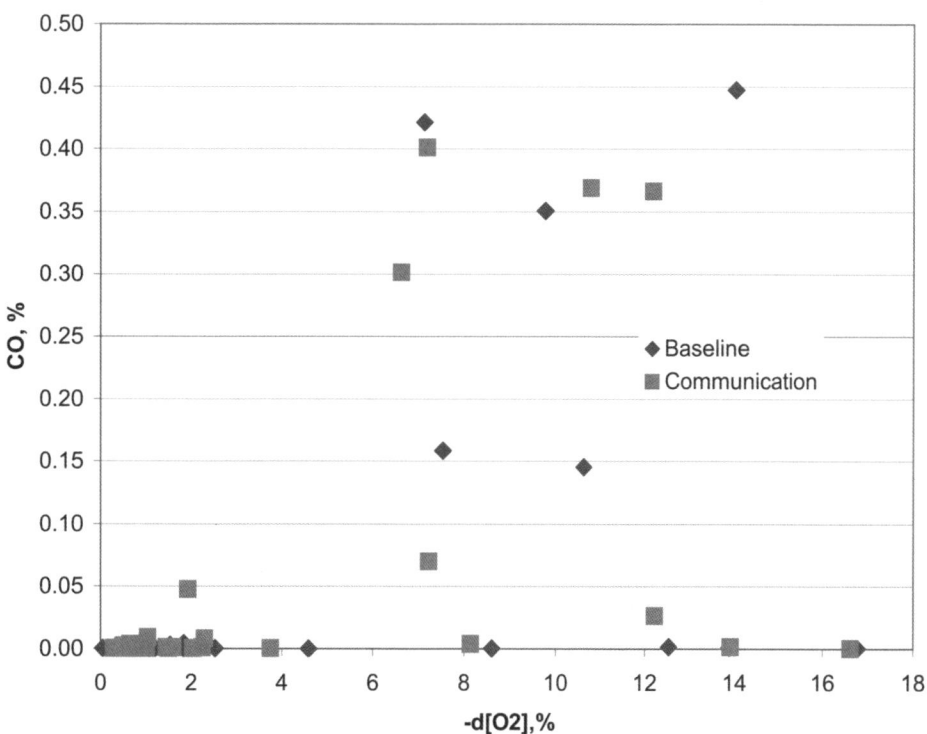

Figure 6. Carbon monoxide (CO) concentration at the Carbondale mine fire as a function of oxygen depletion ($-d[O_2]$) during baseline and communication tests.

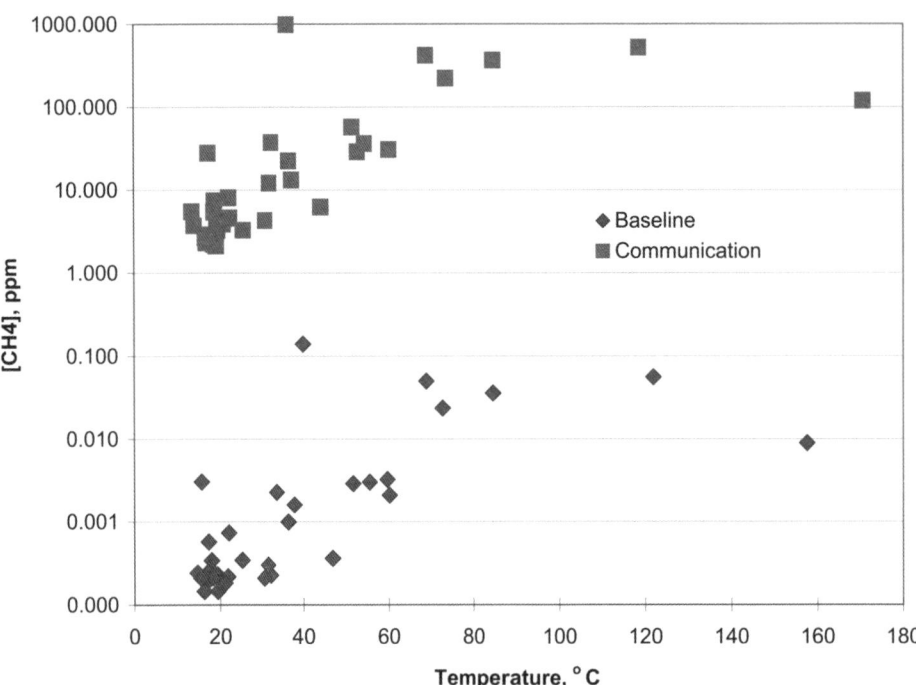

Figure 7. Average methane (CH_4) concentration at the Carbondale site versus underground temperature during baseline and communication tests.

the then-current standard of the law, and the mine operator was released of liability for the mine site.

The mine fire started at an unknown date on the strip-mined area and was first reported in December 1976. At that time, the fire could be seen burning along the highwall and in an entry. Apparently, some of the burning material was removed, and the entry was sealed. This assumption is based on the fact that the entry is no longer visible, and a small pile of burning material was observed near the old mine entry in 1986.

Over a 10 yr period, seven 6.35 cm (2.5 in), sixty-four 20 cm (8 in) boreholes, and four 5 cm (2 in) core holes were drilled at the Large site. Fifty of the 20 cm holes, seven of the 6.35 cm holes, and two core holes were used as monitoring points. The other boreholes were not used because they terminated in solid coal and had been backfilled because they contained more than 0.6 m of standing water or because they had sustained wellbore damage during drilling.

For the majority of boreholes, the mean baseline O_2 concentration was above 12%, a relatively high value for an abandoned mine. The average CO_2 concentration can be correlated with the O_2 depletion (Fig. 8) in hot zones, according to the equation

$$[CO_2] = 0.612(-d[O_2]) + 1.1, R^2 = 0.72. \quad (7)$$

In cold zones, the variation in CO_2 was more random. The CO_2 concentration increased from 0.05% to 15% as the average subsurface temperature increased to less than 125 °C (Fig. 9).

The average baseline CO concentrations exceeded the instrumental detection level at only 10 of the boreholes. In hot boreholes, the CO concentration increased to almost 1% when the O_2 depletion was greater than 15%. The baseline gas composition indicated that combustion was occurring in a relatively oxygen-rich environment at this mine-fire site.

The average CH_4 concentration increased above the background level when the subsurface temperature exceeded 50 °C. At elevated temperatures, the concentration of CH_4 varied between 100 and 1000 ppm (Fig. 10).

Percy Mine Fire

The Percy mine fire in the abandoned Youngstown mine is located in the bituminous region of southwestern Pennsylvania. It had been partially excavated from the outcrop to a depth of 30.5 m; the remaining coal had a 10% dip away from the excavation boundary. Increasing temperatures in monitoring boreholes and the emission of smoke and steam from surface fractures indicated active combustion in the unreclaimed portion of the mine (Kim, 1996).

Based on previous experience, the atmosphere in underground mines is assumed to be in a steady state. Although this baseline condition will vary locally due to combustion, proximity to fresh air, or microbial activity, the composition of air within a given area is generally constant over short periods of time (1–6 mo). When suction is applied to the underground system, changes in gas composition at a particular borehole are detected in air flowing from another area of the mine. Applied suction is assumed to be a temporary perturbation, and the atmosphere is expected to revert to the original composition.

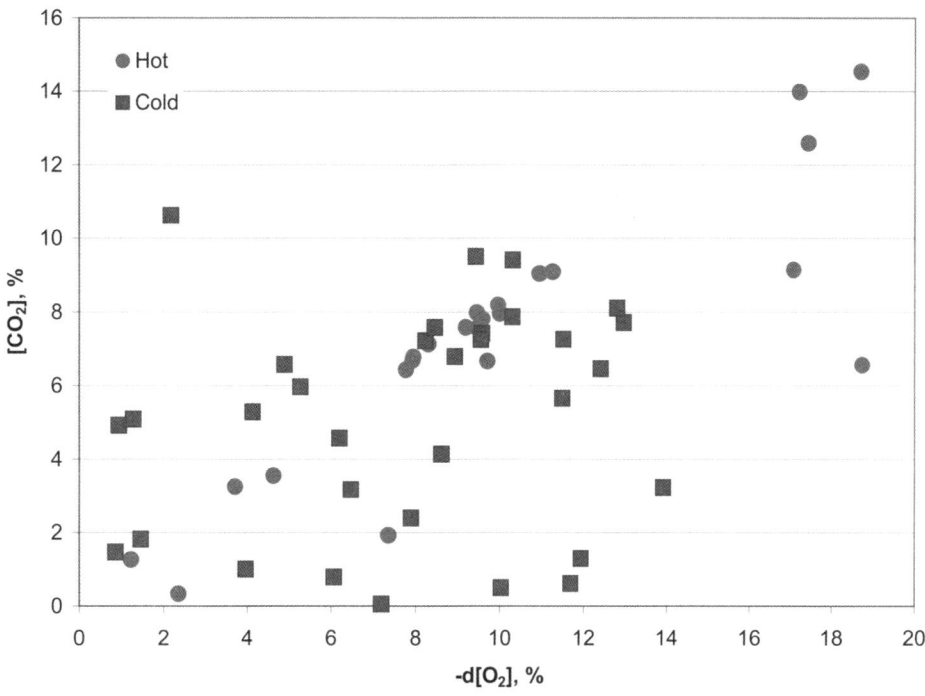

Figure 8. Variation in carbon dioxide (CO_2) concentration as a function of O_2 depletion ($-d[O_2]$) at hot and cold boreholes, Large mine-fire site.

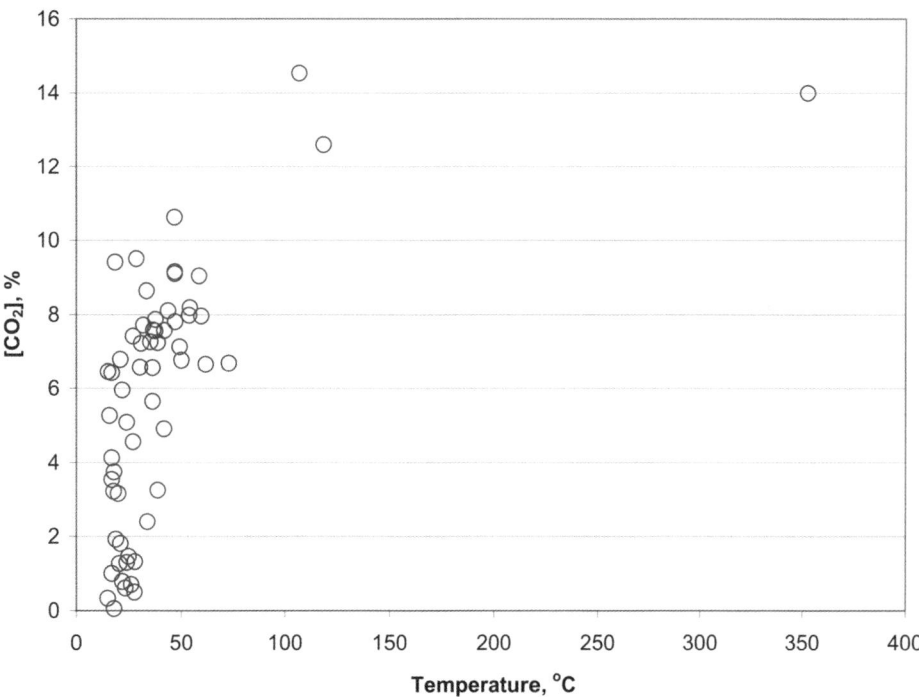

Figure 9. Variation in average carbon dioxide (CO_2) concentration versus average subsurface temperature at the Large mine-fire site.

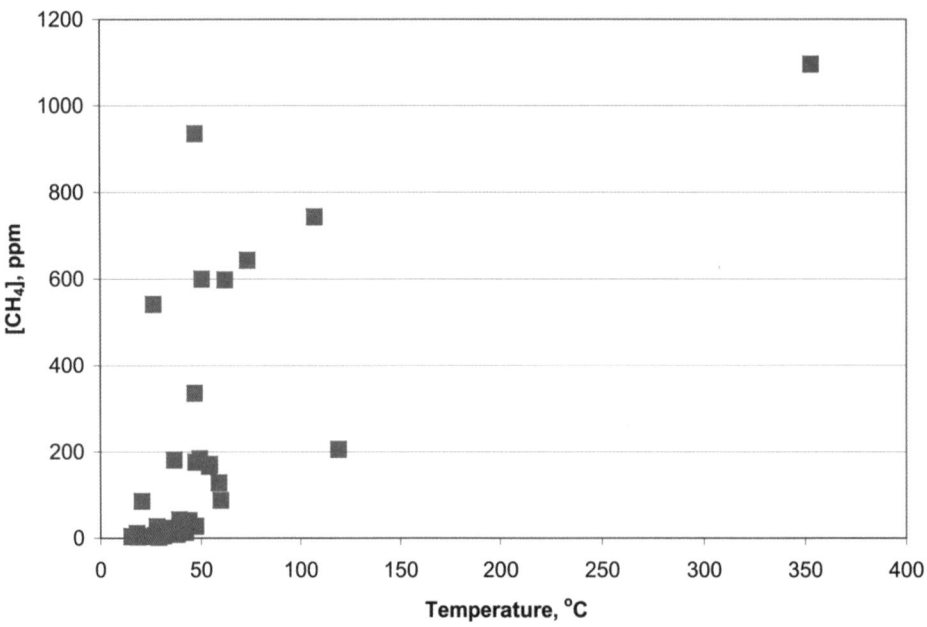

Figure 10. Average methane (CH_4) concentration at the Large mine-fire site versus average subsurface temperature.

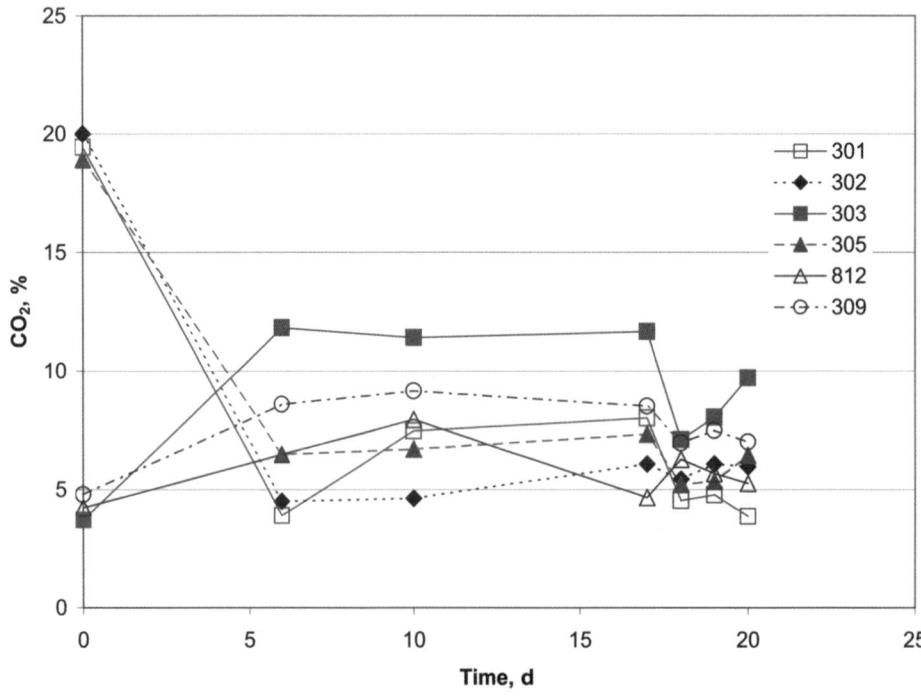

Figure 11. Change in carbon dioxide (CO_2) concentration over the test period at various boreholes at the Percy mine-fire site.

TABLE 5. INITIAL AND AVERAGE CARBON DIOXIDE
CONCENTRATION AND AVERAGE SUBSURFACE
TEMPERATURE AT THE PERCY MINE-FIRE SITE

Borehole number	Initial CO$_2$ (%)	Average CO$_2$ (%)	Borehole temperature (°C)
301	19.44	5.42	15.4
302	20.00	5.44	21.9
303	3.72	9.95	15.4
304	3.68	5.67	15.7
305	18.92	6.25	17.5
306	3.23	6.00	21.6
307	1.82	6.25	27.0
308	4.99	6.88	38.1
309	4.79	7.95	55.4
310	4.56	7.85	88.2
801	5.54	6.09	33.1
802	6.49	11.49	38.2
803	11.05	8.84	33.7
804	3.60	5.18	27.4
805	3.50	5.14	24.9
806	19.66	4.89	23.2
807	3.26	6.58	99.8
808	5.36	6.40	73.1
809	3.29	5.92	56.2
810	5.01	6.08	42.5
811	4.59	6.15	31.9
812	4.23	6.04	32.8
813	4.64	6.10	18.7

This condition did not hold for the Percy mine. There were significant differences in baseline gas composition between the initial concentrations of CO$_2$ and the average concentrations for the rest of the test period (Fig. 11). The concentration decreased at boreholes that had an initially high concentration of CO$_2$ and increased at boreholes where the concentration was initially low. The average initial CO$_2$ concentration over the entire site was 7.2%, and the average concentration was 6.6% for the rest of the test period. Apparently, gas migration within the mine over an extended period of time caused local increases in CO$_2$ concentration. However, neither the initial or long-term CO$_2$ concentrations were correlated to borehole temperature (Table 5). The CO$_2$ concentration was not well correlated with O$_2$ depletion during both baseline and communication tests (Fig. 12).

For normal or uncontaminated air, the value of the ratio N$_2$/O$_2$ was 3.7; higher values are related to O$_2$ depletion. In baseline and communication tests, the N$_2$/O$_2$ ratio indicated increased O$_2$ depletion near the areas of higher subsurface temperature (Fig. 13).

Average CH$_4$ concentration varied to a maximum of 0.13%; it was higher at boreholes with elevated subsurface temperatures (Fig. 14). Very low concentrations of CO were detected at this site, primarily at boreholes that had higher temperatures.

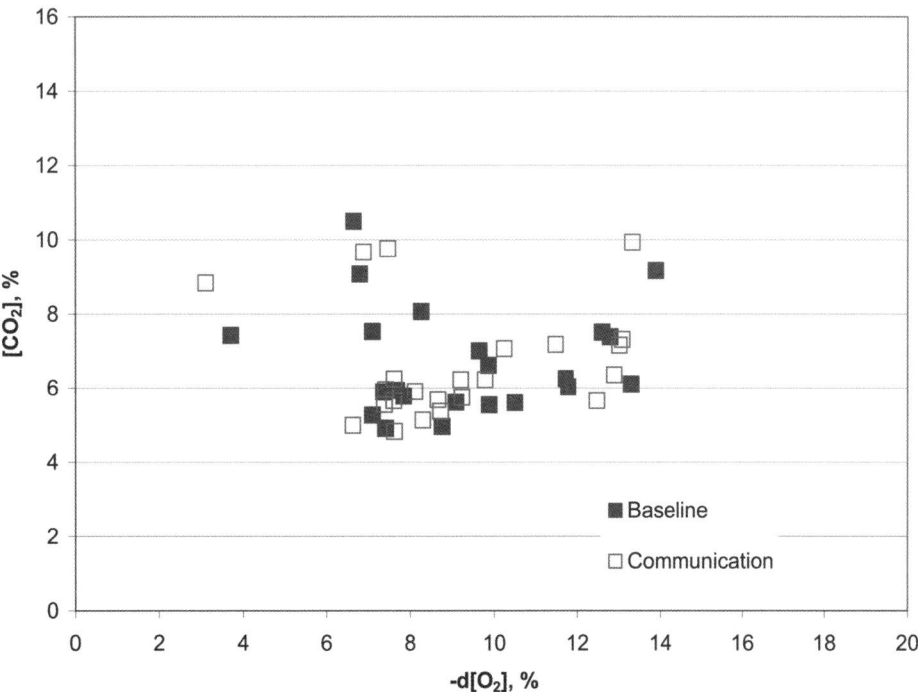

Figure 12. Variation in carbon dioxide (CO$_2$) concentration and oxygen depletion ($-d$[O$_2$]) during baseline and communication tests at the Percy mine-fire site.

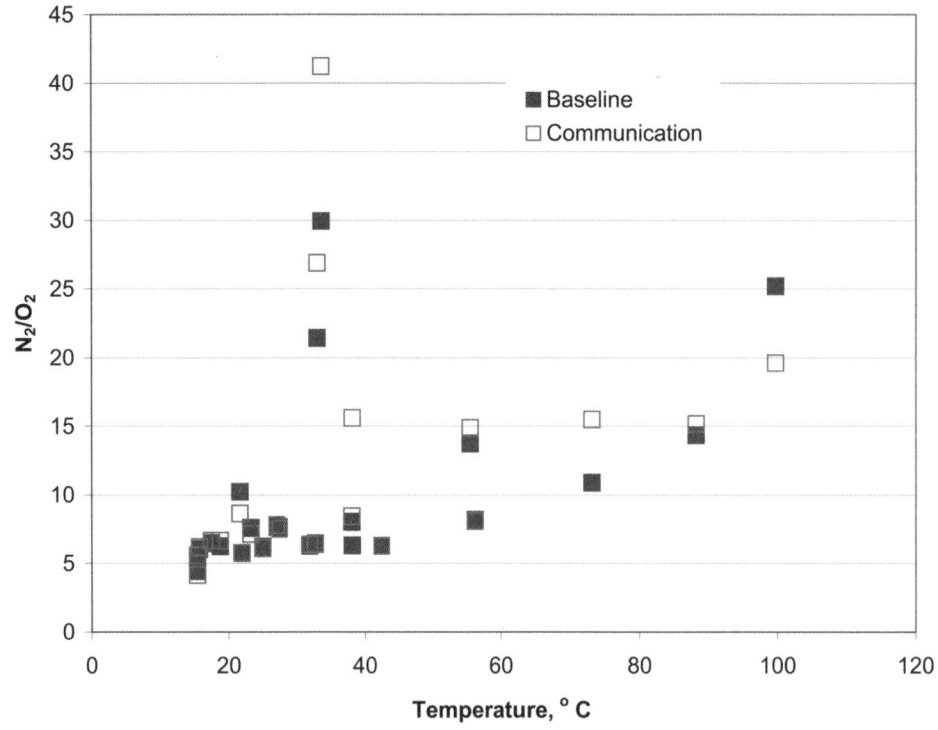

Figure 13. The nitrogen (N_2) to oxygen (O_2) ratio indicates O_2 depletion associated with higher subsurface temperature.

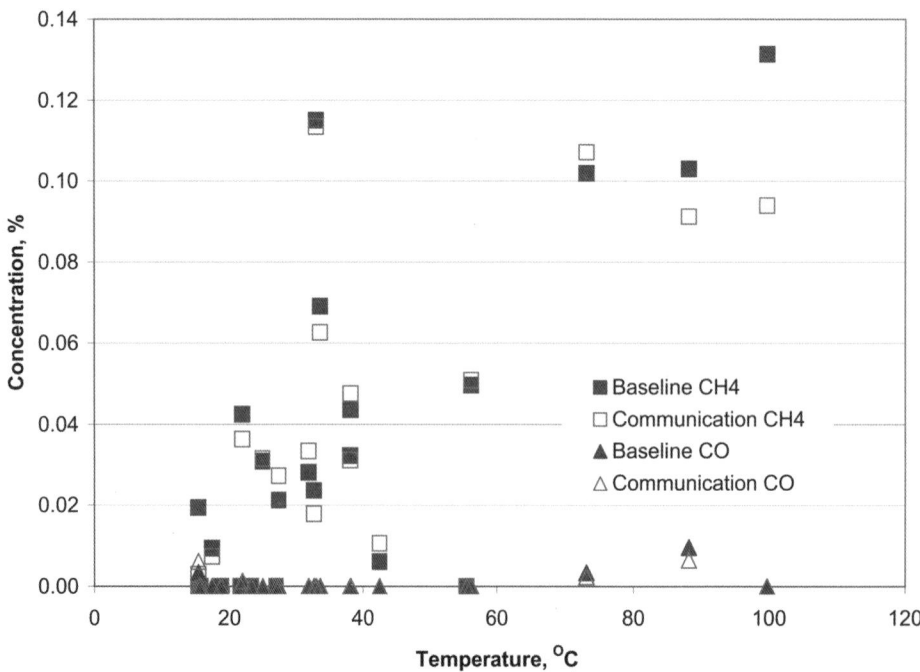

Figure 14. Higher concentrations of methane (CH_4) and carbon monoxide (CO) are detected at boreholes with elevated subsurface temperatures.

SUMMARY

Both laboratory experiments and analysis of mine-fire gases indicate that CO_2 generation is a function of the fixed carbon concentration of the coal, the temperature, and the availability of O_2. Based on these studies, CO_2 production is directly related to O_2 depletion, and its concentration increases as the temperature increases. On average, the CO_2 concentration at a mine fire increases by 0.7% for every 1% decrease in the O_2 concentration. As the O_2 concentration decreases to 2%, the CO_2 concentration approaches an apparent maximum of 15%.

Methane desorption from coal increases with increased temperature. The subsurface concentration of CH_4 increases ~0.001% per 1 °C increase in temperature. The production of CO in mine fires is usually evident only at higher temperatures (>50 °C) and low O_2 concentrations (<12%).

The subsurface generation of CO_2 and CO in mine fires is a function of temperature and O_2 concentration; temperature-dependent desorption controls the emission of CH_4 in mine fires. However, the subsurface concentration of these gases does not necessarily control the contribution of mine fires to atmospheric concentration of greenhouse gases. Physical factors, such as the degree of overburden fracturing and differences between subsurface and ambient temperature and barometric pressure, also influence the rate of gas emission from mine fires. The most serious effect of such fires is usually local, due to the production of CO and particulates. The contribution of underground coal-mine fires to the annual production of greenhouse gases (CO_2 and CH_4) is relatively small but significant. Controlling such fires could have local, regional, and global significance.

ACKNOWLEDGMENTS

The author gratefully acknowledges the support and assistance of R.F. Chaiken (U.S. Bureau of Mines [USBM] and National Institute of Safety and Health [NIOSH], retired), E.M. Burse (USBM and U.S. Department of Energy [USDOE]), L.E. Dalverny (USBM and USDOE, retired), T.R. Justin (USBM, retired), A.M. Kociban (USBM and USDOE, retired), and J.P. Slivon (USBM and USDOE, retired). Financial support for the field work was provided by the Appalachian Regional Office of U.S. Office of Surface Mining Reclamation and Enforcement (OSMRE) and by the Pennsylvania Department of Environmental Protection (PADEP).

REFERENCES CITED

Chaiken, R.F., 1977, Heat Balance in In-Situ Combustion: U.S. Bureau of Mines Report of Investigations 8221, 11 p.

Chaiken, R.F., Brennen, R., Heisey, B., Kim, A.G., Malinka, W., and Schimmel, J., 1983, Problems in the Control of Anthracite Mine Fires: A Case Study of the Centralia Mine Fire: U.S. Bureau of Mines Report of Investigations 8799, 93 p.

Clayton, J.L., 1998, Geochemistry of coalbed gas—A review: International Journal of Coal Geology, v. 35, p. 159–173, doi: 10.1016/S0166-5162(97)00017-7.

Johnson, W., and Miller, G., 1979, Abandoned Coal-Mined Lands: Nature, Extent and Cost of Reclamation: U.S. Bureau of Mines Special Publication 6–79, 29 p.

Kim, A.G., 1977, Laboratory Studies on Spontaneous Heating of Coal: A Summary of Information in the Literature: U.S. Bureau of Mines Information Circular 8756, 13 p.

Kim, A.G., 1995, Relative Self-Heating Tendencies of Coal, Carbonaceous Shales and Coal Refuse: U.S. Bureau of Mines Report of Investigations 9537, 21 p.

Kim, A.G., 1996, Results of Diagnostic Tests at the Percy Mine Fire Site: Pittsburgh, Pennsylvania, U.S. Bureau of Mines Report to the Office of Surface Mining, Reclamation and Enforcement, 40 p.

Kim, A.G., and Chaiken, R.F., 1993, Fires in Abandoned Coal Mines and Waste Banks: U.S. Bureau of Mines Information Circular 9352, 58 p.

Kim, A.G., and Dalverny, L.E., 1994, Mine fire diagnostics at the Large mine site, in Proceedings of the International Land Reclamation and Mine Drainage Conference: Pittsburgh, Pennsylvania, U.S. Bureau of Mines Special Publication SP 06D–94, v. 4, p. 139–147.

Kim, A.G., Justin, T.R., and Miller, J.F., 1992, Mine Fire Diagnostics Applied to the Carbondale, PA, Mine Fire Site: U.S. Bureau of Mines Report of Investigations 9421, 16 p.

McNay, L.M., 1971, Coal Refuse Fires: An Environmental Hazard: U.S. Bureau of Mines Information Circular 8513, 50 p.

OSMRE (U.S. Office of Surface Mining Reclamation and Enforcement), 2005, Abandoned Mine Land Inventory System (AMLIS), Underground Mine Fires: www.osmre.gov/aml/inven/zamlis.htm (accessed 31 March 2005).

Smith, A.F., and Lazarra, C.P., 1987, Spontaneous Combustion Studies of U.S. Coals: U.S. Bureau of Mines Report of Investigations 9079, 28 p.

Stracher, G.B., and Taylor, T.P., 2004, Coal fires burning out of control around the world: Thermodynamic recipe for environmental catastrophe, in Stracher, G.B., ed., Coal Fires Burning Around the World: A Global Catastrophe: International Journal of Coal Geology, v. 59, no. 1–2, p. 7–17.

Su, S., and Agnew, J., 2006, Catalytic combustion of coal mine ventilation air methane: Fuel, v. 85, no. 9, p. 1201–1210, doi: 10.1016/j.fuel.2005.11.010.

MANUSCRIPT ACCEPTED BY THE SOCIETY 7 MARCH 2007

The spontaneous combustion index and its application: Past, present, and future

Sezer Uludağ*
School of Mining Engineering, University of the Witwatersrand, Johannesburg, Private Bag 3, Wits 2050, South Africa

ABSTRACT

The Wits-Ehac Index was developed in 1987 at the School of Mining Engineering, University of Witwatersrand. In this research, several tests were done for a colliery in the Waterberg coalfield using a Wits-Ehac Index apparatus and were compared with the Glasser test. There was a good correlation between the two test types. Another data set was acquired from a risk assessment study of the Witbank coalfield using Geographical Information Systems (GIS). This study used several types of data, including type of coal, overburden thickness, Wits-Ehac Index, and chemical properties of coal. The research concluded that the Kleinkopje area is the most prone to spontaneous combustion, which agrees with historical observations. Improvements have been made on the apparatus that measures the Wits-Ehac Index, with future applications highlighted. Because of increasing spontaneous combustion events in the Witbank coalfield, there is a need for more in-depth research about the process.

Keywords: GIS, risk mapping, spontaneous combustion, Wits-Ehac Index.

INTRODUCTION

Following an experimental study funded by the Explosion Hazard Advisory Committee (EHAC) of the Department of Minerals and Energy, South Africa, M.J. Gouws (1987) developed an index called the Wits-Ehac Index. It uses the crossing-point temperature method and the slope of the differential temperature-time curve and is based on the following formula:

Stage II is the most important part of the curve—it starts from the point when the slope changes to a positive value. Various parts of the curve are shown in Figure 1. Stage II A is much steeper than Stage II B. The crossing-point temperature is the point at which the temperatures of the coal and the inert material are the same; the coal temperature eventually exceeds that of the inert material. Stage II is a good indicator of propensity to spontaneous combustion.

Coal with a high Wits-Ehac Index is more prone to spontaneous combustion than coal with a lower index value (Gouws, 1987). Tests have been conducted on coal samples from various collieries to determine the Wits-Ehac Index as part of a study to assess the spontaneous combustion risk within the Witbank coalfield (Uludağ, 2001), South Africa's premiere coal-producing region.

DESCRIPTION OF THE TEST APPARATUS

The test apparatus that was used for this research consisted of an oil bath, six coal and inert material cell assemblies,

---

*uludag@egoli.min.wits.ac.za

Uludağ, S., 2007, The spontaneous combustion index and its application: Past, present, and future, *in* Stracher, G.B., ed., Geology of Coal Fires: Case Studies from Around the World: Geological Society of America Reviews in Engineering Geology, v. XVIII, p. 15–22, doi: 10.1130/2007.4118(02). For permission to copy, contact editing@geosociety.org. ©2007 The Geological Society of America. All rights reserved.

a circulator, a heater, flow meters for airflow monitoring, air supply compressor, and a computer (Fig. 2).

The oil bath was a 40 L stainless-steel tank with dimensions of 710 mm × 300 mm × 200 mm (Fig. 3). The oil used had a low viscosity and high flash point and breakdown temperatures. Shell Thermia-B was used, which was the best option at the time that the test apparatus was designed. It has a flash point of 210 °C and maximum usable temperature of 320 °C.

The cell assembly (Fig. 4) consisted of six cells. Three of them were used for coal samples; the remaining three cells were filled with inert material, which in this study was calcined alumina. The cell assemblies were oil tight. Each cap that was used for sealing the cell had a platinum resistance thermocouple (PRT) inserted through its middle. The caps had a chimney built into them to release the fumes that were produced in the cell during heating of the coal samples (Fig. 5).

The air that was supplied during testing was directed to the bottom of the cell via a copper tube spiral (Fig. 4). The spiral was designed to be long enough to let the air reach the temperature of the oil before entering the cell. The flow meters controlled the flow of air, which was ~200 mL/min. They were located between the compressor and sample holders and were connected with plastic tubing.

The coal was kept as fresh as possible after it was taken from the mine. This was achieved by storing the coal in a sealed container until the time of preparation of the cell assembly and testing. A glove box, filled with nitrogen, was used for coal preparation. The coal was crushed and sieved to obtain –212 μm (–72 mesh size) coal sample in the glove box. The coal was weighed on an electronic scale, and a 20 g sample was used for each coal cell.

The test environment was kept at an initial temperature of 30 °C. The chimneys on the caps were kept sealed until the temperature stabilized at around 30 °C in all cells.

The test was started by switching on the heater and the air compressor, and data-capture program was initiated on a personal computer. The flow meters were adjusted to the required amount after the oil and cells reached 30 °C. The oil in the steel bath was heated up to a maximum of 200 °C in ~3–5 h. The time required to conduct the experiment depended on the quality of the oil, which deteriorated and caused accumulation of sediment at the bottom of the tank. This increased the viscosity of the oil and, therefore, the heating time of the test. For this reason, the oil had to be changed after several tests. The effect of the heating rate on the Wits-Ehac Index was not established.

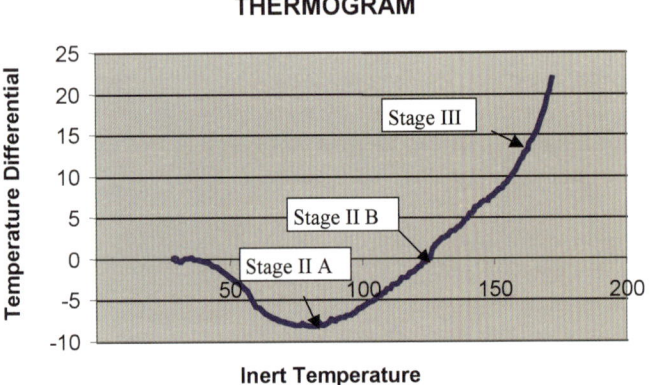

Figure 1. A differential thermal analysis thermogram.

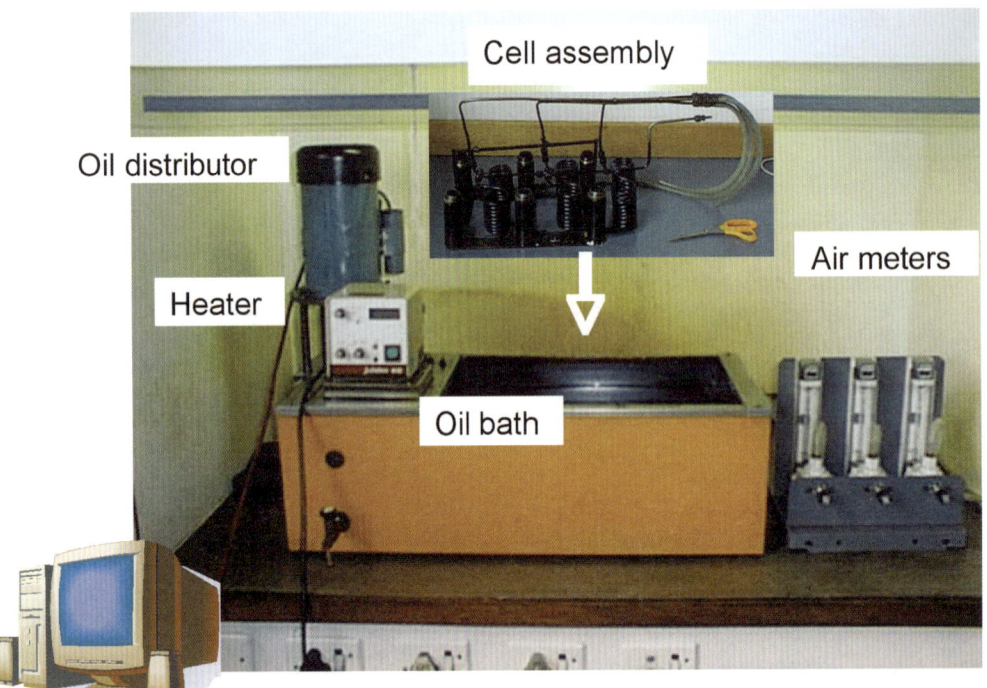

Figure 2. Wits-Ehac Index test apparatus setup.

Figure 3. The oil bath with external oil pump and heater attached.

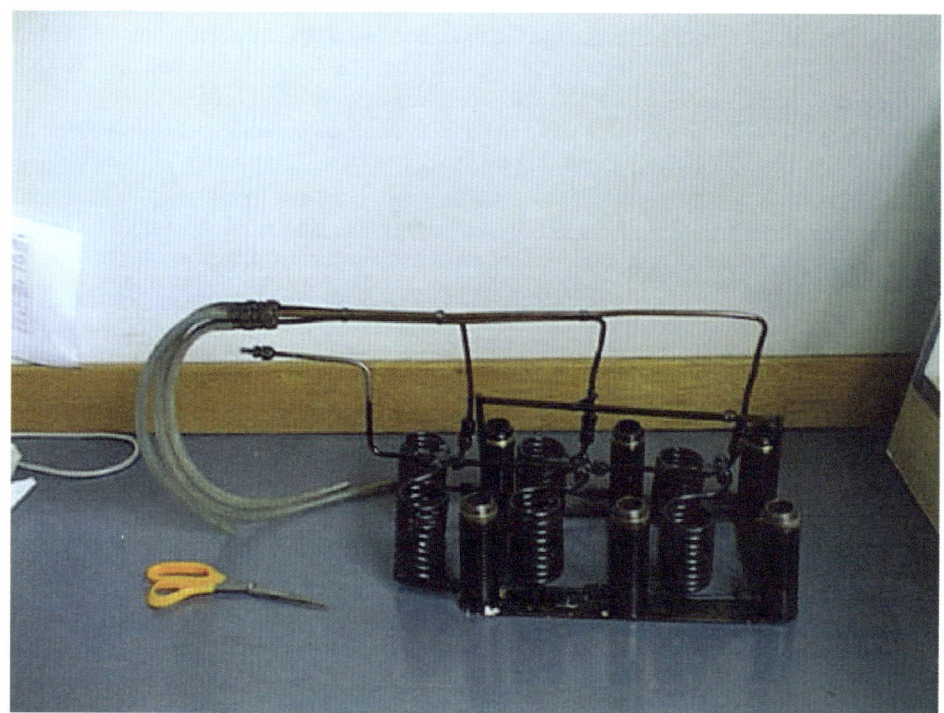

Figure 4. The cell assembly.

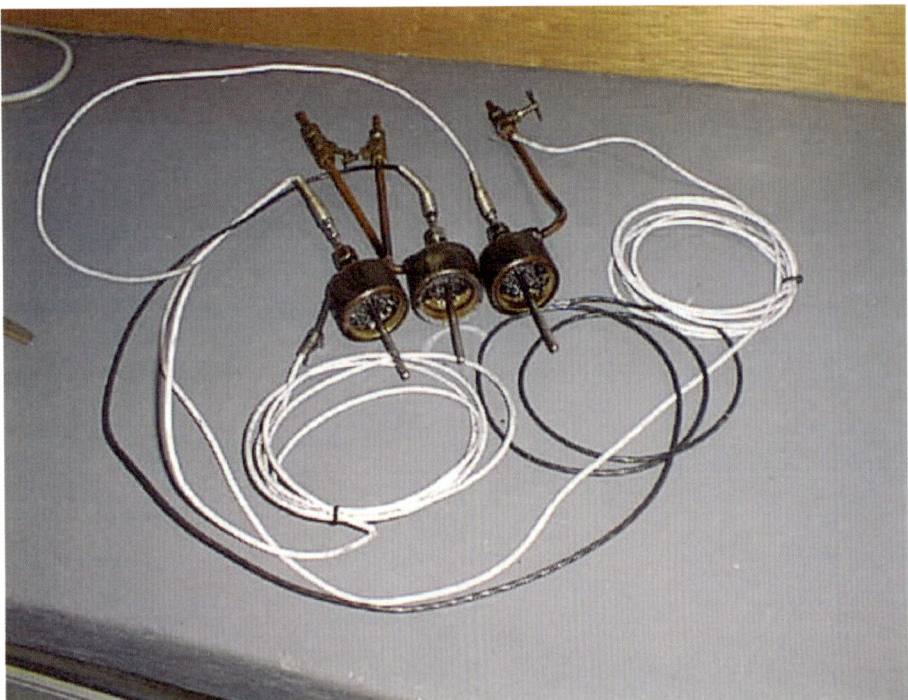

Figure 5. The cell caps with platinum resistance thermocouple (PRT) and the chimney.

DATA PROCESSING

A simple DOS operating computer program written in Basic, called Coaltemp, was used for data capture during testing in the early days, and real-time temperatures could be displayed on the computer screen. Because DOS-based programs can no longer be used with Windows, calculation of the Wits-Ehac Index is now achieved by using a simple template in Excel using the original code written in DBase. After the test ends, all of the data that was captured during the testing is transferred to the Excel template, and calculation of the Wits-Ehac Index is a copy and paste process.

A typical thermogram (Fig. 1) can be obtained by plotting the temperature difference between coal and inert material against the time in minutes. The slopes of the curves are calculated, which, in turn, are used to calculate the Wits-Ehac Index. The index that was calculated for the particular thermogram in Figure 1 is 4.38. The Wits-Ehac Index for South African coalfields ranges from 2 to 7. Coal with an index of less than 3 has a low risk for spontaneous combustion, whereas index values between 3 and 5 represent medium risk. Risk indices of 5–7 represent the highest liability for self-heating.

RECENT CHANGES TO THE APPARATUS

Initially, Sineax V 920 industrial data transmitters were used for converting analog signals to digital. After many years of use, the system was no longer capable of capturing data accurately, and the cables that connected the PRTs and the computer had deteriorated considerably. PRTs are known to give inaccurate data readings, and they need to be calibrated against a normal thermometer. Also, each time a PRT was broken, a new one had to be calibrated with readings from a normal thermometer. The calibration testing took ~3–5 h, and the temperatures of the PRT and mercury thermometer had to be recorded manually during the entire calibration test. This has been eliminated by changing the old PRTs to a more accurate thermocouple type.

A locally manufactured data-capturing system was used to convert analog data to digital data. This was achieved by placing an A/D (analog/digital) card into the computer.

The old Sineax transmitters have been replaced with Dickson temperature data loggers, which are almost one-tenth the size of the previous transmitters. The data logger connects the new thermocouples to the multipoint connector to communicate with software that is installed on the computer.

The new data-capturing system was selected to be compatible with the previous testing method. The data logger (Fig. 6) stores 32,512 sample points with an internal temperature range of $-40\ °F$ to $+176\ °F$ ($-40\ °C$ to $+80\ °C$) and a remote-temperature range of $-238\ °F$ to $+2000\ °F$ ($-150\ °C$ to $+1093\ °C$). The power source of the data logger is a 3 V lithium coin cell battery, which has a lifetime of 5 yr. A multipoint connector (MP100) connects the data logger to the computer by means of a logger interface cable.

The software that records the data and displays it in real time is Windows-based, and data viewing is possible in a few formats (e.g., table, graph). The scale of the graph can be set easily. The whole data-capturing and monitoring system is more user-friendly

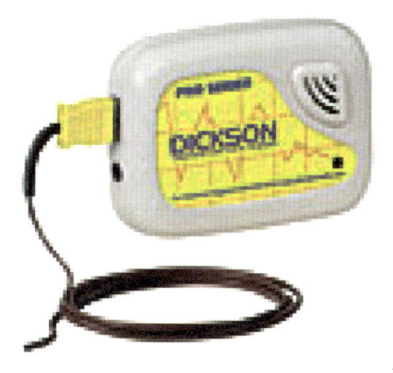

Figure 6. Data logger.

TABLE 1. GLASSER AND WITS-EHAC INDEX RESULTS

Sample name	Glasser	Wits-Ehac I	Wits-Ehac II
Bench 2-12	58	3.59	3.48
Bench 2-35	58	No data	No data
Bench 3-12	92	3.78	3.7
Bench 3-35	92	3.57	3.42
Bench 4-12	122	4.06	4.01
Bench 4-35	112	3.8	3.9
Bench 5	191	4.05	3.97
Bench 6	599	4.37	4.33
Bench 7a	145	4.2	4.3
Bench 7b	151	3.54	3.54
Bench 8	48	3.12	3.08
Bench 9a	325	4.15	4.27
Bench 9b	626	4.23	4.17
Bench 11-1.41	484	4.51	4.35
Bench 11-1.54	359	4.3	4.26

than the earlier one was because there is no longer a need to calibrate the temperature readings.

Another major problem is the quality of the oil that is being used. It deteriorates considerably after several uses and has to be discarded. The flash point of oil is very close to the temperature range requirement for coal testing, which presents a risk. The oil that would be most suitable is high-temperature silicon oil, which has a high flash point, is chemically more stable, and does not melt the plastic materials that are used during preparation of the assembly before testing. Silicon oils are new in South Africa, and there is no local producer of these oils. Means of obtaining this oil should be investigated in the future.

APPLICATION OF THE WITS-EHAC INDEX

Comparison of Wits-Ehac Index and Glasser Tests

Tests were undertaken on several coal samples that were taken from an opencast colliery called Grootegeluk in the Waterberg coalfield, South Africa. The samples obtained are shown in Table 1. Two values of the Wits-Ehac Index were obtained by using two different methods for calculations. The same coal samples also were tested with an oxygen absorption test that was developed by Glasser (Adamski, 2003). The Glasser test is a simple test that is performed by using a small amount of coal at ambient temperatures. The coal is contained in a reaction vessel with a burette, and the absorption of oxygen by the coal is deduced from the reduction of total gas volume. The test is continued until a milliliter of oxygen has been absorbed per gram of coal. The time necessary for absorption is used to estimate the reactivity value, which is expressed as milliliter of oxygen consumed per kilogram per day. A detailed description of the Glasser test can be found on page 138 in the dissertation by Adamski (2003).

The sample names represent the bench number from which they were taken, and they are listed in order of increasing depth. Gouws (1987) questioned whether the oxygen absorption tests are as reliable as the crossing-temperature methods. Figure 7 indicates that there is an acceptable visual correlation between

Figure 7. Comparison of Wits-Ehac Index and Glasser tests.

the Glasser test and the Wits-Ehac Index. It is not reasonable to expect a very close correlation due to a big difference in the way in which the tests are carried out. It also is evident from the test results that as the depth of coal increases, the risk for spontaneous combustion also increases. The Grootegeluk mine bench descriptions are given in Table 2.

Coal near the surface that has degraded over time as a result of exposure to atmospheric changes may absorb more oxygen because of an increase in naturally occurring fractures. Therefore, a low liability of self-heating is measured.

Sample 10 is the minimum value in the graph, which is the result for interlayered sandstone (Table 2). It is believed that the geologic difference of the coal in this area causes the graph to be nonlinear.

When assessing the risk for spontaneous combustion, exclusion of some of the contributors of risk (e.g., type of coal, annual rainfall, groundwater conditions, type of mining method) might result in unrealistic conclusions. It is not certain if this is true, and further research and testing must be conducted.

Mapping Risk by Geographic Information System (GIS) Techniques

According to the literature survey regarding the measurement and assessment of risk for spontaneous combustion, statistical methods that use data in tabular form do not give satisfactory results. It is evident that geological formations should be investigated in their spatial positions because they correlate better in their depositional environment. Based on this fact, Uludağ (2001) attempted to map the spontaneous combustion risk by using a GIS (Geographic Information Systems) tool. The GIS software used was ILWIS (Integrated Land and Water Information System).

GIS can be useful to combine various kinds of data input (e.g., Boolean, numerical, descriptive). The contributing factors of spontaneous combustion are divided as coal properties, geological properties, and environmental properties. The coal factors that increase the risk for spontaneous combustion are listed in Table 3, together with the weighting values that were assigned to them as part of the 2001 research by Uludağ et al. Some of the classification factors used in Table 3 were adopted from Singh et al. (1984). The steps by which a risk map is generated are as follows: (1) Assign weight values to the classes of the parameter maps. (2) Renumber the parameter maps to weight maps. (3) Combine the weight maps into one single risk map. (4) Classify the combined weight map into a final hazard map.

All of the factors are combined into a general risk map by using raster maps that are generated from individual coal factor maps. Raster maps are thematic maps where each pixel is color-coded with a value assigned to it. For this research, most of the contributing factors were assumed to be equally important. There is no direct method to assess the importance of contributing factors.

TABLE 2. GROOTEGELUK MINE COAL ZONES AND MINING BENCHES (ADAMSKI, 2003)

Bench	RD	Thickness	Bench description
1	2.51	16.5	Overburden
2	1.74	13.5	Bright coal
3	1.83	16.0	Bright coal
4	1.86	16.0	Bright coal
5	1.9	16.7	Bright coal
6	1.67	4.2	Dull coal
7a	2.41	5.7	Interburden carbonaceous shale
7b	1.58	1.6	Dull coal
8	2.41	3.9	Interburden carbonaceous shale
9a	1.58	2.8	Dull coal
9b	1.58	5.3	Dull coal
10	2.49	3.9	Interburden sandstone
11	1.52	4.1	Dull coal

Note: RD—relative density.

TABLE 3. RISK RATING OF COAL FACTORS USED IN RISK MAPPING (SINGH ET AL., 1994)

Coal factor	Range	Ratings	Weights
Wits-Ehac	0.0–3.0	Low	0
	3.0–5.0	Moderate	2
	5.0–7.0	High	4
Calorific value	17.81–20.0	Very low	0
	20.0–24.0	Low	1
	24.0–28.0	Moderate	2
	28.0–31.43	High	4
Ash content	<20.0	Low	0
	20.0–30.0	Moderate	1
	>30.0	High	0
Volatile matter	20.5–24	Low	0
	24.0–27.0	Moderate	1
	27.0–33.7	High	2
Moisture	1.70–3.11	Low	2
	3.11–3.85	Moderate	1
	3.85–5.62	High	0
Pyrite	1.21–3.71	Low	0
	3.71–4.62	Moderate	1
	4.62–7.43	High	2
Sulfur	0.24–0.68	Low	0
	0.68–1.0	Moderate	1
	1.00–1.67	High	2
Vitrinite	14.0–30.0	Low	0
	30.0–42.0	Moderate	1
	42.0–62.0	High	2
Fixed carbonate	38.7–48.0	Low	0
	48.0–52.0	Moderate	1
	52.0–59.5	High	2
Reactive inertinite	10.0–14.0	Very low	0
	14.0–18.0	Low	1
	18.0–22.0	Moderate	2
	22.0–26.0	High	3
P_2O_5	0.19–0.68	Very low	0
	0.68–1.0	Low	1
	1.00–1.5	Moderate	2
	1.50–4.17	High	3
Liptinite	2.069–3.7	Low	0
	3.70–4.7	Moderate	1
	4.70–5.4	High	2
	5.40–7.74	Very high	3
Depth factor	<40.0 m	High	2
	40.0–300.0 m	Low	0

In addition, as the number of variables increases in a system, the solution formulae become more complex.

Recently, many researchers have begun to use fuzzy logic techniques to alleviate the problem of uncertainty in systems with many variables. It seems that the risk assessment problem is an ideal case for the use of fuzzy logic techniques. This may be another topic of further research.

The final map is shown in Figure 8. The total risk changes from a minimum of 8 to a maximum of 36. Depending on the level of risk, these numbers can be regrouped into high, medium, or low risk. The validity of the map could be established by only one datum point. A black arrow in Figure 8 shows this area. The Kleinkopje colliery is located there, and it has had historically serious spontaneous combustion problems.

FUTURE APPLICATIONS

The lack of experiments to determine accurate weighting factors decreases the level of confidence of the final risk map. The following should be investigated further against the Wits-Ehac Index:
- Depth of mining;
- the time of exposure of coal to oxygen in the long term (e.g., time of exposure of coal until pillar extraction time, opencast mine, longwall standing time);
- degree of fracturing of coal;
- degree of porosity;
- degree of moisture;
- the effect of particle size and distribution;
- the effect of hardness of coal;
- the rank of coal;
- the effect of macerals and mineralization of coal; and
- the location of the sample (e.g., pillar, face, conveyor belt, roof, floor).

To improve the reliability of testing, the following need to be measured against the Wits-Ehac Index:
- The effect of heating rate of heat-transfer medium;
- the freshness of coal in the short term (e.g., time from its in situ position to laboratory testing);
- the type of heating environment (organic oils, sand, silicon oils, salt baths);
- the method of calculating the slope change in the differential thermal analysis graph; and
- the sample preparation procedure.

By carrying out spontaneous combustion experiments to answer these questions, an expert system could be developed based on the trends that are obtained. A reliable expert system is possible if the limitations of testing and the effects of the variables are understood. This may, in turn, help mining engineers to avoid spontaneous combustion problems.

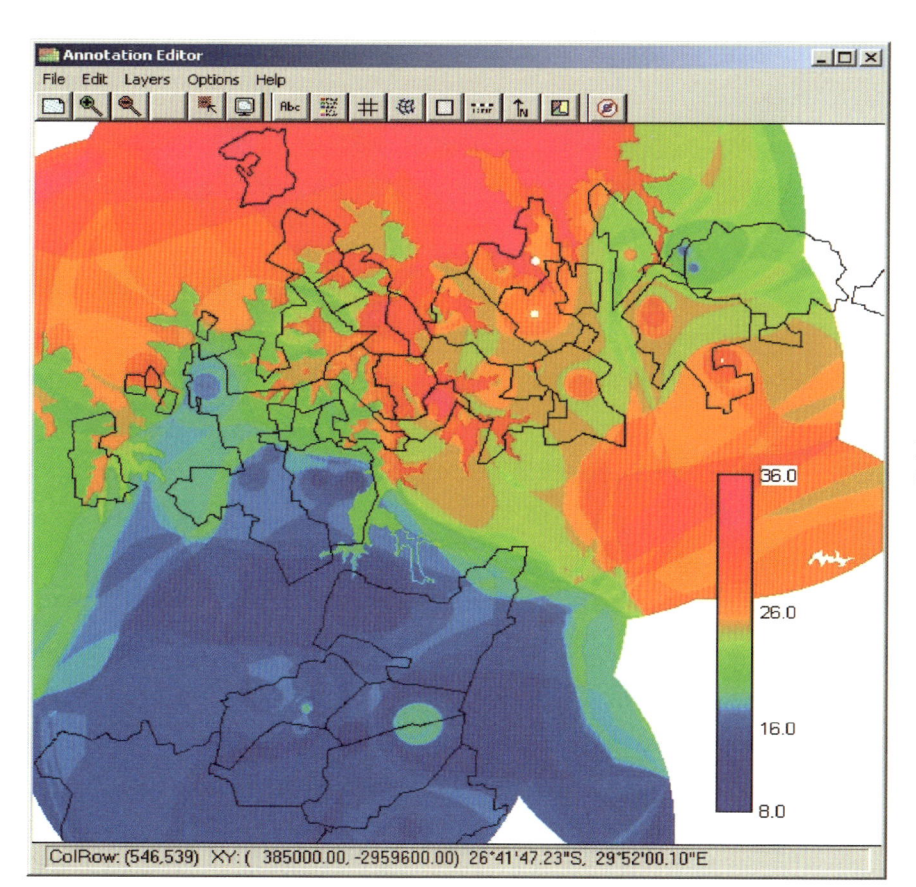

Figure 8. The total risk map for the Witbank coalfield.

CONCLUSIONS

The risk for spontaneous combustion has many contributors, and it is a challenge to accommodate all of the factors. In the past, a few researchers have attempted to assess the risk for spontaneous combustion by using expert systems.

Fuzzy logic applications have been implemented in computer and electrical engineering and control systems. It also has been applied in the mining engineering industry in many ways when uncertainty is common. Fuzzy logic can work with uncertainty better than any other system. Probability-based decision-making can be challenging when the information is nonlinear and there are many variables. Problems like spontaneous combustion could be investigated best by using fuzzy logic systems.

ACKNOWLEDGMENTS

This author thanks the Coaltech, Council of Scientific and Industrial Research, in South Africa for sponsoring the research, H.R. Phillips of the School of Mining Engineering, University of Witwatersrand, and N.H. Eroglu for a comprehensive review. In addition, special thanks are due to G.B. Stracher, East Georgia College, for his encouragement during the preparation of this manuscript.

REFERENCES CITED

Adamski, S.A., 2003, The Prevention of Spontaneous Combustion in Backfilled Waste Material at Grootegeluk Coal Mine [Ph.D. thesis]: Johannesburg, South Africa, University of Witwatersrand, 153 p.

Gouws, M.J., 1987, Crossing Point Characteristics and Differential Thermal Analysis of South African Coals [Master's thesis]: Johannesburg, South Africa, University of Witwatersrand, 165 p.

Singh, R.N., Demirbilek, S., and Turney, M., 1984, Application of spontaneous combustion risk index to mine planning, safe storage and shipment of coal: Journal of Mines, Metals and Fuels, v. 7, p. 347–355.

Uludağ, S., 2001, Assessing Spontaneous Combustion Risk in South African Coal Mines using a GIS Tool [Master's thesis]: Johannesburg, South Africa, University of Witwatersrand, 154 p.

Uludağ, S., Phillips, R.H., and Eroglu, N.H., 2001, Assessing spontaneous combustion risk in South African coal mines using a GIS tool, *in* Unal, E., ed., Proceedings of the 17th International Mining Conference and Exhibition in Turkey: Ankara, Kozan Publisher.

MANUSCRIPT ACCEPTED BY THE SOCIETY 7 MARCH 2007

Geological models of spontaneous combustion in the Wuda coalfield, Inner Mongolia, China

Daiyong Cao*
Xinjie Fan

State Key Laboratory of Coal Resources and Safe Mining, China University of Mining and Technology, Beijing 100083, China

Haiyan Guan

Beijing Remote Sensing Corporation, Shenhua Group, Beijing 100085, China

Chacha Wu
Xiaolei Shi

School of Resource and Safety Engineering, China University of Mining and Technology, Beijing 100083, China

Yuerong Jia

Wuda Mine Ltd., Shenhua Group, Inner Mongolia 016040, China

ABSTRACT

Spontaneous combustion of coal seams is a complicated process that involves complex physical and chemical interactions between the seam and its surrounding environment; the ultimate results are a function of the interplay of internal and external conditions. Based on geologic field investigations and comprehensive analyses, four models of spontaneous combustion for coal were established: (1) a genesis-type model that is based on the coal characteristics that lead to spontaneous combustion; (2) a coal-fires propagation model, namely how spontaneous combustion propagates through a coal seam; (3) a model for the progressive stages and products of a coal fire; and (4) a cross-sectional model of zones that are conducive to spontaneous combustion in a mined-out area. These models provide a scientific and theoretic basis for monitoring and extinguishing the spontaneous combustion of coal.

Keywords: coal, spontaneous combustion, geological models, Wuda coalfield.

*cdy@cumtb.edu.cn

INTRODUCTION

China is a major coal-producing country, and it accounts for about one-third of the annual global output of coal. Coal is one of China's primary energy sources, it provides about three-fourths of the energy that is consumed in the whole country, and it is expected to be its principal energy source for the foreseeable future (Mao and Xu, 1999).

Underground coal fires have been reported from major coal-producing countries, including China, the United States, India, Australia, Canada, South Africa, Russia, and Indonesia (Prakash and Gupta, 1999; Walker, 1999; Stracher and Taylor, 2004; Heffern and Coates, 2004). Northern China has an abundance of shallow and thick coal beds, and extensive underground coal fires exist in the Xinjiang autonomous region, the Ningxia Hui autonomous region, and the Inner Mongolia autonomous region (Fig. 1). The burning of coal induced by spontaneous combustion destroys precious coal reserves, endangers the safety of coal miners, causes desertification, produces noxious or toxic fumes that affect public health, and impairs economic development in the regions where it occurs (Guan et al., 1998; Zhang et al., 1998; Stracher, 2002; Finkelman, 2004). Environmental problems caused by the spontaneous combustion of coal seams were listed on "China's 21st-Century Agenda" by the Chinese government following the "The World Environment and Development" conference that was held in 1992 (Guan et al., 1998; Tan, 2000). The research, monitoring, and governing of coal fires is an academic and technical problem that needs to be solved urgently by scientists in the field of coal science and technology.

The spontaneous combustion of coal seams takes place principally to the north of the Qinling and Kunlun Mountains. This is especially true in the arid northwestern part of China, which extends from 35°N to 50°N and from 73°E to 135°E. As of 1998, there were 56 coal fires, which varied in size from 17 km² to 20 km² (Guan et al., 1998). At the present, eight coal fires have been extinguished and two coal fires are being controlled by fire-fighting methods.

GEOLOGICAL SETTING

The Wuda Coal Mine area is found near Wuhai City, in the Inner Mongolia autonomous region, in the northern section of the Helan Mountains, south of the Ulanbuhe Desert, west of the Yellow River, and east of Bayinshandan. The Wuda Coal Mine area is 16 km long from north to south, 13 km wide from east to west, and extends over a total area of 200 km². The Wuda coalfield, which extends from 39°27′00″N to 39°34′04″N and from 106°34′41″E to 106°38′41″E, is located in the northwestern part of the Wuda Coal Mine area. Its length is 10 km from north to south, its width ranges from ~3 km to 5 km from east to west, and the total area is ~35 km².

The Wuda coalfield consists of coal-bearing strata of Pennsylvanian and Permian ages, including the Pennsylvanian Benxi Formation (C_{2b}), the Pennsylvanian Taiyuan Formation (C_{2t}), the Early Permian Shanxi (P_{1s}) and Xiashihezi (P_{1x}) Formations, and the Late Permian Shangshihezi Formation (P_{2sh}). C_{2t} is the major coal-bearing formation; it consists of sandstone, limestone, mudstone, and mineable coal beds, such as the no. 9, no. 10, no. 12, no. 13, and no. 15 coals. The no. 9 coal formed in a tidal delta-plain setting (Peng and Zhang, 1995; Dai et al., 2003).

Wuda Mining Limited Liability Company, which is part of the Shenhua Group, was established in 1958, and is composed of three coal mines: Wuhushan, Suhaitu, and Huangbaici. The company's designed throughput is 3.9 million tons per year, whereas the actual throughput is 4.1 million tons per year; this is the coking coal production base of the Inner Mongolia autonomous region, and it supplies a large amount of coal for the coking plant and the inhabitants' personal use. In 1961, spontaneous combustion initially was found in the no. 9 and no. 10 coal seams in the Suhaitu mine. From the 1970s to the early 1990s, because of the large number of small mineshafts that was being exploited, spontaneous combustion occurred in most seams, including the no. 1, no. 2, no. 4, no. 6, no. 7, no. 9, no. 10, and no. 12 coal seams. By the end of 1995, 2 million square meters of the surface

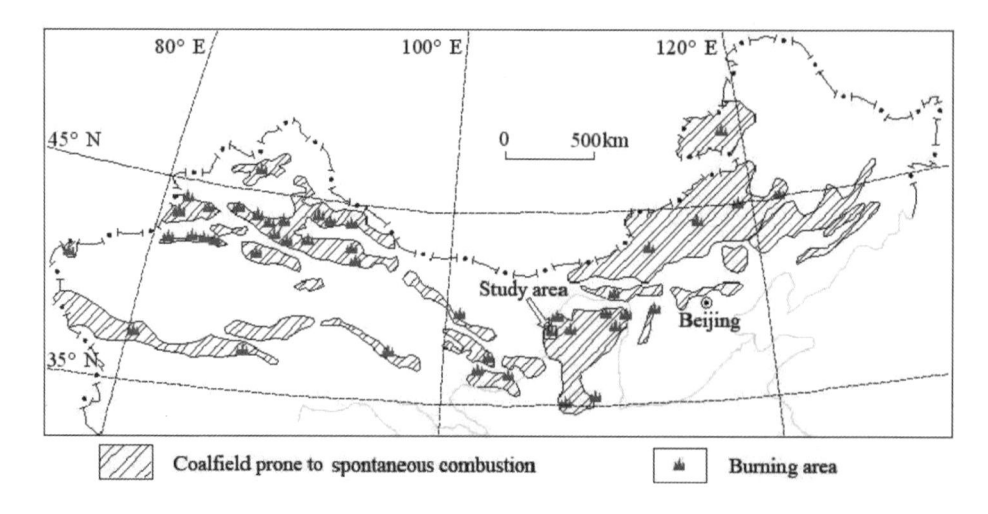

Figure 1. Distribution of coal fires in northern China and location of the study area (modified after Guan et al., 1998).

were affected by coal fires. In July 2002, a thorough investigation of the Wuda coalfield was conducted by Beijing Remote Sensing Corporation (BRSC); it confirmed that there were 26 fire areas burning 3.076 million square meters. In December 2004, spontaneous combustion had coalesced into 16 fire areas of 3.496 million square meters, as noted in the complementary investigation report by BRSC. In 2.5 yr, the fires' area had increased by 0.42 million square meters or nearly 13.7% (Cao et al., 2005).

MODEL OF SPONTANEOUS COMBUSTION GENESIS TYPES

The spontaneous combustion of a coal seam involves complicated interactions among its physics, chemistry, and environment, and it results from the interplay of various internal and external conditions (Wei, 1998; Wang and Zhang, 1990; Erdogan and Vedat, 2002; Cao et al., 2005). The following three sections discuss various internal and external conditions of spontaneous combustion, and based on these conditions, we establish the model of spontaneous combustion genesis types.

Internal Reasons

Various methods of testing and data processing were used to study the internal control factors of coal seam spontaneous combustion in the Wuda coalfield. The relationship between a coal seam's tendency for spontaneous combustion and some physical and compositional factors, such as seam thickness, metamorphic degree (rank), ash content, caloric value, sulfur content, sulfide mineral content, and maceral content, were analyzed. The physical and compositional parameters of Wuda coal are shown in Table 1. Based on the results of cluster analysis testing and statistical data of the Wuda coal, we identified three types of coal seams that are prone to spontaneous combustion because of internal reasons. They are: (1) seams that support combustion because of their high sulfur and sulfide mineral content (no. 9, no. 10, and no. 12 seams); (2) seams that are prone to oxidation because of their great thickness and low metamorphic degree (rank) (no. 1, no. 2, and no. 4 seams); and (3) seams that are transitional between the first two types (no. 6 and no. 7 seams) (Cao et al., 2005).

External Conditions

A coal seam's environmental conditions also have an important effect on its tendency to combust spontaneously. A field investigation revealed several environmental conditions that influence spontaneous combustion: (1) climate (intensity of the solar radiation and rainfall), (2) terrain (flat or rolling), (3) geological setting (structure, attitude of stratum, lithologic composition of the coal roof and floor), and (4) human activities (small mineshafts and state mine excavation). In short, spontaneous combustion of coal is attributed to numerous factors.

The 16 fire areas that were studied were divided into three types according to the locus of fires: mined-out–area fires, small mineshaft fires, and coal outcrop fires (Fan et al., 2006).

Model of Genesis Types

Various internal and external conditions promote spontaneous combustion in the Wuda coalfield, Inner Mongolia. Using three kinds of internal conditions and three kinds of external conditions, we established a model for nine types of coal seam spontaneous combustion that are found in the Wuda coalfield (Table 2).

TABLE 1. PHYSICAL AND COMPOSITIONAL PARAMETERS OF WUDA COAL SEAMS

Coal seam	Thickness (m)	Moisture (%)	Ash (%)	Volatile yield (%)	Caloric value (MJ·kg^{-1})	Total sulfur (%)	Pyrite (%)	Vitrinite (%)	Inertinite (%)
1	2.60	1.08	29.35	33.10	26.42	0.40			
2	4.45	0.99	22.27	31.73	33.30	0.48			
4	4.10	0.82	32.10	33.51	33.71	0.66			
5	0.93	0.94	19.14	33.04	32.50	0.6			
6	1.99	0.95	27.14	30.25	34.13	0.92			
7	2.96	0.94	23.80	31.11	34.40	0.75			
8	1.20	0.95	33.18	34.42	33.00	1.49			
9	2.96	0.73	14.19	31.42	35.55	3.14	2.4	65.6	26.6
10	2.13	0.84	17.08	32.05	35.25	3.02	3.1	64.6	26.4
12	5.01	0.94	23.54	30.02	34.61	2.29	1.9	54.1	25.5
13	1.24	0.86	24.02	29.15	34.12	0.79	0.8	59.2	21.8
14	0.84	0.79	38.12	30.74	33.78	1.23			
15	1.64	1.00	26.05	27.18	34.45	1.69	1.4	50.4	35.6
16	1.22	0.76	36.70	30.36	33.29	1.13			
17	1.03	0.74	33.90	29.87	34.07	1.99			

Note: Italics denote coal seams that are affected by spontaneous combustion. Thickness, moisture, ash, volatile yield, caloric value, and total sulfur were provided by the Department of Geology, Wuda Mining Limited Liability Company. Other data were determined with a microscope.

TABLE 2. SPONTANEOUS COMBUSTION GENESIS TYPES IN WUDA COAL SEAMS

Types from external conditions (i)	Types from internal reasons (j)		
	High-sulfur combustion-supporting ($j = 1$)	Thick layer prone to oxidation ($j = 2$)	Transition ($j = 3$)
Mined-out area fires ($i = 1$)	N_{11}	N_{12}	N_{13}
Small mineshaft fires ($i = 2$)	N_{21}	N_{22}	N_{23}
Coal outcrop fires ($i = 3$)	N_{31}	N_{32}	N_{33}

MODEL OF COAL-FIRE PROPAGATION

By combining high-precision remote-sensing images with geologic field investigations, we found that structural cracks control the development of combustion cracks and collapse cracks, and that combustion cracks expand on the basis of structural cracks and collapse cracks (Fig. 2).

Through field investigation and comprehensive analysis, we established a model of coal-fire propagation. Coal-fire propagation can be divided into three steps. On the two-dimensional plane, the model can be generalized from a point (combustion center) to a line (combustion zone), then from a line to a plane (combustion system). In three-dimensional space, it can be generalized from a spot to a plane, and then to a body. In short, the propagation model represents a process that extends gradually from several combustion points to a large-scale coalfield fire.

Combustion Spot–Combustion Center

Coal fires initiate in one spontaneous-combustion spot (i.e., a combustion center). This spot occurs where the seam has a plentiful supply of oxygen and a favorable environmental condition for heat collection. At this kind of spot, the coal temperature rises gradually to the ignition temperature, and the coal begins to combust. During geologic field investigations, these kinds of spots were easily observable—they exhibited smoking and jetting of gas and precipitation of sulfur, mirabilite, and coal tar (Fig. 3).

Combustion Line–Combustion Zone

Several combustion spots may extend along a crack, including structural cracks and collapse cracks, to form a combustion zone where there is abnormal temperature, smoke, and gas (Fig. 4). Field investigations showed that collapse cracks between coal pillars and mined-out areas typically are the loci where plentiful remaining coal, coal pillars, and favorable-ventilation

Figure 2. Combustion cracks (A) and collapse cracks (B) and their relationship to structure cracks.

Figure 3. Combustion spot with sulfur and mirabilite.

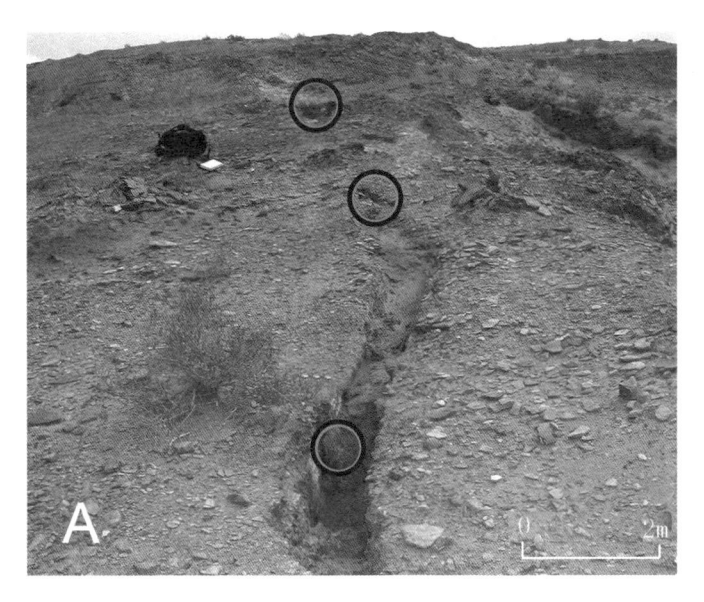

Figure 4. Combustion line formed by several combustion spots. (A) Three combustion spots. (B) Four combustion spots.

conditions can be found. These kinds of combustion spots connect to form a combustion line.

Combustion Plane–Combustion System

Several combustion cracks may cross each other to form a combustion plane (i.e., a combustion system or fires area). Some combustion cracks are parallel, and others are interlaced and cut across each other (Fig. 5).

MODEL OF THE PROGRESSIVE STAGES AND PRODUCTS OF COALFIELD FIRES

In 2004, we conducted three field investigations over 40 d. Based on field observations, comprehensive study, and systematic

Figure 5. Combustion plane–formed combustion cracks interlaced with each other. (A) Polygonal combustion cracks in no. 10 fire area. (B) Interlaced combustion cracks in no. 8 fire area.

TABLE 3. MODEL OF PROGRESSIVE STAGES AND PRODUCTS OF COALFIELD FIRES

Stage	Character and product	Temperature increase	Temperature decrease	Character and product	Stage
VIII. Continuous combustion	Red flame; high-temperature gas	↑ >800 °C	↓	Lava breccias; baked rocks; natural coke	IX. Baked rocks forming
VII. Semicoke combustion	Blue smoke	↑ 501–800 °C	↓	Baked slate; white powder	
VI. General combustion	High-temperature gas; grayer smoke	↑ 301–500 °C	↓	Nacarat baked sandstone	
V. Combustion system forming	Black smoke	↑ 201–300 °C	↓	Saffron baked loess	
IV. Combustion center forming	White and black smoke; heat airflow	↑ 101–200 °C	↓	Sulfur and mirabilite	X. Mineral precipitation
III. Spontaneous combustion	White and black; smoke; coal tar	↑ 61–100 °C	↓		XI. Dormancy
II. Oxidation	White smoke; wet and heat airflow (H_2S)	↑↓ 31–60 °C	↓	White sediment; baked slate	XII. Extinguishment
I. Weathering	Coal body weathered into powder	↑↓ 0–30 °C			

analyses, we established a model of the progressive stages of coal fires (Table 3).

This model divides a coalfield fire into 12 successive stages and two major combustion center formation, combustion system formation, general combustion, semicoke combustion, and continuous combustion. The decreasing temperature phase includes four stages: formation of baked rocks, mineral precipitation, dormancy, and extinguishment. The temperature, combustion character, and products of every stage are different (see Table 3). This model simulates every stage from initiation to extinguishment; however, all coal fires do not follow the full progression. For example, if the heat from oxidation of coal diffuses freely, the oxidation stage may stop, the temperature of coal could decrease back to that of the weathering stage, and spontaneous combustion may not occur. In the dormancy stage, the coal temperature can increase again, and the combustion phenomenon can recur. These two stages are strongly dependent on oxygen supply and ventilation.

MODEL OF ZONES CONDUCIVE TO SPONTANEOUS COMBUSTION IN A MINED-OUT AREA

It is difficult to simulate the development of coal fires in three-dimensional space under the complex conditions of coal mines. Based on the contrast between underground and surface conditions, through the field investigation of eight fire areas (no.

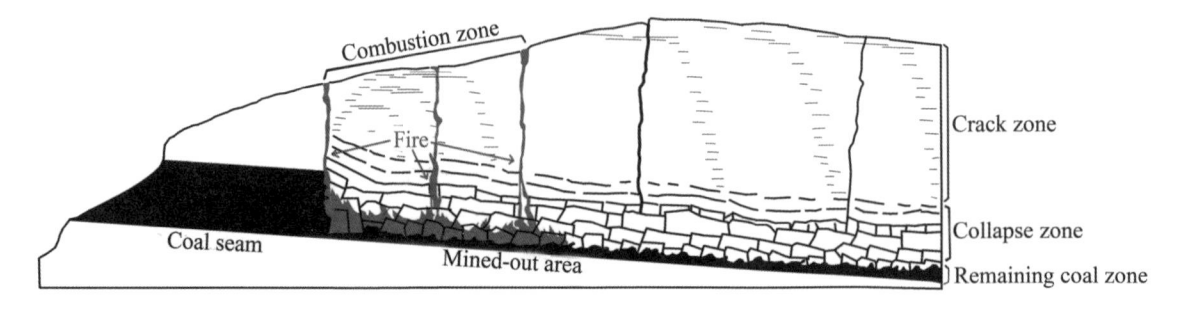

Figure 6. Model of zones that are conducive to spontaneous combustion in a mined-out area.

3, no. 4, no. 5, no. 8, no. 9, no. 10, no. 13, and no. 16), and fires above the 210 Haulage Roadway in the Suhaitu mine, we found that the roof of a coal seam forms the collapse zone, whereas the overlying rocks form the crack zone during the mining of the working face and after mining. Therefore, from top to bottom, three zones can be identified on the geological profile: crack zone, collapse zone, and remaining coal zone. Air can circulate in these three zones. When coal that is left behind after mining has enough oxygen for oxidation activity to occur, the coal temperature rises, reaches ignition temperature, and spontaneous combustion occurs. Smoke and gas that are produced by spontaneous combustion often vent to the surface through collapse and crack zones. Thus, a combustion zone forms in these areas.

On the basis of this analysis, we established a model of zones that are conducive to spontaneous combustion in a mined-out area. The system is composed of four parts: crack zone, collapse zone, remaining coal zone, and combustion zone (Fig. 6). This combustion model simulates the spatial movement of an underground coal fire in a mined-out area.

DISCUSSION

Through systematic geologic field investigations and the study of fires in the Wuda coalfield, Inner Mongolia autonomous region, we have established a spontaneous combustion genesis model, a coal-fire propagation model, a model of the progressive stages of a coal fire, and a model of zones that are conducive to spontaneous combustion in a mined-out area. These models provide a scientific and theoretic basis for coal-fire monitoring, as well for the management and application methods to control such fires (Fig. 7).

The model of the genesis types explains the mechanism of spontaneous combustion of coal seams, which is helpful for forecasting and preventing coal fires; the different types of coal that are prone to spontaneous combustion require different methods of prevention and management. The coal-fire propagation model simulates the propagative law. This model can be taken into account during the course of managing coal fires. The coal-fire progressive stages model simulates the various stages of coal fires from initiation to extinguishment. It also can be applied to the process of controlling coal fires (e.g., estimating the temperature of fire extinguishment, monitoring the temperature of fire areas, gasification application of underground coal fires). The model of zones that are conducive to spontaneous combustion can be applied to monitoring and managing fires in mined-out areas, and it provides a reference for coal mine construction and production in northwestern China.

ACKNOWLEDGMENTS

This research was supported by the National Natural Science Foundation of China (40572092) and the National High

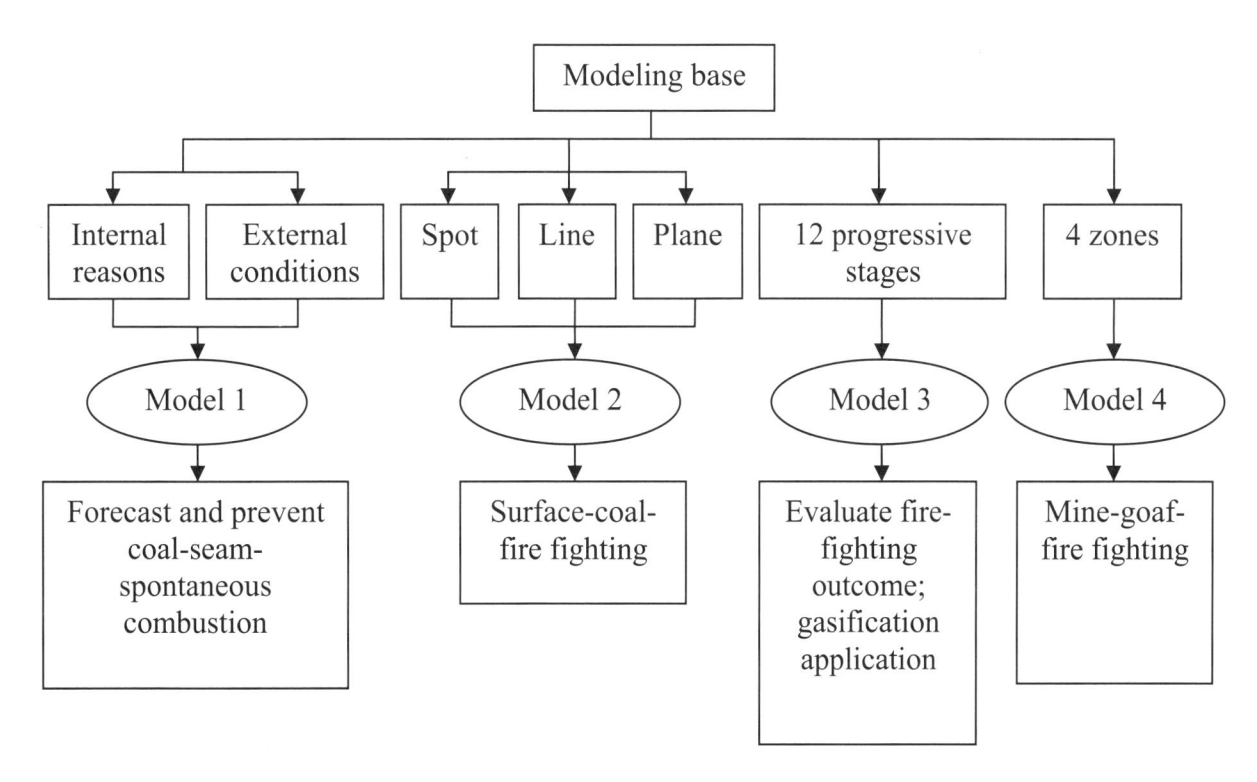

Figure 7. Modeling base and the application of four models: Model 1—spontaneous combustion genesis model; model 2—model of coal-fire propagation; model 3—model of progressive stages of coalfield fires; model 4—model of zones that are conducive to spontaneous combustion in a mined-out area.

Technology Research and Development Program of China (2003AA131100-02). We would like to thank Jianmin Zhang, Jianwei Ma, Mei Wang, Xiaoying Wu, and Yaling Xiang for their constructive discussion. We are grateful to the personnel of the Wuda Mine Ltd., Shenhua Group, who supplied much useful information and support during field investigations. We wish to thank Glenn B. Stracher, Minggao Yu, Shuheng Tang, and Shifeng Dai for their useful comments on the manuscript. We also thank Stacia Spaulding and David Tabet for their revisions of the manuscript.

REFERENCES CITED

Cao, D.Y., Fan, X.J., Shi, X.L., Wu, C.C., and Wei, Y.C., 2005, Analysis of spontaneous combustion internal factors and division of spontaneous combustion types of coal seam in Wuda Coalfield, Inner Mongolia: Journal of China Coal Society, v. 30, no. 3, p. 288–292 (in Chinese).

Dai, S.F., Hou, X.Q., Ren, D.Y., and Tang, Y.G., 2003, Surface analysis of pyrite in the no. 9 coal seam, Wuda Coalfield, Inner Mongolia, China, using high-resolution time-of-flight secondary ion mass-spectrometry: International Journal of Coal Geology, v. 55, p. 139–150, doi: 10.1016/S0166-5162(03)00109-5.

Erdogan, K., and Vedat, D., 2002, Relations between coal properties and spontaneous combustion parameters: Journal of Engineering and Environmental Sciences, v. 26, no. 1, p. 59–64.

Fan, X.J., Cao, D.Y., Shi, X.L., and Wu, C.C., 2006, Controlling factors of spontaneous combustion of coal seams in the Wuda Coalfield, western Inner Mongolia: Geological Bulletin of China, v. 25, no. 4, p. 487–491 (in Chinese).

Finkelman, R.B., 2004, Potential health impacts of burning coal beds and waste banks, in Stracher, G.B., ed., Coal Fires Burning Around the World: A Global Catastrophe: International Journal of Coal Geology, v. 59, no. 1–2, p. 19–24.

Guan, H.Y., van Ganderen, J.L., and Tan, Y.J., 1998, Survey and study on environment of coalfield spontaneous combustion in northern China: Beijing, Publishing House of China Coal Industry, 107 p. (in Chinese).

Heffern, E.L., and Coates, D.A., 2004, Geologic history of natural coal-bed fires, Powder River Basin, USA, in Stracher, G.B., ed., Coal Fires Burning Around the World: A Global Catastrophe: International Journal of Coal Geology, v. 59, no. 1–2, p. 25–47.

Mao, J.H., and Xu, H.L., eds., 1999, The Evaluation and Prediction of Coal Resources in China: Beijing, Science Press, 465 p. (in Chinese).

Peng, S., and Zhang, J., 1995, The coal-bearing Depositional Environment and its Influence on Mining of the Wuda Coalfield: Beijing, Publishing House of China Coal Industry, 99 p. (in Chinese).

Prakash, A., and Gupta, R.P., 1999, Surface fires in Jharia Coalfied, India—Their distribution and estimation of area and temperature from TM data: International Journal of Remote Sensing, v. 20, p. 1935–1946.

Stracher, G.B., 2002, Coal fires: A burning global recipe for catastrophe: Geotimes, v. 47, no. 10, p. 36–37 and 66.

Stracher, G.B., and Taylor, T.P., 2004, Coal fires burning out of control around the world: Thermodynamic recipe for environmental catastrophe, in Stracher, G.B., ed., Coal Fires Burning Around the World: A Global Catastrophe: International Journal of Coal Geology, v. 59, no. 1–2, p. 7–17.

Tan, Y.J., 2000, Disaster and control of spontaneous combustion in coalfield, China: Coalfield Geology and Exploration, v. 28, no. 6, p. 8–10 (in Chinese).

Walker, S., 1999, Uncontrolled fires in coal and coal wastes: London, International Energy Agency Report CCC/16, 72 p.

Wang, X.S., and Zhang, G.S., 1990, Prevention of Coal Mine Fires: Xuzhou, China University of Mining & Technology Press, 256 p. (in Chinese).

Wei, M.H., 1998, Analysis of internal reasons and external conditions about coal seam spontaneous combustion and prevention countermeasure: Coal Mine Safety, v. 2, p. 30–33 (in Chinese).

Zhang, J.M., Ning, S.N., and Cao, Y., 1998, Study of environment influence and governing countermeasure about spontaneous combustion of coal seam in northern China: Disaster Reduction in China, v. 8, no. 1, p. 34–38 (in Chinese).

MANUSCRIPT ACCEPTED BY THE SOCIETY 7 MARCH 2007

Survey of experimental work on the self-heating and spontaneous combustion of coal

Mark I. Nelson*
School of Mathematics and Applied Statistics, The University of Wollongong, Wollongong NSW 2522, Australia

Xiao Dong Chen
Department of Chemical Engineering, Monash University, Clayton, VIC 3800, Australia

ABSTRACT

A wide variety of techniques have been applied to gain insight into the processes that govern the self-heating of coal. These include oxidation mechanisms, ranking the propensity of different coals to self-heat, and the detection and suppression of self-heating. A frequent weakness in the literature about the kinetic data of self-heating systems is the absence of error estimates from regression analysis and the associated constraints on the reliability of the data for modeling. In addition, experimental and numerical work is needed to evaluate the methods used to acquire the kinetic data. Moist coal in coal mines and stockpiles has very different combustion characteristics than those predicted on the basis of dry testing. Consequently, methods for ranking the propensity of coal to spontaneously combust in actual mining conditions need to be developed.

Keywords: spontaneous combustion of coal, self-heating of coal, adiabatic oxidation and testing, coal combustion, coal-mine fires.

1. INTRODUCTION

The propensity of coal to undergo self-heating and spontaneous combustion is a major problem wherever coal is being mined, stored, or transported; it poses problems for both coal producers and users. Accordingly, it has been the subject of extensive fundamental and practical research for well over a hundred years. A succinct overview of the issues relating to the spontaneous combustion of coal and coal-mine fires was provided by Banerjee (1985). Carras and Young (1994) emphasized the causes of self-heating, described mathematical models for modeling self-heating, and assessed the limitations of industrial test methods. Recently, Babrauskas (2003, p. 719–724) provided an overview of the ignition of coal, with a particular interesting description of the historical development of the subject toward the end of the nineteenth century. Our current understanding of the chemical reactions that occur during the low-temperature oxidation of coal was reviewed by Wang et al. (2003b).

In this article, we examine experimental work covering the period 1996–2005. We start in section 1.1 by outlining the economic and environmental consequences of unwanted coal fires and the threat that these fires pose to human life. In section 1.2, we describe the phenomenon that is frequently the reason for coal fires: spontaneous combustion. The connection between spontaneous combustion and coal fires has motivated the ongoing stream of publications relating to the spontaneous combustion of coals.

---

*nelsonm@member.ams.org

Although the emphasis of this paper is experimental work, it will be useful to have an understanding of some of the issues involved in modeling the spontaneous combustion of coal. These are covered in section 2. In particular, in section 2.2, we discuss the simplest theory for the spontaneous combustion of bulk materials subject to self-heating; the model developed by Frank-Kamenetskii (1969). This allows us to gain some elementary insights into the factors governing spontaneous combustion and provides the basis for an experimental method to determine kinetic parameters, the hot-storage test, discussed in section 16. These parameters are required as inputs for any mathematical model of the self-heating process. In section 3, we review some issues dealing with the kinetics of the oxidation of coal at low temperatures.

In sections 4–9, we discuss the spontaneous combustion of coal in various environments. In sections 4 and 5, we consider spontaneous combustion in the context of "natural" coal fires and coal-mine fires, respectively. By a "natural" coal fire, we mean a fire in a coal seam that is not being worked in a coal mine. The distinction between these two categories is very fine, but in view of the extra issues associated with coal-mine fires, such as means of detecting self-heating at an early stage and suppression of combustion, we think that it is useful. Fires in abandoned coal mines are discussed in section 4.1. In section 6, we consider the spontaneous combustion hazard posed by accumulating layers of coal dust. In sections 7 and 8, we discuss hazards associated with stockpiles of coal. Section 7 deals with stockpiles of "fresh" coal, such as those found at power generators. Section 8 deals with stockpiles of the coal waste produced in mining operations. In section 9, we discuss the risk of spontaneous combustion when coal is transported.

In section 10, we discuss various properties of coal that influence its propensity for spontaneous combustion. The inherent moisture content of a coal and the humidity of the air are two properties that exert a strong influence over self-heating in a coal. There are discussed in section 11. Low-rank coals contain large amounts of water that must be removed to make their use economic. However, as discussed in section 11, very dry coals have a high propensity for spontaneous combustion. Thus, low-rank coals must be treated to both decrease their water content and their propensity for self-heating. This is the topic of section 12.

In sections 13–18, we consider experimental techniques for investigating the propensity to spontaneously combust. We start in section 13 by outlining the various experimental procedures that exist for this task. In sections 14–18, we consider some of the more important methods: adiabatic methods in section 14, the crossing point temperature in section 15, the hot-storage test in section 16, the heat release rate method in section 17, and the transient method in section 18.

We draw our review to an end in section 19. In Appendix 1, we define abbreviations used in this review and the nomenclature used in the equations. In Appendix 2, we provide a list of published parameter values for coals, including kinetic values. We hope that this collection of data will be useful in future modeling studies.

1.1. Coal Fires: An Economic and Environmental Hazard and a Threat to Health

The threat of a coal fire is present whenever coal is at or near the surface and is not limited to mined coal; fires can ignite along natural outcrops of coal seams (a "natural" coal fire). The mechanisms that result in coal fires include accidents associated with the coal-mining industry, natural causes, and the spontaneous combustion of coal.

Coal fires in mines, both open and underground, and stockpiles result in a direct loss of valuable resources through the unwanted burning of coal reserves. In addition, there is an indirect loss of coal resources in underground coal-mine fires because access to coal seams may be blocked, and reserves in coal seams not involved in the fire may be devalued. For example, fires in the Jharia Coalfield (India) have resulted in the direct loss of ~37 million tons of coking coal and the indirect loss of 48 million tons, due to flooding or sealing of underground workings (Sinha, 1986). As of 2001, it was estimated that the Wuda coal basin reserves (Inner Mongolia, China) contained 353 million tons of unexploited coal, 100 million tonnes of which have been blocked by coal fires (Stracher et al., 2005). Annual losses to fire, 0.1–0.2 million tons per year, pale into insignificance against the blocked reserves. Furthermore, in any mine with high capital investment, temporary interruptions to mining result in economic losses as equipment and personnel stand idle. Thus, coal fires result in a huge economic loss.

Coal fires are a source of environmental problems, such as acid rain and local/regional atmospheric pollution, through the release of noxious gases and particulate matter. As a consequence of coal fires, topsoil quality is degraded, vegetation coverage is reduced, floral and faunal habitats are destroyed, and the quality of surface and subsurface groundwater significantly decreases. Coal fires can threaten property and surface constructions (such as communication systems, homes, railway lines, roads, and surface installations at a mine), agriculture products (such as crops), and jeopardize the stability of water channels. Coal fires can induce surface subsidence, resulting in the destruction of buildings and surface infrastructure. This may result in the removal of communities endangered by subsidence hazards. Land subsidence caused by fires in the Rukigou Coalfield (Ningxia Province, China) has produced surficial cracks up to several kilometers in length, tens of meters wide, and hundreds of meters deep (Stracher and Taylor, 2004). Coal fires may, in turn, initiate forest and bush fires.

Coal fires release a wide range of airborne pollutants, which, in high enough concentrations, may be detrimental to health through coronary and respiratory diseases—the main causes of death in underground coal fires are due to gas poisoning and suffocation. Carbon monoxide, hydrogen, and volatile organic compounds, particularly methane, are produced by burning coal, as are particulates. Other noxious and harmful gases produced during a coal fire include carbon dioxide, hydrogen sulfide, nitrogen oxides (NO_x), and sulfur oxides (SO_x). Although health risks are known when humans are exposed to individual chemicals, little is known about the health risks in environments containing multiple harmful gases (Freudenstein et al., 2000). Carbon dioxide emission from coal fires also contributes to greenhouse gases in the atmosphere. Thus, coal fires endanger health and safety and cause economic and environmental problems.

1.2. Spontaneous Combustion

1.2.1. What Is Spontaneous Combustion?

Self-heating of coal, that is to say, an increase in the temperature of the coal, occurs when coal is exposed to air at low temperatures. The reaction between coal and air is a solid-gas process that involves the reactions of oxygen at reactive radical sites on the surface of the coal. Low-temperature oxidation of coal, at temperatures <100 °C, is the primary mechanism for self-heating; secondary exothermic processes include the oxidation of pyrite, microbial metabolism, and absorption of water vapor (Banerjee, 1985; Bowes, 1984).

The rate of oxidation is generally very slow at ambient temperature, but it increases exponentially as the temperature rises. If the heat generated by oxidation is not fully dissipated, then it is absorbed by the thermal capacity of the coal, raising the temperature of the coal. Provided that there is an adequate supply of oxygen, an accelerating process can then occur, known as thermal runaway or runaway ignition, in which the release of heat raises the temperature of the coal, which augments the rate of oxidation, which produces more heat, which cannot be dissipated, which increases the temperature of the coal, etc. If this process proceeds unchecked, a runaway ignition event can ensue and subsequently initiate a fire (Banerjee, 1985; Bowes, 1984).

The phenomenon of spontaneous combustion reflects interplay between heat generation and heat loss. Ren et al. (1999, p. 1612) suggested that "spontaneous combustion can thus be defined as an adiabatic oxidation process." This is misleading because heat loss is crucial in defining the phenomenon of spontaneous combustion: any exothermic process will lead to runaway ignition under adiabatic conditions.

When the generated heat is quickly dissipated, spontaneous combustion does not occur. Under these conditions, self-heating leads to a small temperature increase. As outlined in section 7, even these small temperature increases can have adverse technological consequences.

1.2.2. Spontaneous Combustion and Coal Fires

Self-heating of coal, leading to spontaneous combustion, is the most significant cause of fires in coal mines throughout the world (Mahadevan and Ramlu, 1985) and a significant cause of coal fires throughout industries associated with coal mining. Self-heating can occur in underground mines, opencast mines, coal stockpiles (at the mine, at power stations, and at ports), during seaborne transport, and in the disposal of waste materials from coal-using industries in coal waste heaps (Bowes, 1984; Carras and Young, 1994).

The spontaneous combustion of coal has caused widespread underground coal fires in several countries, including China, India, and the United States. Spontaneous combustion in underground mines is especially dangerous because the atmosphere may be inherently explosive, and a highly localized incident of self-heating can initiate an explosion involving the entire mine.

As an illustration of these problems, it has been established that most fires in the Jharia Coalfield (India) originated from spontaneous combustion that occurred either underground or at surface outcrops (Sinha, 1986) and that across all Indian coal mines, 75% of the fires are caused by spontaneous combustion (Singh and Singh, 2004). Up to ten coal fires per year in the Ruhr area of Germany are caused by spontaneous combustion (Pilarczyk et al., 1995). In northern China, underground coal fires are widespread within a region stretching 5000 km east-west and 750 km north-south (Zhang et al., 1998). These fires have started by a combination of spontaneous combustion of surface outcrops, discussed in section 4, and in underground coal mines (Zhang et al., 1998). The cause of most of the mining-related fires is spontaneous combustion (J. Zhang et al., 2004). It is estimated that the fires in northern China consume a vast amount of coal, perhaps as much as 100–200 million tons of underground coal reserves annually, and that, if this estimate is correct, they are responsible for 2%–3% of the world's production of carbon dioxide from burning fossil fuels (Huang et al., 2001).

Stockpiles of coal are prone to spontaneous combustion, especially where large quantities are stored for extended periods. This includes stockpiles of waste materials from mines and industry—the most common cause of burning coal waste piles is the spontaneous combustion of carbonaceous material in the waste materials (Beever, 1982). This point is illustrated by historical data. Surveys in the West Riding of Yorkshire (England) showed that 45 of the county's 153 colliery tips were on fire in 1931 (Sheail, 2005, p. 136) and 78 of 134 heaps visited in January 1937 were on fire (Sheail, 2005, p. 137). Most of these fires were caused by spontaneous combustion. A more recent example is the spontaneous combustion of spoil heaps at the Middelburg Colliery in the Witbank Coalfield (South Africa) following the cessation of mining (Bell et al., 2001).

Fires caused by spontaneous combustion can lead to explosions. In June 1991, an undetected fire in a coal bunker at a coal-fired plant, believed to have started by spontaneous combustion, gave rise to a minor explosion, which ignited coal dust that resulted in a massive explosion (Dantoin et al., 2003). The estimated total cost of this explosion exceeded $4 million (USD).

The results of spontaneous combustion are serious and negative: unwanted economic effects, undesired environmental consequences, and unwished for health problems. To prevent these outcomes, we must understand the processes that lead to runaway self-heating so that the risk of an incident can be reduced by controlling the processes leading to spontaneous combustion.

2. MATHEMATICAL MODELS

In this section, we discuss some issues relating to the mathematical modeling of spontaneous combustion. The points discussed in this section will be relevant in subsequent sections.

We start in section 2.1 by discussing the simplest representation for the reaction rate and heat release in a low-temperature oxidation, an Arrhenius expression, and the extent to which this is justified by experimental data. In section 2.2, we discuss the analysis of self-heating in bulk materials by Frank-Kamenetskii (1969). In section 2.3, we outline some of the physical and chemical processes that are included in more detailed models.

2.1. Rate of Heat Release and Reaction Rate

Although the self-heating of coal is due to a number of exothermic reactions, many studies have shown that the rate of oxygen consumption in the low-temperature oxidation reaction of coal can be expressed by a single Arrhenius expression (Krishnaswamy et al., 1996b)

$$r_{O_2} = A^\dagger \exp\left[-\frac{E}{RT}\right][\rho]^m [O_2]^n. \quad (1)$$

It follows that the rate of heat release is

$$r_h = QA^\dagger \exp\left[-\frac{E}{RT}\right][\rho]^m [O_2]^n. \quad (2)$$

In Equations 1 and 2, Q is the heat of reaction, $A^\dagger$ is the pre-exponential factor (the units of which depend upon m and n), E is the activation energy, R is the universal gas constant, T is the temperature, ρ is the density of the coal, $[O_2]$ is the concentration of oxygen, m is the order of reaction with respect to coal, assumed to be unity, and n is the order of reaction with respect to oxygen.

It is often simply assumed in mathematical models that $n = 1$ (see, for example, Sujanti and Zhang, 1999). A value of $n = 0.63$ ($R^2 = 0.995$) has been reported from experiments in a semi-adiabatic reactor (Smith and Glasser, 2005a), and a value $n = 0.70$ has been reported from experiments in a static isothermal reactor (Smith and Glasser, 2005b).

In investigations into the spontaneous combustion of solid fuels undergoing self-heating, it is commonly assumed that consumption of oxygen is negligible. With the additional assumption that m = 1, the rate of heat release becomes

$$r_h = QA \exp\left[-\frac{E}{RT}\right]\rho_c. \quad (3)$$

Many workers have applied Arrhenius plots to determine the values of the kinetic parameters (A and E) from Equation 3, and some of these studies are reported in sections 16–18. There is one recorded instance of non-Arrhenius behavior for a coal: a Chinese bituminous coal that displayed non-Arrhenius behavior in microcalorimeter experiments over the temperature range 311–338 K (Jones and Newman, 2003).

Studies show that values of A and E only apply to the coal for which they have been obtained—different coals have different values. This seems obvious. However, it has been assumed in some safety literature that *all* coals have the *same* activation energy, for example, in the IMCO self-heating test (International Maritime Consultative Organisation, 1974). This assumption has led to methods for determining safety in shipping coal that are not reliable. This matter is discussed in section 9.1.

Finally, we note that from a chemical engineering perspective, it is more natural to write the pre-exponential factor for a gas-solid reaction occurring within a porous solid in the form

$$A = A^{\dagger\dagger} S_g, \quad (4)$$

where S_g is the effective surface area per kilogram of coal. The amount of surface area of the coal, and therefore the particle size, is now a direct factor in its heating tendency.

2.2. Frank-Kamenetskii Analysis

In this section, we discuss an analysis of self-heating by Frank-Kamenetskii (1969) that has been applied to evaluate the hazards posed by self-heating of many bulk materials (Bowes, 1984), including activated carbon, agricultural wastes, coal milk powder, wood, and wool. In section 2.2.1, we review the mathematical basis of Frank-Kamenetskii's theory and outline some results that are particular useful in practice. In section 2.2.2, we discuss some issues relating to the theory regarding its practical applications and interpretation of experimental results.

2.2.1. Analysis of an Infinite Slab and Some Other Geometries

Although the spontaneous combustion of coal is influenced by many factors, quantitative insight into the behavior of a coal pile is obtained by considering two factors: heat generation due to oxidation, which we discussed in section 2.1, and heat losses to the surroundings. If we assume that, for simplicity, heat transfer in a reactive porous slab of coal is one-dimensional, convective flow can be ignored within the sample, depletion of the reactants (fuel and oxygen) can be neglected up to the time of ignition, and the sample is dry, then an energy balance gives

$$\rho c \frac{\partial T}{\partial t} = \frac{\partial}{\partial x}\left(\lambda \frac{\partial T}{\partial x}\right) + QA\rho \exp\left[-\frac{E}{RT}\right], \quad (5)$$

with boundary conditions

$$T(x = L) = T_a, \quad (6)$$

and

$$\left.\frac{\partial T}{\partial x}\right|_{x=0} = 0, \quad (7)$$

where c is the specific heat capacity, λ is the thermal conductivity, T is the absolute temperature, L is the half-width of the pile, and T_a is the ambient temperature. All physical and chemical properties are assumed to be temperature independent. In Equation 5, the left-hand side is the local rate of enthalpy change in the solid, the first term on the right-hand side is the conductive heat transfer in a porous solid, and the second term is the heat generated by low-temperature oxidation.

The boundary condition given as Equation 6 defines a constant temperature at the solid-gas surface, while the condition given as Equation 7 represents a symmetry condition at the center of the slab.

Under steady-state conditions, the left-hand side of Equation 5 is zero, giving the boundary-value problem

$$\frac{\partial}{\partial x}\left(\lambda \frac{\partial T}{\partial x}\right) = -QA\rho \exp\left[-\frac{E}{RT_a}\right], \quad (8)$$

subject to conditions in Equations 6 and 7.

If we introduce a dimensionless temperature excess, θ, a dimensionless length, x^*, defined in Appendix 1, the boundary-value problem can be rewritten

$$\frac{\partial^2 \theta}{\partial x^{*2}} = -\delta \exp\left[\frac{\theta}{1+\epsilon\theta}\right], \quad (9)$$

subject to the boundary conditions

$$\theta(x^* = 1) = 0, \quad (10)$$

and

$$\left.\frac{\partial \theta}{\partial x^*}\right|_{x^*=0} = 0, \quad (11)$$

where δ, the Frank-Kamenetskii (F-K) parameter, and ϵ, the reduced activation energy, are defined by

$$\delta = \frac{EL^2 \rho Q A}{\lambda R T_a^2} \exp\left[-\frac{E}{RT_a}\right], \quad (12)$$

and

$$\epsilon = \frac{RT_a}{E}. \quad (13)$$

Equation 9 can be further simplified by making the "pre-exponential approximation" that $1 + \epsilon\theta \approx 1$. With this assumption, the solution to the boundary-value problem defined by Equations 9–11 depends only upon the value of the F-K parameter (δ).

Analysis has revealed (Bowes, 1984, chapter 1) the existence of a critical value of δ, δ_{cr}, which for a one-dimensional slab is given by $\delta_{cr} = 0.88$. If $\delta \leq \delta_{cr}$, the system is *subcritical*, i.e., there is a stable steady-state in which there is measurable self-heating of, typically, up to a few tens of degrees above the oven temperature. The *maximum* stable temperature rise that can occur in a slab without ignition corresponds to the case $\delta = \delta_{cr}$ and is given by $\theta_{max} = 1.19$. This gives the maximum safe temperature rise as

$$\Delta T = (T - T_a) = 1.19 \frac{RT_a^2}{E}. \quad (14)$$

Knowledge of this is sometimes useful in the interpretation of experimental results (see, for example, Jones and Littlefair, 1997).

It can also be shown that when $\delta < \delta_{cr}$, there is a second, unstable, steady-state solution. If $\delta > \delta_{cr}$, the system is *supercritical*. The boundary-value problem defined by Equations 9–11 no longer has a solution, and the maximum temperature increases without bound, becoming infinite in finite time—the system is said to "blow up in finite time," or to "exhibit blow up." Physically, this means that the system is unsafe and spontaneous combustion will occur. Thus, there is a sharp transition between subcritical systems, which have a small temperature increase, and supercritical systems, which have an infinitely large temperature increase associated with combustion.

Equation 9 has been solved for different sample geometries, and in each case there is a critical value of the F-K parameter (Bowes, 1984). Some values that are appropriate for self-heating studies involving coal are shown in Table 1. Critical values of the F-K parameter for other geometries are also available (Bowes, 1984).

The analysis reviewed in this section assumed that, on the boundary $x = L$, resistance to external heat transfer to the surroundings is significantly less than by heat conduction within the solid. It is possible to include Newtonian cooling on the boundary $x = L$. This introduces a new parameter, the Biot number ($Bi = hL/\lambda$), where h is the convective heat transfer coefficient at the boundary $x = L$. The boundary condition given as Equation 6 corresponds to the limit $Bi \rightarrow \infty$, which occurs experimentally if conditions within an oven are highly convective. This requires a powerful fan to agitate the air. A critical value of δ exists for any finite value of the Biot number. This theory has also made the "pre-exponential approximation" $(1 + \epsilon\theta \approx 1)$. When this approximation is dropped, there is still a critical value for the F-K parameter, but this must now be found numerically. For example, when $\epsilon = 0.02$, 0.04, and 0.06, δ_{cr} in a slab is 0.898, 0.918, and 0.941, respectively (Chen and Chong, 1995). It is also possible to correct the value of δ_{cr} to take into account fuel consumption, although this correction is not required for a solid substrate such as coal (Jones, 1999a). An evaluation of the critical value of the F-K parameter when these assumptions are relaxed is discussed by Bowes (1984).

TABLE 1. CRITICAL VALUES OF THE F-K PARAMETER δ, δ_{cr}, AND THE MAXIMUM SUBCRITICAL TEMPERATURE INCREASE, θ_{max}, FOR SOME COMMON GEOMETRIES (BOWES, 1984)

Sample geometry	δ_{cr}	θ_{max}
Circular cylinder (height = diameter)	2.844	1.778
Circular cylinder (infinite height)	2.00	1.386
Cube	2.569	1.888
Slab (infinite)	0.878	1.119

Note: δ_{cr}—critical value of δ (the Frank-Kamenetskii or F-K parameter).

2.2.2. The Critical Value of the F-K Parameter: Practical Interpretations

The theory presented in the previous section, which ignores consumption of the reactants, predicts that under subcritical conditions, the temperature increases to a maximum value that is sustained indefinitely. In experimental systems, reaction rates decrease over time due to reactant consumption. Taking this into account, the expected subcritical behavior is that the temperature will rise to a maximum, at most a few tens of degrees above the oven temperature, which is sustained for a finite length of time, during which reactant depletion becomes increasingly significant, before falling back to the oven temperature.

The expected subcritical behavior has been observed in oxidation of a bituminous coal (Jones, 1996b). A quasi steady state was observed for over 4 d with $\Delta T = 10 \pm 1$ (K); the system was still in this steady state at the conclusion of the experiment. It was estimated that no more than 6% of the fuel was consumed during the 4 d steady period—the actual amount of fuel consumed was probably significantly less than this estimate. For other materials, such as cellulosic materials, lignocellulosic materials, and activated carbons, the expected subcritical behavior is not observed. Instead, the measured temperature decreases from its maximum value far sooner than expected, after a comparatively short period on the order of hours, during which reactant consumption is very minor (Jones, 1996b). When the temperature has returned to ambient, a large amount of combustible fuel remains (Jones et al., 1996a). The reasons for this behavior are not fully understood.

For supercritical systems, the temperature rises hundreds of degrees, sometimes over one thousand degrees, above the oven temperature. After reaching a maximum, the temperature falls back to the oven temperature due to reactant depletion. In comparisons of the expected subcritical and supercritical behavior for experimental systems, we see that there is still a clear demarcation between the two types of behavior.

The critical value of δ defines the transition from a subcritical, i.e., safe, pile to a supercritical, i.e., unsafe, pile. If, for instance, all the parameters are known except for the size of a stockpile ($2L$ for a cube or slab), a critical thickness can be calculated from Equation 12 and Table 1, above which a stockpile will inevitably ignite. The length required for a coal pile to ignite under normal storage conditions will depend upon the origin of the coal (principally determined by the values for A, E, and Q). In practice, all the parameters required to calculate δ, from Equation 12, can be readily estimated except for the values for the kinetic parameters (A and E). Methods to obtain these values are discussed in sections 16, 17, and 18. One of these, the hot-storage test discussed in section 16, is based upon the Frank-Kamenetskii analysis.

From Equation 12, we see that spontaneous combustion, i.e., when $\delta > \delta_{cr}$, is favored by conditions that maximize heat generation and minimize heat dissipation. For instance, low thermal conductivity (λ) increases the value of δ. This makes physical sense, since heat accumulation within the pile will increase as the value for λ decreases.

2.3. Modeling

The Frank-Kamenetskii theory, outlined in section 2.2.1, is the simplest spatially structured model that accounts for the phenomena of spontaneous combustion. It provides useful insights into the competing influences of heat production and heat dissipation and provides a means, to be discussed in section 16, of calculating the kinetic parameters (A and E). These parameters are required for more detailed spontaneous combustion models.

The Frank-Kamenetskii theory does not include many factors that are known to be important in the oxidation of coal at low temperatures. These include, but are not limited to: the oxygen concentration in the surrounding atmosphere, the transport of oxygen and moisture within the coal pores by convection and diffusion, gas absorption, the formation of gaseous and solid products, the moisture content of both the surrounding atmosphere and the coal, the evolution of heat by secondary mechanisms, the transport of heat within the coal by convection, the coal-pore structure, the particle size of the coal, and the exposed surface area of the coal. In addition, physical and chemical properties of coal may be a function of both temperature and chemical composition (water moisture, oxygen, reaction products).

There is now an extensive modeling literature investigating how these processes influence self-heating within a coal pile. Early work has been reviewed by Carras and Young (1994).

3. KINETICS

In this section, we review some issues relating to chemical kinetics. We start in section 3.1 by providing a brief overview of the mechanism of low-temperature oxidation. In section 3.2, we discuss the effect of aging and preoxidation on coal. In section 3.3, we discuss the possibility of gaseous species, either produced during the oxidation of coal or occurring naturally as in-seam gases, inhibiting the oxidation rate.

3.1. The Mechanism of Low-Temperature Oxidation

The low-temperature oxidation chemistry of coal has been extensively studied. Although the exact process by which this happens is complex and the mechanisms are not fully understood, some general features of the process are known. The current understanding of the reaction mechanism has recently been reviewed (Wang et al., 2003b), and, accordingly, we provide only a brief overview.

The first step in the process of coal oxidation is the chemisorption of oxygen at active sites on the surface of the coal. At temperatures below 40 °C, the interactions between coal and oxygen are mainly due to physical adsorption and chemisorption of oxygen in coal pores. The next step of the low-temperature oxidation of coal is the formation of unstable surface oxygen-coal complexes, generally through the attack of molecular oxygen on certain aliphatic bonds, usually methylene (CH_2) in the α-position to an aromatic ring or in OR groups, to produce peroxides and hydroperoxides (Mulcahy et al., 1991; Lopez et al., 1998)

$$CH_2 + O_2 \rightarrow CHOOH.$$

At these temperatures there is very little evolution of gaseous products.

At higher temperatures, ~40–70 °C, the unstable oxy-compounds break down to give gaseous products, stable oxygen-coal complexes, and heat. At temperatures below 70 °C, carbon dioxide is the major gaseous product (Wang et al., 2002)—this is a change in view, since early studies indicated that carbon monoxide was the main product below 70 °C. The stable oxy-compounds contain oxygenated functional groups, mainly carbonyls, esters, and carboxylic acids (Garcia et al., 1999). The specific group formed depends upon the functionalities adjacent to the CH_2 group that was attacked in the initial stage of oxidation (Lopez et al., 1998). The overall oxidation rate is significantly more rapid at these temperatures than below 40 °C.

At even higher temperatures, the stable oxy-complexes degrade, generating new active sites for coal oxidation. Accompanying the chemical changes that occur as a consequence of oxidation are changes in the macromolecular structure of the coal (Garcia et al., 1999).

This mechanism does not involve the interaction of oxygen with aromatic rings because these are unreactive due to their chemical properties. Quantum chemical calculations have confirmed that aromatic structures have negligible influence on active aliphatic groups (Shi et al., 2005).

Wang et al. (2003a, 2003c) provided evidence that carbon dioxide and carbon monoxide are produced by two parallel reaction sequences. The general sequence outlined above, which predominantly generates carbon dioxide rather than carbon monoxide, and a second mechanism, called the direct burn-off reaction, which occurs at specific reaction sites in a coal's aliphatic or aromatic structure and results in the direct formation of gaseous products, including carbon dioxide, carbon monoxide, and water. Wang et al. (2003c) found that 70 °C was a critical temperature for the thermal decomposition reactions and for the direct burn-off reaction.

Several studies have found that the oxidation rate of coal increases dramatically as the temperature crosses a threshold of ~70 °C (Beamish et al., 2000, 2001; Chamberlain, 1974) to 80 °C (Mahidin et al., 2003a). In simple representations of the oxidation rate, this means that the values for the kinetic parameters (A and E) differ on either side of the threshold. Historically, this effect was ascribed to a change in the structure of the coal that increased the available surface area. It has also been described as a change from a region in which heat release, below the threshold, is controlled by oxygen adsorption to a region, above the threshold, in which heat release is controlled by coal oxidation (Mahidin et al., 2003a, 2003b).

The inhomogeneity of coal hinders an understanding of the oxidation mechanism. Recently, eight simplified aromatic molecular models have been proposed for coal, each with a different active group connected to one or two aromatic rings (Shi et al., 2005). The oxidation of these structures was analyzed using techniques from quantum chemistry. The reactivity of the active groups was in accord with experimental observations. In the initial stage of oxidation, oxygen was adsorbed by active groups to form a molecule with either a carboxyl group (-COO), a carboxyl acid group (-COOH), or a hydroxyl group (-OH). In six of the compounds, water was released as a product, in one compound carbon dioxide was released, and in one compound no small molecule containing oxygen was liberated. Although the failure of these models to release carbon monoxide was interpreted as indicating a need to improve the molecular structure of the simulated molecules, this is not necessarily the case, because carbon dioxide is believed to be the major gaseous product at low temperatures (Wang et al., 2002).

The development of kinetic models that can be used to predict self-heating will represent an important step forward in developing a better understanding of when, and why, spontaneous combustion occurs (Wang et al., 2003b).

3.2. The Effect of Aging and Pre-Oxidation

3.2.1. General Points

Studies have shown that aging, also known as preoxidation, has a strong negative influence on the oxidation rate (Beamish et al., 2001, 2003; Ren et al., 1999). The lower reactivity of aged coals is due to the removal of active sites by oxidation. The decrease in oxidation rate of aged coals is more noticeable at higher temperatures because the rate of oxidation increases with temperature. The maximum rate of decrease in reaction rate occurs when fresh coal particles are first exposed to oxygen. Weathering also decreases the calorific value of coal (Garcia et al., 1999; Pis et al., 1996).

Schmidt and Elder (1940) proposed that the oxidation rate decreases with time according to the equation

$$k(t) = k_0 t^{b-1}, \quad b < 1. \quad (15)$$

In this equation k_0 should be taken as the reaction rate after one day, on initial testing, since it is not possible to test coal immediately after sample recovery (Beamish et al., 2000). The value for the parameter b is likely to depend upon the storage method used and possibly the rank of the coal (Beamish et al., 2000): values of $b = -0.14$ for particles of size <5 mm (Beamish et al., 2000) and $b = -0.2$ for particles of size <6.4 mm (Schmidt and Elder, 1940) have been reported. It is possible that, because of preoxidation effects, coals that are air-dried for 24 h prior to use in experiments have significantly different reaction rates to coals that are dried using other methods (Beamish et al., 2003). Such a difference would be amplified for coals with smaller particles.

The kinetic parameters of coal that has been stored, or preoxidized, increase with age. This is illustrated in the kinetic parameters for new and aged Kopako coal shown in Appendix 2. As a further illustration of this point, samples stored for two years took two to three times longer to reach "uncontrolled ignition" in adiabatic tests than fresh samples (Beamish et al., 2001). It has been suggested that weathered coals have an increased moisture-holding capacity (Ren et al., 1999).

3.2.2. Experimental Investigations

Garcia et al. (1999) investigated nonisothermal oxidation enthalpies in weathered Colombian coals using DSC (differential scanning calorimetry). The samples were weathered in acrylic boxes designed to reproduce natural weathering as close as possible—samples were exposed to rain, wind, and solar heating. Samples were tested initially and then again after 15, 30, 45, 75, and 105 d of weathering. Progressive weathering decreased the measured enthalpy of oxidation, and the decrease was particularly severe during the first 15 d of weathering.

Pis et al. (1996) investigated the self-heating behavior of fresh and oxidized coals using DTA (differential thermal analysis). Prior to testing, coals were oxidized at 200 °C over different periods of time ranging from 0 to 72 h. For all coals, there was a decrease in the calorific value of the coal with weathering, although for two coals, the variation was small. The variation in the initial stages of weathering was the most significant.

Mondragón et al. (2002) investigated the production of H_2S during the low-temperature oxidation of coal in the presence of elemental sulfur. The most active sites in low-temperature oxidation are the hydrogen atoms in hydro-aromatic structures. Since both elemental sulfur and oxygen compete for these sites, it was suggested that the reaction with sulfur could be used to monitor the deterioration of a sample over time due to aging. Wang et al. (2003c) suggested that the extent of weathering could be investigated by following the ratio of the rates of production of carbon dioxide to carbon monoxide.

3.3. Inhibition

Gaseous reaction products produced during the oxidation of coal have the potential to reduce the rate of oxidation by being adsorbed at active sites in preference to oxygen. This has been investigated in a static isothermal apparatus by adding 1%, which is a typical level in the early stages of the reaction, and 5% concentrations of carbon monoxide to the reactor (Smith and Glasser, 2005b). The rate of oxidation only decreased significantly at a concentration level of 5%, leading to the conclusion that product inhibition is insignificant in the early stages of the reaction.

The presence of seam gases and moisture also inhibits self-heating by blocking access to coal pores. It has been shown that removal of these by degassing increases the risk of coal self-heating (Beamish and Jabouri, 2005; Beamish, 2005a). A "gassy" moist as-mined coal underwent thermal runaway in 8.5 d, starting from an initial temperature of 30 °C. The coal was then allowed to cool and equilibrate at 30 °C. When it was retested, thermal runaway was reached in only 4.25 d (Beamish and Jabouri, 2005). These results suggest that returning moisture to the coal by water infusion following degassing may be a sensible precaution to prevent self-heating in degassed coal seams.

4. NATURAL COAL FIRES

By a "natural" coal fire, we mean a fire in a coal bed that it not associated with an ongoing mining operation. Such fires consume a significant amount of coal. Heffern and Coates (2004) estimated that somewhere between 33 billion and several hundred billion metric tons of coal has burned naturally in the Powder River Basin (North America) over the last four million years, most of which burned within the last two million years. Such coals have been an important source of greenhouse gas emissions in the geological past.

Natural coal fires usually start at a surface outcrop of a coal seam with a certain thickness that has been exposed by erosion. Surface outcrops are formed by geological processes such as erosion by rivers, faulting, and folding (X. Zhang et al., 2004a). Self-heating within the seam may be given a "push" toward ignition by an external source of ignition, such as bush fires, forest fires (Bustin and Mathews, 1982), lightning strikes, or solar heating (Zhang et al., 2003), or it may lead to spontaneous combustion on its own accord. It is also possible for a natural coal fire to start, not at an exposed outcrop, but through the burning of trees that are rooted into a coal bed.

As noted in section 3.1, the oxidation rate of coal rapidly increases in the region of 70–80 °C. Strong solar heating can cause the surface temperature of newly exposed coal seams to reach 75 °C (Zhang et al., 2003), which therefore puts them at high risk of spontaneous combustion. Seams heated to such temperatures by solar-heating can be detected by integrating daytime and nighttime airborne infrared thermal data (Zhang et al., 2003).

The burning of a coal seam several meters in thickness results in a layer of ash on the order of tens of centimeters in thickness, sometimes less (Bustin and Mathews, 1982; Zhang et al., 2003). The pressure on the unsupported overburden rocks causes deep cracks to develop, eventually leading to the subsidence of the overlying strata. The combination of cracks, collapsed areas, and fracturing of the highly burnt overburden rocks, which produce a highly porous medium, allows access of oxygen further underground. Thus, a coal fire started at the surface outcrop can progressively spread deeper into the coal seam (Zhang et al., 2003) until it is extinguished, either through running out of oxygen or by exhausting the accessible coal.

4.1. Fires in Abandoned Coal Mines

Subsidence in abandoned mines, such as the collapse of rooms between pillars, gives raise to crownholes, fracturing, and surface cracks that provide ready passage of air into the mine. This supply of oxygen increases the risk of spontaneous combustion, especially because the fracturing and collapsing of pillars also produce a large surface of coal. Subsidence may also be caused by previous fires deteriorating the rock mass around the coal workings to such an extent that the overlying strata collapse. Burning can also reduce the strength of pillars, leading to their collapse, which in turn accelerates subsidence and exacerbates the problem of spontaneous combustion.

The importance of subsidence in sustaining underground coal fires has been demonstrated numerically in a two-dimensional steady-state model for heat and mass transfer in underground coal fires (Huang et al., 2001). It was found that areas of higher permeability, as represented by areas of fracturing or subsidence in a mine, are required to sustain a convective flow of air that can circulate air to the burning coal surface. Calculations show that the heat conduction is the dominant mode of heat transfer in deeper coal fires, and convection becomes increasingly important as the depth of the fire decreases.

In addition to subsidence, spontaneous combustion at abandoned mines can reduce the quality of groundwater (low pH and high concentration of dissolved solids) seeping from the mine. Fires within a decommissioned mine in the Middelburg Colliery (Witbank Coalfield, South Africa) have enhanced the oxidation of pyrite, leading to the formation of sulfuric acid. Water flowing out of the mine is highly polluted and has a low pH, which has destroyed most of the vegetation within a 3 ha area where seepage occurs (Bell et al., 2001).

5. COAL-MINE FIRES

The initiation of coal-mine fires due to self-heating of coal is an inevitable hazard of mining coal seams and is a worldwide problem. Underground coal fires have been reported in all major coal-producing countries, including: Australia, China, India, Indonesia, Russia, South Africa, and the United States. China, in particular, is notorious for its coal-mine fires. For instance, it is reported that 42 out of 88 coal mines and coal-production bases in the Xinjiang Uygur Autonomous region are severely affected by coal fires, with annual losses estimated at 10–20 million tons (X. Zhang et al., 2004b). Spontaneous combustion remains a threat even in well-regulated coal-mining industries. For instance, during the 1990s, there was one major spontaneous combustion mine explosion in both Australia and New Zealand and several other near misses. More recently, heating in the Southland Colliery (Hunter Valley, Australia) caused ignition, thus forcing the mine to close in December 2003 and sending its owner into receivership.

The study of coal-mine fires applies knowledge about the self-heating and spontaneous combustion of coal to developing improved procedures for the early detection and suppression of self-heating and, in the worst case, the containment and control of any fires that break out.

We start in section 5.1 by discussing some of the origins of fires in coal mines. In section 5.2, we discuss how the hazard from spontaneous combustion can be assessed. In section 5.3, we consider matters once self-heating has led to spontaneous combustion: control and suppression of coal fires. Finally, in section 5.4, we overview the detection and monitoring of coal fires over large areas through the use of surveillance techniques. Such techniques can facilitate the early detection of underground coal fires.

5.1. Origins of Coal-Mine Fires

We noted in section 4 that coal outcrops can ignite through the process of self-heating leading to spontaneous combustion. Alternatively, an external source can "push" a self-heating coal into ignition by providing additional energy. Both of these mechanisms can generate fires in coal mines. Examples of mine-related activities that can act as an additional energy source include: cutting and welding, electric work, the use of explosives, and smoking. Mine-related activities can also increase the propensity for spontaneous combustion without providing an ignition source. For example, the use of explosives and the movement of heavy machinery/vehicles can create fracture and microfractures in coal benches. This increases the permeability of the bench, which increases oxygen circulation through the coal. The oxidation of powdered coal contained in cracks within a coal bench at a depth of greater than half a meter from the surface may lead to the buildup of a significant amount of heat due to poor heat transfer—heat generated by powdered coal closer to the surface, less than one quarter of a meter, leads to a much more limited temperature increase due to heat dissipation by wind currents (Singh and Singh, 2004). These mechanisms can also induce the spontaneous combustion in coal seam joints (Stracher and Taylor, 2004).

A problem wherever coal is processed, including coal mines, is the accumulation of fine-grained loose coals. These accumulations are more susceptible to spontaneous combustion than larger-grained materials. Fires started by the ignition of fine-grained loose coals have been observed to spread to the coal face (Singh and Singh, 2004).

5.2. Assessing the Hazard

In this section, we describe how the hazard from self-heating can be assessed. We start in section 5.2.1 by discussing factors that are particular to a coal mine in determining the propensity to self-heating. Knowledge of these factors and of the state of mining operations allows us to perform a risk assessment for a worked coal seam.

In section 5.2.2, we discuss how self-heating in a mine can be detected, concentrating on the concept of an index gas. The earlier a potential hazard is identified, the longer there is to implement countermeasures to eliminate the risk. Furthermore, the rate of self-heating rapidly accelerates near the point of criticality. In a self-heating experiment using a 2 m adiabatic column containing 41.4 kg of coal, the temperature increase over the first 5 d was on the order 1–2 °C d^{-1} (Beamish et al., 2002). It took ~17 h for the temperature to increase from 110 °C to 150 °C. At the completion of the test, after 7 d, the temperature was rising at a rate of 120°C d^{-1}. If self-heating is detected at this late stage, there may not be time for suppression procedures to be implemented, and ignition may be inevitable.

5.2.1. Before Self-Heating Begins

There are many factors that influence the propensity for spontaneous combustion within a coal mine. These include: properties of the coal, such as rank and pyrite content, which are discussed in section 10, and moisture content, discussed in section 11; geological factors, including environmental, geographic, and hydrological considerations; and, mining practices. Examples of geological factors include (Mahadevan and Ramlu, 1985`; Sensogut and Cinar, 2000): bacterial activity; barometric pressure; caving characteristics; depth of cover; faulting, which allows air to leak into the workings and increase self-heating; geothermal gradient, which is higher in deep mines; hanging-wall conditions; mine temperature; the presence of other materials that are prone to spontaneous combustion, such as carbonaceous shale and poor quality coal that has been deemed not worth mining; seam depth, seam gradient, and seam thickness; and the strength of overlying strata.

Petrographic composition is sometimes identified as a geologic property that influences the propensity to self-heat (Mahadevan and Ramlu, 1985; Sensogut and Cinar, 2000). This is based on the fact that minerals such as exinite and vitrinite are known to promote self-heating. However, in one study, statistical analysis failed to find a relationship between petrographic composition and self-ignition properties (Pilarczyk et al., 1995). Supporting this finding, Garcia et al. (1999, p. 41) contended that "conventional characteristics such as petrographic, ultimate or proximate analysis are of little value" in identifying which coals are liable to spontaneously combust.

Mining practices also govern self-heating (Feng et al., 1973; Mahadevan and Ramlu, 1985; Sensogut and Cinar, 2000). These factors include: airflow humidity, air leakage; coal grades; daily advance; mining method; monitoring method; multiple seam working in close proximity; pillar conditions; seam recovery; ventilation; volume flow rate of oxygen; waste support method; working method; and advance rate.

It is possible to provide a practical risk assessment to predict the propensity of a coal seam to spontaneously combust by taking into account intrinsic and extrinsic factors. The intrinsic factors are a property of the coal that is being mined. The extrinsic factors include the influence of the environment and the geological and operating conditions of the mine.

In the method described by Singh et al. (2002), the intrinsic risk factor was obtained from adiabatic test results, see their Table 9. Where adiabatic data are not available, as might be the case when only borehole samples are available, an empirical relationship based upon a chemical analysis of the coal may be used to estimate the required parameters (IRH [initial rate of heating] and TTR [total temperature raise]) (Singh et al., 2002; Equations 1 and 2). A complete listing of extrinsic factors is given in the original publication (Singh et al., 2002, their Table 2). An example of such a factor is the seam thickness, with ratings: <1.5 m (−1), 1.5–3.0 m (+2), >3 m (+3); and the thick seam slicing method (+8). The values assigned to the intrinsic and extrinsic factors represent an empirical assessment of their contribution to the likelihood of a fire.

The intrinsic and extrinsic factors are added together to give the risk classification shown in Table 2. The extrinsic factors, geological factors, and mining conditions are site specific and can therefore vary between mines and within localities of a mine. Thus, the classification of a particular seam might be "low risk" in one location and "very high risk" in another. An example of this is the classification of coal originating from two adjacent sites of the same seam reported in Pilarczyk et al. (1995): the self-ignition tendency of one is classified by pit-fire statistics as being of medium risk, while the other is classified as high risk. Of course, variation in extrinsic factors cannot be determined in laboratory measurements, which is why risk assessment is based upon the combination of intrinsic and extrinsic factors.

The risk assessment for the seams within a mine provides a foundation on which to devise a risk management plan for spontaneous combustion in a mine, which includes designing both the layout of the mine, including the ventilation system, and an appropriate monitoring system to detect self-heating (see section 5.2.2). Preparation of a risk management plan and a spontaneous combustion management plan are mandatory in New South Wales (Australia) (Singh et al., 2002).

Singh et al. (2002) showed how these ideas are applied by providing a case history analysis of an underground coal mine located in the Upper Hunter Valley (New South Wales, Australia). A risk assessment of the seams was followed by the development of a spontaneous combustion management plan for the one very high-risk coal seam and the one high-risk coal seam.

An alternative approach is the calculation of a risk index, which is described in Mahadevan and Ramlu (1985). The intrinsic risk factor of the coal is given by a liability index, which is obtained from a crossing-point temperature experiment, described in section 15, and by determining the coal's Mahadevan-Ramlu (MR) index, section 15.2. A "mine environment index," an integer from 1 (low) to 4 (very high), is determined by considering coal loss, fissuration, and ventilation pressure differential within the coal mine. The product of the liability index and the mine environment index is the risk index, which is shown in Table 3.

TABLE 2. SPONTANEOUS COMBUSTION RISK ASSESSMENT FOR COAL MINING*

Risk classification	Sum of risk ratings	Incubation period in months
Low risk	1–10	>18
Medium risk	11–20	9–18
High risk	21–40	3–9
Very high risk	>40	0–3

*From Singh et al. (2002, their Table 3).

TABLE 3. RISK INDEX CLASSIFICATION (LIABILITY INDEX × MINE ENVIRONMENT INDEX) FOR COAL MINING*

Risk classification	Risk index
Low risk	0–20
Medium risk	20–40
High risk	>40

*From Mahadevan and Ramlu (1985).

5.2.2. Once Self-Heating Has Started

The prevention of coal-mine fires relies upon the detection of areas in the mine where the temperature is increasing, i.e., areas liable to spontaneously ignite in the future (Banerjee et al., 1970). As noted at the start of section 5.2, early detection of self-heating is essential.

One of the main ways in which this is done is by monitoring the presence of an "index gas" in the mine air. As the temperature of coal increases, it will release gases. Ideally the quantity of emitted index gas is directly related to the temperature of the coal releasing it. The variation of percentage of the index gas in the gases released from the coal as a function of the coal temperature is called a "spontaneous combustion ladder" (Singh et al., 2002). Alternatively, if the index gas consists of two, or more, constitutive gases, then the ratio of these is related in some known manner to the temperature of the coal. In these circumstances, measurements of the index case allow mine operators to quickly identify areas at risk of spontaneous combustion. Before an index can be used, an index value must be decided above which safety precautions are to be implemented.

Whether a particular gas is suitable for use as an index gas depends partly on the chemical makeup of the coal and upon the presence, or absence, of gases trapped within the coal seam. For example, hydrogen cannot be used if it occurs naturally with the in-seam gases, as was the case in Singh et al. (2002). Both carbon dioxide and methane are released from coal seams under normal mine conditions and are therefore not suitable as an index gas.

Popular risk assessment methods include the observation of carbon monoxide emissions, where self-heating is indicated by a large amount of carbon monoxide and the measurement of the carbon monoxide/oxygen deficiency ratio (Graham's ratio). Carbon monoxide is popular as an index gas because numerous investigations have shown that the rate of carbon monoxide production increases with temperature. Furthermore, unlike carbon dioxide, carbon monoxide can only be formed in auto-oxidation reactions. Hence, it is a good indicator of self-heating. A combination of measurements of both carbon monoxide emissions and Graham's ratio has been shown to produce an effective detection system (Singh et al., 2002).

The concentration of an index gas in a mine is used to indicate the stage of the fire. However, the concentration also depends upon variable factors such as the quantity of airflow and the amount of coal involved in the fire. It has been found that the rate of increase in the concentration of carbon monoxide is quite distinct in three stages of the self-heating process (Güyagüler et al., 2003). It has therefore been suggested that the *rate of increase* of carbon monoxide is an important indicator in assessing the likelihood of spontaneous combustion in a real mine. This hypothesis was confirmed by analyzing carbon monoxide data recorded during a mine fire (Güyagüler et al., 2003).

Lu et al. (2004) analyzed the compositions and quantities of gas produced by various coals over a range of temperatures. They found that carbon monoxide and ethene (C_2H_4) could be used as index gases: detection of carbon monoxide (at a level of 1 ppm) at the coal workface indicates that the gob area has reached temperature of 50–64 °C, simultaneous detection of carbon monoxide and ethene (at a level of 0.01 ppm) indicates that the temperature in the gob area has reached 64 °C. Once the concentration of both gases starts to increase regularly, self-heating is developing and countermeasures should be deployed to prevent spontaneous combustion.

Singh et al. (2002) analyzed gas emissions as a function of coal temperature for a coal from a seam at a mine in New South Wales (Australia). Carbon monoxide was released at temperatures of 50 °C and higher. Graham's ratio was found to gradually increase with increasing temperature.

Wang et al. (2002) investigated the low-temperature oxidation of a well-oxidized bituminous coal. The molar ratio of produced carbon monoxide to consumed oxygen increased with temperature. The authors suggested that this ratio could therefore be used both to determine the onset of spontaneous combustion and also as an index for ranking propensity for self-heating for well-oxidized coals.

Any monitoring scheme for spontaneous combustion should involve more than testing for the presence of an index gas. Underground inspections can also include (Singh et al., 2002):

- detecting visual signs of heat, haze, sweating, smell, and smoke;
- checking the effectiveness of ventilation devices, doors, air-crossings, regulators, etc., and
- using handheld instrumentation for detecting gases such as: carbon dioxide, carbon monoxide, hydrogen, methane, and oxygen.

5.3. Control and Suppression of Coal Fires

Once initiated, underground coal fires have the potential to burn for a long time. For instance, two fires in the Northern Coalfield (New South Wales, Australia) have been burning for over 20 yr (McNally, 2000). Fires in the Liu Huangou Coalfield (Xinjiang Province, China) have been burning for over 20 yr and possibly for as long as 40 yr (Stracher and Taylor, 2004). A coal seam fire at Aldridge Creek (British Columbia, Canada) was started in 1936 by a forest fire and was still burning almost 50 yr later (Bustin and Mathews, 1985). A coal fire due to spontaneous combustion was noticed at the Middelburg Colliery (Witbank Coalfield, South Africa) in 1947 during decommissioning of the mine, and the fire is still burning despite attempts to extinguish it (Bell et al., 2001). In the Jharia Coalfield (India), subsurface fires caused by mining-induced spontaneous combustion have been burning for over five decades and are still spreading (Sinha, 1986). Fires in coal mines in northern China have been burning for several hundred years (de Boer et al., 2001).

The costs of containing, controlling, and extinguishing a fire can be so high that the only course of action, particularly for underground fires, is to let the fire burn itself out—this is the case for numerous fires in Pennsylvania (USA) (Stracher and Taylor,

2004). However, sometimes these costs are insubstantial compared with the value of the retrieved coal, including the release of coal that would otherwise be blocked. This has been the case for several coal fires in India (Sinha, 1986).

Although it may be economically viable to extinguish a large fire within a mine, the priority is to prevent matters reaching this stage by taking appropriate action as soon as the fire is identified. In section 5.3.1, we discuss the mechanisms that are available to fight coal fires. One possibility, particularly for self-heating events that have not yet produced a fire, is the application of a fire retardant. Research along these lines is reviewed in section 5.3.2.

5.3.1. Combating Coal Fires

Several techniques have been applied to control, or isolate, coal-mine fires. These techniques include (Sinha, 1986; Stracher and Taylor, 2004):

- **Trench cutting:** A trench, which may be backfilled with earth, is cut in the direction of propagation of the fire. If the trench is wide enough, it provides a firebreak that the fire is unable to cross.
- **Filling of galleries:** Fires can travel ahead of the coal face in underground mines through open galleries due to a chimney effect. To prevent this, galleries are filled with inert materials delivered through boreholes from the surface.
- **Hydraulic sand stowing:** Similar to trench cutting, this provides an inert barrier that the fire is unable to cross.
- **Inert gas injection:** Injection of an inert gas reduces the oxygen concentration in the fire area, lowering the rate of oxidation and the rate of spread of the fire. It may possibly extinguish the fire. The infusion of carbon dioxide and nitrogen has proven successful in combating mine fires (Sinha, 1986).
- **Injection of foam/water:** Once a barrier has been constructed between an unburning coal seam and an approaching fire, foam or water is circulated through the barrier through boreholes drilled into the barrier. The aim is to prevent the fire from crossing the barrier by keeping the barrier saturated with water.
- **Injection of slurry and ash:** A water-mud slurry can be injected through fractures and cracks or through holes drilled into underground shafts. The aim is to smother the flames.

It is sometimes possible to exert some control over the spread of a fire by manipulating the ventilation within the mine. Ventilation control can also improve visibility in selected areas by reducing fumes. In order to be used, numerical methods that analyze air-current optimization must arrive at correct solutions quickly. The use of genetic algorithms to determine the optimal ventilation through the mine during a fire, with the aim of producing the best conditions for the safe evacuation of miners and increasing safety in salvaging equipment and for fighting the fire, has been reported (Zhong et al., 2003).

Successful application of one, or more, of the procedures listed above will contain a fire, which will eventually burn itself out. To directly extinguish a fire, the following measures may be adopted (Sinha, 1986):

- **Digging:** The burning coal seam and the overlying strata are dug out and water, or community sewage (Stracher and Taylor, 2004), is applied to provide cooling. This works best for a small fire. For large fires, besides the size of the area to be dug out, there are technical difficulties, such as the availability of sufficient water and stabilization of the area immediately in front of the working face (Sinha, 1986).
- **Surface sealing:** Where the fire surface is too large to dig out, the fire may be deprived of oxygen by sealing the surface. Sinha (1986) suggested that alluvium be laid in three layers, each 40 cm thick, over the entire fire area and at least 10 m beyond the fire boundary. For fires in opencast mines, it has been suggested to surface-seal burning coal by applying a one meter layer of an inert material, typically soil (Stracher and Taylor, 2004).

These mechanisms should be compared with those used to extinguish fires in coal waste heaps listed in section 8.

Although the methods for control and extinction can be effective, there are numerous instances in which they have proven ineffective because operators have underestimated the size of the problem they are dealing with or they have applied countermeasures inadequately (Bell et al., 2001; Bustin and Mathews, 1982; Sinha, 1986). For instance, trenches may not be dug long enough or wide enough. Water injection is ineffective if it merely drains away. Bell et al. (2001) discussed a case in which water injection was applied for a matter of weeks, whereas it may have been required for years to put out a fire. Bell et al. (2001) also noted that the application of water potentially has two drawbacks: the release of carbon monoxide and hydrogen produced from water-gas reactions; and, the release of the heat of condensation may increase the temperature in the mine.

5.3.2. Fire Retardants

In areas where self-heating has been detected, but a fire has not yet started, the application of a fire retardant may eliminate the danger. Lu et al. (2004) showed that injection of MEA-1A retardant, a water-soluble high-molecular compound, into a coal sample sealed some of its surface pores and hindered the penetration of oxygen into the coal. Following the identification of a self-heating event at the Longgu mine (North China), a MEA-1A retardant solution was sprayed in the gob area once a day. Two days later, the index gas concentrations were down and the risk of spontaneous ignition had been eliminated.

To prevent spontaneous combustion and fire in open-case mines, Singh and Singh (2004) developed a mechanized spraying devise for spraying a fire-protective coating material over freshly exposed coal benches, coal stacks, and worked galleries.

The coating materials cover the coal surface and reduce the contact with air. The spraying device can deliver the suppressant up to 20 m height over the exposed coal surface.

Colaizzi (2004) described a cellular (foam-containing) grout suppressant, which is highly flowable and has high heat resistance, and the ways in which the suppressant *could* be used, such as spraying the material onto exposed coal surfaces after completion of mining. However, the paper provided no references to work where the suppressant had been used, either in a real fire or in a laboratory test. The product is commercially available.

5.4. Detecting Fires Using Large-Scale Surveillance Techniques

Remote-sensing techniques are the ideal tool to monitor coal fires in isolated, inaccessible areas, particularly when the coalfields cover large areas, such as in the coalfields of northeast China (J. Zhang et al., 2004) and the Jharia Coalfield of India (Gupta and Prakash, 1998). Many remote-sensing techniques have been used to detect and monitor coal fires using images obtained from satellites (microwave, optical, thermal) and aircraft (infrared, optical, aerial photography, and thermal). These techniques work by identifying regions of surface anomalies. For instance, subsurface fires may dry the topsoil covering the burning coal seam, causing its reflectance to increase relative to adjacent areas where the fire is not burning and the soil is wetter, and rendering the soil barren of vegetation. Areas of dry and barren soil can be detected on satellite sensor images (Gupta and Prakash, 1998). The existence of such areas is therefore a possible indicator of an underground coal fire.

In isolation, each technique provides only partially detection, depending upon parameters such as the depth of the fire, the time of the data acquisition, the topographical relief, and prevailing weather conditions. Combinations of data from multiple techniques provide a more comprehensive picture, in terms of completeness, robustness, and improved reliability, especially when used with ancillary data such as geological and topographical maps (X. Zhang et al., 1998).

For instance, combined daytime and nighttime thermal data may provide the detailed information required to combat underground fires by indicating the area, depth, spreading direction, and the speed of fire propagation (X. Zhang et al., 2003). These data are still best obtained by plane, because the resolution of the Landsat-5 thermal data is limited by its low spatial resolution; a coal fire smaller than 50 m^2 cannot be detected (X. Zhang et al., 1997). Airborne data can be used to monitor hot spots on a much smaller scale and are therefore still important for the early detection of small coal fires (X. Zhang et al., 2003; X. Zhang et al., 2004b).

Integration of airborne thermal infrared data from 8–12.5 μm nighttime, 8–12.5 μm daytime, and 3–5 μm daytime bands reveals more detailed structures of surface thermal anomalies than would be obtained by any one band in isolation (X. Zhang et al., 2004b). Nighttime data are the best at detecting low-temperature thermal anomalies, because background temperatures caused by solar heating are suppressed. Daytime 3–5 μm data are best at detecting small, subpixel-sized (data resolution is 7.5 m × 7.5 m), high-temperature thermal anomalies. Thus, daytime 3–5 μm data can be used to identify newly started underground fires and the cores of older underground fires. Nighttime data do not reveal temperature gradients in a burning region, and this is where 8–12.5 μm daytime data can be utilized. When combined with geological data for coal seams, integration of these three sets of data allows the spreading direction of the fire to be determined.

Combinations of techniques are sometimes required to distinguish between areas where fires are currently burning and areas where fires have burnt in the geological past. For instance, areas of thermally altered sedimentary rock with enhanced magnetic properties are a by-product of coal fires. These rocks form due to the melting of overlying rocks, due to the high temperatures, in excess of 1100 °C, produced by the combustion of coal seams. These temperatures are higher than the Currie point, and, on cooling, the rocks record the geomagnetic direction of that point in time. However, aeromagnetic detection of areas of magnetic anomalies cannot distinguish between rocks that are being burnt by current fires or fires that were active in the geological past. When combined with other techniques, magnetic anomalies can be used to identify areas where coal fires are currently burning (X. Zhang et al., 1998). The mapping of the extent and depth of extinct coal fires from the geological past is of independent interest as a method to refine estimates of the paleorelease of carbon dioxide (de Boer et al., 2001; X. Zhang et al., 2004a). Aeromagnetic exploration also has the potential to be a tool to aid coal exploration (de Boer et al., 2001).

Combinations of methods also allow the right method to be deployed at the right time. For example, although both aerial photography and airborne thermal infrared data are very useful in detecting and studying underground coal fires, they are very costly to use for monitoring purposes. If potential sites of underground fires can be detected using other methods, then these methods can be deployed to provide detailed coverage on an "as-needed" basis, rather than as a speculative tool.

6. THE SPONTANEOUS COMBUSTION OF COAL DUST

Virtually all mining operations produce coal dust as either a product or by-product. When this dust accumulates on heated surfaces, there is an enhanced risk of spontaneous combustion. The combination of self-heating in a dust layer and an energy input from a hot surface can lead to damage to machinery or, in the worst case, to a spontaneous combustion event that provides the energy to initiate a dust explosion. In coal mining, dust can collect on bearings, conveyor belt idlers, motor housings, and rolls (Reddy et al., 1998). Coal dust can also gather in "dead areas," such as corners. In facilities like coal mills, the dust in dead areas is progressively dried. As outlined in section 11.1, a pile of dried coal dust is a significant combustion hazard should a moist airflow occur: in a coal mill, such a scenario can lead to an explosion (Ren et al., 1999).

Little quantitative data exist on the behavior of pulverized coal. Air humidity has been identified as an important factor in determining whether self-heating progresses rapidly or not (Ren et al., 1999). In this study, smaller particles were shown to be more reactive than larger particles.

Reddy et al. (1998) studied the ignition of two coal dusts using a hot-plate apparatus. Dust layers of thicknesses 5, 10, 15, 20, 25, and 50 mm were used. For a given thickness of dust, they determined the lowest temperature of the hot plate at which thermal runaway occurred—the self-ignition, or layer, temperature of the dust layer. The layer ignition temperature decreased as layer thickness increased. Experiments were also carried out when the coal was mixed with one of two inert materials (dolomite and limestone) at loadings of 40%, 50%, 60%, 65%, and 70% additive. For sufficiently high additive loadings, ignition was not observed experimentally. However, the mixing of inert additives with coal dust is not a feasible mechanism for preventing self-heating of dust layers on hot surfaces.

The experimental data were analyzed using an extension of the Frank-Kamenetskii model discussed in section 2.2 and the modifications by Thomas and Bowes (1961), which incorporate both Newtonian cooling and asymmetric heating from the hot plate. The experiment was modeled as an infinite plane slab with thickness $2L$, where one face of the slab is in perfect thermal contact with the hot plate at a temperature T_p, and the other face is losing heat by Newtonian cooling to an atmosphere at ambient temperature. This model is given by

$$\lambda \frac{\partial^2 T}{\partial x^2} = -QA\rho \exp\left[-\frac{E}{RT}\right], \quad (16)$$

with boundary conditions

$$T(x=0) = T_p, \quad (17)$$

and

$$-\lambda \frac{\partial T}{\partial x}\bigg|_{x=2r} = h(T_s - T_a), \quad (18)$$

where L is the half-thickness of the dust layer, T_s in the surface temperature of the dust layer in contact with ambient temperature, and h is the surface heat-transfer coefficient. (The Frank-Kamenetskii model is recovered as the limiting case in which the Biot number $[hL/\lambda]$ approaches infinity and for which $T_p = T_a$.) The model allowed the parameters E and AQ to be calculated from experimental data and provided reliable predictions of the layer ignition temperature beyond the range of experimentation for samples without additives.

For the experiments using inert materials, the changes in the thermal conductivity (λ) and heat of reaction (Q) with additive loading were taken into account. The model predictions for systems with additives were in reasonable agreement with experimental data at loadings up to 50 wt% additive (Pittsburgh coal admixed with rock dust) and 60 wt% additive (Prince coal admixed with dolomite). At higher loadings, there was a significant divergence between theory and experiment: criticality was predicted by the model but was not observed experimentally. This shows that the standard F-K model is not applicable for materials that have a low heat of reaction.

Theories and data on the ignition of single coal particles and coal dust clouds have been reviewed by Babrauskas (2003, p. 360–364). He concluded that "The state of research must be assessed as being incomplete and contradictory at this point. Not only are widely disparate values reported, but even qualitative trends seem to depend greatly on the details of the experimental procedure adopted."

7. COAL STOCKPILES

In this section, we consider self-heating and spontaneous combustion in stockpiles of "fresh" coal. There have been few large-scale experimental investigations into the behavior of stockpiles of the size found in coal storage yards due to the expense of running such a test and the length of time it can take to run one experiment. Despite these drawbacks, there have been some well thought experiments that have provided insightful data.

The larger the stockpile, the greater is the risk of spontaneous combustion. However, for large stockpiles, even subcritical heating can lead to significant economic losses. Thus, we start in section 7.1 by considering self-heating and spontaneous combustion in coal stockpiles. In section 7.2, we discuss procedures that can be implemented to reduce self-heating. No procedure provides a 100% guarantee of safety, and in section 7.3, we consider the identification of areas within a stockpile in which self-heating is occurring. If identified early enough, appropriate procedures can be applied to eliminate the risk. Finally, in section 7.4, we consider an incident at a large stockpile at which noxious chemicals that were produced as a by-product of self-heating were released into a populated area.

7.1. Self-Heating and Spontaneous Combustion in Large Coal Stockpiles

It is often necessary to store coal in large stockpiles over long periods of time, for example, at power generators and at mines that export coal. Such stockpiles can undergo subcritical self-heating that reduces both the calorific value and the weight of stored coal (Fierro et al., 1999b). This reduces the financial value of the stored coal. Furthermore, during self-heating, progressive changes occur in the coal's chemical and physical properties that have undesirable consequences, such as a loss of coking properties (e.g., the degree of plasticity of coking coals during heating; Mondragón et al., 2002), a reduction in liquefaction yield, and a reduction in solvent extraction. These factors make the coal less suitable for technological utilization. The change in coal properties due to subcritical heating is known as weathering, or preoxidation, and was discussed in section 3.2.

In addition to self-heating, secondary mechanisms for mass and calorific loss include (Fierro et al., 1999b): partial or total coal combustion, removal of small coal particles due to wind and rain (leaching), volatilization due to pyrolysis, and incomplete recovery of the pile during removal. The significance of leaching and dragging is difficult to estimate experimentally (Fierro et al., 1999b). Should self-heating reach a stage where the stockpile needs to be cooled, water can be applied. However, this increases the water content of the coal and reduces its heating value. It should be noted that when stockpiled coal is exposed to water, whether through rain or the application of a water spray, some organic and inorganic matter, such as calcium, magnesium, and sulfur (Nakajima et al., 2005), leach out. This may cause a small decrease in the calorific and weight value of the coal, but it also produces an environmental hazard. Since self-heating can increase the local temperature inside a stockpile to above 100 °C, the application of a water spray to reduce localized self-heating results in hot water running through the stockpile, which increases leaching (Nakajima et al., 2005).

There is also the ever-present threat of spontaneous combustion. In addition to the possibility of spontaneous combustion due to the size of the pile, variation in the ambient temperature due to solar heating can also induce ignition. Should the stockpile ignite, the shear amount of coal involved (hundreds or even thousands of tons) presents huge problems in extinguishing the fire. Stockpiles must therefore be continuously monitored because it can take several months before a hot spot appears within the pile.

Sensogut and Ozdeniz (2005) recorded the temperature at 22 locations within a stockpile over a 75 d period. Over this time frame, the stockpile was subcritical. A statistical model was developed that predicted the temperature behavior of stockpiled coal based upon the air temperature, atmospheric pressure, moisture content, and time. This model was thought to be useful for mine management.

In order to develop reliable practices that minimize self-heating in large stockpiles, mathematical models are required. These models require knowledge of the kinetic parameters (A and E) and the equilibrium water content. Methods to obtain the former are discussed in sections 16–18; the importance of the latter is discussed in section 11.

7.2. Preventing Self-Heating in Stockpiles

7.2.1. General Considerations

To reduce economic losses, self-heating must be minimized. This will happen if the heat-generating capacity of the coal is reduced by, for example, establishing high heat losses. A variety of measures have been tried to reduce self-heating, including: minimizing the angle of the slopes of the stockpile (reducing wind access into the pile) (Fierro et al., 1999a), compaction of the pile (which reduces voidage) (Fierro et al., 1999a; Smith et al., 1988), protection of the coal pile through covering it with an inert material (Fierro et al., 1999a; Glasser and Bradshaw, 1990), making the atmosphere inert (Smith et al., 1988), and the use of natural or artificial wind barriers around the perimeter of a stockpile (to reduce airflow through the stockpile) (Borges and Viegas, 1988; Cai et al., 1983; Fierro et al., 1999a). The advantages and disadvantages of these approaches have been discussed by Glasser and Bradshaw (1990).

Compaction of the pile reduces the pore volume between coal particles, which decreases air ventilation through the stockpile, and is an effective way to reduce permeation of air. A loosely stored pile may have a voidage of 25% or 30%. This may be reduced to 10% through compacting with mechanical means. Compaction also reduces the permeation of water into the pile, and changes the thermal conductivity of the pile. Compaction can significantly increase the safety of a stockpile for *reactive* coals, but it can have the *opposite* effect for unreactive coals (Brooks et al., 1988a; Kok et al., 1989).

Stockpiles are safe in the two extremities of sufficiently low air circulation and sufficiently high air circulation. In the former, the oxidation rate is limited by the supply of oxygen and only a minor amount of self-heating occurs—clearly, if there is no airflow through the stockpile, then once the oxygen within the pile is consumed, the oxidation rate will be zero. In the latter case, heat is removed quicker than it is generated, and the temperature of the stockpile approaches the air temperature—this is called a ventilated pile (U.S. Department of Energy, 1994). Between these two extremes, there are two scenarios. In the first case, self-heating inevitably leads to spontaneous combustion. In the second case, there are two possibilities: either limited self-heating occurs or spontaneous combustion occurs. Which possibility happens depends upon the initial conditions of the problem—this is a thermal explosion of the "second kind" (Bowes, 1984).

Thus, one mechanism to reduce self-heating is to control the flow of air through the pile. This has the danger that if it is incorrectly implemented, the risk of spontaneous combustion increases. This happens if the airflow rate is sufficiently "high" to support heat generation by maintaining a high oxygen concentration while being at the same time sufficiently "low" so that heat is not dissipated.

It has been noted that fires in stockpiles almost always occur on the windward side of the pile (Fierro et al., 1999a). At a given wind velocity, coal piles with gentler slopes are less susceptible to spontaneous combustion (Fierro et al., 1999a).

The wind velocity through the pile (ventilation) and the porosity of the pile (voidage) have been identified in modeling work as being key parameters for controlling self-heating in stockpiles (Fierro et al., 2001). Significantly, voidage was found to be the more important because it controls the effects of wind velocity. The crucial role played by voidage is shown in the observation that hot spots did *not* appear in simulations of stockpiling over 400 d at a voidage of 10%, whereas at a voidage of 25% hot spots appeared in a few hours.

It has been observed that aged and preoxidized coals have a higher initial rate of heating in adiabatic tests than fresh coals (Ren et al., 1999). This suggests that freshly mined coals should neither be processed nor stored with weathered coals because the more rapid initial heat release from the aged coals might provide sufficient energy to ignite the fresh coal. This mechanism may explain the observation that "when a new pile of coal is laid on an existing weathered pile, fires occur in the plane of contact between the two piles" (Fierro et al., 1999a, p. 24).

Where two coals have the same mixture of particle sizes, the propensity for spontaneous combustion of the reactive coal can be reduced by blending it with a less reactive coal or by increasing its ash content. However, a coal blend containing a mixture of sizes throughout its volume is more vulnerable to self-heating than one in which the different sizes are segregated throughout its volume (Nugroho et al., 2000a). The effect is very nonlinear, and the presence of a small amount of either a finely crushed coal or a very reactive coal can lead to a large increase in the propensity to self-ignite. Finally, where a stockpile contains coal that has been added at different times, a "first-in, first-out" policy should be used when coal is removed (U.S. Department of Energy, 1994).

7.2.2. Experimental Studies

The efficiencies of four mechanisms in reducing losses (the coefficient of total losses, weight losses, and calorific value losses) in stockpiles of a "highly dangerous" coal have been investigated by constructing five large stockpiles (of weight 2000–3000 t) (Fierro et al., 1999a, 1999b). One pile was unprotected and therefore acted as a reference. One safety measure was applied to each of the remaining piles: periodic compaction, use of a low-angle slope in the prevailing wind direction, protection of the stockpile with an artificial wind barrier, and coverage of the stockpile with an inert barrier (ash-water slurry of average thickness between 3 and 4 cm). Since the inert barrier was impermeable, the effect of the ash layer was to produce a stockpile with an effective voidage of 0%. In order to maximize the reduction of wind pressure by the wind screen, the screen design was optimized using wind tunnel tests. With the exception of the ash-water slurry pile, the piles were also studied by infrared tomography (Fierro et al., 1999b).

Although the coefficient of total loses per year is an important economic figure, the studies reported in Fierro et al. (1999a, 1999b) are the first reported measurements of this parameter for large stockpiles. The efficiency of each mechanism at reducing losses was evaluated after 190 d for the slurry pile and 270 d for the other piles. Loss coefficients were calculated by the difference between the initial and final data, reported on a yearly basis, though the authors noted that the most interesting phase in the evolution of the coal piles was the first 50 d. The total loss in the reference pile was 19.5% per year.

Although there were difficulties associated with handling the large amounts of coal involved, the results showed that each safety measure decreased the self-heating of the coal. The most effective measure was the ash-water slurry (coefficient of total losses 3.1% per year), followed by the wind screen (6.1%). The pile protected by the wind barrier showed no signs of self-heating, except for a period during which the pile was unprotected due to the wind screen being blown over by a strong storm. An assessment of the efficiencies of the protection systems that incorporated the costs of their construction and maintenance showed that the ash-water slurry was the most economic. This is an attractive technique because the slurry could be taken from the power station at which the stockpiles were built. This method also has favorable scale-up characteristics. The experimental results matched the predictions of a mathematical model regarding the time at which self-heating became significant (to within an order of magnitude), the location of spontaneous combustion, and the magnitude of calorific losses (Fierro et al., 2001).

A mathematical model was used to model heat transfer due only to atmospheric heating, i.e., without self-heating. Combining measured surface temperatures from infrared thermography with predicted temperatures from the model allowed the energy produced in the interior of the stockpile to be assessed. This enabled the *calorific loss* to be estimated. The estimated value was the same order of magnitude as the value determined experimentally. It is interesting to note that the temperatures calculated from the model were not valid for the first few days after falling rain.

7.3. Identifying Areas of Self-Heating

Although infrared thermography is a suitable technique for measuring radiation emitted from coal at temperatures at around 300 K, very little work has been reported in the literature regarding its use (Fierro et al., 1999b; Kok et al., 1989). In an early study (Kok et al., 1989), the method was unsuccessful in detecting self-heating in a coal pile. In a more recent study, surface temperatures of a stockpile measured by an infrared thermographic camera were in good agreement with those measured by thermocouples (Fierro et al., 1999b). In order to obtain good results, infrared measurements have to be restricted to the hours of the day in which the surface temperature does not change rapidly. Infrared thermography was also deemed very efficient at detecting hot spots in the coal.

Where hot spots are identified near the surface of a stockpile, they can be excavated or are exposed to the atmosphere to permit cooling (U.S. Department of Energy, 1994).

7.4. Self-Heating in a Stockpile: A Case Study

Freudenstein et al. (2000) reported an incident in which gaseous substances were released by a 7000 t stockpile at an opencast mine. The coal contained high sulfur content. It is known that when such coals burn, they can release sulfur dioxide, which may react with water to form hydrogen sulfide and sulfuric acid. Hydrogen sulfide can also form if there is insufficient oxygen present for complete oxidation. The health effects of sulfur dioxide, acid aerosols, and particulates, all likely combustion products in this case, include the acute irritation of airways and the aggravation of asthma and chronic airway disease (Committee on the Medical Effects of Air Pollutants, 1997).

Sulfur gases from stockpiles can be readily detected by the nose, especially after rainfall (Bell et al., 2001). Complaints from villagers living downwind from the stockpile included strong unpleasant odors, predominantly of a tar-like smell, although a sulfurous "rotten-egg" smell was also described, upper airway and skin irritation, and exacerbations of asthma and eczema.

Freudenstein et al. (2000) highlighted deficiencies in how the problem was tackled by local and health authorities:

- Measurements of sulfur dioxide in the air at the mine carried out by one of the environmental health departments used equipment that had a threshold for detection 100 ppm higher than the level at which theoretical health effects are known to exist (125 ppb).
- No measurements were made of gases that could provide evidence of a fire inside the stockpile: carbon dioxide, carbon monoxide, hydrogen, and methane.
- Although the district fire brigade had equipment that could detect elevated temperatures within the stockpile, "involvement of the fire brigade for this purpose was not considered at the time of the incident" (Freudenstein et al., 2000, p. 43).
- Measurements of the sulfur dioxide and sulfite levels at the stockpile were taken by a private company only after health-related complaints had stopped and could not be used to estimate either peak or personal exposure.
- "Individual complaints were not followed up to substantiate or refute a casual link between the air pollution and the complaints of ill health" (Freudenstein et al., 2000, p. 43).

To prevent these types of problems from reoccurring in the future, the authors recommended the creation of a national database in England of companies and public bodies that have the necessary facilities to provide a rapid measurement of relevant airborne pollutants. A worrying fact to come out of this paper is that "national data on the frequency of such incidents do not exist" (Freudenstein et al., 2000, p. 41).

8. COAL WASTE HEAPS

In this section, we consider self-heating and spontaneous combustion in stockpiles of waste materials associated with coal-related industries. Waste heaps containing appreciable amounts of unprocessed coal, which is not an uncommon situation, are prone to self-heating: of 163 coal-mine fires identified by the Indian Department of Coal, 44 were due to the spontaneous combustion of quarry overburden containing carbonaceous matter (Sinha, 1986). In section 8.1, we consider self-heating and spontaneous combustion in coal waste heaps. In section 8.2, we outline methods that have been used to combat fires in coal waste heaps. In section 8.3, we discuss the recovery of usable fuel from the waste materials in the heap. Finally, in section 8.4, we consider a historical perspective on the dangers associated with fires at colliery spoil heaps by considering attitudes of British mine owners in the 1930s.

8.1. Self-Heating and Spontaneous Combustion in Coal Waste Heaps

Coal mining and cleaning operations associated with coal technology generate a large amount of solid waste: 6×10^6 t annually in Spain (Alonso et al., 2002), over 10^8 t annually in China (Liu et al., 1998), and more than 450×10^6 t annually in the Kemerovo region in Siberia (Sidenko et al., 2001). The total coal-mine waste in China was estimated at more than 1.6×10^9 t in 1994 (Liu et al., 1998). The product in waste heaps consists of a variety of rock types that have been extracted during mining operations, including varying amounts of coal that was not separated during processing, either intentionally, because it was deemed a low-quality coal, or unintentionally. The fraction of usable coal may be sufficient to consider reworking the dump: coal contents in the range of 28%–32% have been reported (Alonso et al., 2002).

The appreciable amount of coal in waste heaps means that they are at risk of spontaneous combustion, a risk that may be exacerbated by the small size of the coal fines. Even a poor-quality coal that has been rejected for commercial purposes, such as one with a high ash content, can pose a spontaneous combustion risk (Beamish and Blazak, 2005). The most common reason for waste heaps to ignite is the spontaneous combustion of the remaining coal. Older heaps, associated with abandoned mines, are particularly at risk of spontaneous combustion because these are often poorly compacted compared with their modern-day counterparts. Waste heaps that catch fire are then at risk of collapsing as the fire undermines the heap (Singh et al., 2002).

Self-heating can lead to environmental problems such as the production of smoke and fumes and the pollution of runoff water. It has been estimated that the combustion of 1 t of typical mine waste produces: 99.7 kg CO, 0.61 kg H_2S, 0.03 kg NO_x, 0.84 kg SO_2, and 0.45 kg smoke (Liu et al., 1998). Thus, self-heating and spontaneous combustion of coal waste heaps results in a serious environmental problem.

The risk of spontaneous combustion can be reduced by using an appropriate design for the heap. The methods that are used to prevent self-heating in stockpiles, discussed in section 7.2, are relevant. Another approach is to reduce the size of the waste heap by mixing the material in the waste heap with inert materials and using the composite material for building construction, brick making, and road construction.

8.2. Combating Waste Heap Fires

A variety of techniques can be applied to control, or isolate, waste heap fires. These include (Liu et al., 1998):

- **Deep-layer grouting:** A deep trench is dug into the burning waste heap which is filled with grout. This is the most common and effective method for extinguishing burning waste tips in China.

- **Digging:** The burning layers and the overlying material are dug out, and water is applied from a high-pressure water spray to provide cooling. This is an effective mechanism for extinguishing a fire, especially during early combustion.
- **Leveling and compressing:** In this method, equipment is used to level and compress the surface of the waste tip, reducing airflow into the burning layer. It is not suitable for extinguishing burning waste tips because of the risk to workers and equipment.
- **Surface sealing:** Where the fire surface is too large to dig out, the fire may be deprived of oxygen by sealing the surface.
- **Spraying:** Sealant materials or fire-protective coating materials are sprayed onto the top of the waste tip to seal air pores, cutting off the contact of air with the coal surface.
- **Grouting:** Cementitious grout is injected into the top of the waste tip to fill air pores and restore structural integrity.
- **Controlling:** Whereas the other methods seek to extinguish a fire, this method seeks to enhance it by inserting wind pipes into the burning layer. The heat that is produced is collected and used for other purposes.

Surface sealing, spraying, and grouting are methods that seal the surface of the waste tip, preventing oxygen flow into the pile, and consequently extinguishing any fire. The most frequently used covering materials are lime and clay. These methods are not effective for large waste tips because the covering agents do not penetrate into deep burning layers. These mechanisms should be compared against those used to extinguish fires in coal mines listed in section 5.3.

In China, many waste heaps contain a sizeable fraction of sulfurous iron ore, which frequently is the main cause of spontaneous combustion (Liu et al., 1998). Removal of this ore is therefore an effective mechanism for preventing the risk of spontaneous combustion. The recovered ore is processed to produce either sulfurous acid or fine sulfurous sand.

8.3. Energy Recovery

The presence of a sizeable coal fraction in a waste heap is a waste of energy resources. The recovery of usable fuel from these waste materials, which means maximizing recovery of organic matter while reducing recovery of ash, is desirable on both economic and safety grounds, since it reduces the risk of spontaneous combustion. In addition, the recovery of a usable product reduces the amount of waste that requires disposal.

Alonso et al. (2002) investigated the production of an efficient fuel by agglomeration of coal fines waste with colza oil. This process allowed the recovery of substantial amounts of energy, with maximum organic recovery in the range 75%–85%, and reduced the weight of disposed material by over 50%.

Where the recovered product had a high ash content, 39% in one instance, it could be blended with a coal of lower ash content to produce a usable fuel.

8.4. Self-Heating in Colliery Spoil Heaps: Views from the Past

Sheail (2005) recounted the increasing protest in the UK during the 1930s as a consequence of the damage to public health and amenity by the spontaneous combustion of colliery spoil heaps. Governments of the day claimed that industry could be encouraged to reduce pollution through extinguishing current fires and preventing future fires without the need of legislation. Coal owners claimed that there was no "radical cure" for the problem of burning heaps, and the Chairman of the Mining Association dismissed the very idea of a public health risk.

At the start of World War II, the need for civil-defense precautions against enemy bombers forced both the government and the mine owners to concede to comprehensive action in suppressing and preventing such fires. The outlay of taxpayers' money seemed to eliminate many of the previous objections by industry.

9. TRANSPORT OF COAL

Power-generating companies commonly import coal, and worldwide trade in coal amounts to some 300 million tons per year (Singh et al., 2002). Special care must be exercised in the transportation of coal to avoid spontaneous combustion in ship holds, and it is therefore important to have a method that appraises the safety of any coal that is to be shipped.

Bulk materials are commonly transported in cubic containers with sides of 3 m. The maximum mean ambient temperature that a container in a shipping hold will be exposed to for a reasonable amount of time during transport is 38 °C (311 K) (Bowes and Cameron, 1971; Cameron and MacDowall, 1972). Using Frank-Kamenetskii theory, section 2, a coal can be safely stored during transport if

$$\delta = \frac{EL^2 \rho Q A}{\lambda R T_a^2} \exp\left[-\frac{E}{RT_a}\right] \leq \delta_{cr} = 2.569, \quad (19)$$

where $L = 1.5$ m and $T_a = 311$ K. It has been suggested that a "safety margin" can be built into the calculation by assuming instead that the maximum ambient temperature is $T_a = 323$ K (Cameron and MacDowall, 1972, p. 1014). The required calculation is straightforward once all the constants are known. However, the value of the activation energy (E) is typically known to within ±5% (Jones, 1999a), and since this parameter appears within an exponential, it is necessary to take note of this uncertainty in the calculation (Jones, 1999a).

The kinetic parameters A and E could be obtained from the hot-storage test described in section 16. However, that is not the basis on which international standards have been developed. The methods that have been used are described in section 9.2. Our account of these methods is heavily based upon the descriptions provided by Jones (Jones et al., 1998; Jones, 2000a, 2000c).

One feature of supercritical systems, i.e., when $\delta > \delta_{cr}$, is that the time to ignition can be very long. On this basis, it has been suggested that coals that are judged unsafe to ship based on propensity for spontaneous combustion may be safe to ship in practice if their ignition times are much greater than the estimated transport times. This is discussed in section 9.2.

The dangers associated with shipping coal were well known to seamen in the nineteenth century and feature in a biographical account (*Youth*) of the voyage of a coal-bearing barque from England to Bangkok in the 1880s by Joseph Conrad (Conrad, 1902). An exegesis of this account, as it relates to the spontaneous combustion of coal, has been published in Jones (2002). Notable aspects of this account include the following: Most damage to the ship was caused *not* by spontaneous combustion but by a subsequent dust explosion. After the explosion, the barque was towed by a steamer, a much faster vessel. This had the unfortunate effect of significantly enhancing the rate of combustion due to the increased air supply.

In addition to the danger of spontaneous combustion due to self-heating, an additional hazard in the transport of bituminous coals is the slow release of molecular hydrogen. Over the course of transport, hydrogen concentrations in the dead space above the coal can reach levels close to the lower explosion limit for hydrogen in air (Jones, 1997, p. 51–52).

9.1. Is It Safe to Ship Coal: Subcritical or Supercritical?

In this section, we describe methods that address the issue as to whether a self-heating bulk solid shipped, or stored, in a 3 m cube will be subcritical or supercritical. In section 9.1.1, we describe various international standards that have been developed since the early 1970s. It turns out that each of these has a fundamental flaw. In section 9.1.2, we outline some newer proposals, seeking to overcome the flaws in the older methods.

9.1.1. Old International Standards

As noted earlier, the kinetic parameters A and E can be determined from the hot-storage test, described in section 16. However, this is a time-consuming method, and the standard test method that was developed in the 1970s, the IMCO test (International Maritime Consultative Organisation, 1974), did not use this approach. The IMCO test was developed from experimental work, which suggested that the activation energies for carbonaceous materials were very similar (Bowes and Cameron, 1971; Cameron and MacDowall, 1972). In designing the IMCO standard, it was assumed that carbonaceous materials had the *same* activation energy. This allows a simple test to be developed to distinguish between nonhazardous and hazardous materials.

In the IMCO test, the material to be transported is packed into a 10 cm cubic gauze container and is placed in a recirculating air oven at 140 °C for 24 h (Jones, 2000c). The temperature at the center of the sample is recorded by a thermocouple. If the sample fails to ignite, the material is judged safe to ship in a 3 m cube, while if it ignites, it is judged unsafe to ship in a 3 m cube. The advantage of the IMCO method is that it is a "single-point" test that gives a definitive answer within 24 h. What is its disadvantage?

The theoretical basis for the IMCO test is Equation 19. It can be shown that criticality in a 10 cm cube at 140 °C (413 K) is equivalent to criticality in a 3 m cube at 38 °C (311 K) provided that the activation energy of the combustible material is $E = 77$ kJ mol^{-1} (Jones, 1996a). If a "safety margin" is built into the prediction by assuming that the maximum ambient temperature is 50 °C (323 K), the assumed activation energy is $E = 90$ kJ mol^{-1} (Jones, 2000c). However, self-heating materials can have activation energies over the range 50–150 kJ mol^{-1} (Jones, 2000c). The assumption of a single activation energy for all materials to which the test is applied is the inherent flaw in the method. This leads to the conclusion that "if, as is usually the case, the test is applied in total ignorance of the activation energy, it cannot be expected to give a reliable result" (Jones, 2000c, p. 69).

Assuming a maximum ambient temperature of 311 K and providing that all the constants are known, Equation 19 can be used to determine if a particular material is safe to transport in a 3 m cube. It may happen that materials that fail the IMCO test are also found to be unsafe to ship using Equation 19—this was found to be the case for six Scottish coals with known values for all parameters (Jones et al., 1998). That this is coincidental is illustrated by the following example (Jones, 2000c):

Consider a substrate X, for which the critical ambient temperature for a 10 cm assembly is 150 °C and for which the kinetic parameters are $A = 77$ s^{-1} and $E = 70$ kJ mol^{-1}.

Consider a substrate Y, for which the critical ambient temperature for a 10 cm assembly is 130 °C and for which the kinetic parameters are $A = 1.8 \times 10^7$ s^{-1} and $E = 110$ kJ mol^{-1}.

Suppose that for both substrates, $Q = 35$ MJ kg^{-1}, $\lambda = 0.1$ W m^{-1} K^{-1}, and $\rho = 500$ kg m^{-3}, which are typical values for carbonaceous materials.

According to the standard test procedures, it is safe to ship material X and unsafe to ship material Y. *However*, a calculation using Equation 19 shows that for a 3 m cube at 50 °C (323 K), the critical half-length for substrate X is 0.83 m, signifying a cube of 1.7 m, and for substrate Y, it is 2.4 m, signifying a cube of side 4.8 m. Thus, in fact, it is unsafe to ship material X in a 3 m cube, whereas it is safe to ship material Y in a 3 m cube: a reversal of what the standard says. This example shows that a substrate that is subcritical in a 10 cm cube at 140 °C can be supercritical in a 3 m cube at 50 °C and vice-versa, thoroughly demolishing the IMCO test. Jones noted that "this does not augur at all well for the traditional approach" (Jones, 2000d, p. 70).

In the mid-1990s, a provisional International Organization for Standardization (ISO) standard was proposed for assessing the propensity for spontaneous heating of carbonaceous materials (International Organization for Standardization, 1994). This method was also based on the assumption that the failure of a 10 cm cube to ignite at 140 °C indicates safety for shipping purposes in a 3 m cube (Jones et al., 1998). It therefore has the same disadvantages as the IMCO test (Jones et al., 1998; Jones, 2000d).

The United Nations recommendations for the transportation of goods susceptible to self-heating (United Nations, 1995), described in (Jones, 2000a), involves tests of various container sizes to produce various classifications of hazard. For instance, a material that fails the IMCO test can be retested in a 10 cm cube at either 373 K or 393 K. If the material passes the retest, i.e., it does not ignite, then it is "safe" to store or transport the material in either a cube of volume 450 L (side 0.77 m), if it passed the test at 373 K, or a cube of volume 3 m³ (side 1.44 m), if it passed the test at 393 K. It can be shown that both of these criteria assume that the activation energy of the combustible material is $E = 87$ kJ mol^{-1} (Jones, 2000a). This test is simply a variant of the IMCO test and, accordingly, has the same disadvantages as the IMCO test (Jones et al., 1998; Jones, 2000d).

We conclude this section by making the, by now, obvious remark that test procedures that predict the propensity to combust based upon an assumed activation energy are unreliable.

9.1.2. New Standards

In section 9.1.1, it was shown that the IMCO (International Maritime Consultative Organization, 1974), ISO (International Organization for Standardization, 1994), and UN (United Nations, 1995) test procedures lead to unreliable predictions because they use an arbitrary activation energy. This has led to new procedures being proposed to assess the propensity for spontaneous combustion of combustible solids in which the value of the activation energy is determined as part of the test procedure (Jones, 1998c, 2000c). The first of these methods (Jones, 1998c) uses heat-release-rate measurements made with a microcalorimeter. The second of these methods (Jones, 2000c) augments the hot-storage test, described in section 16, with the heat-release method, described in section 17, and is outlined in section 17.3. Both methods represent fundamental improvements in assessing the propensity of hazardous materials to spontaneously combust during transport (Jones, 1999b).

9.2. Time to Ignition

It has been suggested that it can be safe to ship materials under conditions that are nominally supercritical, provided that the anticipated shipping time is less than the time to ignition of the assembly (Jones, 1999c, 2000d). The time to ignition, in seconds, under adiabatic condition is given by (Boddington et al., 1983)

$$t_{ad} = \frac{RT_R^2}{E} \frac{c}{QA} \exp\left[\frac{E}{RT_R}\right], \quad (20)$$

where T_R is the initial temperature of the reactants. The assumptions of adiabacity mean that this is a lower bound on ignition times. If the calculated ignition time is higher than the shipping time, a prac-

tical upper bound on which is three weeks (Jones, 2000d), then "there is no hazard in practical terms" (Jones, 2000d, p. 1561).

Table 4 shows the calculated critical ambient temperature for six coals stored in a 3 m cube (Jones, 2000d). For each coal, this temperature is lower than the maximum ambient temperature that a container will be exposed to during shipping, 311 K. On this basis, none of these coals should be transported as a 3 m cube assembly if the coal is supercritical. However, for two of the coals, the calculated ignition time is much longer than any shipping time and these might therefore be deemed safe to transport.

10. PROPERTIES THAT INFLUENCE THE PROPENSITY TO COMBUST

The propensity of coal to spontaneously combust is determined by many factors, which can be divided into two main types: properties of the coal (intrinsic factors) and the environment/storage conditions (extrinsic factors). Examples of these factors are given in sections 2.3 and 5.2.1. In this section, we are interested in the intrinsic properties of the coal, particularly in how they influence the propensity for self-heating. The spontaneous combustion behavior of coal is a combined product of many factors, and it is difficult to isolate one factor for experimental study. In Table 5, we list experimental studies that have attempted to systematically investigate how coal properties influence the propensity to combust. Earlier work studying the influence of intrinsic properties upon the oxidation rate of coals at low temperatures has been reviewed in Banerjee (1985) and Carras and Young (1994).

In section 10.1, we discuss a regression analysis that relates the propensity for combustion of a coal to its intrinsic properties. In section 10.2, we discuss the influence of inorganic additives, in section 10.3, the influence of particle size and surface area, and in section 10.4, we discuss reaction rates and kinetic parameters. Finally, in section 10.5, we group together some miscellaneous properties.

10.1. Regression Analysis

As noted earlier, many factors influence the oxidation process, and it is difficult to determine their relative effects by varying one factor while keeping the others constant. A popular

TABLE 4. CALCULATED TIME TO IGNITION UNDER ADIABATIC CONDITIONS FOR SIX SCOTTISH COALS

Coal	E (kJ mol^{-1})	A (s^{-1})	$(T_a)_{cr}$	t_{ad} (d)
Rosslynell	54	9	256	2
Dalquhandy 1	74	4×10^3	285	11
Killoch 6015	93	1×10^6	302	73
Killoch 5561	78	3×10^3	301	71
Dalquhandy 2	50	1	262	4
Killoch 5736	57	15	268	4

Note: The value of $(T_a)_{cr}$ is for a 3 m cube. Other parameter values: $Q = 25$ MJ kg^{-1}, $T_R = 300$ K, $c = 1260$ J g^{-1} K^{-1}. $(T_a)_{cr}$ data are from Jones (1998a); all other data are from Jones (2000d).

TABLE 5. SYSTEMATIC EXPERIMENTAL INVESTIGATIONS INTO HOW INTRINSIC COAL
PROPERTIES INFLUENCE THE COMBUSTION CHARACTERISTICS OF COAL

	Humidity	Coal moisture content	Oxygen concentration	Particle size	Aging
Beamish et al. (2001)					X
Garcia et al. (1999)					X
Kadioğlu and Varamaz (2003)		X			
Küçük et al. (2003)	X	X		X	
Nugroho et al. (2000a, 2000b)				X	
Pis et al. (1996)					X
Ren et al. (1999)	X	X		X	X
Smith and Glasser (2005b)	X	X	X	X	

Note: X indicates that the property was studied.

approach has therefore been to test a variety of coals using some measurement of self-heating tendency and then to apply a regression analysis on coal components such as ash content, calorific value, density, fixed carbon content, iron content (total), iron content (nonpyritic), moisture content, pyrite content, sulfur content (organic), sulfur content (total), volatile matter content, etc. Examples of such correlations appear in Mazumdar (1996) and Singh et al. (2002).

Smith and Glasser (2005b) measured the initial rate of oxidation for seventy coals at an ambient temperature (23 °C) and pressure (625 mm Hg) chosen to represent typical conditions at South Africa's main coal-shipping terminal. A regression analysis was undertaken to relate the measured initial rate of oxidation, as determined on an ash-free basis, for these seventy coals to their composition, as determined by petrographic, proximate, and ultimate analyses. Twenty-four regressors were examined. An R^2 value close to unity was obtained by the following equation on 66 samples

$$\log(\text{rate}) = 0.89 (\text{IM})^{0.14} (\text{VM})^{0.43}, \quad (21)$$

where IM is the inherent moisture content, and VM is the volatile matter content of the coal. The importance of these quantities is that VM represents the reactive component of the sample, while IM acts as a proxy for the surface area of the coal. Whatever transform of the rate was taken and regardless of whether a one-, two-, or three-parameter model was used, the most significant regressor was *always* VM content, followed by the IM content.

Other studies have also identified the importance of one, or both, of these regressors.

- Practical experience with the Jhingurdah seam (Singrauli, India) has established that the coals with greater susceptibility to spontaneous combustion are characterized by a high moisture content and high volatile matter content (Singh and Singh, 2004).
- Correlation coefficients were developed between four susceptibility indices (based upon critical air-blast analysis, the crossing-point temperature, differential scanning calorimetry, and differential thermal analysis) and the combined influence of ash, moisture, and volatile content for 31 Indian coal samples (Panigrahi and Sahu, 2004). Correlation coefficients ranged from between $R^2 = 0.76$ and $R^2 = 0.85$.
- Analysis of coal samples from six Indian coalfields in relation to their petrography and rank showed that propensity to spontaneously combust, using the crossing-point temperature discussed in section 15, can be predicted either by the volatile matter content of the coal or its calorific value (Behera, 2004).
- Coals especially prone to self-heating and spontaneous ignition in the Northern Coalfield (New South Wales, Australia) are characterized by one or more of the following: significant sulfide content, high moisture content, and high volatile matter content (McNally, 2000).
- It has also been found that the ignition temperature of coals increases as the volatile content decreases (Kreulen, 1948).

In contrast to these findings, Pilarczyk et al. (1995) found no correlation between volatile content and tendency for self-ignition for German coals.

10.2. The Influence of Inorganic Additives

The effects of inherent and added inorganic matter on the low-temperature oxidation and spontaneous combustion of a Victorian brown coal were investigated by Sujanti and Zhang (1999, 2000), Watanabe and Zhang (2001), and Zhang and Sujanti (1999). The critical ambient temperature, above which thermal runaway occurs, was found using the hot-storage test, described in section 16, for: the raw coal, water-washed coal, acid-washed coal, and acid-washed coal doped with a variety of additives. Additives were introduced into the coal by ion-exchange (Watanabe and Zhang, 2001; Zhang and Sujanti, 1999) and bulk-loading (wet mixing) (Sujanti and Zhang, 1999, 2000; Watanabe and Zhang, 2001). The critical ambient temperatures of the acid-washed and water-washed coals were higher than that of the raw coal, indicating that removal of the inherent inorganic matter from coal reduces its propensity to spontaneously combust. The critical ambient temperature of the acid-washed coal was used as a baseline, and additives were identified as inhibitors (promoters) if they lowered the critical ambient temperature. The effect of the additive (promotion/inhibition) did not depend upon the method of application (Watanabe and Zhang, 2001). The findings of this research are summarized in Table 6. There is no clear trend on whether cations or anions play the most important role in influencing the spontaneous combustion characteristics of the coal.

Copper acetate was found to have a stronger promoting influence on self-heating when it was ion-exchanged into the coal ($T_{a,cr}$ = 127.5 ± 0.5 °C) than when it was bulk mixed ($T_{a,cr}$ = 143.5 ± 0.5 °C) (Watanabe and Zhang, 2001). The lower reactivity of the ion-exchanged coal was due to its significantly lower surface area and micropore volume (Watanabe and Zhang, 2001). The change in surface area and volume was due to the absorbed Cu^{2+}, which provided a resistance to oxygen diffusion within the coal pores and reduced the number of available active sites. Wet-mixing potassium acetate into the coal resulted in a lower surface area than when the additive was ion-exchanged into the coal (Watanabe and Zhang, 2001).

The effect of additive loading was also investigated for calcium acetate and sodium acetate (Sujanti and Zhang, 2000; Watanabe and Zhang, 2001; Zhang and Sujanti, 1999). The critical ambient temperature remained unchanged as the loading of calcium acetate increased (Sujanti and Zhang, 2000; Watanabe and Zhang, 2001), although the inhibiting effect was stronger when the additive was bulk mixed with the coal (Watanabe and Zhang, 2001). The critical ambient temperature decreased as the loading of sodium acetate increased. The inhibition effects of calcium acetate (Zhang and Sujanti, 1999) and sodium acetate (Watanabe and Zhang, 2001) are mainly physical in nature—the additive increases resistance to oxygen diffusion by blocking the coal pore structure.

The effect of additive loading was investigated for bulk-loaded potassium chloride and sodium acetate by Sujanti and Zhang (1999). The inhibiting/promoting effects of these additives were enhanced with an increase in additive loading.

Low-temperature oxidation kinetics of the coal samples were determined using the transient heating method (Sujanti and Zhang, 2000; Watanabe and Zhang, 2001; Zhang and Sujanti, 1999), described in section 18, and using an adiabatic energy balance (Sujanti and Zhang, 1999). The calculated reaction rates showed a general trend: the lower the critical ambient temperature, the higher the sample reactivity. Kinetic data indicated that ion-exchanged copper acetate and potassium acetate affected the reaction rate primarily through chemical means (Watanabe and Zhang, 2001).

10.3. Particle Size and Surface Area

The total surface area of a coal particle is given by the sum of its external and internal surface areas. The former is insignificant for particles with a large effective internal surface area. Conversely, for particles with low effective internal surface areas, the external surface area plays a key role during low-temperature oxidation (Akgün and Arisoy, 1994; Palmer et al., 1990).

The internal surface area of a particle depends upon its size, structure, and porosity. Smaller particles have a larger surface

TABLE 6. THE EFFECT OF INORGANIC ADDITIVES ON THE CRITICAL AMBIENT TEMPERATURE OF A VICTORIAN BROWN COAL IN THE HOT-STORAGE TEST

Additive	Promoter	Inhibitor	No effect
Ammonium chloride (NH_4Cl)			[a]
Calcium acetate [$Ca(Ac)_2$]		[b]	[c] [d]
Calcium carbonate ($CaCO_3$)	[a] [b]		
Calcium chloride ($CaCl_2$)		[a] [b] [c] [d]	
Calcium hydroxide [$Ca(OH)_2$]	[b]		
Copper acetate [$Cu(Ac)_2$]	[b] [c] [d]		
Copper carbonate ($CuCO_3$)	[b]		
Magnesium acetate [$Mg(Ac)_2$]		[a] [b] [c] [d]	
Magnesium carbonate ($MgCO_3$)		[b]	
Montan powder (*)		[a]	
Potassium acetate (KAc)	[a] [b] [c] [d]		
Potassium carbonate (K_2CO_3)	[b]		
Potassium chloride (KCl)		[a]	
Pyrite (FeS_2)	[a]		
Sodium acetate (NaAc)	[a] [b] [c] [d]		
Sodium carbonate (Na_2CO_3)	[b]		
Sodium chloride (NaCl)		[a] [b] [c] [d]	
Sodium hydroxide (NaOH)		[b]	
Sodium nitrate ($NaNO_3$)			[a]

Note: Additives were bulk-loaded (wet mixing) (a, b, c) and ion-exchanged (c, d) into the coal samples. Key: [a] Sujanti and Zhang, 1999; [b] Sujanti and Zhang, 2000; [c] Watanabe and Zhang, 2001; [d] Zhang and Sujanti, 1999.
*Montan powder is calcium chloride powder + 3% sodium dodecyl sulfate.

area per unit volume of the coal particle (Akgün and Arisoy, 1994) and a lower diffusional resistance. Other factors being equal, this means that smaller particles have a higher oxidation rate, and release more heat, than larger particles. Particle size is therefore an important factor in determining whether self-heating leads to spontaneous combustion (Akgün and Arisoy, 1994; Carras and Young, 1994; Krishnaswamy et al., 1996b). Studies have shown that the propensity for spontaneous combustion increases as the coal particle size decreases (Kadioğlu and Varamaz, 2003; Küçük et al., 2003; Nugroho et al., 2000a; Ren et al., 1999). In addition to the increased oxidation rate of smaller particles, the rate of heat transfer between particles increases as their size decreases (Ren et al., 1999).

The rate of oxidation is found to increase with decreasing particle size until a critical diameter is reached. Above the critical value, the oxidation rate is approximately proportional to the cube root of the external surface area (Küçük et al., 2003). Below the critical value, further decreases in the particle size have little effect on the oxidation rate (Carpenter and Sergeant, 1966b; Kaji et al., 1985; Palmer et al., 1990; Ren et al., 1999). The critical diameter corresponds to a size at which oxygen penetrates the particles without experiencing any mass-transfer resistance (Ren et al., 1999). The critical diameter depends upon factors such as the reaction conditions and the porosity of the coal. It varies significantly between coals. Critical values of 50 µm (Palmer et al., 1990), between 138 and 387 µm (Carpenter and Sergeant, 1966a, 1966b), and 1000 µm (Kaji et al., 1985) have been reported. Chamberlain and Hall (1973) found no significant variation in the oxidation rate for particle sizes of <210 µm and 250–450 µm; these sizes may have been below the critical diameter of their coal. It has been reported that the surface area of lower-rank coals, and consequently their reactivity, is independent of particle size (Nugroho et al., 2000a). It has also been suggested that the surface area of a particle is reduced by low-temperature oxidation (Kaji et al., 1985; Swann et al., 1974). The effect is particularly noticeable in low-rank coals, where a reduction of up to 30% has been measured following oxidation at 35 °C (Swann et al., 1974).

Nugroho et al. (2000a) studied the effect of particle size and the physical structure of coals (including pore size and surface area) on the self-heating of high- and low-rank Indonesian coals and their blends. The effects of particle size were very significant for high-rank coals: the surface area of one higher-rank coal increased almost fivefold as the particle diameter decreased from 1.5 mm to 0.06 mm. However, the effects of particle size were almost negligible for low-rank coals as the surface area hardly changed. It was also concluded that a coal bed with mixed sizes was more vulnerable to spontaneous combustion than one with segregated sizes. However, the authors' experimental results do not support this conclusion. Rather, they show that a coal pile with a mixture of large and small particles is more vulnerable to spontaneous combustion than one containing only large particles.

In addition to changing the reaction rate, the particle size also affects the permeability of the stockpile. It has been suggested that for *stockpiles*, permeability exerts more control on the overall reaction rate than particle size (Krishnaswamy et al., 1996a, p. 359). Such considerations do not apply to experiments on a sufficiently small scale. Thus, in small-scale experiments, beds containing smaller particles have a stronger propensity to spontaneously combust than those containing larger particles. However, this trend is reversed in sufficiently large beds (Nugroho et al., 2000b). The relationship between particle size and permeability explains why there is a greater tendency to self-heat in stockpiles in which segregation of coal particles has occurred: the larger particles allow more air to enter the stockpile, which reacts with the smaller particles that have a higher surface area (U.S. Department of Energy, 1994).

10.4. Reaction Rates and Kinetic Constants

10.4.1. Reaction Rates

Smith and Glasser (2005a) measured the rates of reaction for nine coals, ranging from lignite to anthracite, in a semi-adiabatic reactor. The rate of reaction was determined for at least two temperatures and on an ash-free basis. They obtained initial reaction rates that varied between 2.3×10^{-5} and 3.6×10^{-4} mol O_2 min^{-1} g^{-1} over the temperature range 57.6–79.77 °C. Smith and Glasser (2005b) measured the rates of reaction for seventy coals in a static isothermal apparatus. The initial reaction rates varied over two ranges of magnitude between 2.34×10^{-6} and 4.36×10^{-4} mol O_2 min^{-1} g^{-1} at an ambient temperature of 23 °C.

10.4.2. Kinetic Constants

In the simplest models for the low-temperature oxidation of coal, as described in section 2.1, it is assumed that heat is released in a single-step oxidation reaction. There has therefore been considerable interest in determining appropriate values for the kinetic parameters: the activation energy (E) and the pre-exponential factor (A).

Smith and Glasser (2005a) measured the activation energy for eight coals, ranging from lignite to anthracite, in a semi-adiabatic reactor. Their results are given in Table 7. They commented that although there is only a minor range in variation for the activation energies, the rates of reaction vary by one order of magnitude.

TABLE 7. APPARENT ACTIVATION ENERGIES, PRE-EXPONENTIAL FACTORS, AND OXIDATION HEAT FOR EIGHT COALS OVER THE TEMPERATURE RANGE 50–80 °C

Sample	E (kJ mol^{-1})	A (m^3 min^{-1} g^{-1})	Q (kJ mol^{-1} O_2)
14	35	2.61×10^1	364
15	38	2.67×10^1	398
17	55	4.92×10^3	426
18	31	2.05×10^{-1}	339
19	75	4.92×10^6	409
21	72	2.37×10^6	287
26	49	3.33×10^2	306
34	91	6.55×10^8	437

Note: From Smith and Glasser (2005a).

Beamish et al. (2003) found rate parameters for five coals using adiabatic methods. Their results are in Appendix 2. The activation energies increase as the rank of the coal increased: 55 kJ mol^{-1} (sub-bituminous C), 61 kJ mol^{-1} (lignite), and 73–83 kJ mol^{-1} (high-volatile B bituminous). The increase in activation energy with an increase in rank has also been observed by other authors (Carpenter and Giddings, 1964; Kuchta et al., 1980; Smith and Lazzara, 1987). These results contrast with the findings of Nugroho et al. (2000a), who found that the activation energies for two volatile Indonesian bituminous coals ($E = 51$ and 53 kJ mol^{-1}) were lower than those for a low-rank sub-bituminous coal ($E = 76$ kJ mol^{-1}) and lignite ($E = 89$ kJ mol^{-1}). In the work reported in Nugroho et al. (2000a), the samples were air dried for over 24 h before testing. It is possible that preoxidation effects, discussed in section 3.2, increased the activation energy of the more reactive coals (Beamish et al., 2003). A decrease in coal reactivity at 45 °C following air drying has been noted by Unal et al. (1992). Preoxidation effects are noticeable in the results presented for Kopako coal, a sub-bituminous C coal, in Appendix 2.

Sujanti and Zhang (1999) determined kinetic parameters, using the heat-release rate method described in section 17, for cylinders with height (cm)/diameter (cm) ratios of 27/9.8, 15/8, 12/6, and 8/4. The apparent activation energy showed little variation between the four reactors: $105.49 \leq E$ (kJ mol^{-1}) ≤ 107.37. Although no regression errors were provided, this variation is almost certainly within the statistical uncertainty in each calculation. The values of A varied over an order of magnitude. It was suggested that preoxidation effects during the course of the reaction prior to detection of thermal runaway, on the order of one to two hours, could have been more significant for the smaller reactors, causing a loss of reactivity and consequently lower values for A. It was noted that the critical ambient temperature of the reactors increased linearly with the reactor-specific external surface area. This was taken to indicate the "important role of external heat transfer in a spontaneous combustion process" (Sujanti and Zhang, 1999, p. 556). Nugroho et al. (2000a) also noted that, in tests using the transient method, described in section 18, while the measured activation energy was independent of basket size, the parameter AQ was lower for smaller basket sizes.

Nugroho et al. (2000a, 2000b) determined the values for the product of the pre-exponential factor and the exothermicity (AQ) and the activation energy (E) for a variety of Indonesian coals, including coals of different particle sizes, and noted that there was a general correlation between higher values for E corresponding to higher values in the product AQ. This is illustrated in Nugroho et al. (2000b, their Figure 3). It was also noted that these parameters, AQ and E, increased with decreasing particle size, and the influence of particle size was more prominent in the pre-exponential factor than in the activation energy (Nugroho et al., 2000a).

Kinetic values, obtained using other methods, are discussed in sections 16 and 18.

As noted in section 9.1.1, standard test methods involve the assumption of common activation energies for self-heating materials. The results quoted in this section, and elsewhere in this review, such as Appendix 2, clearly show that this assumption is invalid.

10.5. Miscellaneous

In this section, we discuss various properties of coal: coal rank, section 10.5.1, heat capacity, section 10.5.2, heat of reaction, section 10.5.3, the oxygen content of the coal, section 10.5.4, and pyrite content, section 10.5.5.

10.5.1. Coal Rank

The propensity of coal to self-heat and to spontaneously combust tends to increase with decreasing rank. Thus, lignite and sub-bituminous coals are more prone to spontaneous combustion than bituminous coals and anthracites (Kuchta et al., 1980). It needs to be reinforced that the relationship with rank is a *tendency*; not an absolute fact—studies suggesting that this is a fact tend to be based on a limited rank range (Beamish, 2005a).

It has been found that as rank decreases, the moisture content, oxygen content, and internal surface area of a coal all increase (Michalski et al., 1990). Each of these effects acts to increase the oxidation rate, which increases the propensity for spontaneous combustion. The importance of surface area and oxygen content is discussed in sections 10.3 and 10.5.4, respectively. As noted in section 10.1, moisture content may act as a proxy for surface area.

10.5.2. Heat Capacity

Smith and Glasser (2005a) measured the heat capacities of nine coals, ranging from lignite to anthracite, in a semi-adiabatic reactor. Typically, the heat capacity was determined at two temperatures. Mean values of eight of the coals ranged between 1520 and 1729 J kg^{-1} K^{-1}. Australian lignite provided an outlier with mean heat-capacity 2128 J kg^{-1} K^{-1}. This coal had very high moisture content, "which alone could be responsible for a high specific heat" (Smith and Glasser, 2005b, p. 1154).

Beamish et al. (2003) calculated the heat capacities of five coals, ranging in rank from lignite to high-volatile bituminous. Their values ranged between 1399 and 1608 J kg^{-1} K^{-1}.

10.5.3. Heat of Reaction

Smith and Glasser (2005a) measured the heat of reaction for nine coals, ranging from lignite to anthracite, in a semi-adiabatic reactor, and eight of their values are provided in Table 7. Typically, the heat of reaction was determined at two temperatures. The temperature-averaged values ranged from 283 to 437 kJ mol^{-1} O$_2$. There was no obvious correlation between petrographic, proximate, and ultimate properties of the samples and their heat of reaction.

10.5.4. Oxygen Content of the Coal

Coals with higher oxygen content have a higher self-heating tendency (U.S. Department of Energy, 1994; Schmidt, 1945). Nugroho et al. (2000a) found that, generally, the propensity for spontaneous combustion increases with increasing oxygen percentage in the coal. This effect becomes less significant for smaller particles. Low-rank coals have relatively high oxygen contents (Garcia et al., 1999), which partly account for their increased reactivity as opposed to high-rank coals.

10.5.5. Pyrite Content

As shown in Table 6, the presence of inorganic impurities in coal can promote self-heating. Pyrite (FeS_2) has long been known to enhance the risk of spontaneous combustion. It does so by accelerating the reaction process (Banerjee, 1985, p. 7; Durie, 1991, p. 267; Carras and Young, 1994; Glasser and Bradshaw, 1990). Pyrite has a twofold action. It first catalyzes the oxidation reaction. Second, in moist air, pyrite is itself oxidized, which provides a secondary heat source and accelerates the process of self-heating. Unsurprisingly, pyrite is listed as a promotor of self-heating in Table 6.

11. MOISTURE AND HUMIDITY

The influence of moisture and humidity on the self-heating of coal has been a subject of investigation for many years. They can act to either enhance or moderate self-heating, and they are important factors that influence the extent to which self-heating occurs (Banerjee, 1985; Beamish and Jabouri, 2005; Bowes, 1984). The temperature of the coal will increase due to self-heating until a plateau is reached, at which the temperature is temporarily stabilized. In this region, heat generated by oxidation is used to vaporize the moisture in the coal (Beamish and Jabouri, 2005). Once all the moisture has been vaporized, the temperature increases rapidly. At the other extreme, dry materials can readily ignite following the sorption of water. Thus, dry coal, whether in a mine or during storage, should not be kept in a damp place because this can promote self-heating (Ren et al., 1999). Similarly, dry deposits of coal particles in coal mills become particularly susceptible to spontaneous ignition, through adsorption of moisture, under conditions of high humidity (Ren et al., 1999). It is also recommended that dry and wet coals should be stored separately (Ren et al., 1999).

In section 11.1, we discuss heat-transfer processes that are associated with the adsorption and desorption of water moisture. In section 11.2, we discuss experiments where the main emphasis is on the moisture content of the coal, whereas in section 11.3, we discuss experiments where the main emphasis is on atmospheric humidity. Since the adsorption and desorption of water vapor is an equilibrium process, a phrase such as "high water content" really means a high water content relative to the humidity of the experiment. In section 11.4, we discuss the production of water vapor during oxidation, and in section 11.5, we make a few comments regarding modeling.

The effect of moisture transfer (evaporation, condensation, diffusion, and convection) on the overall heat balance, described in section 11.1, is well understood. However, the effect of moisture on the rate of coal oxidation—does water have a catalytic affect?—is not fully understood.

11.1. Heat Transfer by Adsorption and Desorption of Water Moisture

Under normal circumstances, coal, whether in a mine or a stockpile, contains moisture that is in equilibrium with the humidity of the surrounding atmosphere. Consequently, there is no net heat transfer due to adsorption and desorption. Under nonequilibrium conditions, this is no longer the case, and adsorption and desorption processes have a strong controlling effect. This was first realized in the 1950s (Stott and Baker, 1953; Berkowitz, 1957).

When dry air flows over moist coal, desorption of water occurs. This is an endothermic process that decreases, or restricts, the coal temperature. At the other extreme, when moist air flows over a dry coal, adsorption of water occurs. This is an exothermic process, and it releases the heat of wetting, about 2261 J kg (H_2O)$^{-1}$. This can increase the coal temperature and enhance self-heating.

The heat released by water adsorption is particularly significant for dry coals. Practical experience has revealed that in stockpiles, hot spots occur frequently after rain. For very dry coal, the problems are very serious. If almost all of the surface and inherent moisture has been removed, then moisture from saturated air condenses not only onto the external surface of the coal but throughout its internal pore structure. This can rapidly release a tremendous amount of heat: very dry coals can ignite by water sorption. It has been noted in China that, in the rainy season, when the atmospheric temperature is high, heavy rainfall can lead to explosions at waste heaps where the surface is already on fire (Liu et al., 1998). The risk of explosions has led to a reluctance to use water to cool heatings in stockpiles (Ren et al., 1999).

Moist air has a higher thermal conductivity and heat capacity than dry air (Greankoplis, 1983, p. 205; Reid et al., 1987, p. 491). The flow of a moistened air transfers more heat to a coal pile than that of a drier air; this is a secondary heat-transfer effect. In large coal systems, "dry" and "wet" regions can both occur. Another ancillary heat-transfer effect is that heat released in a "wet" area increases the local temperature, which in turn increases heat flow into a dry area, enhancing its rate of oxidation (Schmal et al., 1985).

11.2. Moisture Content of Coal

The moisture content of a coal may be bound to active sites on the surface of coal particles or it may be loosely bound, "free water," filling a portion of the coal pores. The mechanisms by which moisture content affects self-heating include the following:

- **Heat sink:** When water is present, the heat released by oxidation is used to evaporate the moisture. Significant self-heating only starts to occur in an area once the moisture in that area has been evaporated. This point is nicely shown in the temperature-time trajectories reported by Vance et al. (1996, their Fig. 4). These show a rapid increase in the heating rate as moisture is removed from the sample.

- **Preventing oxidation by "chemical" occupation of active sites:** The adsorption and desorption of water changes the number of active sites that are available for the oxidation process to occur (Akgün and Arisoy, 1994; Nordon et al., 1979). Moisture desorption increases the number of active sites available for oxidation, increasing the oxidation rate (Banerjee et al., 1970; Walker, 1967).
- **Preventing oxidation by reducing access to smaller pores:** Adsorbed moisture can provide a resistance to oxygen diffusion within the coal pores. Thus, removal of this water exposes more surface area for oxidation, increasing the oxidation rate.
- **Preventing oxidation by "physical" occupation of active sites:** In areas of the coal matrix that are occupied by free water, the oxidation rate becomes negligible because, before oxidation can occur, oxygen has to first dissolve in the water and then to diffuse to an active site. The solubility of oxygen in water is low, and the diffusion rate of oxygen in water is four orders of magnitude lower than that of oxygen in air (Panaseiko, 1974).

11.2.1. Moisture Content and Internal Surface Area

As described in section 10.1, a detailed regression analysis has been carried out on the initial rate of oxidation for 70 coals (Smith and Glasser, 2005b). The inherent moisture content, which ranged from 0.7% to 9.1%, was found to be the second most significant regressor, behind the volatile matter content, in predicting the oxidation rate: the oxidation rate was an increasing function of moisture content in this range. The rate of adsorption and desorption of moisture was stated to be slow compared to the rate of oxygen adsorption because of the low moisture content of saturated air (3.8% at 23 °C). Thus, the observed increase in reaction rate was not due to heating effects. Although it is possible that inherent moisture plays a catalytic role, it may act as a proxy for the surface area of the coal. If this is the case, a wetter coal could have a larger number of vacant active sites than a drier coal and, consequently, a higher initial oxidation rate. The relationship between increased inherent (equilibrium) moisture content and an increased tendency for coals to self-ignite is well established (U.S. Department of Energy, 1994).

The importance of moisture content as a predictor of self-heating has been noted by Mazumdar (1996), who correlated the self-heating temperature of coal with the volatile content and moisture content. It was suggested that moisture acts as "an index of the widely varying quality of volatiles" (Mazumdar, 1996, p. 647). It was also noted that high volatile coals are "highly amenable to classification by their capacity moisture content only" (Mazumdar, 1996, p. 647). We return to this point in section 11.2.2.

Taking these factors into account, the findings of Smith and Glasser (2005b) suggest that, other things being equal, the maximum oxidation rate of an intrinsically "wetter" coal should be higher than that of an intrinsically "drier" coal.

11.2.2. Moisture Content and Propensity to Self-Ignition: The Catenary Effect

Although many investigations have emphasized the importance of moisture during the self-heating of coal, the effect of moisture on the oxidation rate, other than its influence on the local temperature through adsorption and desorption, is not fully understood.

At relatively low temperatures, an increase in free moisture can increase the rate of heating (Bhattacharyya et al., 1968); low-rank coals with 5–10 wt% moisture undergo oxidation more rapidly than completely dried samples (Clemens and Matheson, 1996). As the moisture content is increased from "low" to "high," the oxidation rate is maximized at a median moisture content (Chen and Stott, 1993; Sondreal and Ellman, 1974; Vance et al., 1996). Other evidence of a catenary effect includes the existence of a global minimum when either the self-heating temperature (Mazumdar, 1996) or the crossing-point temperature (Banerjee, 1985, p. 45–47; Mazumdar, 1996) of a range of coals is plotted as a function of their moisture content. It should be noted that there is a lack of data points for "wetter" coals in the figures contained in Mazumdar (1996).

Vance et al. (1996) investigated the self-heating of sub-bituminous coals under adiabatic conditions over the temperature range 40–140 °C. The moisture content of the coal was varied between 4.5 wt% and 15.7 wt%. The time to reach temperatures of 80 °C and 140 °C was found to be a catenary function of the moisture content, which was minimized at ~7 wt%. The maximization of the oxidation rate was more pronounced at temperatures below 80 °C.

Sondreal and Ellman (1974) showed that as the moisture content of lignite increased toward ~20 wt%, so did the oxidation rate. Over this value, the oxidation rate decreased as the moisture content increased. These trends became less conspicuous as the temperature approached 70 °C. Chen and Stott (1993) investigated the effect of moisture below 70 °C. They showed that, for a sub-bituminous coal, the maximum oxidation rate occurred when it contained between 7 wt% and 17 wt% moisture.

A consequence of the catenary effect is that the oxidation rate of a relatively dry coal is enhanced by addition of "small" amounts of moisture. In fact, a "slightly" wet coal should be more reactive than its dried form. Why does water moisture have a catenary effect? It has been suggested that the reaction rate increases because water has a catalytic effect on oxidation rates (Akgün and Arisoy, 1994; Arisoy and Akgün, 1994; Marinov, 1977; Nordon et al., 1979; Nordon and Bainbridge, 1979; Nordon, 1979; Saranchuk et al., 1978). It has been proposed that the formation of a coal-moisture-oxygen intermediate product is responsible for the catalytic effect (Moxon and Richardson, 1985). Such a moisture-sorption reaction would be exothermic, which may also enhance the oxidation rate (Mulcahy et al., 1991). It has also been suggested that at low temperatures (30 °C), tightly bound water cannot act directly as a catalyst, but instead that it generates radical sites in the coal that are more reactive than those available in a fully dried coal (Clemens and Matheson, 1996).

In contrast, it has also been proposed that the adsorption of moisture on the coal *inhibits* the reaction rate (Akgün and Arisoy, 1994; Buckley, 1994; Ogunsola and Mikula, 1991). The proposed mechanism is: water molecules compete with oxygen to form hydrogen bonds to oxygen-containing functional groups attached to aromatic rings. If a water-functional group is formed, then the water molecule shields the functional group from attack by oxygen molecules.

Finally, Nordon and co-workers (Nordon, 1979; Nordon and Bainbridge, 1979; Nordon et al., 1979) found that the activation energy for the oxidation of coal and char was independent of moisture content, indicating a hindrance effect of water molecules (which would lower the pre-exponential factor or rate constant, since it represents the frequency of the oxygen molecules interactions with the active sites).

11.2.3. Moisture Content and Propensity for Self-Ignition: Noncatenary Effects and Other Studies

A number of studies have not shown a catenary effect as the moisture content is varied. This may be because a catenary relationship is not universal in coals or that the range of moisture values used was too small. It has even been suggested that the catenary effect is an artifact of the test methods, which may have altered the coal (Bhat and Agarwal, 1996).

Li and Skinner (1986) reported that the reactivity of a sub-bituminous coal progressively decreased as the moisture content was increased from a perfectly dry sample to a raw wet sample.

Kadioğlu and Varamaz (2003) reported that the propensity for spontaneous combustion of two Turkish lignites was reduced as their moisture content was increased. It was also found that if moist coals were air-dried, their propensity for self-heating increased as the drying time increased, i.e., the samples became more reactive as their moisture content was decreased.

Ren et al. (1999) tested a coal under both dry coal/saturated air and saturated (wet) coal/saturated air conditions. The former was more reactive. In fact, the temperature of the wet coal/saturated air system was observed to initially decrease due to evaporation of moisture from the coal.

Lu et al. (2004) investigated the effect of injecting water into a coal sample (5%) that was heated in an oven. The water inhibited coal oxidation during the early stages of the experiment, but at temperatures of over 140 °C, the rate of change of temperature was greater in the water-coal sample than in a coal-only sample. (The coal-only sample contained 4.8% moisture). It was suggested that in the water-coal system, the added water might have had a catalytic action.

11.2.4. Moisture Content and Propensity for Self-Ignition: The Effect of the Drying Method

It is known that the extent and method of drying changes the concentration and nature of active sites within coal (Buckmaster and Kudynska, 1992; Carr et al., 1995; Dack et al., 1983). It is therefore unsurprising that the oxidation rates of dried coals have been found to depend upon the drying method (Chen and Stott, 1993; Unal et al., 1992; Vance et al., 1996). It is believed that different drying methods damage the solid structure of the coal to different extents, thereby inducing a variation in the oxidation rate. Chen and Stott (1993) found that a near-equilibrium drying procedure, which has very low drying rates, gave different trends for the influence of moisture on oxidation rates from those obtained using vacuum drying. A clear difference in oxidation rates depending upon the choice of drying method was also shown in Vance et al. (1996).

11.3. Humidity

As discussed in section 11.1, air humidity is an important factor in governing whether the oxidation of a dry coal will progress into rapid self-heating. For example, the self-heating risk is higher when coal is exposed to humidified air rather than dry air, or saturated oxygen rather than dry oxygen (Guney and Hodges, 1969; Smith and Lazzara, 1987; Stott, 1960; Vance et al., 1996). On the other hand, it has been suggested that in stockpiles, humidity variations in the air at the surface are insignificant when the surface temperature is below 50 °C "because humidity only plays a role at temperatures above 50 °C" (Fierro et al., 2001, p. 126).

Mukherjee and Lahiri (1957) showed that at temperatures above 70 °C, the oxidation of coals occurs more readily in dry rather than in moist air. This may be linked to the change in the mechanism of coal oxidation that occurs at about 70 °C (Berkowitz, 1985). Below 70 °C, acid functions and peroxides are generated during coal oxidation, and a higher moisture content is believed to promote these functions (Berkowitz, 1985). At higher temperatures, peroxides form only transiently or not at all. Thus, it is possible that humidity plays a promoting role at lower temperature and an inhibiting role at higher temperatures.

Küçük et al. (2003) found that the propensity of Turkish lignite to self-heat increased with decreasing humidity of the air. This result was interpreted as showing the occurrence of a reaction-inhibiting mechanism.

11.4. The Production of Water Vapor

Once the physically and chemically adsorbed water, and any free water, has been removed from the coal matrix, any water that is liberated is due to chemical changes in the structure of the coal as a result of oxidation and/or gas-phase oxidation processes (Vorres et al., 1992).

There is no consensus in the literature regarding the trend in the production of water vapor during oxidation. Karsner and Perlmutter (1982) reported a small increase in water production over the temperature range from 175 to 300 °C, whereas Swann and Evans (1979) reported the opposite trend over the temperature range from 35 to 70 °C. Wang et al. (2002, p. 591) noted that once the physically adsorbed water had been removed from a well-oxidized coal, the amount of water produced displayed "a direct relationship with temperature during coal oxidation."

11.5. Modeling

We have seen throughout this section that under nonequilibrium conditions, adsorption and desorption processes exert a strong influence over the self-heating process. There is considerable evidence that adsorbed water influences the rate of oxidation by acting as a catalyst. Thus, any general mathematical model for self-heating in coal should include these processes. However, many modeling studies have only investigated the self-heating of *dry* coals (e.g., Bradshaw et al., 1991; Brooks et al., 1988b; Hull et al., 1997). Models of "dry coal" experiments can be justified; however, as in many experimental techniques, the first step is to remove the moisture by drying.

Because many of the principles underlying the behavior of wet coals are not fully understood, there is ongoing discussion on the appropriate way to model their self-heating. For instance, Schmal et al. (1985) assumed that the air flowing through a coal bed is fully saturated with water. This assumption has been criticized as being unrealistic (Chong and Chen, 1999), especially in regions of the coal bed that have become very dry and where the local temperature has increased over 100 °C.

Thus, there remains scope for the development of mathematical models to further analyze the influence of inherent moisture content and humidity on the self-heating behavior of coal.

12. UPGRADING OF LOW-RANK COAL

Low-rank coals, such as brown coals, lignite, and sub-bituminous, are found in abundant deposits all over the world. For example, 86% of the estimated 58 billion tons of coal resources in Indonesia are classified as lignite and sub-bituminous coals (Umar et al., 2005), and most of the estimated 150 billion tons of coal resources in Mongolia are thought to be low-rank brown coal (Avid et al., 2004). The estimated worldwide reserves of low-rank coals are comparable to those of high-rank coals. Consequently they are increasing in importance as an energy resource.

Unfortunately, as discussed in section 10.5.1, low-rank coals are more susceptible to spontaneous combustion than high-rank coals. This imposes restrictions on both their storage in stockpiles and in their transportation. Almost all low-rank coals have high moisture contents, reaching 60 wt% for some brown coals, and consequently low calorific values. Coals with such high water contents must be used close to the mining site since their high transportation cost per thermal unit of coal makes it uneconomic to transport them. These restrictions hinder their effective utilization. The heating value of a low-rank coal can be greatly increased by drying, often to values similar to those of high-rank coals. For instance, the heating value of a sub-bituminous coal from the United States increased when it was dried from 19.3 MJ kg^{-1} wet basis, to 28.0 MJ kg^{-1} (Sato et al., 2004). When dried coal is ground into small particles, a very efficient fuel is obtained. However, dried low-rank coals are highly susceptible to oxidation, moisture re-absorption, and therefore to spontaneous combustion (Carras and Young 1994; Chen and Stott, 1993; Ogunsola, 1993; Swann and Evans, 1979), and this susceptibility imposes restrictions on their storage and transportation. These problems can be overcome by a combination of dewatering, to reduce moisture content and increase calorific value, and thermal stabilization by upgrading, to prevent self-heating and combustion during long-term storage and long-distance transportation. Methods to upgrade low-rank coals have been developed since the 1920s (Mahidin et al., 2003b).

Nakagawa et al. (2004) hydrothermally treated a brown coal at 300 °C. The moisture content decreased from 1.31 kg/kg dry matter to 0.59 kg/kg dry matter, while the calorific value increased from 25.8 to 27.8 MJ kg^{-1} dry matter. The susceptibility to spontaneous combustion of the raw and upgraded coals was assessed using a temperature-programmed reaction technique (TGA), from which it was concluded that the propensity for spontaneous combustion was reduced by the treatment process. The wastewater contained dissolved organic compounds, the concentrations of which increased with increasing process temperature. The organic compounds in the wastewater were completely gasified under hydrothermal conditions using a Ni-supported carbon catalyst at 350 °C under 20 MPa pressure, producing a combustible gas rich in methane and hydrogen.

Sato et al. (2004) investigated the thermal liquid-phase upgrading of low-rank coal in a 200 mL autoclave at 380–440 °C under an initial nitrogen pressure of 2 MPa. The change in heating values for three coals is shown in Table 8. The susceptibility to spontaneous combustion of the raw coal and upgraded product was assessed using an adiabatic spontaneous ignition test and by thermogravimetry and differential thermal analysis (TG-DTA) data. The upgraded coals did not ignite in the adiabatic ignition test. The ignition temperatures measured by thermogravimetry were higher for the upgraded coals than for the raw coals. Thus, the treatment process reduced the propensity to spontaneously combust.

Avid et al. (2004) applied the dewatering and upgrading technique used in Sato et al. (2004) to three Mongolian low-rank coals. The change in heating values is shown in Table 8. The heating value of the upgraded coal was higher when coal tar was used as the solvent compared to t-decalin. The upgrading process significantly increased the spontaneous ignition temperature measured by thermogravimetric analysis.

Umar et al. (2005) applied the upgraded brown coal (UBC) process developed by Kobe Steel Ltd. to an Indonesian low-rank

TABLE 8. HEATING VALUES OF RAW AND UPGRADED PRODUCT ON A DRY BASIS

Coal	Raw coal	Upgraded product	Solvent
Bagabuur-2A [a]	26.3	32.9	Coal tar
Shiviee-Ovoo [a]	21.2	28.8	Coal tar
Tsaidam-Nuur [a]	22.7	33.1	Coal tar
Buckskin [b]	28.0	31.3	t-decalin
Taiheiyo [b]	27.8	27.8	t-decalin
Yallourn [b]	25.0	33.6	t-decalin

Heating value (MJ kg^{-1})

Note: [a]—Avid et al. (2004); [b]—Sato et al. (2004).

coal. The coal was processed at a rate of 100 kg d⁻¹ at a temperature of 150 °C and a pressure of 350 kPa using kerosene as the process media. The total moisture content of the coal decreased significantly, from 24.2% before upgrading to 0.3% after upgrading. The dewaterization process increased the specific energy from 5467 kcal kg⁻¹ to 6279 kcal kg⁻¹. The susceptibility to spontaneous combustion of the raw and upgraded coals was assessed from CPT and DTA-TG data. It was concluded that the upgraded coal was less likely to ignite than the raw material. In order to prevent an increase in the moisture content after processing, a low sulfur wax residue was added to the upgraded coal.

Mahidin et al. (2003a, 2003b) developed an upgrading method that combines vacuum drying with a tar-coating treatment. This process can upgrade a low-rank coal to a product that is comparable to a bituminous coal, particularly in the moisture, volatile matter, and carbon contents. In the vacuum-drying stage, almost all of the moisture content is removed (up to 94% at 200 °C), and the volatile matter is significantly reduced. The specific surface area is also reduced from ≈18 m² g⁻¹ for the raw coal to ≈6 m² g⁻¹ (Mahidin et al., 2003b). In the second stage, tar is added and the coal-tar mixture is heated to 270–300 °C. During the heating and cooling process, tar is vaporized and deposited onto the surfaces of the coal, sealing the surface by plugging macropores in the coal. This further decreased the specific surface area to ≈2–4 m² g⁻¹, depending upon the process conditions.

The kinetic parameters of oxygen adsorption were obtained using DTA-TG techniques (Mahidin et al., 2003a). The activation energy for oxygen adsorption increased as the maximum temperature of the second processing stage was increased. The low-temperature oxidation of the raw and upgraded coals was investigated using an adiabatic reactor (Mahidin et al., 2003a, 2003b). The upgraded product had a reduced susceptibility toward spontaneous combustion. Analysis of the adiabatic results showed that the activation energy (E) and the parameter grouping (QA/c) were both higher for the upgraded products than for the untreated coal.

The risk of spontaneous combustion can be reduced by spraying a coal immediately after drying with certain liquid petroleum hydrocarbons. One such hydrocarbon is No. 6 fuel oil, which is readily available, inexpensive, and works well as a suppressant. However, this agent is potentially damaging to human health because, depending upon how it is blended, it can contain polynuclear aromatic hydrocarbons (PNAs) that are known carcinogens. Furthermore, it is known that inhaled particles can increase the potency of PNAs adsorbed onto the particles when the particles are in the human lung (Dalbey et al., 1998).

Dalbey et al. (1998) investigated the carcinogenic potential of coal particles treated with No. 6 fuel oil by comparing the treated coal with petroleum coke treated in a similar way that had been tested by chronic inhalation in monkeys and rats. The authors concluded that the treated coal would have no greater carcinogenic potential than petroleum coke.

13. EXPERIMENTAL TECHNIQUES: INTRODUCTION

A large number of test methods have been used to investigate the self-heating of coal (Banerjee, 1985; Berkowitz, 1985; Carras and Young, 1994), each of which has its own advantages and disadvantages. In this section, we provide a short introduction to such methods. Several of the panoply of methods are listed in section 13.1. The more "important" of these methods are discussed in sections 14–18. Recent investigations using the remaining methods are outlined in section 13.2. In section 13.3, we discuss "medium-scale" experimental investigations.

13.1. Experimental Procedures

Many experimental techniques use "small" samples, typically on the order of 100–200 g and sometimes smaller. The main disadvantage of such tests is that the conditions under which the coals are tested are far removed from the self-heating conditions experienced in a stockpile or coal seam. For instance, in large assemblies of coal, it is the energy released from self-heating that drives the ignition process. In small assemblies of coal, the energy released from self-heating is insufficient for spontaneous combustion to occur, and an external heat source must be used to drive self-heating into spontaneous combustion. A more concrete problem is that there is no rigorous way to obtain from small-scale experiments the propensity for spontaneous combustion in large-scale systems. It is therefore difficult to extrapolate from their results to predict the behavior of a coal in large-scale systems. The main advantage of small-scale test methods is the rapidity at which they provide results.

Coal producers and users favor methods that provide a well-defined "index," based upon a measurement or measurements made during a test. The index can then be used to rank a set of coals in relative order with regard to their propensity to self-heat. Such a ranking should correlate with the known heating behavior of the coals. There is an emphasis on test methods that are straightforward and in which the "index" is determined rapidly. This requires a test that uses a small sample. Samples used in such tests are almost always "fresh." Thus, the rankings obtained cannot necessarily be applied to situations in which coals are either being stored or transported because in these circumstances, aging effects, described in section 3.2, are important.

Where test methods are being used to rank the propensity of coals to self-heat within mines, e.g., when they are used to provide a risk hazard assessment as described in section 15.2.1, it is important that precautions be taken to minimize preoxidation, section 3.2, before samples are tested. An example of the care that is required is illustrated by the collection and preparation of samples in Beamish and Blazak (2005): samples were placed in plastic cling wrap that was then sealed with layers of aluminum foil and masking tape before being frozen on-site; they were transported to the laboratory in an insulated container full of ice; they were then stored at the laboratory in a freezer. Samples can also be stored as soon as they are collected in an inert atmosphere and prepared in the laboratory, prior to any experiment, under an inert atmosphere. Such precautionary measures were not undertaken in many early investigations (Mahadevan and Ramlu, 1985).

The following methods have been used in major coal-mining countries to evaluate the spontaneous combustion of coal (Singh et al., 2002):

1. adiabatic methods (Banerjee, 1985), described in section 14;
2. adiabatic oxidation using the R70 technique, described in section 14.2; and
3. crossing-point temperature method (Banerjee, 1985), also known as the ignitability method, described in section 15.

Other experimental techniques that have been used to investigate various aspects of self-heating include:

- differential scanning calorimetry (DSC), described in section 13.2.1;
- differential thermal analysis (DTA) and thermal gravimetry (TG), described in section 13.2.2;
- isothermal methods, described in section 13.2.3; and
- determination of minimum self-ignition temperatures, described as part of the hot-storage test in section 16.

In the absence of a universally accepted criterion for characterizing and classifying coals with respect to their propensity to self-heat, many research groups have developed their own methods and techniques. An example of this is the laboratory test method developed by the Institute for Cokemaking and Fuel Technology in Essen (Germany) (Pilarczyk et al., 1995).

13.2. Miscellaneous Methods

13.2.1. Differential Scanning Calorimetry (DSC)

Many experimental techniques may be considered to be "isothermal," in the sense that the environment temperature is constant even though the temperature of the sample changes. DSC is a nonisothermal method, in which the environment temperature is usually increased at a constant rate. In DSC, two pans, one containing a coal sample and the other empty, are maintained at the same temperature, and the difference in the amount of heat required to achieve this is measured as a function of the changing environment temperature. This allows endothermic and exothermic processes occurring within the coal, which are related to self-heating processes, to be identified. Sample sizes are typically on the order of 10 mg, and operating conditions ensure that there are no diffusion limits for oxygen to reach the coal surface (Garcia et al., 1999).

DSC has been used to investigate the effect of aging upon oxidation enthalpies of coals weathered over a period of up to 105 d (Garcia et al., 1999). The onset temperature, defined as the interception with the temperature axis of the tangent to the enthalpy profile at 290 °C, has been suggested as a suitable index to define the propensity of coals to self-heat, as it increases with aging in a more systematic way than the oxidation enthalpies.

DSC has also been used to investigate the role of moisture in the self-heating of low-rank coals (Clemens and Matheson, 1996) and to obtain kinetic parameters for low-temperature coal oxidation (Malow and Krause, 2004).

13.2.2. Differential Thermal Analysis (DTA) and Thermal Gravimetry (TG)

DTA can be used to study the self-heating behavior of coals by following the evolution of heat during an experiment. DTA has been used to the study the self-heating behavior of fresh and oxidized coals (Pis et al., 1996). DTA-TG curves have been used to characterize the propensity for spontaneous combustion of low-rank coals and their upgraded solid products (Avid et al., 2004; Mahidin et al., 2003a, 2003b; Sato et al., 2004; Umar et al., 2005). TG curves have been used to characterize the propensity for spontaneous combustion for an upgraded brown coal (Nakagawa et al., 2004).

13.2.3. Isothermal Methods

Smith and Glasser (2005b) developed a static isothermal method to measure the rate of reaction of coal under conditions in which the oxygen concentration is effectively constant. The apparatus is simple, and the authors express the hope that it could be used to obtain comparable data from other laboratories.

Wang et al. (2002, 2003a, 2003c) investigated the gaseous products formed during low-temperature oxidation of a bituminous coal in an isothermal flow reactor. Their technique allowed a complete mass balance to be obtained. An *increase* in the coal sample mass, on the order of 1% of the initial sample mass, was observed in the majority of experiments (Wang et al., 2003a). The increase in mass was due to the formation of surface oxygen-coal complexes in the coal pores. It was also noted that "the rates of oxygen consumption and the formation of carbon oxides decreased with time" (Wang et al., 2002, p. 586). The ratio of the rate of production of carbon dioxide to carbon monoxide in preoxidized coals was initially high but decreased to a steady state that was independent of the size of the coal particles and the oxygen concentration, but which depended upon the reactor temperature (Wang et al., 2003c). The results reported in these papers led to a new mechanism being proposed for the low-temperature oxidation of coal (Wang et al., 2003b, 2003a, 2003c).

13.3. Medium-Sized Investigations

Medium-scale experiments, defined as experiments using between 40 and 1000 kg of coal, have the advantages of being more akin to "real" self-heatings than small-scale experiments because no external heat source is required to push forward self-heating, and they provide results considerably quicker than true large-scale experiments, in weeks rather than months. A further advantage of medium-sized investigations over experiments using smaller amounts of coal is that the tested coal samples can incorporate a range of particle sizes and moisture states that more closely resemble "real" conditions.

Chen (1991) built a "full-scale experiment apparatus," which was two meters long and 0.3 m in diameter, that was used to investigate the propensity of New Zealand coals to combust (Stott and Chen, 1992). A modified version of Chen's column has been developed at the University of Queensland (Beamish et al., 2002). This one has a length of two meters, an effective diameter of 0.19 m, and a carrying capacity of 62 L. Depending upon packing density, it can hold between 40 and 70 kg of coal (Hancock et al., 2005). The use of computer-controlled heating elements enables the large column to be operated adiabatically. A test was run using 41.4 kg of a sub-bituminous A coal, and thermal runaway occurred after 19 d from a starting temperature of 19–22 °C (Beamish et al., 2002). The column was loaded with two layers, where the lower layer was more reactive. As moisture was removed by heating from the lower layer, it moved to the higher layer, where some of it was adsorbed. This accounted for some of the initial temperature increase in the upper region. This was a nice observation, and it showed an effect that would be insignificant in the common small-scale experiments. Experiments show that the higher the moisture content, the longer it takes to reach thermal runaway (Beamish and Jabouri, 2005).

The processes that occur during self-heating are difficult to visualize. To overcome this, an animation program has been developed at the University of Queensland to visualize temperature and gas data recorded during operation of the two-meter column (Hancock et al., 2005). Temperature measurements are made at eight equally spaced thermocouples located in the center of the column and running along its length. The concentrations of carbon dioxide, carbon monoxide, ethane, ethylene, hydrogen, methane, and oxygen are measured exiting the reactor, as well as Graham's ratio. The animation shows the hot spot migrating upward during a 19 d experiment. Such animations have the potential to be used for training mining personnel, particularly regarding the recognition of the importance of gas sampling frequency in tracking hot spot development.

In recent work, the University of Queensland column has been used to investigate the development of the hot spot and off-gas evolution patterns (Beamish and Jabouri, 2005; Beamish, 2005b). These tests have shown that coals in seams that have been "degassed" and dried reach thermal runaway (defined as a temperature >150 K) quicker than seams that have not been gas-drained. Analysis of gas data shows that substantial amounts of hydrogen are produced during self-heating at temperatures below 80 °C. Gas data also reveal that the Graham's ratio is much higher for the degassed coal than the gassy coal. This is an important consideration when gas-monitoring results are used to assess the risk of spontaneous combustion in areas of a mine that have been gas-drained.

14. ADIABATIC METHODS

Adiabatic methods, sometimes called adiabatic oxidation methods, encompass a variety of experimental procedures. They are considered to be particularly good at simulating the initial stages of in situ heating of coal, during which heat losses are thought to be negligible. They have become a popular way to determine the propensity of a coal to self-heat, and it has been suggested that they provide the *best* method to assess the self-heating propensity of a coal (Singh et al., 2002).

In section 14.1, we provide a general description of such methods. In sections 14.2 and 14.3, we describe recent work using adiabatic methods. In the former, we discuss the R_{70} method, which is used in Australia for the commercial evaluation of coal samples, and in the latter, we discuss other approaches. In section 14.4, we describe how kinetic parameters (A and E) can be obtained from temperature measurements made in adiabatic experiments.

14.1. General Description of Test Method

Adiabatic test methods are carried out in reaction vessels that have been designed to minimize heat loss, and, ideally, to act as if they are perfectly insulated. The reaction vessel is placed inside an adiabatic oven, such as in the R_{70} self-heating test described in section 14.2 or in the investigations of Mahidin et al. (2003a, 2003b, Sato et al. (2004), Singh et al. (2002), and Vance et al. (1996), or an oil bath (Ren et al., 1999). In order to maintain adiabatic operation, the temperature of the oven is controlled automatically to equal that of the coal in the reaction vessel; consequently, there is no heat transfer between the two. This requires the use of a computer to monitor the temperature of the coal and to control a heating element in the oven or within the oil bath. Oxygen or air flows through the reaction vessel after being preheated to the oven temperature. The self-heating rate is measured by monitoring the rise in temperature in the reaction vessel. These measurements are used to rank the relative self-heating tendencies of coals. Although the method was first proposed as a means to evaluate the propensity of coals to self-heat in 1924 (Davis and Byrne, 1924), it did not become technically feasible until the 1970s.

Different implementations of the adiabatic method use different classifications, and one such classification is shown in Table 9. In this table, the initial rate of heating is defined as the tangent at the starting point of the temperature-time trajectory, and the total temperature rise is the difference between the maximum temperature recorded and the initial temperature. Although the propensity for spontaneous combustion can be ranked using the initial rate of heating (IRH) and total temperature rise (TTR) values, it has also been suggested that the temperature-time trajectory be taken into account (Ren et al., 1999). The initial temperature of the experiment was taken to be 40 °C in Ren et al. (1999), which is a typical temperature in a British coal mine. An initial temperature of 60 °C would be relevant to conditions

TABLE 9. CRITERIA IN DETERMINING THE RISK FACTORS FOR THE ADIABATIC OXIDATION METHOD FROM SINGH ET AL. (2002)

Risk classification	Initial rate of heating (°C/h)	Total temperature rise (°C)	Risk index
Very high	>2.0	>7.0	8
High	1.2 to 2.0	4.5 to 7.0	4
Medium	0.6 to 1.2	2.5 to 4.5	2
Low	<0.6	0 to 2.5	1

Note: Used in Güyagüler et al. (2003), Ren et al. (1999), and Singh et al. (2002). The risk index is the intrinsic factor input into the risk assessment calculation of a coal seam described in Section 1 of this paper.

TABLE 10. CLASSIFICATION OF COALS ACCORDING TO THE R_{70} SELF-HEATING INDEX (MOREBY, 1997)

Risk classification	R_{70} (°C/h)
High risk	>0.8
Low risk	<0.5
Medium risk	0.5–0.8

Note: (°C/h)—rate of heating.

within a coal mill (Ren et al., 1999); samples tested at both 40 °C and 60 °C were more reactive at the latter, and the TTR values were much higher at 60 °C than at 40 °C.

The main problem with adiabatic test methods is ensuring that adiabatic conditions are maintained. This requires very tight control over heat losses and temperature control down to small fractions of a degree difference during the early stages. This is very difficult to obtain (Smith and Glasser, 2005a). To overcome these problems, a "semi-adiabatic" reactor has been developed in which a low rate of heat transfer occurs under controlled conditions (Smith and Glasser, 2005a). Allowance is made for heat transfer in the mathematical model, using parameter values, heat capacity, heat of reaction, reaction rate, and activation energies, determined from experimental data. Smith and Glasser (2005a, p. 1159) noted that if the heat losses are too large, then they "mask the effect of the heat generated by the reaction." However, if they are too small, "then the temperature rises may be too large," which invalidates some of the modeling assumptions and significantly reduces the accuracy of parameter values derived from the test method.

14.2. The R_{70} Method

14.2.1. Description of the R_{70} Method

The R_{70} self-heating rate index, introduced by Humphreys (Humphreys, 1981; Humphreys et al., 1981), is a measure of a coal's propensity to self-heat and is defined as the reciprocal of the time taken in an adiabatic oven for the temperature to reach 70 °C from a starting temperature of 40 °C. The units of R_{70} are °C/h. Once the temperature increases past ~70 °C, the rate of temperature increase dramatically escalates as self-heating leads to thermal runaway.

The testing procedure, as currently applied at the University of Queensland, involves several steps (Beamish et al., 2005). In the first of these, 150 g of the crushed coal, <212 μm, is dried at 100 °C under nitrogen for ~16 h. The coal is then cooled to 40 °C under a nitrogen atmosphere, whereupon it is transferred into an adiabatic oven where it is allowed to equilibrate under a nitrogen flow. When the temperature has stabilized, oxygen is passed through the container at 50 mL min^{-1}.

Table 10 shows the classification of coals according to their R_{70} self-heating rate index. This index has been used in Australia since 1981 by the Australian Coal Industry Research Laboratories (ACIRL) to commercially assess the propensity of Australian and New Zealand coals to self-heat (Beamish et al., 2002). Tests carried out at the University of Queensland have resulted in the development of a large database of R_{70} values (Beamish et al., 2005). The classification scheme shown in Table 10 is therefore out-dated, and it is in the process of being replaced by a five-tier classification system that is semilogarithmic (Beamish, 2005, personal commun.).

14.2.2. Investigations Using the R_{70} Method

Beamish et al. (2000, 2001, 2003) measured temperature-time curves over the temperature range from 40 °C to 180 °C. This range allows the self-heating process to be followed at various stages of generation and shows the time taken to reach "uncontrolled ignition." The variation in measured R_{70} values for dried coals has been shown to be less than ±5% of the average value for tests performed consecutively over a three to four day period and is due to the oven consistently maintaining adiabatic conditions over sustained testing periods (Beamish et al., 2000). Repeatability of R_{70} values to within ±5% was confirmed in a later study (Beamish and Blazak, 2005). The measured variations could have resulted from differences in particle-size distribution, and/or minor coal compositional variations, between samples. The time to reach thermal runaway has been defined as the time taken to teach a temperature of 140 °C (Blazak et al., 2001).

Over longer time periods, aging effects, see section 3.2, are noticeable, even for samples kept under controlled storage (Beamish et al., 2000). This preoxidation effect was described by the equation $R_{70}(t) = at^b$, where t is the time in days since the initial testing, which is assumed to be day one, a is the R_{70} value on initial testing, and b is a constant that depends on the particle size and storage method. Due to improvements in the way that samples are protected once they are collected on-site, outlined in section 13.1, it is now possible to test samples in the laboratory at effectively time $t = 0$. This will lead to a more accurate assessment of the propensity for spontaneous combustion of coal seams.

R_{70} values show a rank dependence (Beamish, 2001, personal commun.): sub-bituminous coals have values in the range $14.91 \leq R_{70}$ (°C/h) ≤ 33, a lignite sample has a value of $R_{70} = 7.76$ (°C/h), and high-volatile bituminous A and B coals have values in the range $0.31 \leq R_{70}$ (°C/h) ≤ 8. This trend has been attributed to changes in the internal surface area of the coal, which governs the number of available sites for oxidation to take place.

Using the method described in section 4, kinetic parameters, listed in Table A2 were determined over the temperature range 70–140 °C (Beamish et al., 2003). A plot of self-heating rate index (R_{70} on the y axis) against Suggate rank (Suggate, 2000) (S_r on the x axis) showed a local maximum, i.e., the greatest propensity to spontaneous combustion, at a value of $S_r \sim 6.2$. This value of the Suggate rank corresponded to a local minimum in the plot of activation energy (E) against Suggate rank. On either side of the critical Suggate rank, reactivities, and R_{70} values, decrease significantly.

The Suggate rank is calculated using either the calorific value and volatile matter of the coal (based on a dry mineral matter- and sulfur-free basis) or the atomic O/C and H/C ratios of the coal (on a mineral matter–free basis). These parameters are usually available because they are obtained during coal exploration. Thus, to ensure a true rank comparison using the Suggate rank, R_{70} values for dry coals have to be converted to a mineral-free basis. A formula for this conversion has been developed (Humphreys et al., 1981) and has been used to construct a correlation curve (Beamish, 2005a). Thus, in principle, an initial risk assessment can be obtained by estimating the R_{70} value of a new coal based upon its Suggate rank. However, the R_{70}–Suggate rank correlation curve is on a mineral-free, moisture-free basis, and this can create a misleading impression of the relative self-heating rates of two coals in a coal seam, or when stockpiled, since, in both cases, the coal is neither mineral-free nor dry (Beamish, 2005a) and both of these parameters strongly affect the R_{70} value (Beamish and Blazak, 2005; Beamish, 2005a; Beamish et al., 2005). Thus, R_{70} values of dried coals are now plotted versus ash content, which is closely related to the mineral matter content of the coal. On such a figure, the different coal ranks separate themselves out (Beamish et al., 2005, their Fig. 1).

The R_{70} value significantly decreases with increasing mineral matter content, as defined by the ash content of the coal, while the time to thermal runaway increases with increasing ash content (Beamish and Blazak, 2005; Blazak et al., 2001). As an example, it has been shown that the thermal runaway time for two coals with approximately the same R_{70} values differed by 6 h, where the coal with a lower ash content ignited first (Beamish et al., 2005, their Fig. 2). The moderating effect of mineral content is ascribed to its role as a heat sink (Beamish and Blazak, 2005), and it also acts by diluting the "fuel" component of the coal.

A seam profile for the Callide Coalfield (Queensland, Australia) has been drawn based upon R_{70} values for samples obtained from a borecore. These show how the self-heating index varies with the depth of the borehole, and, subject to the provisions outlined above, it enables a risk assessment to be made within the seam. This allows regions that are more prone to spontaneous combustion to be designated prior to the start of mining. In turn, this allows the development of an appropriate spontaneous combustion management plan, as outlined by Singh et al. (2002).

Six borehole samples were analyzed, and a second-order polynomial curve was obtained by plotting R_{70} value (on the y axis) against percent ash content, dry basis, (on the x axis) (Beamish and Blazak, 2005). A seventh sample was obtained by blending two of the original samples. The R_{70} value for the blended sample fell on the curve. From this data point, it was concluded that R_{70} values of blended samples can be predicted solely on the basis of their ash content. It was suggested that this curve could be applied to reduce the potential of spontaneous combustion within a stockpile, since it allows the R_{70} value of blended coals to be predicted.

In the standard test procedure, R_{70} values are found for dry coal. In Beamish and Hamilton (2005), the effect of moisture on the R_{70} values of two isorank coals was investigated. The R_{70} value increased dramatically as the moisture content was decreased below 40%–50% of the moisture-holding capacity of the coal; the moisture content of the two coals was 17.0% and 18.6% on an as-received basis. In fact, it was not possible to induce self-heating in sufficiently wet samples. Reinforcing the results reported in Beamish and Blazak (2005), the R_{70} value of isorank coals with similar moisture content was found to decrease as mineral content increased.

14.3. Investigations Using Adiabatic Methods

Adiabatic techniques have become a standard method to measure the propensity of a coal to self-heat. As such, they can be used to investigate how the combustion characteristics of a coal are modified by other factors. For instance, they have been used to investigate the effect of water content and the influence of two drying methods on the heating rate of coal (Vance et al., 1996). They have also been used to compare the combustion characteristics of low-rank coals and their upgraded products (Mahidin et al., 2003a, 2003b; Sato et al., 2004) and to investigate the influence of the moisture content of the air, particle size, and aging on self-heating (Ren et al., 1999).

Güyagüler et al. (2003) analyzed samples from 16 Turkish lignite seams using both an adiabatic method and a crossing-point temperature method, described in section 15. The results were in close agreement with each other and allowed the progress of a self-heating event in a mine to be predicted. In addition to measuring temperature, the concentrations of carbon dioxide, carbon monoxide, nitrogen, and oxygen were measured during the experiment.

14.4. Kinetic Data from Adiabatic Tests

In an adiabatic test, kinetic data for *dry* coal is obtained using the equation

$$\rho c \left(\frac{\partial T}{\partial t} + v \nabla T \right) = k \nabla^2 T + \rho Q A \exp\left[-\frac{E}{RT}\right]. \quad (22)$$

Under adiabatic operating conditions, heat loss by conduction is negligible ($k \nabla^2 T \approx 0$). When the flow velocity through the coal bed is sufficiently small, convection due to the inflow is also negligible ($v \nabla T \approx 0$). Under these circumstances, Equation 22 reduces to

$$\rho c \frac{\partial T}{\partial t} = \rho Q A \exp\left[-\frac{E}{RT}\right]. \quad (23)$$

Mahidin et al. (2003a, 2003b) checked the assumption that convection due to the inflow was negligible. They assumed that $v_r = v_z$ and $v_o = v_z$ and estimated that the axial heat-convection term ($v_z \nabla T$) was much smaller than the heat-generation term.

Equation 23 can be rearranged to give

$$\ln\frac{\partial T}{\partial t} = \ln(QA/c) - (E/RT). \quad (24)$$

Thus, plotting $\ln(\partial T/\partial t)$ versus $1/T$ produces a straight line of slope $-E/R$ and intercept of $\ln(QA/c)$. This method was used by Beamish et al. (2003) and Mahidin et al. (2003a, 2003b). In Mahidin et al. (2003a), one set of kinetic values was obtained for temperatures below 80 °C and a second set for temperatures above 80 °C. The activation energies determined from adiabatic measurements below 80 °C were in reasonable agreement with values determined for oxygen adsorption using TGA-TA techniques, and it was therefore concluded that at low temperatures, oxidation is limited by oxygen adsorption.

15. CROSSING-POINT TEMPERATURE

In section 15.1, we provide a general description of the crossing-point temperature (CPT) method. In section 15.2, we discuss how data obtained from a crossing-point temperature experiment is used to obtain an index that can be used to rank a coal's propensity for spontaneous combustion. Finally, in section 15.3, we describe recent work using the crossing-point temperature method.

It has been suggested that, due to its simplicity, the CPT method is suitable to be a standard technique for the study of mine fires (Sensogut and Cinar, 2000).

15.1. General Description of Test Method

In the CPT method, two reactors are placed in a programmed oven or temperature bath. Each reactor contains on the order of 30–60 g of coal (30 g, Sensogut and Cinar, 2000; 40 g, Umar et al., 2005; 60 g, Kadioğlu and Varamaz, 2003, and Küçük et al., 2003). Nitrogen flows through one reactor, and either air or nitrogen flows through the second. Both gas streams pass through a drier and are preheated before entering the furnace. The temperature of the furnace is increased at a constant rate that raises the ambient temperature to which the coal samples are exposed. (Sometimes only one reactor is used.)

The temperatures of the furnace and the two reactors are recorded continuously from ambient temperature to combustion. The crossing-point temperature is the temperature at which the thermocouples in the second reactor, measuring the local air temperature, and furnace are equal. The CPT indicates the susceptibility of the coal to spontaneous combustion—the lower the CPT, the higher the risk of spontaneous combustion. The temperature increase in the coal bed above that of the oven is due to the heat released in exothermic oxidation reactions. A more detailed description of the method is provided in Küçük et al. (2003).

Flow-rates of 40 cm³ min⁻¹ (Sensogut and Cinar, 2000), 50 cm³ min⁻¹ (Kadioğlu and Varamaz, 2003; Küçük et al., 2003), 80 cm³ min⁻¹ (Umar et al., 2005), and 200 cm³ min⁻¹ (Güyagüler et al., 2003) have been used. Küçük et al. (2003) investigated how the gas flow rate affected the measured combustion characteristics: a coal with a CPT of 160 °C at a flow rate of 50 cm³ min⁻¹ did not have a CPT at flow rates of either 25 or 100 cm³ min⁻¹. At the lower flow rate, heat release was not high enough to sustain combustion due to insufficient supply of oxygen, while at the higher flow rate, dissipation of heat dominated over heat accumulation in the reactor. Heating rates of 1/60 K min⁻¹ (Panigrahi and Sahu, 2004), 1/30 K min⁻¹ (Banerjee, 1985), 1/2 K min⁻¹ (Sensogut and Cinar, 2000; Umar et al., 2005), and 1.0 K min⁻¹ (Kadioğlu and Varamaz, 2003; Küçük et al., 2003) have been used.

Instead of defining the CPT by comparing the temperature of the coal against the surrounding medium, it is possible to put an inert material in the second reactor and to define the CPT by comparing the temperature of the coal against the inert material (Gouws and Wade, 1989). When the CPT apparatus is setup in this manner, a differential thermal analysis thermogram can be obtained in which the temperature difference between the coal and the inert material (on the y axis) is plotted as a function of the inert temperature (on the x axis). This figure has three stages. In stage I, the temperature difference decreases from 0 °C to a local minimum. In stage II, the temperature difference increases approximately linearly. In stage IIa, the temperature difference is negative, while in stage IIb it is positive. In stage III, the sample has entered runaway ignition, and the temperature difference–inert temperature trajectory is very steep. The CPT is the transition point between stage IIa and stage IIb.

15.2. Evaluating the Data

Various indices have been used to evaluate CPT data. An index is only useful if an increase in the index corresponds to an increased propensity for spontaneous combustion. The FCC (Feng Chakravorty Cochrane) index is based on the average heating rate between 110 and 220 °C and is given by Feng et al. (1973) as

$$\text{FCC Index} = \frac{(\text{Average heating rate between 110 to 220C})}{\text{Crossing Point Temperature}} \times 1000 \; (\text{min}^{-1}). \quad (25)$$

The lower limit (110 °C) was chosen so that there was a minimal influence on the self-heating rate due to the evaporation of moisture. The higher limit (220 °C) was chosen so that the exit of volatile matter from the coal had a minimal influence on the self-heating rate. This definition was criticized by Mahadevan and Ramlu (1985) because the index is undefined for coals that have a CPT greater than 220 °C. Gouws and Wade (1989) used an experimental system in which the crossing-point temperature was defined against an inert reference rather than the surrounding medium. The CPTs measured using this definition were lower than those commonly reported using the traditional definition. Five of the samples tested had a CPT of lower than 110 °C, the lower limit in the FCC index. Furthermore, all coals had started runaway ignition by the upper limit of 220 °C (region III on a differential thermal analysis thermogram)—this region is not thought to be relevant to studies on low-temperature self-heating. The limits of 110 °C and 220 °C were therefore criticized as being arbitrary; it was suggested that the limits should be defined by the behavior of the coal. Gouws and Wade (1989) put forward a modified FCC index given by

$$FCC_{mod} = \frac{\text{stage II slope}}{\text{CPT}}, \qquad (26)$$

where "stage II slope" is the average heating rate over stage II of a differential thermal analysis thermogram. This is a more appealing definition because the arbitrary limits over which the heating rate is determined are removed. Instead, the limits are the start and end temperature over which the coal naturally undergoes steady self-heating.

In order to eliminate arbitrary prescribed temperature limits, Mahadevan and Ramlu (1985) divided the temperature-time trajectory into three regions: between ambient temperature and the "perceptible self-heat generation point," or "inflexion-point temperature" (IPT); from the IPT to the crossing point; and from the crossing point to active combustion. The "inflexion point" was not defined in a rigorous way, but loosely speaking, it is the point at which the temperature starts to increase more rapidly with time than linearly. The Mahadevan and Ramlu (MR) index is defined by

$$\text{MR index} = \frac{\dfrac{\text{time to reach the CPT}}{\text{heating rate at the CPT}} \times \dfrac{\text{average heating rate between the IPT and CPT}}{\text{time to reach the IPT}} \times 10. \qquad (27)$$

This definition has been criticized because an increase in the MR index was found not to necessarily correspond to an increase in liability to self-heat (Gouws and Wade, 1989). In Gouws and Wade (1989), coal sample 25 was deemed to be less liable to self-heat than coal sample 12 on the grounds that it had a higher CPT and a lower average stage IIA slope. The criticism of the MR index stemmed from the fact that coal sample 12 had a lower index than that of coal sample 25. However, this analysis could in turn be criticized because the definition of "liability to self-heat," higher CPT and lower average stage IIA slope, is arbitrary and not related to any fundamental properties of self-heating. Despite the criticism of the MR index, it was found that it agreed "reasonable well" with the FCC_{mod} index, defined by Equation 26, when rankings were compared over a set of 58 coals (Gouws and Wade, 1989).

Gouws and Wade (1989) proposed an index that requires a differential thermal analysis thermogram to be obtained. This index uses two measurements, the CPT and the average stage II slope, and is given by

$$\text{WITS} - \text{EHAC index} = 500 \times \frac{\text{stage II slope}}{\text{CPT}}. \qquad (28)$$

Thus, a coal with a higher CPT or a low average heating rate is rated as having a low propensity to spontaneously combust. An analysis of 58 coals showed that the WITS-EHAC and FCC_{mod} indices ranked the samples in the same order. However, a comparison of Equations 26 and 28 will show that this result is unsurprising.

Gouws and Wade concluded their article with the following comment, which applies to many indices used to rank coals: "Although the new liability index adequately identifies the spontaneous combustion potential of South African coals, it nevertheless suffers from the same disadvantage as all simple and composite liability indices—it is empirical in nature. It still remains to test the new liability index against a mathematically consistent kinetic model…" (Gouws and Wade, 1989, p. 85).

Coal samples using the FCC, MR, and WITS-EHAC indices are classified as being low, medium, and high risk, as indicated in Table 11.

Umar et al. (2005) use a liability index defined by

$$I_t = \frac{1}{20} \cdot \frac{T_2 - T_1}{\text{CPT}}, \qquad (29)$$

to classify the susceptibility of coals to spontaneous combustion. In this index, T_1 is the coal-bed temperature 10 min before the crossing point, and T_2 is the coal-bed temperature 10 min after the crossing point.

15.3. Investigations Using the Crossing-Point Temperature Method

The crossing-point temperature method has become a standard method used to measure the propensity of a coal to self-heat and to investigate how the combustion characteristics of a coal are modified by other factors. For instance, it has been used to characterize the propensity for spontaneous combustion of upgraded brown coals (Umar et al., 2005). It has also been

TABLE 11. LIABILITY INDICES INDICATING RISK GROUPS FOR THE CROSSING-POINT TEMPERATURE METHOD (GÜYAGÜLER ET AL., 2003)

Risk classification	FCC [a]	MR [b]	WITS-EHAC [c]
Low	0 to 5	0 to 10	0 to 2.5
Medium	5 to 10	10 to 20	2.5 to 5
High	>10	>20	>5

Note: [a] Feng et al. (1973); [b] Mahadevan and Ramlu (1985); [c] Gouws and Wade (1989). FCC—Feng Chakravorty Cochrane index, MR—Mahadevan and Ramlu index.

used to investigate the influence upon spontaneous combustion characteristics of air humidity (Küçük et al., 2003), coal moisture content (Kadioğlu and Varamaz, 2003; Küçük et al., 2003), and particle size (Küçük et al., 2003). An investigation using both the CPT method and an adiabatic method by Güyagüler et al. (2003) is mentioned in section 14.3.

Sensogut and Cinar (2000) used the crossing-point method to investigate the susceptibility to combustion of coal samples from the Ermenek District in Turkey. They concluded that the main cause for spontaneous combustion in the collieries was due to the environmental conditions within the mine rather than the innate susceptibility of the coal in the seams, which was classified as being of "low risk."

16. THE HOT-STORAGE TEST

The hot-storage test, also known as the basket test, the F-K method, the steady-state approach, the oven-heating test, and the wire-mesh basket test, has been widely used to study the self-heating characteristics of solid materials (Bowes, 1984). The method was initially developed to investigate the self-heating and ignition of activated carbons (Bowes and Cameron, 1971; Cameron and MacDowall, 1972), and it has provided the basis for several international standards (International Maritime Consultative Organization, 1974; International Organization for Standardization, 1994; United Nations, 1995).

In section 16.1, we provide a general description of the method, and in section 16.2, we describe recent work using the hot-storage test.

16.1. Description of the Hot-Storage Test

In the hot-storage test, particles of the self-heating material are placed into a gauze container of known size, which is then suspended in a fan-assisted oven of given ambient temperature. The temperature at the center of the container is measured using a thermocouple. From the maximum temperature increase, it is easy to decide, as described in section 2.2.2, if the system is subcritical or supercritical. The experimental setup must enforce two conditions in order to apply the theory from section 2.2.1 (Jones, 1999b): there must be purely conductive heat transfer within the material undergoing self-heating, and there must be good convective heat transfer within the oven so that the surface temperature of the sample is maintained at ambient temperature. The last condition is not strictly necessary as the theory can be developed to allow for less-efficient convective cooling on the boundary, but it simplifies the analysis of the experimental data.

16.1.1. Review of Theory

The hot-storage test is based on the theoretical model of Frank-Kamenetskii (1969) that was described in section 2.2. Recall that there is a critical value of the F-K parameter δ, δ_{cr}, which distinguishes between subcritical and supercritical systems. The F-K parameter is defined by

$$\delta = \frac{EL^2 \rho QA}{\lambda R T_a^2} \exp\left[-\frac{E}{RT_a}\right]. \quad (30)$$

The value of δ_{cr} depends upon the container geometry, which is usually either cubical (Bowes and Cameron, 1971; Cameron and MacDowall, 1972; Jones and Littlefair, 1997; Nugroho et al., 1998, 2000a, 2000b) or cylindrical (Malow and Krause, 2004; Sujanti et al., 1999; Sujanti and Zhang, 1999, 2000; Watanabe and Zhang, 2001; Zhang and Sujanti, 1999). Values of δ_{cr} are given in Table 1. Evaluating Equation 30 at the critical value and taking logs leads to the equation

$$\ln\left(\frac{\delta_{cr} T_{a,cr}^2}{L^2}\right) = \ln\left(\frac{\rho EQA}{R\lambda}\right) - \frac{E}{R} \cdot \frac{1}{T_{a,cr}}. \quad (31)$$

Equation 31 is the basis for determinations of the kinetic parameters using the hot-storage test.

16.1.2. Applying the Theory

In order to apply Equation 31, we need to know the critical ambient temperature, the threshold between subcritical and supercritical behavior, at which thermal runaway occurs ($T_{a,cr}$) for a number of container sizes (L).

For a fixed sample size (L), a series of experiments is carried out, each using a fresh sample of the material, to bracket the critical ambient temperature, which is very sharply defined by the marked differences in the maximum temperature increase between the two regions, to some predetermined degree of accuracy. Critical ambient temperatures have been determined to within ±0.5 K (Nugroho et al., 1998; Sujanti et al., 1999; Sujanti and Zhang, 2000; Watanabe and Zhang, 2001; Zhang and Sujanti, 1999), ±2.0 K (Jones, 1996a; Malow and Krause, 2004; Nugroho et al., 2000a, 2000b), and ±2.5 K (Sujanti and Zhang, 1999). A critical temperature of, for example, $T_{a,cr} = 127.5 \pm 0.5$ K means that a sample failed to ignite with an ambient temperature of 127 °C, but ignited with an ambient temperature of 128 °C.

This procedure is repeated for containers of different sizes. For instance, Nugroho et al. (2000a) used cubical baskets of lengths 35, 40, 50, 60, 70, and 80 mm. Plotting the experimental results in the form of $\ln(\delta_c T a^2_{a,cr}/L^2)$ versus $1/T_{a,cr}$ gives a straight line of slope $-E/R$. The value of A, or the value for QA when the value for Q is unknown, is found from the y intercept.

The reproducibility of critical ambient temperatures is very good (Sujanti et al., 1999; Sujanti and Zhang, 1999, 2000; Watanabe and Zhang, 2001). In Watanabe and Zhang (2001), the authors state that the same critical ambient temperature was obtained for repeat tests, whereas Zhang and Sujanti (1999) give a variation of ±1 K. Errors in determining the critical ambient temperature will lead to uncertainty in both the x value ($1/T_{a,cr}$) and the y value [$\ln(\delta_c T a^2_{a,cr}/L^2)$]. Irreproducibility is due to factors such as the variation in the equilibrium moisture content of the coal, which depends upon atmospheric humidity, and variations in the oven temperature over the duration of an experiment. The latter have been reported to be ±1 K (Chong et al., 1996; Jones et al., 1996b; Nugroho et al., 2000a) and ±2 K (Chong et al., 1995; Nugroho et al., 1998). The accuracy of activation energies measured using this technique has been reported as ±5% (Jones and Wake, 1990).

The advantage of this method is that it is relatively small scale, and, once enough data points have been determined to obtain a straight-line fit, it can be used to estimate critical ambient temperatures for self-ignition of larger quantities. The disadvantage of the hot-storage test is that is a very time consuming (Chen and Chong, 1995) because an iterative procedure is required to determine the critical ambient temperature for a given sample container. Thus, several experiments are required to obtain one data point. This process must then be repeated to obtain data points for sample containers of different sizes. Measuring the critical ambient temperature for five container sizes requires a week or more of experimental work (Jones, 1999b). It is for this reason that the hot-storage test has not been directly used as the basis for determining if self-heating materials can be safely stored during transport; test methods that have been used for this purpose are described in section 9. Jones pointed out another potential disadvantage of the method pertaining to its use in forensics; there might be insufficient quantities of the "suspect material" to run enough tests to determine the kinetic constants (Jones, 1999b).

16.2. Investigations Using the Hot-Storage Test

Sujanti et al. (1999) calculated kinetic parameters using both the hot-storage method and the transient method described in section 18. Their results are summarized in Table 12.

The hot-storage test has been used to evaluate the effects of inert inorganic matter and additives upon the propensity for spontaneous combustion of a Victorian brown coal (Sujanti and Zhang, 1999, 2000; Watanabe and Zhang, 2001; Zhang and Sujanti, 1999). The critical ambient temperature for a fixed container size was used to determine if an additive promoted or suppressed combustion. These investigations are reported in section 10.2. Oxidation kinetics were estimated using the transient method (Sujanti and Zhang, 2000; Zhang and Sujanti, 1999), described in section 18, and the heat-release method (Sujanti and Zhang, 1999), described in section 17.

Nugroho et al. (1998, 2000a, 2000b) used the hot-storage test to investigate the propensity for spontaneous combustion of air-dried, high- and low-rank Indonesian coals and their blends. Critical ambient temperatures and the activation energy (E) and the product of the pre-exponential factor and the exothermicity (AQ) were determined using the hot-storage test (Nugroho et al., 1998, 2000a, 2000b), the heat-release method (Nugroho et al., 1998), and the transient method (Nugroho et al., 1998, 2000a, 2000b). These results are summarized in Tables 13 and Appendix 2.

TABLE 12. COMPARISON OF KINETIC DATA FOR VICTORIAN BROWN COALS OBTAINED USING THE HOT-BASKET TEST AND THE CROSSING-POINT TEMPERATURE METHOD USING SEVEN DIFFERENT CYLINDERS*

Dimensions Diameter (mm)	Height (mm)	Distance between thermocouples, Δr (mm)	E (kJ mol^{-1})	A (s^{-1})
25.0	25.0	3.0	108.0	3.22×10^7
40.0	40.0	6.0	117.5	4.61×10^8
60.0	60.0	7.0	114.8	1.96×10^8
75.0	75.0	10.0	100.6	3.91×10^6
40.0	260.0	6.0	101.2	6.08×10^6
60.0	120.0	7.0	111.3	1.21×10^8
95.0	47.5	12.0	94.7	6.93×10^6
Hot-storage test			101.7	6.32×10^6

*From Sujanti et al. (1999).

TABLE 13. COMPARISON OF KINETIC DATA FOR INDONESIAN COALS OBTAINED BY VARIOUS METHODS

Coal and method	E (kJ mol^{-1})	QA (J kg^{-1} s^{-1})	R^2	Reference
KPC-Prima—FK	55 ± 4			[a]
KPC-Prima—TM	56 ± 4			[a]
KPC-Prima—HRRM	54 ± 2			[a]
Prima—FK	90 ± 5	1.88×10^{12}	0.9943	[b]
Prima—TM	90 ± 4	4.48×10^{12}	0.9957	[b]
Pinang—FK	62 ± 5	$2.13 \pm 0.17 \times 10^8$	0.9843	[c]
Pinang—TM	57 ± 4	$2.25 \pm 0.16 \times 10^8$	0.9983	[c]
South Bangko—FK	72 ± 2			[a]
South Bangko—TM	78 ± 7			[a]
South Bangko—HRRM	75 ± 2			[a]
Tanjung Enim—FK	56 ± 5			[a]
Tanjung Enim—TM	52 ± 1			[a]
Tanjung Enim—HRRM	69 ± 9			[a]

Note: [a] Nugroho et al. (1998); [b] Nugroho et al. (2000a); [c] Nugroho et al. (2000b). FK—Frank-Kamenetskii parameter; HRRM—heat-release-rate method; TM—transient method.

Jones and Littlefair (1997) investigated the self-heating of coal filter cake in a 10 cm cube basket. The expected demarkation between subcritical and supercritical behavior was not observed. At oven temperatures of 393 K and 396 K the system was clearly subcritical, with maximum temperature increases of 22 K and 38 K, respectively. At an oven temperature of 458 K, the system was clearly supercritical, with a maximum temperature increase of 610 K. However, at oven temperatures between 408 K and 445 K, maximum temperature increases of between 80 K and 498 K were recorded. These temperatures are lower than usually observed in supercritical experiments, yet the experiments were supercritical because all the combustible material was depleted by the time the temperature returned to ambient. It was suggested that the unusual results were due to catalysis by inorganic materials present in the filter cake—it would have been interesting to see the results of experiments in which the filter cake had been acid washed to remove inorganic matter. In the absence of a clear demarkation between subcritical and supercritical regions, it was noted that "there is no non-arbitrary way of distinguishing between ignition and failure to ignite" (Jones and Littlefair, 1997, p. 1167). Given sufficient data, point sensitivity analysis could be used to identify a critical value in a well-defined, nonarbitrary manner (Varma et al., 1999).

17. HEAT-RELEASE-RATE METHOD

The heat-release-rate method (HRRM) was introduced in Jones (1996b) and Jones et al. (1996a, 1996b). It assumes that there is no heat transfer at the geometric center of a sample when the temperature there is equal to the oven temperature. The advantage of the HRRM method, over the hot-storage test, is that one experiment leads to one data point on the kinetic plot, as opposed to the hot-storage test, where several experiments are required to obtain one data point. Furthermore, all data points can be obtained using one container size, whereas several container sizes are required in the hot-storage test. The HRRM is therefore very time-effective compared to the hot-storage test.

In section 17.1, we describe the HRRM. In section 17.2, we discuss theoretical issues relating to the validity of the HRRM. In section 17.3, we describe recent work using the HRRM. Investigations in which the HRRM is benchmarked against other methods for determining kinetic parameters are discussed in section 18.4.1.

17.1. Description of the HRRM

In section 17.1.1, we describe the HRRM. In section 17.1.2, we discuss the accuracy of measurements made by thermocouples during the HRRM.

17.1.1 Analysis of Data from the HRRM

In the HRRM, as in the hot-storage test, a gauze container is loaded with particles of the self-heating material, and it is placed in a fan-assisted oven of given ambient temperature. A cubical container with sides of 10 cm is typically used (Jones, 1996b; Jones et al., 1996b, 1998). Where the HRRM differs from the hot-storage test is how the data are analyzed.

Consider the standard model

$$\rho c \left(\frac{\partial T}{\partial t} \right) = k \nabla^2 T + \rho Q A \exp\left[-\frac{E}{RT} \right]. \quad (32)$$

Suppose that at some time t, the crossover time t_{hrm}, the temperature at the geometric center of the sample x_{gc}, the crossover position, is equal to the oven temperature. In the HRRM, it is *assumed* that at this time and position, thermal conduction at the center of the material becomes zero. Then, at the crossover position, this equation reduces to

$$\left(\frac{\partial T}{\partial t} \right)_{(x_{gc}, t_{hrm})} = \frac{QA}{c_p} \exp\left(-\frac{A}{RT_a} \right) \quad (33)$$

or

$$\ln\left(\frac{\partial T}{\partial t} \right)_{(x_{gc}, t_{hrm})} = \ln\left(\frac{QA}{c_p} \right) - \frac{E}{R} \cdot \frac{1}{T_a}. \quad (34)$$

In Equation 34, T_a is the oven temperature. For convenience, we refer to the temperature at the center where it crosses over the oven temperature as the crossing-point temperature.

Equation 34 is the basis of the HRRM for estimating oxidation kinetics. In one experiment, the rate of change of temperature with time at the crossover position and crossover time, $([\partial T/\partial t]_{x_{gc}, t_{hrm}})$, is determined at the crossing-point temperature (T_a)—this measurement is easily obtained from the temperature trace of the thermocouple. The experiment is repeated for different values of the oven temperature, giving data points for the rate of temperature rise at the crossing temperature. Each experiment uses a fresh sample.

The values of the parameters E and A, or the product AQ if the value for Q is unknown, are then estimated from Equation 34 by plotting the logarithm of the transient heating at the crossing-point temperature ($\ln[\partial T/\partial t]_{x_{gc}, t_{hrm}}$) versus the reciprocal of the crossing (oven) temperature $1/T_a$. Note that Equation 33 can also be written as

$$\rho c_p \left(\frac{\partial T}{\partial t} \right)_{T_a} = QA \exp\left(-\frac{E}{RT_a} \right), \quad (35)$$

which means that the rate of heat generation can be determined at the crossing point.

17.1.2. Thermocouple Issues

Experimental errors that occur in the HRRM are associated with temperature measurements made by the thermocouples. According to Malow and Krause (2004), there is an uncertainty of ~2 K in thermocouple measurements, which leads to an uncertainty in the value of the transient heating term $(\partial T/\partial t)_{(x_{gc}, t_{hrm})}$ of

~5%. Variations in the oven temperature over the duration of an experiment would be on the order of ~1 K, as they are in the hot-storage test (Jones et al., 1996b). Experimental data can therefore be analyzed assuming that most of the experimental errors are in the value of the transient heating term (Jones et al., 1998).

Jones et al. (1998) noted that when a thermocouple of the type used in their experiments experiences a step change in temperature of ~100 °C, which can occur for supercritical samples, there is a response time of ~1 s. This could potentially pose a problem in determining the transient heating term at the crossing point. However, since the temperature measurements were made every 2.5 s, the recorded temperature should represent the true temperature.

17.2. Discussion of Theory

In the HRRM, as in the hot-storage test, a "cold sample" is placed into a "warm" oven. There is therefore a heating period in which the sample acclimates to the oven temperature. The HRRM is based upon the belief that during this period, the extent of reaction is negligible, and that the temperature profile throughout the sample is flat. Therefore "at the time when the sample temperature is at or very close to the oven temperature, there can be no heat transfer between oven and sample; therefore heat gain by the sample is due solely to its own reaction with oxygen" (Jones et al., 1996a, p. 1734). At the crossover time, "the temperature profile of the sample at this stage will be flat; therefore the reaction rate will be the same at all places" (Jones, 1996b, p. 161).

The HRRM assumption seems to have been first articulated by Bowes, who stated that when the temperature at the geometric center is "in the vicinity of the oven temperature...heat transfer between [substrate] and the oven atmosphere is clearly minimal" (Bowes, 1984, p. 189).

No mathematical reasoning has been put forward to support the conjecture that at the crossing point, there is no heat conduction, and that the temperature profile is flat throughout the sample. The validity of the latter part of the conjecture has been questioned by Gray et al. (1992). Nugroho et al. (2000a, p. 2986) commented that the HRRM suffers from "not taking account of the temperature gradient through the sample" at the crossing point. Having compared activation energies obtained using the hot-storage test, the HRRM, and the transient method, discussed in section 18, Nugroho et al. (1998) preferred to use the hot-storage test and the transient method in subsequent work (Nugroho et al., 2000a, 2000b).

It has been suggested that the soundness of the HRRM can be determined from seeing whether data plotted according to Equation 34 produce a straight line (Jones et al., 1996a). In practice, data obtained using the HRRM do give a good straight line. It is concluded from this that if heat conduction *is* nonzero at the crossing point, then it must either be negligible in comparison with the heat-release term or, if non-negligible, that it is significant but *constant* across the range of oven temperatures used, so that when logarithms are taken, it becomes absorbed into the constant term (Jones et al., 1996a).

A comparison of activation energies determined using the HRRM and the hot-storage test is given in Tables 13 and 14, which show that the accuracy of activation energies determined using the HRRM varies.

We conjecture that the flatter the temperature profile at the crossing point, the more accurate the estimates using the HRRM. At present, there seems to be no way of knowing whether a given material will have a "more flatter" or "less flatter" profile and, consequently, whether the HRRM will provide "more accurate" or "less accurate" estimates of kinetic parameters.

The HRRM assumes that there is an infinite Biot number, so that the temperature at the surface is the oven temperature. It would be interesting to know how the calculated values of the kinetic parameters change as the Biot number increases toward infinity. This could easily be investigated numerically.

17.3. Investigations Using the HRRM

17.3.1. General Investigations

The HRRM has been used to examine properties of Scottish coals (Jones, 1996b; Jones et al., 1996b, 1998). In Jones et al. (1996b, 1998), kinetic values, shown in Table A2, were determined. In Jones (1996b), the HRRM was used to calculate values for the rate of heat generation for a bituminous coal under subcritical conditions. This investigation is noticeable because a wet sample was tested; most investigations use *dry* coal. The coal used had a moisture content of 16.4%. At an oven temperature of 349 K, moisture was only lost after 13–17 h in the oven.

Sujanti and Zhang (1999) carried out experiments using a cylindrical reactor with a height of 8 cm and diameter of 4 cm placed inside an oven. Air passed through the reactor at a rate of

TABLE 14. ACTIVATION ENERGIES (kJ mol $^{-1}$) DETERMINED FOR A VARIETY OF SUBSTRATES USING VARIOUS METHODS

Substrate	F-K	TM	HRRM	OFW (DSC)	KAS (DSC)
Cork dust	107 ± 10	111 ± 26	134 ± 34	130 ± 16	126 ± 16
Detergent powder	102 ± 30	112 ± 24	165 ± 40	104 ± 5	99 ± 5
German lignite coal dust	102 ± 20	99 ± 19	116 ± 25	115 ± 8	111 ± 7
Riboflavin	151 ± 37	127 ± 27	151 ± 26	404 ± 147	416 ± 153

Note: Data are from Malow and Krause (2004). F-K—Frank-Kamenetskii parameter; HRRM—heat-release-rate method; TM—transient method. OFW and KAS are differential scanning calorimeter (DSC) methods.

100 mL min^{-1} (N.T.P. [normal temperature and pressure]). Four thermocouples, each with a precision of ±0.1 K, were inserted inside the reactor to monitor the temperature along the reactor axis. Having established that there was negligible heat transfer along the reactor axis, the thermocouples recorded the same temperature during the experiments; the HRRM was applied when the measured thermocouple temperature was equal to the oven temperature. Kinetic parameters were determined for the raw coal, water-washed coal, acid-washed coal, and acid-washed coal doped with 11 additives.

17.3.2. A New Test Method to Evaluate Propensity for Self-Ignition

Jones (1998b) has shown how the HRRM can be applied to estimate the activation energy from a single criticality data point obtained using the hot-storage test described in section 16. Suppose that it is known that when T_a = 355 K, a sample of a given size is subcritical, whereas when T_a = 359 K, the sample is supercritical. The critical ambient temperature is therefore $T_{a,cr}$ = 357 ± 2 K, and the HRRM is applied at this temperature. So, we have

$$\left[\frac{\partial T}{\partial t}\right]_{T_{a,cr}} = \frac{QA}{c}\exp\left[-\frac{E}{RT_{a,cr}}\right], \quad (36)$$

where the quantity on the left-hand side of the equation is measured from the thermocouple data. Then

$$\delta_{cr} = \frac{EL^2\rho QA}{\lambda RT_{a,cr}^2}\exp\left[-\frac{E}{RT_{a,cr}}\right],$$

the critical value of the F – K parameter

$$= \frac{EL^2\rho}{\lambda RT_{a,cr}^2}\cdot\left[\frac{\partial T}{\partial t}\right]_{T_{a,cr}},$$

which gives

$$E = \frac{\delta_{cr}\lambda RT_{a,cr}^2}{L^2\rho c}\cdot\frac{1}{\left[\frac{\partial T}{\partial t}\right]_{T_{a,cr}}}, \quad (37)$$

using Equation 36.

A comparison of values obtained using this procedure and the full HRRM is shown in Table 15. The single-point method estimates the activation energy to within 10%–15%. This method has also been applied to determine the activation energy of wood sawdust (Jones, 2000b).

This abbreviated procedure provides the basis of a proposed test method to evaluate the safety of transporting a self-heating material (Jones, 2000c). In this method, a critical ambient temperature is found for a sample of fixed size. From this measurement, an estimate of the activation energy is obtained using the method outlined above. The pre-exponential factor is then calculated using Equation 12 and the appropriate value for δ_{cr}. The critical size of a cube containing the self-heating material at a given ambient temperature is then determined, again using Equation 12. The proposed test is a little more time-consuming than the traditional IMCO test, described in section 19.1.1, because several oven testings are required, rather than one, but it is still sufficiently rapid that it could be used as a test procedure. However, it has the great advantage that it provides a criticality estimate based upon an estimate of the kinetic parameters rather than in ignorance of them.

18. THE TRANSIENT METHOD

The transient method (TM), introduced in Chen and Chong (1995) and Chong et al. (1995, 1996), is usually referred to as the "crossing-point temperature method," but in this article, we avoid this phrase in order to prevent confusion with the entirely different crossing-point temperature method described in section 15. To increase the potential for confusion, the HRRM described in section 17 has also been described as being a "crossing-point method."

The TM is based upon the observation that when a subcritical experiment is carried out using the hot-storage test, there is a time at which the heat-conduction term vanishes at any position near the geometric center of the container. The advantage of the transient method is that only one experiment is required to obtain one data point on the kinetic plot, as opposed to the hot-storage test, where series of experiments are required to obtain one data point. Furthermore, all data points can be obtained using one container size, whereas several container sizes are required in the hot-storage test. This substantially reduces the time required to determine kinetic parameters, making it very time-effective compared to the hot-storage test. A further advantage of the transient method is that the method is not dependent upon obtaining a large Biot number for heat transfer within the oven.

In section 18.1, we describe how the idea of a crossing-point temperature (CPT) is used to analyze data from the TM so as to obtain kinetic parameters. In section 18.2, we discuss the theoretical underpinnings of the TM to answer the following questions: why might we expect a CPT to exist? Under what conditions can

TABLE 15. COMPARISON OF ACTIVATION ENERGIES (kJ mol^{-1}) FOR TWO COALS USING THE FULL HEAT-RELEASE METHOD AND THE HEAT-RELEASE METHOD APPLIED TO A SINGLE CRITICALITY POINT DETERMINED USING THE HOT-STORAGE TEST

Coal	Full HRRM [a]	Single data point + HRRM [b]
Dalquhandy	54 ± 5	68
Rosslynlee	74 ± 4	66

Note: [a] Jones et al. (1996b); [b] Jones (1998b). HRRM—heat-release-rate method.

a CPT be measured? The central practical issues in applying the TM revolve around the placement of the thermocouples within the container and the accuracy of their temperature measurements. These matters are discussed in section 18.3. In section 18.4, we discuss experimental investigations using the TM.

18.1. Description of the TM

Consider the transient-energy equation for the standard model

$$\rho c \frac{\partial T}{\partial t} = k \nabla^2 T + \rho Q A \exp\left[-\frac{E}{RT}\right]. \quad (38)$$

Suppose that at some time $t = t_{tm}$, the crossover time, and at some position, x_{tm}, the crossover position, the thermal conduction term becomes zero. Then, at this position and time, the energy equation reduces to

$$\left(\frac{\partial T}{\partial t}\right)_{(x_{tm}, t_{tm})} = \frac{QA}{c_p} \exp\left(-\frac{A}{RT_o}\right) \quad (39)$$

or

$$\ln\left(\frac{\partial T}{\partial t}\right)_{(x_{tm}, t_{tm})} = \ln\left(\frac{QA}{c_p}\right) - \frac{E}{R} \cdot \frac{1}{RT_o}. \quad (40)$$

In Equation 40, T_0 is the temperature at the crossover time and crossover position where the thermal conduction term has vanished, with the restriction that $T_0 > T_a$. This temperature is known as the crossing-point temperature (CPT).

Equation 40 is the basis for estimating oxidation kinetics using the TM. In one experiment, the crossing-point temperature (T_0) and the rate of change of temperature at the crossover position at the crossover time ($[\partial T/\partial t]_{x_{tm}, t_{tm}}$) are measured. The experiment is repeated at different values of the oven temperature, giving a sequence of data points ($[\partial T/\partial t]_{x_{tm}, t_{tm}}, T_0$). Once a suitable data set has been obtained, the values of the parameters E and AQ are then estimated from Equation 40 by plotting the logarithm of the transient heating at the crossing-point temperature $\left[\ln\left(\frac{\partial T}{\partial t}\right)_{(x_{tm}, t_{tm})}\right]$ versus the reciprocal of the crossing point temperature $1/T_0$. This is a straight line with slope $-E/R$ and y-intercept $\ln(AQ/c_p)$. Note that, unlike the hot-storage test, the analysis of the experimental data does not require an infinite Biot number at the gas-surface boundary.

Note that if the position x_{tm} is at a point of symmetry within the body, so that $\nabla T = 0 \forall t$, then at the crossover time, the temperature profile at this position is locally flat.

Sample containers used have either been cubical (Chong et al., 1995, 1996; Nugroho et al., 1998, 2000a, 2000b) or cylindrical (Chen and Chong, 1995; Chong et al., 1996; Malow and Krause, 2004; Sujanti et al., 1999; Sujanti and Zhang, 2000; Watanabe and Zhang, 2001; Zhang and Sujanti, 1999).

18.2. Why and When Is There a CPT?

The theory that the TM uses to analyze data *is* correct; Equation 39 *must* hold at the crossover position when the local conduction term is zero. But under which circumstances does a CPT occur? Consider the case of a one-dimensional slab: At time $t = 0$, a "cold" slab is placed into a "warm" oven. The temperature of the slab is uniform, and the conduction term is everywhere zero. As time increases, heat penetrates into the slab by diffusion from the boundaries. The temperature of the slab is higher near the boundaries and lower in the middle. Therefore, heat flows into the center of the slab so that the heat-conduction term is positive at the center. Conversely, as the system approaches its steady state, the temperature is highest at the center of the sample and lowest at the boundaries. Therefore, heat flows away from the center of the slab, and the heat-conduction term is negative at the center. There must therefore be a time at which the heat-conduction term at the center is zero. Numerical simulations showing the developing temperature profiles within a slab are presented in Chen and Chong (1995). These simulations reveal that at the geometric center, for a given material, a container of fixed size, and fixed oven temperature, the CPT is unique (to within ±0.01 K), not depending upon the initial, uniform temperature of the material (Chen and Chong, 1995).

In order to obtain a CPT, the oven temperature must neither be too low nor too high. If the oven temperature is higher than the critical ambient temperature of the system, a suitable CPT may not be measured (Chen and Chong, 1995; Nugroho et al., 1998). If the oven temperature is too low, then the amount of self-heating at the CPT is insufficient to cause a measurable temperature rise, and a CPT cannot be identified (Chen and Chong, 1995). A CPT could not be detected for wood sawdust when the oven temperature was below 170 °C (Chong et al., 1995), for whole milk powder when the oven temperature was below 130 °C (Chong et al., 1996), and for skim milk powder when the oven temperature was below 140 °C (Chong et al., 1996). The uniform initial temperature must also not be too high, otherwise the sample will ignite, and a suitable CPT may again not exist (Chen and Chong, 1995). Within the range of acceptable oven temperatures, the CPT is typically at most a few degrees above the oven temperature (Nugroho et al., 1998).

18.3. Issues with Thermocouples

In order to use the TM, we need to determine the time at which the heat-conduction term at the crossover position vanishes. If the crossover position is at the geometric center of the container, then when the heat-conduction term vanishes, the temperature profile will be "locally flat." Thus, in order to determine the crossover temperature, we can either calculate the heat-conduction term by discretization or seek to identify when the temperature profile becomes locally flat.

The problem of knowing where to locate the thermocouples so as to determine the crossover temperature most accurately has been raised as a potential disadvantage of the TM (Nugroho et al., 1998). An analysis of how the values of the kinetic parameters depend upon the choice of location of the thermocouples remains to be done and would be a very useful addition to the literature. In the absence of such an analysis, we review the experimental placement of thermocouples within cylindrical and cubical samples in sections 18.3.1 and 18.3.2, respectively.

In section 18.3.3, we discuss the accuracy with which measurements of the CPT and the rate of change of temperature at the crossover position can be made.

18.3.1. Location of Thermocouples in a Cylindrical Container

Suppose that the cylinder is sufficiently long so that only radial heat conduction is significant and that three thermocouples are placed in a radial direction. The heat-conduction term at the crossover position vanishes when

$$\frac{1}{r}\frac{\partial}{\partial r}\left(r\frac{\partial T}{\partial r}\right) = 0 \quad (41)$$

or in finite-difference form

$$\frac{T_3 - 2T_2 + T_1}{(\Delta r)^2} + \frac{1}{r_2}\frac{T_3 - T_1}{2(\Delta r)} \approx 0, \quad (42)$$

where T_2 is the temperature at the crossover position (located at radial position r_2) and T_1 and T_3 are the temperatures of thermocouples placed at distance r on either side of the crossover position. The crossover temperature occurs when $T_1 = T_3 = T_2 = T_0$. If the crossover position is at the geometric center of the sample (so that $r_2 = 0$) then $T_3 = T_1$ by symmetry, and only two thermocouples are required. When these two thermocouples record the same temperature ($T_1 = T_2 = T_0$), there is no heat conduction and the temperature profile is locally flat. Note the crossover position need not be located at the geometric center of the sample, although this has the advantage of reducing the number of thermocouples from three to two.

A very useful question to ask is: where should the thermocouples be placed? This question has not been thoroughly investigated in the literature. Numerical simulations could cast light on this issue by investigating how the estimates of kinetic parameters vary as a function of the thermocouple location. As Δr decreases, the discretization becomes a better approximation to the heat-conduction term. Thus, one would expect the errors to increase as Δr increases.

In Chen and Chong (1995), the sample container was a cylinder with diameter and height of 6 cm. Three thermocouples were inserted into the container: one at the geometric center, the second 7 mm (in the radial direction) from the first, and the third 7 mm (in the radial direction) from the second on the same horizontal and vertical cross-sectional plane. The crossing-point temperature was determined in three ways:

1. **Flat temperature profile #1:** When the temperatures at the first and second thermocouples were equal.
2. **Flat temperature profile #2:** When the temperatures at the first and third thermocouples were equal.
3. **Heat-conduction calculation:** The heat-conduction term was approximated by the following discretization

$$\frac{\partial^2 T}{\partial x^2} \approx \frac{T(3) - 2T(2) + T(1)}{(\Delta x)^2},$$

where $T(i)$ is the temperature measured by the ith thermocouple, and $\Delta x = 0.007 \pm 0.001$ m is the distance between the thermocouples. Thus, the crossover position was the location of the second thermocouple.

The CPT was obtained using each method for four initial temperature profiles. These results are shown in Table 16. The average errors calculated by each method were $\pm 0.6\%$, $\pm 0.7\%$, and $\pm 3.3\%$, respectively (based on the average temperature). The errors using the first two methods are very small and confirm that the CPT is independent of the uniform initial sample temperature, as is suggested by numerical simulations.

The larger errors using method three were ascribed to a combination of the errors in the thermocouple readings (± 0.2 °C) and error in the location of the thermocouples (± 1 mm) (Chen and Chong, 1995). At the CPT, the temperature difference between the three thermocouples was on the order of ± 0.5 °C, so a measurement error of ± 0.2 °C is significant. However, it should be noted that the discretization used for the heat-conduction term was incorrect (cf. Equation 42). Thus, the first two methods determined the CPT at the origin ($r = 0$), whereas the third method calculated the CPT at the location $r = 0.007$.

Chong et al. (1996) carried out experiments using stainless-steel mesh baskets: two cubical and two cylindrical containers were used. The purpose of these experiments was to see if the shape and size of the container affected the test results. No such influence was found. The first thermocouple was fixed at the geometric center of the sample, while the second was placed in the location indicated in Table 17. The crossover temperature was taken to be when the two thermocouples recorded the same temperature.

In Sujanti and Zhang (2000), Watanabe and Zhang (2001), and Zhang and Sujanti (1999), a cylindrical container with diameter and height of 2.5 cm was used. One thermocouple was placed at the geometric center of the sample, the second thermocouple was located 3 mm from the first, and the third thermocouple was located 3 mm from the second along the cylinder radius. The heat-conduction term was discretized using Equation 42. Thus, the crossover position was at a radial position 3 mm from the center of the cylinder. In Sujanti and Zhang (2000), it is noted that the choice $\Delta r = 3$ mm achieved the most accurate measurement of the radial derivative.

Sujanti et al. (1999) carried out experiments using seven cylinders of varying dimensions, four cylinders with equal diameter and height and four with an equal volume. Three thermocouples

TABLE 16. CROSSING-POINT TEMPERATURES OBTAINED EXPERIMENTALLY DURING THE SELF-HEATING OF WOOD SAWDUST AS A FUNCTION OF THE INITIAL UNIFORM TEMPERATURE PROFILE

IC (K)	Method 1	Method 2	Method 3
274	214.5	214.8	216.0
297	214.8	214.1	211.1
298	215.7	213.5	212.8
311	214.8	216.0	218.3

Note: Data are from Chen and Chong (1995). IC—initial condition.

TABLE 17. DISTANCE BETWEEN THE THERMOCOUPLE CENTERS (Δx) USED IN CHONG ET AL. (1995)

Basket	Width/diameter (cm)	Height (cm)	Δx (cm)
Cube A	5	5	0.8
Cube B	6	6	0.8
Cylinder A	6	6	0.5
Cylinder B	4	12	0.4

were inserted in each reactor separated by a distance Δr along a radius of the cylinder. The first thermocouple was placed at the geometric center of the sample. The distance Δr was optimized to achieve the most accurate measurement of the radial temperature and its derivative. Thus, the crossover position was at a radial position Δr from the center of the cylinder. The dimensions of the cylinders and the values for Δr are given in Table 12.

Malow and Krause (2004) carried out experiments using a cylinder in which the diameter and height were equal ($V = 400$ cm^3). The crossover position was located at the geometric center of the sample, and a thermocouple was positioned at half the distance between the center and the container surface. Tests were not carried out to investigate if the calculated kinetic parameters were sensitive to the location of the second thermocouple.

18.3.2. Location of Thermocouples in a Cubical Container

The heat-conduction term in a cubical container vanishes when

$$\frac{\partial^2 T}{\partial x^2} + \frac{\partial^2 T}{\partial y^2} + \frac{\partial^2 T}{\partial z^2} = 0. \quad (43)$$

Suppose that the crossover point is the center of the cube. Then, by symmetry, the heat conduction term vanishes when

$$\frac{\partial^2 T}{\partial x^2} = 0 \quad (44)$$

or in finite-difference form when

$$\frac{T_3 - 2T_2 + T_1}{(\Delta r)^2} \approx 0. \quad (45)$$

In this equation, T_2 is the temperature at the center of the cube, while T_1 and T_3 are the temperatures of thermocouples placed at a distance r on either side of the center. By symmetry, $T_1 = T_3$, so that only two thermocouples are required, and the CPT occurs when $T_1 = T_2 = T_0$.

Recall from section 17.1.1 that all investigations to date using the HRRM have been carried out in a cubical container. The HRRM can be considered to be a limit of the TM in which the distance between the thermocouples (Δr) is the half-length of the cube ($\Delta r = L$). Under an infinite Biot number assumption, the crossing-point temperature then becomes the oven temperature. Note that the HRRM assumes that at the crossing point, the flat profile extends all the way across the sample. If this is true, then the TM and HRRM methods would have the same crossover temperature and the values of the kinetic parameters identified using the two methods would be the same.

Chong et al. (1995) carried out experiments using cubical stainless-steel meshes with sides of 50, 60, and 70 mm. Three thermocouples were placed within the container to obtain a temperature profile along a line. The first thermocouple was placed at the center of the container, the second 4 mm from the center, and the third 4 mm away from the second. The CPT was defined to be the temperature at which the temperatures at the first and third thermocouple were equal.

Nugroho et al. (1998, 2000a, 2000b) carried out experiments using a cubical wire-mesh basket with sides of 50 mm. Initially, five thermocouples were placed within the basket to obtain a temperature profile along a line. One thermocouple was placed at the center of the container, the second was placed 5 mm away from the center, the third was placed 5 mm away from the second, the fourth was placed 7.5 mm away from the third, and the fifth was placed 7.5 mm away from the fourth. The crossing-point temperature was defined as the temperature at which the temperatures at the first and third thermocouples were equal (Nugroho et al., 1998). In subsequent experiments (Nugroho et al., 2000a, 2000b), only two thermocouples were used: one in the center of the basket and one 10 mm away from the center.

Nugroho et al. (1998) noted that a disadvantage of the TM is that the crossing point must be at the geometric center of a cubical container; otherwise, the heat-conduction term is not necessarily zero when the temperatures of the thermocouple located at the crossing point and its nearest neighbor are zero. Strictly speaking, this is not a problem from a theoretical perspective because six thermocouples could be placed around the crossing-point thermocouple to estimate the heat flux in all three coordinate directions. However, there are practical objections to the use of seven thermocouples. An interesting numerical study would be to quantify how small errors in the placement of the crossing-point thermocouple are reflected in errors in the calculated values of the kinetic parameters.

18.3.3. Accuracy of Thermocouple Measurements

How accurate are the temperature measurements made by thermocouples? What are the associated errors in determining the CPT and the rate of change of temperature at the crossing point?

Errors in thermocouple measurements have been estimated at ±0.2 K by Chen's group (Chen and Chong, 1995; Chong et al., 1996), ±0.5 K in the work of Nugroho et al. (2000a, 2000b), and ±2 K by Malow and Krause (2004). Errors in the CPT have been estimated at ±0.5 K (Nugroho et al., 2000a, 2000b) and ±1 K (Chong et al., 1996). Errors in the transient heating rate at the crossover position and crossover time have been estimated at ±5% (Malow and Krause, 2004), ±10% (Chong et al., 1996), and ±0.04 K min^{-1} (Nugroho et al., 2000a, 2000b).

How do these errors accumulate in the calculation of kinetic parameters? Nugroho et al. (2000b) estimated that the errors in the kinetic parameters E and AQ were smaller than 10%.

18.4. Investigations Using the Transient Method

The foundations of the TM were laid in Chen and Chong (1995), which was primarily concerned with establishing the conditions under which a CPT exists. The majority of the results in this paper were obtained by numerical simulation; however, experiments using wood sawdust as the self-heating material were also reported. The numerical and experimental results showed that the CPT was independent of initial, uniform temperature profile. The method was then applied to obtain the kinetic parameters (AQ and E) of untreated and treated wood sawdust (Chong et al., 1995) and whole and skim milk powder (Chong et al., 1996). In Chong et al. (1996), R^2 values were found for Equation 40 to be: 0.894 (whole milk powder below 145 °C), 0.969 (whole milk powder above 145 °C), and 0.989 (skim milk powder).

In section 18.4.1, we describe experiments in which kinetic parameters have been obtained using the TM and at least one other method. In section 18.4.2, we discuss points arising from other investigations.

18.4.1. Benchmarking the TM Method against Other Methods

In this section, we review investigations that have determined kinetic parameters using the TM method and one (or both) of the hot-storage test method and the HRRM. Nugroho et al. (1998, 2000a, 2000b) determined the kinetic parameters for Indonesian coals using a combination of the transient method, the heat-release method, and the hot-storage test. In Nugroho et al. (1998), the TM gave good results for the activation energy for all three coals tested, whereas the HRRM gave good estimates for two of the coals. Therefore in Nugroho et al. (2000a, 2000b), only the TM and the hot-storage test were used. Good agreement was obtained between the two methods. These results are shown in Table 13.

Sujanti et al. (1999) determined values of the kinetics parameters for the low-temperature oxidation of a Victorian brown coal using seven cylindrical wire-mesh reactors of different size using the TM and the FK methods. A good straight-line fit was found using the TM method, although a value for R^2 and regression errors were not provided. Kinetic constants from both methods predicted the critical thickness of a layer of coal deposit (infinite slab) to within an acceptable accuracy. It was suggested that the good agreement between the methods means that the TM can replace the F-K method in future investigations.

The authors did not address the issue as to why the measured values of the kinetic parameters varied with the geometry of the cylinder. If we take the values determined in the hot-storage test as being "correct," then some of the values reported in Table 12 would fall within the error of uncertainty inherent in the hot-storage test. A numerical study would be useful to estimate the extent to which variation in kinetic parameters is due the placement of the thermocouples and the extent to which the variation is caused by the assumption that the cylinder is infinite in length, i.e., that only heat conduction in the radial direction is important.

Malow and Krause (2004) determined activation energies for the exothermic reaction of a German lignite coal dust, a cork dust, a detergent powder, and vitamin B2 using self-heating methods (TM, F-K, HRRM) and differential scanning calorimeter methods (KAS, OFW). For the HRRM and TM experiments, the sample container was a cylinder of equal height and diameter with a total volume of 400 cm^3. Only four data points were used to determine the kinetic values using the hot-storage test. From these results, it appears that the TM was accurate for three out of the four materials, whereas the HRRM was accurate for only one of the four materials.

18.4.2. Other Investigations

Nugroho et al. (2000a) investigated the effect of the basket size upon the measured values of the kinetic oxidation parameters. They found that the value of the activation energy was independent of the size of the basket used, but that there was a slight reduction in the product AQ as the basket size was reduced. In Nugroho et al. (2000b), the same authors noted that a plot of $\ln(AQ)$ against E for a variety of coals of different ranks and particle sizes produced an excellent linear plot, where a higher value of E corresponded to a higher value of $\ln(AQ)$. However, a R^2 value was not quoted. Based upon a theoretical analysis of their experimental results, a new risk index was proposed to evaluate the propensity for spontaneous combustion. This classification is based upon the rate of temperature change at the crossing-point temperature measured in a 50 mm cubical basket at an ambient temperature of 400 K (Table 18).

The TM has been used to determine the oxidation kinetics of a Victorian brown coal in the presence of inorganic matter and

TABLE 18. RISK INDEX FOR COAL SELF-IGNITION PROPOSED BY NUGROHO ET AL. (2000b)

Risk classification	T (K min^{-1})
Low	$0.0 < T < 0.2$
Medium	$0.2 < T < 0.4$
High	$0.4 < T < 0.6$
Very high	$T > 0.6$

Note: $T = (\partial T/\partial t)(x_{tm}, t_{tm})$, measured in a 50 mm cube at an oven temperature of 50 K.

additives (Sujanti and Zhang, 2000; Watanabe and Zhang, 2001; Zhang and Sujanti, 1999). The reaction rates showed a general trend: the lower the critical ambient temperature in a hot-storage test, the higher the reaction rate of the sample.

19. CONCLUSIONS

In this article, we have reviewed self-heating and spontaneous combustion of coal, concentrating on experimental work. We have also reviewed the Frank-Kamenetskii steady-state model for the self-heating of bulk solids, which is the simplest model for the spontaneous combustion of a coal stockpile. This method underpins experimental methods that are used to test the safety of transporting hazardous materials. All mathematical models require values for the kinetic parameters (A, or AQ, and E). These can be obtained from the hot-storage test, which is based upon the steady-state model, or either the heat-release-rate method or the transient method, which both use a transient formulation of the Frank-Kamenetskii model. The HRRM and the TM provide a much quicker way to obtain kinetic parameters than the traditional hot-storage test. Where comparisons have been made among the three methods, the HRRM and the TM have been, on the whole, accurate. The TM has a stronger theoretical basis than the HRRM. However, experimental work comparing the two methods is limited, and further work is required both experimentally and through numerical simulations, to identify the relative strengths of these two methods. Are there circumstances under which either method is intrinsically more reliable than the other? Considering the TM on its own, a useful investigation would be to determine how the errors in the TM depend upon the location of the thermocouples. To date, all experimental investigations using a cylindrical container have assumed that it is infinitely long—it would be useful to know how "long" a cylindrical reactor has to be for this to be a good approximation.

In all three methods, kinetic parameters are obtained by plotting the straight line $y = mx + c$ and identifying the measured values of m and c with various physical and chemical parameters. Beamish et al. (2003, p. 669) noted that "Having reliable values of E and A will enhance the prediction of the propensity of coals to combust spontaneously." No experimental method is 100% accurate, and there will always be an error in the estimated value of the kinetic parameters. However, one would not know this from many of the values that are reported in the literature! In many papers that determine kinetic parameters, either no error estimates from regression analysis are provided (Beamish et al., 2003; Mahidin et al., 2003a, 2003b; Reddy et al., 1998; Smith and Glasser, 2005a; Sujanti et al., 1999; Sujanti and Zhang, 1999, 2000) or only partial regression errors are reported (Jones et al., 1996b; Jones et al., 1998; Malow and Krause, 2004; Nugroho et al., 1998, 2000a). In the latter case, an error estimate is typically provided for the activation energy (E) but not for the pre-exponential factor (A). The only paper we found in which an uncertainty was provided for both kinetic parameters was Nugroho et al. (2000b). This appears to be a problem not just in coal research—an unwillingness to quote regression errors has been noted in the reporting of kinetic parameters for skim milk powder (Chong et al., 1996). It is unfortunate that R^2 values and regression errors are not published when experimental data are analyzed. Thus, the reliability of published kinetic values cannot be assessed.

There has been a limited amount of experimental work carried out on large stockpiles, due to the expense and time-consuming nature of such work. The work reported by Fierro et al. (1999a, 1999b, 2001) has set a new standard, showing the advantages that accrue when experimental work of this size is undertaken in conjunction with mathematical modeling. Careful design of such experiments allows the validity of modeling assumptions to be carefully investigated. We believe that it may be possible to apply the transient method to a large stockpile to measure the reactivity of coal near the boundary.

We have not reviewed modeling work in this paper. We note that almost all papers dealing with the spontaneous combustion of coal model large stockpiles. Little has been published on smaller-scale test methods, perhaps out of a belief that the steady-state Frank-Kamenetskii model says all that needs to be said. However, there remain issues to be addressed. For example, it has been suggested that the effect of preoxidation/aging effects can be significant in hot-storage tests (Sujanti and Zhang, 1999; Nugroho et al., 2000a). This issue can be explored through mathematical modeling. Another issue that is often not investigated in modeling studies is the influence of particle size upon the oxidation rate—this is known to be important from experimental studies, see section 3.

The role of moisture on the self-heating of coal is complex, and much work is required before it will be fully understood. Many experimental techniques for ranking the propensity of coals to spontaneously combust use dried samples. Yet coal in coal mines and in stockpiles, but particularly in coal mines, may be wet and have very different combustion characteristics. For example, the R_{70} value of a coal dropped by 50% for a 6% increase in moisture from the dry state of the test (Beamish and Hamilton, 2005). In a related study, the R_{70} value of a sub-bituminous coal was 20 times higher than that of a high volatile A bituminous coal on a dry mineral matter–free basis, but only 2–3 times greater when the effects of mineral and as-received moisture content were taken into account (Beamish, 2005a). R_{70} values obtained in small-scale tests for dry coal have been shown to have limited usefulness in making a risk assessment for coal that is being mined, because the mining conditions, i.e., moisture content and seam-gas concentration, significantly modify the self-heating behavior of the coal (Beamish, 2005b).

Methods that use dry coals, the norm for small-scale tests, can lead to conclusions being drawn about the propensity for in situ self-heating that differ dramatically from what might be observed in a mine. In particular, the delaying effects of in-seam gases and moisture are not recognized in small-scale experiments. This has led to the suggestion that, in order to gain a more reliable indication of the propensity for coals to self-heat, a bulk-scale test needs to be devised that can mimic conditions closer to those that occur in mines (Beamish, 2005a; Beamish et al., 2005). The University of Queensland's 2 m column is an example of such a test.

There is evidence that, in addition to heat-transfer effects through condensation and evaporation, water directly affects the oxidation rate, section 11.2.2, and the oxidation rate is maximized at "median" concentrations. This effect is missing from all experiments using dry coal and from experiments on wet coal that are carried out at elevated temperatures where the moisture is driven off, as might occur, for example, in the hot-storage test. This strengthens the call for a test that incorporates the effect of moisture content and poses a question mark against the validity of test methods that rank the relative propensity of dry coals. For the case of bagasse, a self-heating material, it is known that dry test results are unreliable for "wet" materials: "maximum safe stockpile heights" for the storage of bagasse based upon oven-heating tests are well in excess of what is known to be safe from industrial experience (Jones, 2001).

Studies about the self-heating and spontaneous combustion of coal start with fundamental research questions relating to issues such as the kinetics of oxidation and the role of moisture, both as a heat-source/heat-sink and as an agent influencing the chemical mechanism. Mathematical models are used to obtain kinetic parameters and to predict the behavior of large stockpiles.

Our understanding of the chemistry is used when index gases are identified. Engineering principles are applied when monitoring schemes are devised and implemented to improve mine safety and when coal-mines fires are brought under control and suppressed. Advances in any fundamental area have manifold implications throughout the field.

Some of the fundamental processes that govern self-heating in coal also influence self-heating in other industrial processes, such as those involving the hot-air drying of food powders. Powders such as dairy products and flour react exothermically upon contact with oxygen. When they are dried, layers of deposit can build up inside and on the drying chamber walls, presenting a self-heating hazard that can cause dust explosions. The milling process is also a heat-generation process, and self-heating can again lead to an explosion if there is insufficient ventilation. There are also issues regarding the storage of grain particles in packed beds since self-heating can occur in humid conditions; e.g., when food contains oil, the oxidation of lipids is influenced by water content.

Similar self-heating processes can happen during industrial textile drying or laundry operation; oil content is also a promoting factor for self-heating in these processes. Forestry litters, which are combustible, have exhibited self-heating in laboratory testing. Thus, self-heating of litter can trigger a fire of major consequence provided that the litter has sufficient thickness. From an ecological perspective, such fires may be "useful." However, reported fires in forests are more related to lightning or undesirable human "interventions." Mining of combustibles other than coal may have self-heating issues, but so far there are few literature reports on this.

We have seen that when the heat generated from low-temperature oxidation is greater than the heat dissipated, self-heating will lead inexorably to spontaneous combustion, and spontaneous combustion is a potential hazard wherever coal is being mined, stored, or transported. Our ever-expanding insight into the mechanisms that lead to self-heating can be applied to reduce, but not eliminate, the risk. We end this review with a quote from an era in which there was little understanding of the phenomenon of self-heating, but in which it was an ever-present threat: "From the day we sailed, the *Titanic* was on fire, and my sole duty, together with eleven other men, had been to fight that fire. We had made no headway against it," John Dilley, RMS Titanic Fireman (U.S. Senate, 1912, quoted in Stracher and Taylor, 2004, p. 7).

ACKNOWLEDGMENTS

We thank the following for their help in answering our questions during the course of this work: Basil Beamish (University of Queensland, Australia), Bogdan Dlugogorski (University of Newcastle, Australia), Robert Essenhigh (Ohio State University, USA), Clifford Jones (University of Aberdeen, Scotland), and Glenn B. Stracher (East Georgia College, USA).

APPENDIX 1. NOMENCLATURE

TABLE A1. NOMENCLATURE

Abbrev. or variable	Definition	Units
CPT	Crossing-point temperature	
DSC	Differential scanning calorimetry	
DTA	Differential thermal analysis	
FCC	Feng Chakravorty Cochrane index	
F-K	Frank-Kamenetskii	
HRC	High-rank coal	
HRRM	Heat-release-rate method.	
IC	Initial condition (assumed to be a uniform temperature profile)	K
IM	Inherent moisture content of a coal	
IRH	Initial rate of heating	°C/h
IPT	Inflection-point temperature	°C
LIT	Layer ignition temperature	
LRC	Low-rank coal	
MR	Mahadevan and Ramlu index	
PNA	Polynuclear aromatic hydrocarbons	
TG	Thermal gravimetry	
TM	Transient method	
TTR	Total temperature rise	°C
UBC	Upgraded brown coal	
UQ	University of Queensland	
VM	Volatile matter content of a coal	
A	Pre-exponential factor	s^{-1}
$A^{\dagger}$	Pre-exponential factor	Appropriate units
$A^{\dagger\dagger}$	Pre-exponential factor	$kg\ s^{-1}\ m^{-2}$
Bi	Biot number $Bi = hL/\lambda$	—
E	Activation energy	$J\ mol^{-1}$
L	Half-width of a coal pile	m
$[O_2]$	Concentration of oxygen	Appropriate units
Q	Oxidation heat of coal	$J\ kg^{-1}$
R	Universal gas constant $R = 8.314\ J\ mol^{-1}\ K^{-1}$	
S_g	Effective surface area of coal per kilogram of coal	$m^2\ kg^{-1}$
T	Temperature	K
T_a	Ambient, or oven, temperature	K
T_R	The initial temperature at which coal is assembled prior to transportation	K
T_0	Crossover temperature	K
c	Specific heat capacity	$J\ kg^{-1}\ K^{-1}$
h	Convective heat transfer coefficient at the boundary $x = L$	$W\ m^{-2}\ K^{-1}$
m	Order of oxidation reaction with respect to coal	—
n	Order of oxidation reaction with respect to oxygen	—
r_h	Rate of heat release	$J\ m^{-3}\ s^{-1}$
r_{O_2}	Rate of oxidation	Appropriate units
t	Time	s
t_{tm}	Crossover time in the HRRM	s

(*continued*)

TABLE A1. NOMENCLATURE (*continued*)

Abbrev. or variable	Definition	Units
t_{hrm}	Crossover time in the HRRM	s
x	Distance	m
x_{gc}	The geometric center of a sample	m
x_{tm}	The crossover position in the TM	m
x^*	Dimensionless distance $x^* = x/L$	—
δ	Frank-Kamenetskii parameter $\delta = \dfrac{EL^2 \rho QA}{\lambda RT_a^2} \exp\left[-\dfrac{E}{RT_a}\right]$	—
δ_{cr}	Critical value of the Frank-Kamenetskii parameter	—
ε	Reduced activation energy $\varepsilon = \dfrac{RT_a}{E}$	—
θ	Dimensionless temperature excess $\theta = \dfrac{E}{RT_a^2}(T - T_a)$	—
λ	Thermal conductivity	$W\ m^{-1}\ K^{-1}$
ν	Linear airflow rate	$J\ m^{-3}\ s^{-1}$
ρ	Bulk density of coal	$kg\ m^{-3}$

APPENDIX 2. PARAMETER VALUES

In this appendix, we record parameter values published for various coals. We hope that this collection will be useful for future modeling studies.

TABLE A2. PUBLISHED PARAMETER VALUES FOR VARIOUS COALS

Coal	AQ ($J\ s^{-1}\ kg^{-1}$)	A (s^{-1})	E ($kJ\ mol^{-1}$)	Q ($MJ\ kg^{-1}$)	c ($J\ kg^{-1}\ K^{-1}$)	λ ($W\ m^{-1}\ K^{-1}$)	ρ ($kg\ m^{-3}$)
Dalquhandy (Scotland) [a]	—	4×10^3	74 ± 4	25	1260		530
Dalquhandy (Scotland) [b, c]	—	1	50			0.143	
Killoch 5561 (Scotland) [b, c]	—	1×10^3	78			0.143	
Killoch 5736 (Scotland) [b, c]	—	15	57			0.143	
Killoch 6015 (Scotland) [b, c]	—	1×10^6	$93 \pm 5\%$	25		0.143	669
Kopako (NZ) [d]		1.01×10^2	55	22.08	1515		
Kopako (NZ, 56 days later) [d]		2.690×10^3	66	22.08	1515		
KPC-Prima (Indonesia) [e]		See Table 15		31.89			725
Lignite coal dust (Germany) [f]		See Table 17		22.90		0.0813	433
New Vale (NZ) [d]		2.37×10^2	61	22.11	1608		
Pinang (Indonesia) [g]		See Table 15		29.0	1250	0.11	720 ± 5
Pittsburgh (USA) [h]	3.04×10^9	—	65.4	33.08		0.1	492
Plateau (NZ) [d]		64.852×10^3	83	30.48	1399		
Prince (USA) [i]	7.0×10^{12}	—	95.7	31.3		0.1	477
Prima (Indonesia) [i]		See Table 15		31.9		0.11	725 ± 6
Renown (NZ) [d]		7.23×10^2	61	24.17	1468		
Rosslynlee (Scotland) [a, j, c]	—	9	54 ± 5	24.7	1260	0.143	745
South Bangko (Indonesia) [e]		See Table 15		22.97			635
Strongman (NZ) [d]		4.870×10^3	73	31.88	1453		
Tanjung Enim (Indonesia) [e]		See Table 15		28.7			620
VBC (Australia) [k]		See Table 16		22.9	1132	0.23	1350

Note: References: [a] Jones et al. (1996b); [b] Jones et al. (1998); [c] Jones (1998a); [d] Beamish et al. (2003); [e] Nugroho et al. (1998); [f] Malow and Krause (2004); [g] Nugroho et al. (2000b); [h] Reddy et al. (1998); [i] Nugroho et al. (2000a); [j] Jones (1996a); [k] Sujanti et al. (1999).

REFERENCES CITED

Akgün, F., and Arisoy, A., 1994, Effect of particle size on the spontaneous heating of a coal stockpile: Combustion and Flame, v. 99, no. 1, p. 137–146, doi: 10.1016/0010-2180(94)90085-X.

Alonso, M.I., Valdés, A.F., Martínez-Tarazona, R.M., and Garcia, A.B., 2002, Coal recovery from fines cleaning wastes by agglomeration with colza oil: A contribution to the environment and energy preservation: Fuel Processing Technology, v. 75, p. 85–95, doi: 10.1016/S0378-3820(01)00233-8.

Arisoy, A., and Akgün, F., 1994, Modelling of spontaneous combustion of coal with moisture content included: Fuel, v. 73, no. 2, p. 281–286, doi: 10.1016/0016-2361(94)90126-0.

Avid, B., Sato, Y., Maruyama, K., Yamada, Y., and Purevsuren, B., 2004, Effective utilization of Mongolian coal by upgrading in a solvent: Fuel Processing Technology, v. 85, p. 933–945, doi: 10.1016/j.fuproc.2003.10.010.

Babrauskas, V., 2003, Ignition Handbook: Principles and Applications to Fire Safety Engineering, Fire Investigation, Risk Management and Forensic Science: Issaqyah, Washington, Fire Science Publishers, 1116 p.

Banerjee, S.C., 1985, Spontaneous Combustion of Coal and Mine Fires: Rotterdam, Holland, A.A. Balkema, 168 p.

Banerjee, S.C., Banerjee, B.D., and Chakravorty, R.N., 1970, Rate studies of aerial oxidation of coal at low temperatures (30°C to 170°C): Fuel, v. 49, no. 3, p. 324–331, doi: 10.1016/0016-2361(70)90024-4.

Beamish, B.B., 2005a, Comparison of the R_{70} self-heating rate of New Zealand and Australian coals to Suggate rank parameter: International Journal of Coal Geology, v. 64, p. 139–144, doi: 10.1016/j.coal.2005.03.012.

Beamish, B.B., 2005b, Laboratory-scale assessment of hot spot development in bulk coal self-heating, in Gillies, A.D.S., Proceedings of the Eighth International Mine Ventilation Congress: Melbourne, The Australasian Institute of Mining and Metallurgy, p. 355–359.

Beamish, B.B., and Blazak, D.G., 2005, Relationship between ash content and R_{70} self-heating rate of Callide coal: International Journal of Coal Geology, v. 64, no. 1–2, p. 126–132, doi: 10.1016/j.coal.2005.03.010.

Beamish, B.B., and Hamilton, G.R., 2005, Effect of moisture content on the R_{70} self-heating rate of Callide coal: International Journal of Coal Geology, v. 64, no. 1–2, p. 133–138, doi: 10.1016/j.coal.2005.03.011.

Beamish, B.B., and Jabouri, I., 2005, Factors affecting hot spot development in bulk coal and associated gas evolution, in Proceedings of the Coal 2005 Conference: Melbourne, Australia, The Australian Institute of Mining and Metallurgy, p. 187–193.

Beamish, B.B., Barakat, M.A., and St. George, J.D., 2000, Adiabatic testing procedures for determining the self-heating propensity of coal and sample aging effects: Thermochimica Acta, v. 362, no. 1–2, p. 79–87, doi: 10.1016/S0040-6031(00)00588-8.

Beamish, B.B., Barakat, M.A., and St. George, J.D., 2001, Spontaneous-combustion propensity of New Zealand coals under adiabatic conditions: International Journal of Coal Geology, v. 45, no. 2–3, p. 217–224, doi: 10.1016/S0166-5162(00)00034-3.

Beamish, B.B., Lau, A.G., Moodie, A.L., and Vallance, T.A., 2002, Assessing the self-heating behaviour of Callide coal using a 2-metre column: Journal of Loss Prevention in the Process Industries, v. 15, p. 385–390, doi: 10.1016/S0950-4230(02)00020-7.

Beamish, B.B., St. George, J.D., and Barakat, M.A., 2003, Kinetic parameters associated with self-heating of New Zealand coals under adiabatic conditions: Mineralogical Magazine, v. 67, no. 4, p. 665–670, doi: 10.1180/00 26461036740125.

Beamish, B.B., Blazak, D.G., Hogarth, L.C.S., and Jabouri, I., 2005, R_{70} relationships and their interpretation at a mine site, in Proceedings of the Coal 2005 Conference: Melbourne, Australia, The Australian Institute of Mining and Metallurgy, p. 183–185.

Beever, P.F., 1982, Understanding the problems of spontaneous combustion: Fire Engineers Journal, v. 42, no. 126, p. 38–39.

Behera, P., 2004, Spontaneous combustion in coals—A panacea: Journal of the Geological Society of India, v. 63, no. 2, p. 158–170.

Bell, F.G., Bullock, S.E.T., Hälbich, T.E.J., and Lindsay, P., 2001, Environmental impacts associated with an abandoned mine in the Witbank Coalfield, South Africa: International Journal of Coal Geology, v. 45, p. 195–216, doi: 10.1016/S0166-5162(00)00033-1.

Berkowitz, N., 1957, On the differential thermal analysis of coal: Fuel, v. 36, no. 3, p. 355–373.

Berkowitz, N., 1985, The Chemistry of Coal: Amsterdam, The Netherlands, Elsevier, 513 p.

Bhat, S., and Agarwal, P., 1996, The effect of moisture condensation on the spontaneous susceptibility of coal: Fuel, v. 75, p. 1523–1532, doi: 10.1016/0016-2361(96)00121-4.

Bhattacharyya, K.K., Hodges, D.J., and Hinsley, F.B., 1968, The influence of humidity on the initial stages of the spontaneous heating of coal: The Mining Engineer, v. 126, p. 274–284.

Blazak, D.G., Beamish, B.B., Hodge, I., and Nichols, W., 2001, Mineral Matter and Rank Effects on the Self-Heating Rates of Callide Coal: Brisbane, Queensland Mining Industry Health and Safety Conference, Queensland Mining Council, p. 347–350.

Boddington, T., Feng, C.G., and Gray, P., 1983, Thermal-explosion and times-to-ignition in systems with distributed temperatures. 1. Reactant consumption ignored: Proceedings of the Royal Society of London, v. 385, no. 1789, p. 289–311.

Borges, A.R., and Viegas, D.X., 1988, Shelter effect on a row of coal piles to prevent wind erosion: Journal of Wind Engineering and Industrial Aerodynamics, v. 29, no. 1–3, p. 145–154, doi: 10.1016/0167-6105(88)90153-5.

Bowes, P.C., 1984, Self-Heating: Evaluating and Controlling the Hazards: Amsterdam, The Netherlands, Elsevier, 500 p.

Bowes, P.C., and Cameron, A., 1971, Self-heating and ignition of chemically activated carbon: Journal of Applied Chemistry and Biotechnology, v. 21, p. 244–250.

Bradshaw, S., Glasser, D., and Brooks, K., 1991, Self-ignition and convection patterns in an infinite coal layer: Chemical Engineering Communications, v. 105, p. 255–278.

Brooks, K., Bradshaw, S., and Glasser, D., 1988a, Spontaneous combustion of coal stockpiles—An unusual chemical reaction engineering problem: Chemical Engineering Science, v. 43, no. 8, p. 2139–2145, doi: 10.1016/0009-2509(88)87095-7.

Brooks, K., Svanas, N., and Glasser, D., 1988b, Evaluating the risk of spontaneous combustion in coal stockpiles: Fuel, v. 67, p. 651–656, doi: 10.1016/0016-2361(88)90293-1.

Buckley, A.N., 1994, A survey of the application of X-ray photoelectron spectroscopy to flotation research: Colloids and Surfaces, v. 93, p. 159–172, doi: 10.1016/0927-7757(94)02906-7.

Buckmaster, H.A., and Kudynska, J., 1992, Dynamic in situ 9 ghz c.w.-e.p.r. low-temperature oxidation study of selected Alberta coals. 4. Influence of moisture on hv bituminous coal: Fuel, v. 71, no. 10, p. 1147–1151, doi: 10.1016/0016-2361(92)90096-7.

Bustin, R.M., and Mathews, W.H., 1982, In situ gasification of coal, a natural example: History, petrology, and mechanics of combustion: Canadian Journal of Earth Sciences, v. 19, no. 3, p. 514–523.

Bustin, R.M., and Mathews, W.H., 1985, In situ gasification of coal, a natural example: Additional data on the Aldridge Creek coal fire, southeastern British Columbia: Canadian Journal of Earth Sciences, v. 22, p. 1858–1864.

Cai, S., Chen, F.F., and Soo, S.L., 1983, Wind penetration into a porous storage pile and use of barriers: Environmental Science & Technology, v. 17, no. 5, p. 298–305, doi: 10.1021/es00111a011.

Cameron, A., and MacDowall, J.D., 1972, The self heating of commercial powdered activated carbons: Journal of Applied Chemistry and Biotechnology, v. 22, p. 1007–1018.

Carpenter, D.L., and Giddings, D.G., 1964, The initial stages of oxidation of coal with molecular oxygen. 1. Effect of time, temperature and coal rank on rate of oxygen consumption: Fuel, v. 43, p. 247–266.

Carpenter, D.L., and Sergeant, G.D., 1966a, Initial stages of oxidation of coal with molecular oxygen. 3. Effect of particle size on rate of oxygen consumption: Fuel, v. 45, no. 4, p. 311.

Carpenter, D.L., and Sergeant, G.D., 1966b, Initial stages of oxidation of coal with molecular oxygen. 4. Accessibility of internal surface to oxygen: Fuel, v. 45, no. 6, p. 429.

Carr, R.M., Jumagai, H., Peake, B.M., Robinson, B.H., Clemens, A.H., and Matheson, T.W., 1995, Formation of free radicals during drying and oxidation of a lignite and a bituminous coal: Fuel, v. 74, no. 3, p. 389–394, doi: 10.1016/0016-2361(95)93472-P.

Carras, J.N., and Young, B.D., 1994, Self-heating of coal and related models, application and test methods: Progress in Energy and Combustion Science, v. 20, p. 1–15, doi: 10.1016/0360-1285(94)90004-3.

Chamberlain, E.A.C., 1974, Spontaneous combustion of coal: Colliery Guardian, p. 79.

Chamberlain, E.A.C., and Hall, D.A., 1973, The ambient temperature oxidation of coal in relation to the early detection of spontaneous heating: The Mining Engineer, May, p. 387–397.

Chen, X.D., 1991, The Spontaneous Heating of Coal—Large Scale Laboratory Assessment Supporting Theory [Ph.D. thesis]: Christchurch, New Zealand, University of Canterbury, 197 p.

Chen, X.D., and Chong, L.V., 1995, Some characteristics of transient self-heating inside an exothermically reactive porous solid slab: Process Safety and Environmental Protection, v. 73, no. B2, p. 101–107.

Chen, X.D., and Stott, J.B., 1993, The effect of moisture content on the oxidation rate of coal during near-equilibrium drying and wetting at 50°C: Fuel, v. 72, no. 6, p. 787–792, doi: 10.1016/0016-2361(93)90081-C.

Chong, L.V., and Chen, X.D., 1999, A mathematical model for the self-heating of spray-dried food powders containing fat, protein sugar and moisture: Chemical Engineering Science, v. 54, p. 4165–4178, doi: 10.1016/S0009-2509(99)00115-3.

Chong, L.V., Shaw, I.R., and Chen, X.D., 1995, Thermal ignition kinetics of wood sawdust by a newly devised experimental technique: Process Safety Progress, v. 14, no. 4, p. 266–270, doi: 10.1002/prs.680140409.

Chong, L.V., Shaw, I.R., and Chen, X.D., 1996, Exothermic reactivities of skim and whole milk powders as measured using a novel procedure: Journal of Food Engineering, v. 30, p. 185–196, doi: 10.1016/S0260-8774(96)00043-X.

Clemens, A.H., and Matheson, T.W., 1996, The role of moisture in the self-heating of low-rank coals: Fuel, v. 75, no. 7, p. 891–895, doi: 10.1016/0016-2361(96)00010-5.

Colaizzi, G.J., 2004, Prevention, control and/or extinguishment of coal seam fires using cellular grout, in Stracher, G.B., ed., Coal fires Burning Around the World: A Global Catastrophe: International Journal of Coal Geology, v. 59, no. 1–2, p. 75–81.

Committee on the Medical Effects of Air Pollutants, 1997, Handbook on Air Pollution and Health: London, UK, Department of Health, 144 p.

Conrad, J., 1902, Youth, a Narrative, and Two Other Stories: Edinburgh, Scotland, W. Blackwood and Sons, 375 p.

Dack, S.W., Hobday, M.D., Smith, T.D., and Pilbrown, J.R., 1983, Free radical involvement in the oxidation of Victorian brown coal: Fuel, v. 62, no. 12, p. 1510–1512, doi: 10.1016/0016-2361(83)90123-0.

Dalbey, W.E., Blackburn, G.R., Roy, T.A., Sasaki, J., Krueger, A.J., and Mackerer, C.R., 1998, Use of a surrogate aerosol in a preliminary screening for the potential carcinogenicity of coal coated with No. 6 fuel oil: American Industrial Hygiene Association Journal, v. 59, p. 90–95.

Dantoin, B., Hossfeld, R., and McAtee, K., 2003, Converting from funnel flow to mass flow: Power, v. 147, no. 8, p. 61.

Davis, J.D., and Byrne, J.F., 1924, An adiabatic method for studying spontaneous heating of coal: Journal of the American Ceramic Society, v. 7, p. 809–816, doi: 10.1111/j.1151-2916.1924.tb18175.x.

de Boer, C.B., Dekkers, M.J., and van Hoof, T.A.M., 2001, Rock-magnetic properties of TRM carrying baked and molten rocks straddling burnt coal seams: Physics of the Earth and Planetary Interiors, v. 126, p. 93–108, doi: 10.1016/S0031-9201(01)00246-1.

Durie, R.A., 1991, The Science of Victorian Brown Coal: Structure, Properties and Consequences for Utilization: Oxford, UK, Butterworth-Heinemann, 750 p.

Feng, K.K., Chakravorty, R.N., and Cochrane, T.S., 1973, Spontaneous combustion—A coal mining hazard: CIM (Canadian Mining and Metallurgical) Bulletin, v. 66, no. 738, p. 75–84.

Fierro, V., Miranda, J.L., Romero, C., Andrés, J.M., Arriaga, A., Schmal, D., and Visser, G.H., 1999a, Prevention of spontaneous combustion in coal stockpiles—Experimental results in coal storage yard: Fuel Processing Technology, v. 59, p. 23–34, doi: 10.1016/S0378-3820(99)00005-3.

Fierro, V., Miranda, J.L., Romero, C., Andrés, J.M., Pierrot, A., Gómez-Landesa, E., Arriaga, A., and Schmal, D., 1999b, Use of infrared thermography for the evaluation of heat loss during coal storage: Fuel Processing Technology, v. 60, p. 213–229, doi: 10.1016/S0378-3820(99)00044-2.

Fierro, V., Miranda, J.L., Romero, C., Andrés, J.M., Arriaga, A., and Schmal, D., 2001, Model predictions and experimental results on self-heating prevention of stockpiled coals: Fuel, v. 80, p. 125–134, doi: 10.1016/S0016-2361(00)00062-4.

Frank-Kamenetskii, D.A., 1969, Diffusion and Heat Transfer in Chemical Kinetics (2nd edition): New York, Plenum Press, 574 p.

Freudenstein, U., Crowley, D., and Welch, F., 2000, Chemical incident management: Gaseous emissions from a stockpile of coal: Public Health, v. 114, p. 41–44, doi: 10.1016/S0033-3506(00)00307-3.

Garcia, P., Hall, P.J., and Mondragon, F., 1999, The use of differential scanning calorimetry to identify coals susceptible to spontaneous combustion: Thermochimica Acta, v. 336, p. 41–46, doi: 10.1016/S0040-6031(99)00183-5.

Glasser, D., and Bradshaw, S.M., 1990, Spontaneous combustion in beds of coal, in Cheremisinoff, N., ed., Handbook of Heat and Mass Transfer. 4. Advances in Reactor Design and Combustion Science: Houston, Gulf Pub., p. 1071.

Gouws, M.J., and Wade, L., 1989, The self-heating liability of coal: Predictions based on composite indices: Mining Science and Technology, v. 9, p. 81–85, doi: 10.1016/S0167-9031(89)90797-4.

Gray, B.F., Little, S.G., and Wake, C.G., 1992, The prediction of a practical lower bound for ignition delay times and a method of scaling times-to-ignition in large reactant masses from laboratory data, in Twenty-Fourth Symposium (International) on Combustion: Pittsburgh, Pennsylvania, The Combustion Institute, p. 1785–1791.

Greankoplis, C.J., ed., 1983, Transport Processes: Momentum, Heat and Mass: Boston, Massachusetts, Allyn and Bacon, Inc., 538 p.

Guney, M., and Hodges, D.J., 1969, Adiabatic studies of the spontaneous heating of coal: Part 2, Colliery Guardian, v. 217, p. 173–177.

Gupta, R.P., and Prakash, A., 1998, Reflectance aureoles associated with thermal anomalies due to subsurface mine fires in the Jharia Coalfield, India: International Journal of Remote Sensing, v. 19, no. 14, p. 2619–2622, doi: 10.1080/014311698214415.

Güyagüler, T., Karpuz, C., and Bağci, S., 2003, The spontaneous combustion characteristics of Turkish lignite and correlation of the self-heating process with the actual fire: CIM (Canadian Mining and Metallurgical) Bulletin, v. 96, no. 1070, p. 75–79.

Hancock, M., Kizil, M.S., and Beamish, B.B., 2005, Computer animation of hot spot development in bulk coals as an aid for training coal miners, in Proceedings of the Coal 2005 Conference: Melbourne, Australia, The Australian Institute of Mining and Metallurgy, p. 243–248.

Heffern, E.L., and Coates, D.A., 2004, Geologic history of natural coal-bed fires, Powder River Basin, USA in Stracher, G.B., ed., Coal Fires Burning Around the World: A Global Catastrophe: International Journal of Coal Geology, 59, no. 1–2, p. 25–47.

Huang, J., Bruining, J., and Wolf, K.-H.A.A., 2001, Modeling of gas flow and temperature fields in underground coal fires: Fire Safety Journal, v. 36, p. 477–489, doi: 10.1016/S0379-7112(01)00003-0.

Hull, A.S., Lanthier, J.L., Chen, Z., and Agarwal, P.K., 1997, The role of diffusion of oxygen and radiation on the spontaneous combustibility of a coal pile in confined storage: Combustion and Flame, v. 110, p. 479–493, doi: 10.1016/S0010-2180(97)00088-6.

Humphreys, D.R., 1981, A Study of the Propensity of Queensland Coals to Spontaneous Combustion [M.S. thesis]: Brisbane, Australia, The University of Queensland, 108 p.

Humphreys, D., Rowlands, D., and Cudmore, J.F., 1981, Spontaneous combustion of some Queensland coals, in Hargraves, A.J., ed., Ignitions, Explosions & Fires: Melbourne, Australia, The Australasian Institute of Mining and Metallurgy, p. 5-1–5-19.

International Maritime Consultative Organisation, 1974, Self-Heating Test for Carbon: International Maritime Consultative Organization, Dangerous Goods Code Supplement 1974–5, Amendment 11–75, 83 p.

International Organisation for Standardization, 1994, Brown Coals and Lignites—Testing of Self-Ignition: Draft Proposal: Geneva, International Organisation for Standardization.

Jones, J.C., 1996a, A single-point estimation of the combustion rate expression for a bituminous coal: Journal of Fire Sciences, v. 14, p. 325–330.

Jones, J.C., 1996b, Steady behaviour of long duration in the spontaneous heating of a bituminous coal: Journal of Fire Sciences, v. 14, no. 1–2, p. 159–166.

Jones, J.C., 1997, Topics in Environmental and Safety Aspects of Combustion Technology: Caithness, Scotland, Whittles Publishing, 192 p.

Jones, J.C., 1998a, Letter to the editor: Fuel, v. 77, no. 13, p. 1521.

Jones, J.C., 1998b, A means of obtaining a full kinetic rate expression for the oxidation of a solid substrate from a single criticality data point: Fuel, v. 77, no. 14, p. 1677–1678, doi: 10.1016/S0016-2361(98)00098-2.

Jones, J.C., 1998c, Towards an alternative criterion for the shipping safety of activated carbons: Journal of Loss Prevention in the Process Industries, v. 11, p. 407–411, doi: 10.1016/S0950-4230(98)00024-2.

Jones, J.C., 1999a, Calculation of the Frank-Kamenetskii critical parameter for a cubic reactant shape from experimental results on bituminous coals: Fuel, v. 78, p. 89–91, doi: 10.1016/S0016-2361(98)00116-1.

Jones, J.C., 1999b, Developments in test procedures for propensity towards spontaneous heating, in Proceedings of the International Conference on Fire Safety: Sissonville, West Virginia, Product Safety Corporation, v. 27, p. 57–64.

Jones, J.C., 1999c, Towards an alternative criterion for the shipping safety of activated carbons, Part 4: The role of times to ignition: Journal of Loss Prevention in the Process Industries, v. 12, p. 533–535, doi: 10.1016/S0950-4230(99)00018-2.

Jones, J.C., 2000a, Commentary on the UN test for spontaneous heating of solid substances: Journal of Loss Prevention in the Process Industries, v. 13, p. 177–178, doi: 10.1016/S0950-4230(99)00077-7.

Jones, J.C., 2000b, A means of obtaining a full kinetic rate expression for the oxidation of a solid substrate from a single criticality data point. Part 3: Possible application to cellulosics: Letter to editor: Fuel, v. 79, p. 717–718, doi: 10.1016/S0016-2361(99)00219-7.

Jones, J.C., 2000c, A new and more reliable test for the propensity of coals and carbons to spontaneous heating: Journal of Loss Prevention in the Process Industries, v. 13, p. 69–71, doi: 10.1016/S0950-4230(99)00055-8.

Jones, J.C., 2000d, On the role of times to ignition in the thermal safety of transportation of bituminous coals: Fuel, v. 79, p. 1561–1562, doi: 10.1016/S0016-2361(00)00003-X.

Jones, J.C., 2001, On the extrapolation of results from oven heating tests for propensity to self-heating: Combustion and Flame, v. 124, p. 334–336, doi: 10.1016/S0010-2180(00)00204-2.

Jones, J.C., 2002, A scientific exegesis of "Youth" by Joseph Conrad: Journal of Fire Sciences, v. 20, no. 4, p. 343–345, doi: 10.1177/073490402762574776.

Jones, J.C., and Littlefair, J., 1997, Novel behaviour in the self-heating of coal filter cake: Fuel, v. 76, no. 12, p. 1165–1167, doi: 10.1016/S0016-2361(97)00117-8.

Jones, J.C., and Newman, S.C., 2003, Non-Arrhenius behaviour in the oxidation of two carbonaceous substrates: Journal of Loss Prevention in the Process Industries, v. 16, no. 3, p. 223–225, doi: 10.1016/S0950-4230(02)00115-8.

Jones, J.C., and Wake, G.C., 1990, Measured activation energies of ignition of solid materials: Journal of Chemical Technology and Biotechnology (Oxford, Oxfordshire: 1986), v. 48, p. 209–216.

Jones, J.C., Chiz, P.S., Koh, R., and Matthew, J., 1996a, Continuity of kinetics between sub- and supercritical regimes in the oxidation of a high-volatile solid substrate: Fuel, v. 75, no. 15, p. 1733–1736, doi: 10.1016/S0016-2361(96)00150-0.

Jones, J.C., Chiz, P.S., Koh, R., and Matthew, J., 1996b, Kinetic parameters of oxidation of bituminous coals from heat-release rate measurements: Fuel, v. 75, no. 15, p. 1755–1757, doi: 10.1016/S0016-2361(96)00159-7.

Jones, J.C., Henderson, K.P., Littlefair, J., and Rennie, S., 1998, Kinetic parameters of oxidation of coals by heat-release measurement and their relevance to self-heating tests: Fuel, v. 77, no. 1–2, p. 19–22, doi: 10.1016/S0016-2361(97)00155-5.

Kadioğlu, Y., and Varamaz, M., 2003, The effect of moisture content and air-drying on spontaneous combustion characteristics of two Turkish lignites: Fuel, v. 82, no. 13, p. 1685–1693.

Kaji, R., Hishinuma, Y., and Nakamura, Y., 1985, Low temperature oxidation of coals: Effects of pore structure and coal composition: Fuel, v. 64, no. 3, p. 297–302, doi: 10.1016/0016-2361(85)90413-2.

Karsner, G.G., and Perlmutter, D.D., 1982, Model for coal oxidation kinetics. 1. Reaction under chemical control: Fuel, v. 61, p. 29–34, doi: 10.1016/0016-2361(82)90289-7.

Kok, A., Bolt, N., and Jelgersma, J.H.N., 1989, Spontaneous heating and stockpiling losses of coal: Kema Scientific and Technical Reports, v. 7, no. 2, p. 63–89.

Kreulen, D.J.W., 1948, Elements of Coal Chemistry: Rotterdam, Holland, Niugh and van Ditmar, 204 p.

Krishnaswamy, S., Agarwal, P.K., and Gunn, R.D., 1996a, Low-temperature oxidation of coal. 3: Modelling spontaneous combustion in coal stockpiles: Fuel, v. 75, no. 3, p. 353–362, doi: 10.1016/0016-2361(95)00249-9.

Krishnaswamy, S., Bhat, S., Gunn, R.D., and Agarwal, P.K., 1996b, Low-temperature oxidation of coal. 1: A single-particle reaction-diffusion model: Fuel, v. 75, no. 3, p. 333–343, doi: 10.1016/0016-2361(95)00180-8.

Kuchta, J.M., Rowe, V.R., and Burgess, D.S., 1980, Spontaneous Combustion Susceptibility of US Coals: U.S. Bureau of Mines Report of Investigation 8474, 37 p.

Küçük, A., Kadioğlu, Y., and Gülaboğlu, M.S., 2003, A study of spontaneous combustion characteristics of a Turkish lignite: Particle size, moisture of coal, humidity of air: Combustion and Flame, v. 133, no. 3, p. 255–261, doi: 10.1016/S0010-2180(02)00553-9.

Li, Y.H., and Skinner, J.L., 1986, Deactivation of dried subbituminous coal: Chemical Engineering Communications, v. 49, p. 81–98.

Liu, C., Li, S., Qiao, Q., Wang, J., and Pan, Z., 1998, Management of spontaneous combustion in coal mine waste tips in China: Water, Air, and Soil Pollution, v. 103, p. 441–444, doi: 10.1023/A:1004922620264.

Lopez, D., Sanada, Y., and Mondragon, F., 1998, Effect of low-temperature oxidation of coal on hydrogen-transfer capability: Fuel, v. 77, no. 14, p. 1623–1628, doi: 10.1016/S0016-2361(98)00086-6.

Lu, P., Liao, G.X., Sun, J.H., and Li, P.D., 2004, Experimental research on index gas of the coal spontaneous at low-temperature stage: Journal of Loss Prevention in the Process Industries, v. 17, p. 243–247, doi: 10.1016/j.jlp.2004.03.002.

Mahadevan, V., and Ramlu, M.A., 1985, Fire risk rating of coal mines due to spontaneous heating: Journal of Mines, Metals and Fuels, v. 8, p. 357–362.

Mahidin, O.Y., Nakata, Y., and Usui, H., 2003a, Improvement of devolatilization and control of low-temperature oxidation by vacuum drying and tar coating treatments of low-rank coal: Journal of Chemical Engineering of Japan, v. 36, no. 7, p. 769–775, doi: 10.1252/jcej.36.769.

Mahidin, O.Y., Usui, H., and Okuma, O., 2003b, The advantages of vacuum-treatment in the thermal upgrading of low-rank coals on the improvement of dewatering and devolatilization: Fuel Processing Technology, v. 84, no. 1–3, p. 147–160.

Malow, M., and Krause, U., 2004, The overall activation energy of the exothermic materials of thermally unstable materials: Journal of Loss Prevention in the Process Industries, v. 17, p. 51–58, doi: 10.1016/j.jlp.2003.09.002.

Marinov, V.N., 1977, Self-ignition and mechanisms of interaction of coal with oxygen at low temperatures. 1: Changes in the composition of coal heated at constant rate to 250°C in air: Fuel, v. 56, no. 2, p. 153–157.

Mazumdar, B.K., 1996, On the correlation of moist fuel ratio of coal with its spontaneous combustion temperature: Fuel, v. 75, no. 5, p. 646–648, doi: 10.1016/0016-2361(96)90002-2.

McNally, G.H., 2000, Geology and mining practice in relation to small subsidence in the Northern Coalfield, New South Wales: Australian Journal of Earth Sciences, v. 47, p. 21–34, doi: 10.1046/j.1440-0952.2000.00761.x.

Michalski, S.R., Winschel, L.J., and Gray, R.E., 1990, Fires in abandoned coal mines: Bulletin of the Association of Engineering Geologists, v. 27, p. 479–495.

Mondragón, F., Ruíz, W., and Santamaría, A., 2002, Effect of early stages of coal oxidation on its reaction with elemental sulphur: Fuel, v. 81, p. 381–388, doi: 10.1016/S0016-2361(01)00143-0.

Moreby, R., 1997, Dartbrook coal—Case study, in Proceedings of the 6th International Mine Ventilation Congress: Pittsburgh, Pennsylvania, The Society of Mining, Metallurgy and Exploration Inc., p. 39–45.

Moxon, N.T., and Richardson, S.B., 1985, Development of a calorimeter to measure the self-heating characteristics of coal: Coal Preparation, v. 2, p. 79–90.

Mukherjee, P.N., and Lahiri, A., 1957, Brennstoff-Chemie, v. 38, p. 55.

Mulcahy, M.F.R., Morley, W.J., and Smith, I.W., 1991, Combustion, gasification and oxidation, in Durie, R.A., ed., The Science of Victorian Brown Coal: Structure, Properties and Consequences for Utilization: Oxford, England, Butterworth-Heinemann, p. 359–463.

Nakagawa, H., Namba, A., Böhlmann, M., and Miura, K., 2004, Hydrothermal dewatering of brown coal and catalytic hydrothermal gasification of the organic compounds dissolving in the water using a novel Ni/carbon catalyst: Fuel, v. 83, p. 719–725, doi: 10.1016/j.fuel.2003.09.020.

Nakajima, T., Kanda, T., Fukuda, T., Takanashi, H., and Ohki, A., 2005, Characterization of eluent by hot water extraction of coals in terms of total organic carbon and environmental impacts: Fuel, v. 84, p. 783–789, doi: 10.1016/j.fuel.2004.11.017.

Nordon, P., 1979, A model for the self-heating reaction of coal and char: Fuel, v. 58, no. 6, p. 456–464, doi: 10.1016/0016-2361(79)90088-7.

Nordon, P., and Bainbridge, N.W., 1979, Some properties of char affecting the self-heating reaction in bulk: Fuel, v. 58, no. 6, p. 450–455, doi: 10.1016/0016-2361(79)90087-5.

Nordon, P., Young, B.C., and Bainbridge, N.W., 1979, The rate of oxidation of char and coal in relation to their tendency to self-heat: Fuel, v. 58, no. 6, p. 443–449, doi: 10.1016/0016-2361(79)90086-3.

Nugroho, Y.S., McIntosh, A.C., and Gibbs, B.M., 1998, Using the crossing point method to assess the self-heating behaviour of Indonesian coals, in Twenty-Seventh Symposium (International) on Combustion: Pittsburgh, Pennsylvania, The Combustion Institute, p. 2981–2989.

Nugroho, Y.S., McIntosh, A.C., and Gibbs, B.M., 2000a, Low-temperature oxidation of single and blended coals: Fuel, v. 79, p. 1951–1961, doi: 10.1016/S0016-2361(00)00053-3.

Nugroho, Y.S., McIntosh, A.C., and Gibbs, B.M., 2000b, On the prediction of thermal runaway of coal piles of differing dimension by using a correlation between heat release and activation energy, in Twenty-Eighth Symposium (International) on Combustion: Pittsburgh, Pennsylvania, The Combustion Institute, p. 2321–2327.

Ogunsola, O.I., 1993, Thermal upgrading effect on oxygen distribution in lignite: Fuel Processing Technology, v. 34, no. 1, p. 73–81, doi: 10.1016/0378-3820(93)90062-9.

Ogunsola, O.I., and Mikula, R.J., 1991, A study of spontaneous combustion characteristics of Nigerian coals: Fuel, v. 70, no. 2, p. 258–261, doi: 10.1016/0016-2361(91)90162-4.

Palmer, A.D., Cheng, M., Goulet, J.-C., and Furimsky, E., 1990, Relation between particle size and properties of some bituminous coals: Fuel, v. 69, no. 2, p. 183–188, doi: 10.1016/0016-2361(90)90171-L.

Panaseiko, N.P., 1974, The influence of moisture on the low-temperature oxidation of coals: Solid Fuel Chemistry, v. 8, p. 21.

Panigrahi, D.C., and Sahu, H.B., 2004, Application of hierarchical clustering for classification of coal seams with respect to their proneness to spontaneous heating: Transactions of the Institution of Mining and Metallurgy, Section A, Mining Technology, v. 113, no. 2, p. A97–A106.

Pilarczyk, E., Leonhardt, P., and Wanzl, W., 1995, Characterization of coals with respect to their self-ignition tendency, in Pajares, J.A., and Tascón, J.M.D., eds., Coal Science and Technology: Proceedings of the Eighth International Conference on Coal Science, v. 24, p. 497–500.

Pis, J.J., de la Puente, G., Fuente, E., Morán, A., and Rubiera, F., 1996, A study of the self-heating of fresh and oxidized coals by differential thermal analysis: Thermochimica Acta, v. 279, p. 93–101.

Reddy, P.D., Amyotte, P.R., and Pegg, M.J., 1998, Effect of inerts on layer ignition temperatures of coal dust: Combustion and Flame, v. 114, p. 41–53, doi: 10.1016/S0010-2180(97)00286-1.

Reid, R.C., Prausnitz, J.M., and Poling, B.E., eds., 1987, The Properties of Gases and Liquids: New York, McGraw-Hill, Inc., 741 p.

Ren, T.X., Edwards, J.S., and Clarke, D., 1999, Adiabatic oxidation study on the propensity of pulverised coals to spontaneous combustion: Fuel, v. 78, p. 1611–1620, doi: 10.1016/S0016-2361(99)00107-6.

Saranchuk, V.I., Galushko, L. Ỹa., Paschenko, L.V., and Luk'yanenko, L.V., 1978, Influence of water on the low-temperature oxidation of coal, Solid Fuel Chemistry, v. 12, no. 91, p. 9–12.

Sato, Y., Kushiyama, S., Tatsumoto, K., and Yamaguchi, H., 2004, Upgrading of low rank coal with solvent: Fuel Processing Technology, v. 85, p. 1551–1564, doi: 10.1016/j.fuproc.2003.10.023.

Schmal, D., Duyzer, J.H., and van Heuven, J.W., 1985, A model for the spontaneous heating of coal: Fuel, v. 64, p. 963–972, doi: 10.1016/0016-2361(85)90152-8.

Schmidt, L.D., 1945, Changes in coal during storage, in Lowry, H.H., ed., Chemistry of Coal Utilization, Volume 1: New York, John Wiley and Sons, Inc., p. 627–676.

Schmidt, L.D., and Elder, J.L., 1940, Atmospheric oxidation of coal at moderate temperatures: Engineering and Industrial Chemistry, v. 32, no. 2, p. 249–256, doi: 10.1021/ie50362a021.

Sensogut, C., and Cinar, I., 2000, A research on the tendency of Ermenek district coals to spontaneous combustion: Mineral Resources Engineering, v. 9, no. 4, p. 421–427.

Sensogut, C., and Ozdeniz, A.H., 2005, Statistical modelling of stockpile behaviour under different atmospheric conditions—Western Lignite Corporation (WLC) case: Fuel, v. 84, p. 1858–1863, doi: 10.1016/j.fuel.2005.03.027.

Sheail, J., 2005, 'Burning bings': A study of pollution management in mid-twentieth century Britain: Journal of Historical Geography, v. 31, p. 134–148, doi: 10.1016/j.jhg.2004.04.001.

Shi, T., Wang, X., Deng, J., and Wen, Z., 2005, The mechanism at the initial stage of the room-temperature oxidation of coal: Combustion and Flame, v. 140, p. 332–345, doi: 10.1016/j.combustflame.2004.10.012.

Sidenko, N.V., Gieré, R., Bortnikova, S.B., Cottard, F., and Pal'chik, N.A., 2001, Mobility of heavy metals in self-burning waste heaps of the zinc smelting plant in Belovo (Kemerovo Region, Russia): Journal of Geochemical Exploration, v. 74, p. 109–125, doi: 10.1016/S0375-6742(01)00178-9.

Singh, R.N., Shonhardt, J.A., and Terezopoulos, N., 2002, A new dimension to studies of spontaneous combustion of coal: Mineral Resources Engineering, v. 11, no. 2, p. 147–163, doi: 10.1142/S0950609802000938.

Singh, R.V.K., and Singh, V.K., 2004, Mechanised spraying device—A novel technology for spraying fire protective coating material in the benches of opencast coal mines for preventing spontaneous combustion: Fire Technology, v. 40, no. 4, p. 355–365, doi: 10.1023/B:FIRE.0000039163.07393.96.

Sinha, P.R., 1986, Mine fires in Indian coalfields: Energy, v. 11, no. 11–12, p. 1147–1154, doi: 10.1016/0360-5442(86)90051-4.

Smith, A.C., and Lazzara, C.P., 1987, Spontaneous Combustion Studies of US Coals: Pittsburgh, Pennsylvania, U.S. Bureau of Mines Report of Investigation 9079, 28 p.

Smith, A.C., Miron, Y., and Lazzara, C.P., 1988, Inhibition of Spontaneous Combustion of Coal: Washington, D.C., U.S. Bureau of Mines Report of Investigation 9196, 22 p.

Smith, M.A., and Glasser, D., 2005a, Spontaneous combustion of carbonaceous stockpiles. Part I: The relative importance of various intrinsic coal properties and properties of the reaction system: Fuel, v. 84, p. 1151–1160, doi: 10.1016/j.fuel.2004.12.004.

Smith, M.A., and Glasser, D., 2005b, Spontaneous combustion of carbonaceous stockpiles. Part II: Factors affecting the rate of the low-temperature oxidation reaction: Fuel, v. 84, no. 9, p. 1161–1170, doi: 10.1016/j.fuel.2004.12.005.

Sondreal, E.A., and Ellman, R.L., 1974, Laboratory determination of factors affecting storage of North Dakota lignite: U.S. Bureau of Mines Report of Investigation 7887, 91 p.

Stott, J.B., 1960, Influence of moisture on the spontaneous combustion of coals: Nature, v. 188, no. 54, p. 54–55, doi: 10.1038/188054a0.

Stott, J.B., and Baker, O.J., 1953, Differential thermal analysis of coal: Fuel, v. 32, no. 4, p. 415–427.

Stott, J.B., and Chen, X.D., 1992, Measuring the tendency of coal to fire spontaneously: Colliery Guardian, v. 240, no. 1, p. 9–16.

Stracher, G.B., and Taylor, T.P., 2004, Coal fires burning out of control around the world: Thermodynamic recipe for environmental catastrophe, in Stracher, G.B., ed., Coal Fires Burning Around the World: A Global Catastrophe: International Journal of Coal Geology, v. 59, no. 1–2, p. 7–17.

Stracher, G.B., Prakash, A., Schroeder, P., McCormack, J., Zhang, X., van Dijk, P., and Blake, D., 2005, New mineral occurrences and mineralization processes: Wudu coal-fire gas vents of Inner Mongolia: The American Mineralogist, v. 90, p. 1729–1739, doi: 10.2138/am.2005.1671.

Suggate, R.P., 2000, The rank (s_r) scale: Its basis and its applicability as a maturity index for all coals: New Zealand Journal of Geology and Geophysics, v. 43, p. 521–553.

Sujanti, W., and Zhang, D.-K., 1999, A laboratory study of spontaneous combustion of coal: The influence of inorganic matter and reactor size: Fuel, v. 78, p. 549–556, doi: 10.1016/S0016-2361(98)00188-4.

Sujanti, W., and Zhang, D.-K., 2000, Investigation into the role of inherent inorganic matter and additives in low-temperature oxidation of a Victorian brown coal: Combustion Science and Technology, v. 152, p. 99–114.

Sujanti, W., Zhang, D.-K., and Chen, X.D., 1999, Low-temperature oxidation of coal studied using wire-mesh reactors with both steady-state and transient methods: Combustion and Flame, v. 117, no. 3, p. 646–651, doi: 10.1016/S0010-2180(98)00139-4.

Swann, P.D., and Evans, D.G., 1979, Low-temperature oxidation of brown coal. 3: Reaction with molecular oxygen at temperatures close to ambient: Fuel, v. 58, no. 4, p. 276–280, doi: 10.1016/0016-2361(79)90136-4.

Swann, P.D., Allardice, D.J., and Evans, D.G., 1974, Low-temperature oxidation of brown coal. 1: Changes in internal surface due to oxidation: Fuel, v. 53, no. 2, p. 85–87, doi: 10.1016/0016-2361(74)90060-X.

Thomas, P.H., and Bowes, P.C., 1961, Thermal ignition in a slab with one face at a constant high temperature: Transactions of the Faraday Society, v. 57, no. 11, p. 2007, doi: 10.1039/tf9615702007.

Umar, D.F., Daulay, B., Usui, H., Deguchi, T., and Sugita, S., 2005, Characterization of upgraded brown coal (UBC): Coal Preparation, v. 25, p. 31–45, doi: 10.1080/07349340590927350.

Unal, S., Wood, D.G., and Harris, I.J., 1992, Effects of drying methods on the low temperature reactivity of Victorian brown coal to oxygen: Fuel, v. 71, no. 2, p. 183–192, doi: 10.1016/0016-2361(92)90007-B.

United Nations (UN), 1995, Recommendations for the Transportation of Dangerous Goods: Manual of Tests and Criteria (2nd edition, Section 33.3.1.6): Test Method of Self-Heating of Substances: New York, United Nations.

U.S. Department of Energy, 1994, Primer on Spontaneous Heating and Pyrophoricity: Washington, D.C., U.S. Department of Energy, 89 p.

U.S. Senate, 1912, Titanic Disaster. Senate report no. 806, pursuant to Senate Resolution 283, 62nd Congress, 2nd session: Washington, D.C., U.S. Government Printing Office.

Vance, W.E., Chen, X.D., and Scott, S.C., 1996, The rate of temperature rise of a subbituminous coal during spontaneous combustion in an adiabatic device: The effect of moisture content and drying methods: Combustion and Flame, v. 106, p. 261–270, doi: 10.1016/0010-2180(95)00276-6.

Varma, A., Morbidelli, M., and Wu, H., 1999, Parametric Sensitivity in Chemical Systems: Cambridge, UK, Cambridge University Press, 342 p.

Vorres, K.S., Wertz, D.L., Malhotra, V., Dang, Y., Joseph, J.T., and Fisher, R., 1992, Drying of Beulah-Zap lignite: Fuel, v. 71, no. 9, p. 1047–1053, doi: 10.1016/0016-2361(92)90113-3.

Walker, I.K., 1967, The role of water in spontaneous combustion of solids: Fire Research Abstracts and Reviews, v. 9, no. 1, p. 5–22.

Wang, H., Dlugogorski, B.Z., and Kennedy, E.M., 2002, Examination of CO_2, CO and H_2O formation during low-temperature oxidation of a bituminous coal: Energy & Fuels, v. 16, p. 586–592, doi: 10.1021/ef010152v.

Wang, H., Dlugogorski, B.Z., and Kennedy, E.M., 2003a, Analysis of the mechanism of the low-temperature oxidation of coal: Combustion and Flame, v. 134, p. 107–117, doi: 10.1016/S0010-2180(03)00086-5.

Wang, H., Dlugogorski, B.Z., and Kennedy, E.M., 2003b, Coal oxidation at low temperatures: Oxygen consumption, oxidation products, reaction mechanism and kinetic modelling: Progress in Energy and Combustion Science, v. 29, p. 487–513, doi: 10.1016/S0360-1285(03)00042-X.

Wang, H., Dlugogorski, B.Z., and Kennedy, E.M., 2003c, Pathways for production of CO_2 and CO in low-temperature oxidation of coal: Energy & Fuels, v. 17, p. 150–158, doi: 10.1021/ef020095l.

Watanabe, W.S., and Zhang, D.-K., 2001, The effect of inherent and added inorganic matter on low-temperature oxidation reaction of coal: Fuel Processing Technology, v. 74, p. 145–160, doi: 10.1016/S0378-3820(01)00237-5.

Zhang, D.K., and Sujanti, W., 1999, The effect of exchangeable cations on low-temperature oxidation and self-heating of a Victorian brown coal: Fuel, v. 78, p. 1217–1224, doi: 10.1016/S0016-2361(99)00036-8.

Zhang, J., Wagner, W., Prakash, A., Mehl, H., and Voigt, S., 2004, Detecting coal fires using remote sensing techniques: International Journal of Remote Sensing, v. 25, no. 16, p. 3193–3320, doi: 10.1080/01431160310001620812.

Zhang, X.M., van Genderen, J.L., and Kroonenberg, S.B., 1997, A method to evaluate the capability of Landsat-5 TM band 6 data for sub-pixel coal fire detection: International Journal of Remote Sensing, v. 18, p. 3279–3288, doi: 10.1080/014311697217080.

Zhang, X.M., Cassells, C.J.S., and van Genderen, J.L., 1998, Multi-sensor data fusion for the detection of underground coal fires: Geologie en Mijnbouw, v. 77, p. 117–127, doi: 10.1023/A:1003526100075.

Zhang, X., van Genderen, J.L., Guan, H., and Kroonenberg, S., 2003, Spatial analysis of thermal anomalies from airborne multi-spectral data: International Journal of Remote Sensing, v. 24, no. 19, p. 3727–3742, doi: 10.1080/0143116031000095925.

Zhang, X., Kroonenberg, S.B., and de Boer, C.B., 2004a, Dating of coal fires in Xinjiang, north-west China: Terra Nova, v. 16, p. 68–74, doi: 10.1111/j.1365-3121.2004.00532.x.

Zhang, X., Zhang, J., Kuenzer, C., Voigt, S., and Wagner, W., 2004b, Capability evaluation of 3–5 μm and 8–12.5 μm airborne thermal data for underground coal fire detection: International Journal of Remote Sensing, v. 25, no. 12, p. 2245–2258, doi: 10.1080/01431160310001618112.

Zhong, M., Xing, W., Weicheng, F., Peide, L., and Baozhi, C., 2003, Airflow optimizing control research based on genetic algorithm during mine fire period: Journal of Fire Sciences, v. 21, no. 2, p. 131–153, doi: 10.1177/0734904103021002003.

MANUSCRIPT ACCEPTED BY THE SOCIETY 7 MARCH 2007

A laboratory study of a reactive surface layer for the prevention of spontaneous combustion

Rufaro Kaitano
David Glasser*
Diane Hildebrandt
School of Process and Materials Engineering, University of the Witwatersrand, Johannesburg, Private Bag 3, Wits 2050, South Africa

ABSTRACT

Spontaneous combustion commonly occurs in storage and waste dumps from coal mining. It has been suggested that this can be prevented by the use of a protective layer (possibly compacted) of a reactive material. As the coal ages (becomes oxidized), the coal reactivity toward oxygen decreases and the oxygen profile changes. Analyses of the oxygen concentration profiles in a 3-m-long, 20 cm inside diameter plastic column filled with coal were performed for a period of up to 8 mo. The results were modeled, and they can be adequately fitted to a simple diffusion/reaction model.

Keywords: spontaneous combustion, coal oxidation, coal reactivity.

INTRODUCTION

The oxidation of carbonaceous matter in coal is primarily responsible for the spontaneous combustion in coal dumps. In order to monitor coal dumps and predict the likelihood of spontaneous combustion, a large amount of work has been done by scientists and engineers D. Schmal, D.H. Duyzer, and J.W. van Heuven (Schmal et al., 1985).

There is also a large amount of literature that focuses on understanding and modeling spontaneous combustion by Brooks and Glasser (1986), Brooks et al. (1988b), and K.K. Bhattacharyya (1972). There is a review article by Glasser and Bradshaw (1990) that covers many of the issues.

There has also been a suggestion (Glasser and Bradshaw, 1990) that a useful way to protect a dump that has the potential to burn is to cover it with a layer of a relatively fine reactive coal. The purpose of this layer is to scavenge out the oxygen (while itself not catching fire) before the oxygen can reach the material that could combust. The potential of this theory will not be explored further in this paper, but the interested reader can refer to the previous publications where the theory is explained.

Many of the experimental techniques that are used to study coal oxidation rates are really only suitable for use in research laboratories because they are too complex and so are not directly useful to the people who build and maintain large stockpiles and waste dumps.

It is therefore of use to devise a very simple apparatus that can be used in a mine laboratory and that will give information about the actual run-of-mine material that the practitioner is working with. Furthermore, it is not only the initial reactivity of the material that is of interest but also how it changes with time (degree of oxidation). The apparatus described in this paper is an attempt

---

*david.glasser@wits.ac.za

to devise such a piece of equipment. This apparatus would be ideal for practical testing applications. Indeed this approach has actually been successfully used to protect dumps of material that previously had spontaneously combusted (Adamski, 2003).

EXPERIMENTAL METHODS

Sample Preparation

Coal samples were obtained from New Vaal Colliery in the Orange Free State, South Africa. The mine used to be an underground mine, which was later turned into an opencast mine. The coal can be classified as low-grade bituminous. This is based on the fact that the ash content is very high, 40% on average, and the calorific values are in the region of 1.3MJ kg^{-1}. The particle size was ~1 cm on average. The samples were collected from the piling stacks as the coal came from the crushing and washing section of the mine. The two samples were collected ~8 mo apart. The coal samples were put into polythene bags, into which dry ice was sprinkled to both lower the temperature and to exclude oxygen so as to minimize oxidation prior to storage. Once the coal was taken to the laboratory, it was stored under nitrogen. The coal used in these experiments was not being considered for use as a reactive layer but was merely used to test the apparatus and the proposed model.

Equipment

The apparatus consisted of a plastic pipe that was 3 m in height, with a 20 cm inside diameter, and that was closed at the bottom and open to the atmosphere at the top. The apparatus was placed in a laboratory where the temperature varied in the range 15–25 °C. Gas samples were obtained from sampling points at 10 cm intervals along the column and were analyzed for CO, CO_2, O_2, and N_2. For the analysis, a custom-made gas chromatograph (GC) was used. A thermal conductivity detector (TCD) coupled with a data acquisition unit was used to generate peak data, which were stored on floppy disks.

Separation of all gases could not be achieved by the use of a single column; hence, two columns attached to a four-port valve were used. The valve made it possible to connect the columns in series or bypass one with the carrier gas, depending on the option required. The columns used a Molecular Sieve 5A (1.5 m, 1/8″ inner diameter column) to separate O_2, N_2, and CO peaks and a Porapaq Q (1.2 m, 1/8″ inner diameter column) to trap CO_2 and subsequently analyze it. The temperature of operation was maintained between 20 °C and 22 °C to achieve maximum sensitivity of the instrument. The carrier gas was hydrogen at a pressure of 500 kPa and flow rate of 20 mL/min.

Experimental Procedure

The coal samples were taken out of the storage cupboard, where they had been stored under nitrogen, and quickly charged into the reactor, which was positioned close to the gas chromatograph. A precision 1.0 mL pressure lock syringe was used for sampling from the reactor. Samples of 0.5 mL were successively withdrawn in descending order from the highest point of the reactor and injected into the gas chromatograph. Just after starting the experiment, no CO_2 was expected since the reaction was still in the early stages of oxygen adsorption. There was thus no need for selective elution of the gases for the first week after initiation of the experiment. The samples were allowed to pass through both the Porapaq Q and Molecular Sieve for the first 7 d. Thereafter, the valve was used for all samples as the chance of getting CO_2 as one of the products increased. The valve was turned after 2.5 min from injection to bypass the carrier gas around the molecular sieve and then after 5.5 min to join the columns in series again.

RESULTS

The concentration profile was initially monitored on a daily basis; then the frequency was reduced until it was monitored twice per month. From the results obtained, the concentration of oxygen was calculated at varying heights. The results obtained were averaged over a period of 1 mo. This was to help eliminate the scatter in the results. In any event, the profiles changed very slowly.

Because the rate of reaction at room temperature was low and the columns had a small radius, we would expect the heat losses from the walls of the vessel to be such that the contents of the column would essentially remain at room temperature.

The averaged results were then plotted as a function of height. The graphical results for the two runs on different samples of coal done at different times are shown in Figures 1 and 2. From these results, it can be seen that, as the coal ages, oxygen penetration into the coal increases with time such that there is a general tendency for the oxygen concentration to increase with time at each measuring point. No carbon monoxide was detected during the course of the experiment. Traces of carbon dioxide were detected up to a height of 100 cm from the bottom from the fourth month until the end of the experiment, but the occurrence did not follow a consistent pattern, and no specific trend was noted.

For the first three months, there did not seem to be much change in the reactivity of the coal, as can be seen by the change in concentration with height, which does not vary much with time. The variation in the oxygen profile from the fourth month on seems to increase. This can be seen from the results in both Figures 1 and 2.

MATHEMATICAL MODELING

There has been much work in the literature regarding the search for an expression for the kinetics of gaseous oxygen with coal—the work by Itay (1983), and Smith (1993) can be cited as examples. It seems that the concentration dependence is effectively somewhere between half and first order. Further-

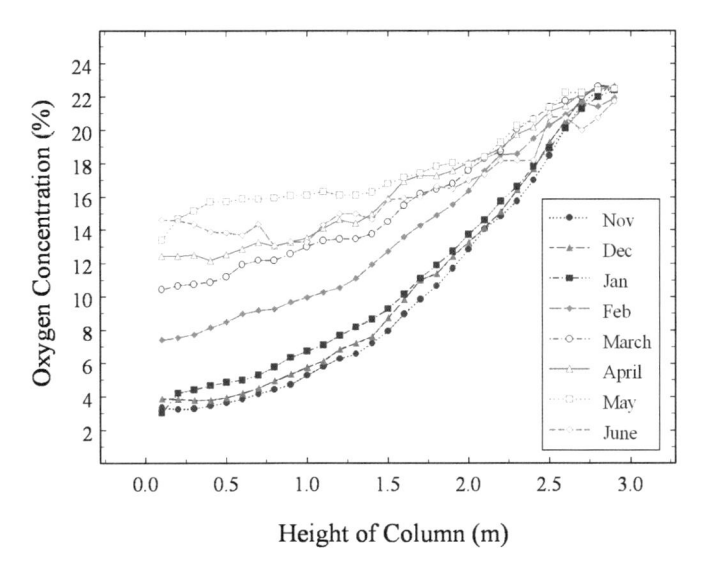

Figure 1. Oxygen concentration vs. height for different months: run 1.

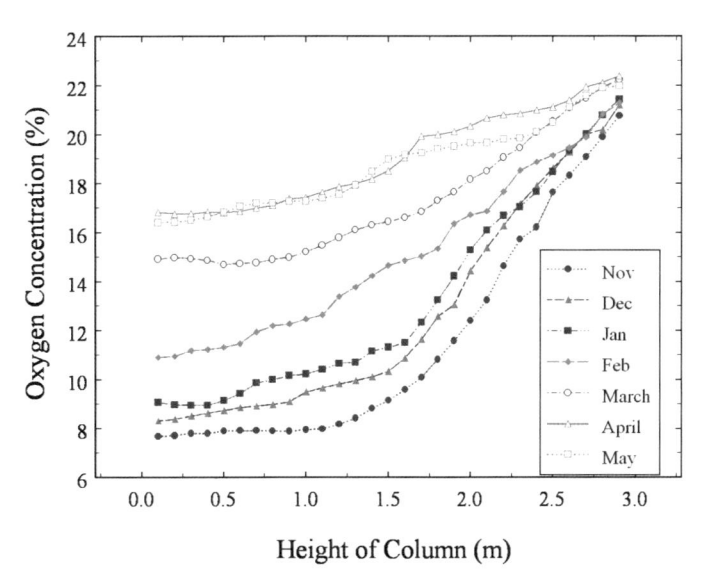

Figure 2. Oxygen concentration vs. height for different months: run 2.

more, as the coal reacts with oxygen, its rate decreases, and this is referred to as the aging effect. Finally, this rate is also quite sensitive to the moisture content of the coal. Thus, it is quite difficult to come up with an explicit rate expression that describes all these phenomena.

For the present, we do not need an explicit expression for the reaction rate of the coal with oxygen except to note that the coal deactivates with degree of oxidation (Itay, 1983; Smith, 1993). Because the reactivity declines with the degree of oxidation, the coal particles that react faster will deactivate more quickly. Thus, there is a tendency for particles that, for whatever reason, start off with different rates to tend toward similar reaction rates. This result will prove useful in trying to model the experimental results.

When the experiment is started, the distribution of coal reactivities can be assumed to be independent of height in the column, as the coal is uniformly distributed. However, because of the diffusion resistance in the bed, the oxygen toward the bottom tends to get depleted faster, as seen in the curves for November (Fig. 1) and December (Fig. 2), respectively. Thus, the reaction rate here will become lower than at the surface because of the dependence of the rate on the oxygen concentration. However, for the reasons discussed already, the rate of reaction at the surface will decay more rapidly than that deeper in the bed, which will tend to equalize the reaction rates. The overall effect of this behavior is that the assumption that the rate of reaction at every depth is eventually approximately the same becomes reasonable; that is, we can use a single rate that does not vary as a function of depth at every time interval. This rate will however change with time. Furthermore, because the coal properties change very slowly, while the gas-phase concentration profile is set up fairly quickly with time, we can make the pseudo-steady-state assumption—namely, we can model the concentration profile as a function of height only at each point in time.

The consequence of this is that at a given point in time, we can assume that the rate behaves as if it were zero-order with respect to oxygen. If we make this assumption, we are in a position to integrate the differential equation describing the behavior in the column, and we can test, using the experimental results, how good this assumption is. If this assumption appears to work well, we are then in a position to plot this average rate as a function of time and see what the aging effect is. We need to also note in doing this that as the coal ages, it will also be drying out, and this could also affect the rate of reaction. Finally, the apparatus is in an open laboratory, and so the average ambient temperature will tend to vary with time, again affecting the rates.

MODEL

The following assumptions were made in the mathematical model in this study:

(1) The material reacts with a zero-order reaction with respect to oxygen, and an average reaction rate over the whole bed is used at each point in time.

(2) The material reactivity decays with time (more accurately, with degree of oxidation and perhaps water concentration).

(3) The reaction conditions are isothermal (the bed of oxygen-absorbing material is kept at room temperature, and the heat of reaction is readily dissipated).

(4) The coal reactivity changes very slowly relative to the speed with which the gas-phase concentration profile sets itself up (pseudo–steady-state assumption).

(5) Fick's law of diffusion was assumed.

Thus, using the pseudo-steady-state assumption, that is the accumulation terms are negligible at a particular time t, we can write a mass balance. The rate of change of concentration with respect to height is given by:

$$D(d^2c/dx^2) = r, \quad (1)$$

where x is the height from the closed end of the bed, r is the average rate of oxidation at time t, and D is the diffusion coefficient and c is the gas-phase oxygen concentration in mole fraction units. To solve this equation, the following boundary condition has to be taken into consideration:

$$\text{at } x = 0, \, dc/dx = 0. \quad (2)$$

That is to say, there is no diffusion of oxygen across the sealed bottom of the bed. The solution under these conditions is:

$$\therefore c = (r/2D)x^2 + b. \quad (3)$$

The other boundary condition, which will not used here is:

$$\text{at } x = L, \, c = 0.21. \quad (4)$$

That is, the ambient oxygen concentration is always assumed to be 21%.

One can test the model by plotting the measured concentration c versus the square of the height, x^2, at a given time t. This has been done for both run 1 in Figure 3 and for run 2 in Figure 4.

As can be seen, the results fit the model reasonably well, and so this tends to vindicate the assumptions we have made. We note from the model Equation 3 that the slope of the lines is given by $r/2D$. Now, since the diffusion coefficient (D) should not vary with time, if we plot this slope versus time, we are effectively seeing how r, that is the average reaction rate, varies with time. This has been done in Figure 5 for both runs 1 and 2. Note that both of these graphs show the same type of S-shaped curve though the material in run 2 is ~30% less reactive. One could of course estimate the value of the diffusion coefficient, but as it remains constant, this is not necessary for the present purpose.

Now recall that r is the average reaction rate in the bed. The value of r in Equation 3 should also be that at the surface of the bed where the coal is exposed to the oxygen concentration in pure air. Even though this was in pure air, it is somewhat different from the situation found in other experiments where both oxygen and moisture concentrations were kept constant (Itay, 1983; Smith, 1993). In particular, other workers found a steady decline in rate, not this S shape. Now it is known (Itay, 1983;

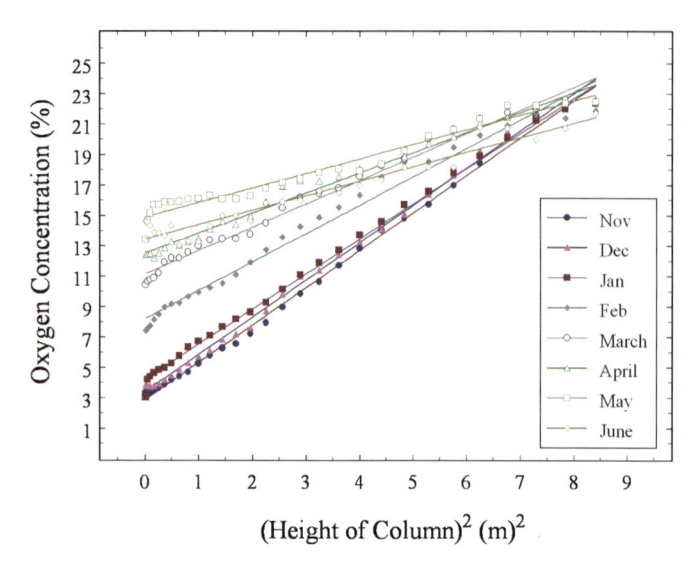

Figure 3. Test of the model for run 1.

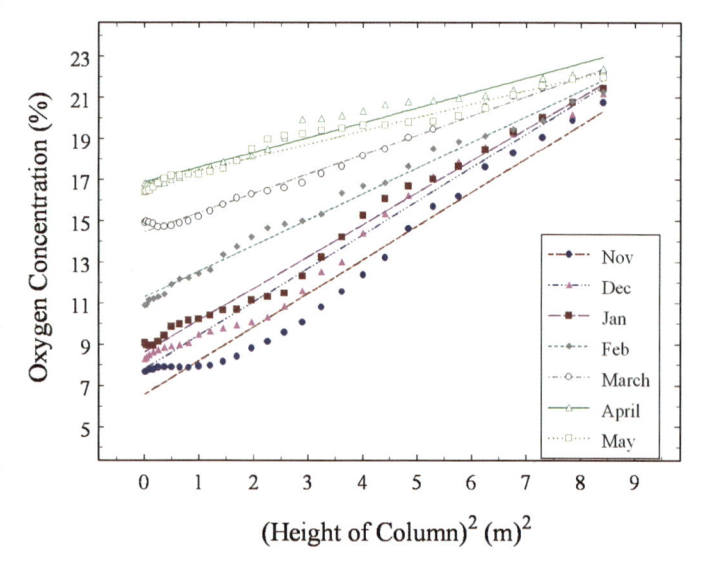

Figure 4. Test of the model for run 2.

Sondreal and Ellman, 1974) that the moisture content can have a large effect on the reaction rate. We also know that the initial coal was very moist because it came from the coal-washing plant, so we can speculate that during the first few months, the coal bed was drying out. It has also been shown that the reaction rate is actually depressed for large moisture contents, it goes through a maximum at intermediate values, and it then falls off as the coal dries even further (Sondreal and Ellman, 1974). We can thus speculate that during the first flat phase in Figure 5, the moisture loss and aging effects were acting in opposite directions, resulting in a fairly flat section. This was followed by a period in which the rate dropped rapidly because the effects of moisture loss and aging were in the same direction. In the flatter, final section, the coal moisture had reached equilibrium with the air and only the

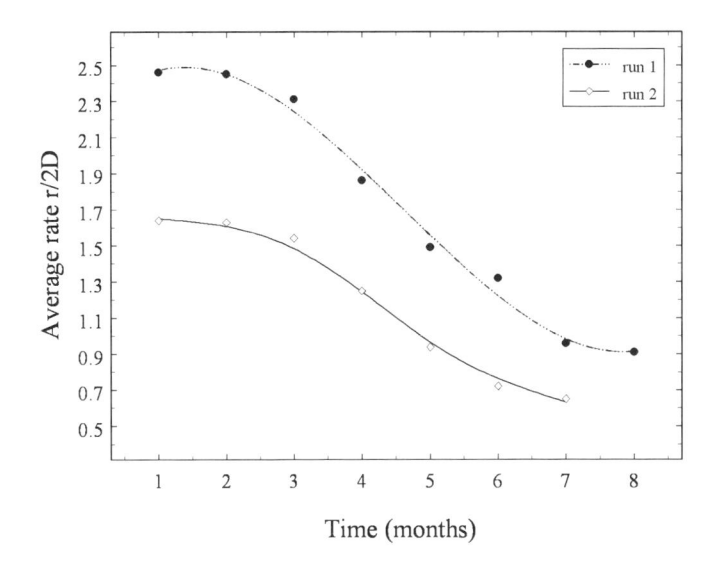

Figure 5. Variation of oxygen absorption rate parameter with time for runs 1 and 2.

aging effect was observed. Clearly, to verify this explanation, more work will need to be done.

The two coal samples used in the experiments were collected in the same way and from the same mine but at different times. The reason for the difference in reactivity as seen in Figure 5 is not clear. We speculate that it represents the natural variation between coal mined from different parts of the mine.

DISCUSSION

The apparatus fulfilled its purpose to measure the reactivity of the carbonaceous material with air and to see how this changed with time. While a particular coal was used in these experiments, there is no reason to believe that the apparatus could not be used for a very wide range of materials that undergo slow oxidation at room temperature.

If a gas chromatograph is not available to do the oxygen analysis, one could easily use a classical wet chemical method similar to those that were formerly used in standard combustion product analyses. This is because experiments (Itay, 1983; Smith, 1993) have shown that the only significant reaction of coal at room temperature is the absorption of oxygen.

In order to use a reactive layer to cover a bed of material that is liable to spontaneously combust, it is important to ensure that the layer will not undergo or initiate spontaneous combustion itself. What this means in practice is that we first need to ensure that natural convection does not occur, and this can be done by ensuring that the Rayleigh number is sufficiently small (Brooks et al., 1988a). This condition is essentially determined by the properties of the packed material, such as size, voidage, and thermal conductivity, and the air properties, but not its reactivity.

Once this is done, we need make sure that the layer cannot spontaneously combust due to conduction alone. How this is done is also outlined in Brooks et al. (1988). This condition depends on the rate of reaction at room temperature, the sensitivity of the reaction rate to temperature, and the heat of reaction, as well as the properties mentioned above.

It has been found (Smith and Glasser, 2005a, 2005b) that the sensitivity of the reaction rate to temperature (Arrhenius factor, activation energy) and the heat of reaction do not vary much from coal to coal. The factor that shows the biggest variation among different materials is the reactivity at room temperature, and this is what is measured in this apparatus. These results should enable scientists to estimate unknown parameters needed in the calculations suggested herein.

An apparatus based on these ideas has been used by Adamski (2003) to test various materials from the mine for their use as a covering at the Grootegeluk Mine in South Africa. Based on the results that were obtained, decisions were made on how to cover the waste material that had a history of spontaneous combustion as the opencast mine was backfilled.

A practical implication of this work is that in a less compact carbonaceous bed that is covered by a compacted layer of a different particle size, oxygen can diffuse into the bed from the top or any other section that is in contact with air. This oxygen will however be scavenged by the compacted layer if it has the correct properties (high reactivity, small particle size, and thick enough layer so that it can prevent oxygen reaching the uncompacted material).

As the coal in the compacted layer ages (reaction rate falls with time) at a given depth, the concentration of oxygen increases with time. Thus, as time increases, oxygen will eventually get into the material that is being protected. As a result, heat generation leading to spontaneous combustion cannot be ruled out. However, a more reactive and more compact layer than that used in this experiment could initially limit the oxygen penetration even more than shown in these experiments. However, as the coal ages, the oxygen penetration will increase with time so that a reactive layer covering a bed of coal can never be a long-term (permanent) solution for preventing spontaneous combustion.

It has also been suggested that a layer of fine sand or clay on the surface of a coal dump could also be used to stop spontaneous combustion. This however may not be true. The layer has the property of slowing the rate of oxygen ingress into the coal layer, but at the same time it slows the dissipation of any heat that is produced. Which of these factors predominates will decide whether this layer will help or actually make matters worse. This situation has been thoroughly analyzed by Raftopoulos (1993).

CONCLUSIONS

This apparatus constitutes a simple experimental technique for measuring the reactivity of coal or waste material and how it changes with time. The apparatus (or a larger equivalent) can easily handle run-of-mine material directly.

The fact that the experiments were carried out under conditions that were very close to those for a typical dump makes it an important diagnostic tool in evaluating what really happens in a coal dump.

In fact, the technique could be seen as a very simple, general technique for evaluating the reactivity of real coal samples. Because there is a strong interaction between many factors, such as moisture content and aging, it is probably less useful as a technique for fundamental studies.

A simple pseudo-steady-state model was developed to describe the diffusion of oxygen into a bed of reacting coal. The model basically depends on the fact that as the coal absorbs oxygen, its reactivity decreases. The experimental data correlate reasonably well with the model, and the results yielded curves that fit the model quite well. This suggests that the assumptions made in the model are reasonable.

ACKNOWLEDGMENTS

The financial support (a bursary for Kaitano) of the Foundation of Research Development (FRD) of South Africa is gratefully acknowledged. The suggestions of the reviewers are also gratefully acknowledged and have led to significant improvements to the paper.

REFERENCES CITED

Adamski, S.A., 2003, The Prevention of Spontaneous Combustion in Back-Filled Waste Material at Grootegeluk Coal Mine [Ph.D. thesis]: Johannesburg, South Africa, University of the Witwatersrand, 153 p.

Bhattacharyya, K.K., 1972, Role of desorption of moisture from coal in its spontaneous heating: Fuel, v. 51, no. 3, p. 214, doi: 10.1016/0016-2361(72)90084-1.

Brooks, K., and Glasser, D., 1986, A simplified model of spontaneous combustion in coal stockpiles: Fuel, v. 65, p. 1035–1041, doi: 10.1016/0016-2361(86)90163-8.

Brooks, K., Balakotaiah, V., and Luss, D., 1988a, Effect of natural convection on spontaneous convection of coal stockpiles: American Institute of Chemical Engineers Journal, v. 34, no. 3, p. 353–365.

Brooks, K., Svanas, N., and Glasser, D., 1988b, Evaluating the risk of spontaneous combustion in coal stockpiles: Fuel, v. 67, p. 651–656, doi: 10.1016/0016-2361(88)90293-1.

Glasser, D., and Bradshaw, S., 1990, Spontaneous combustion in beds of coal, in Cheremisinoff, N.J., ed., Handbook of Heat and Mass Transfer. 4: Advances in Reactor Design and Combustion Science: Houston, Gulf Publishing, p. 1071–1148.

Itay, M., 1983, The Low Temperature Oxidation of Coal: Its Kinetics and Implications for Spontaneous Combustion [Ph.D. thesis]: Johannesburg, South Africa, University of the Witwatersrand, 262 p.

Raftopoulos, E., 1993, Multiplicity Features of Simple Models for Spontaneous Combustion in Coal Stockpiles [Ph.D. thesis]: Johannesburg, South Africa, University of the Witwatersrand, 415 p.

Schmal, D., Duyzer, D.H., and van Heuven, J.W., 1985, A model for the spontaneous heating of coal: Fuel, v. 64, p. 963–972, doi: 10.1016/0016-2361(85)90152-8.

Smith, M.A., 1993, A Laboratory Study of the Low Temperature Oxidation of Coal [Ph.D. thesis]: Johannesburg, South Africa, University of the Witwatersrand, 277 p.

Smith, M.A., and Glasser, D., 2005a, Spontaneous combustion of carbonaceous stockpiles. Part I: The relative importance of various intrinsic coal properties and properties of the reaction system: Fuel, v. 84, p. 1151–1160.

Smith, M.A., and Glasser, D., 2005b, Spontaneous combustion of carbonaceous stockpiles. Part II: Factors affecting the rate of the low-temperature oxidation reactions: Fuel, v. 84, p. 1161–1170, doi: 10.1016/j.fuel.2004.12.005.

Sondreal, E.A., and Ellman, R.C., 1974, Laboratory determination of factors affecting storage of North Dakota lignite: U.S. Bureau of Mines Report BM-RI-7887, 91 p.

MANUSCRIPT ACCEPTED BY THE SOCIETY 7 MARCH 2007

The origin of gas-vent minerals: Isochemical and mass-transfer processes

Glenn B. Stracher*

Division of Science and Mathematics, East Georgia College, Swainsboro, Georgia 30401, USA

ABSTRACT

Minerals derived from coal-fire gas and found encrusting vents and ground fissures form by isochemical or mass-transfer processes. Isochemical processes include sublimation and gas-liquid–solidification (GLS). Evidence for sublimation includes elements found in both the gas and minerals as well as the occurrence of euhedral crystals. Textural evidence for GLS includes drip and flow condensates.

Mass-transfer processes are more complex and include gas-altered substrate (GAS), gas-liquid–altered substrate (GLAS), gas-liquid–precipitation (GLP), and gas reaction ± liquid-solidification (GRLS) processes. During GAS, the depositional substrate is altered by reaction with one or more components of coal-fire gas, whereas during GLAS, the substrate is altered by reaction with a liquid solution. Evidence for GLAS includes vesicles and minerals with ions acquired from the underlying substrate. GRLS is characterized by reaction among select-gas components, ultimately responsible for mineralization. Differentiating GAS, GLAS, and GRLS from one another is difficult if the mineralized by-products of combustion are dissolved by water or windswept from an altered substrate. Amorphous bulb-like masses at the end of elongated crystals are evidence for GLP.

Minerals reported to be coal-fire sublimates on the basis of textural evidence may actually be the reaction products of select-gas components exhaled from a vent or fissure. Reactions involving coal-fire gas components have never been identified. Additional mineral forming reactions based on isochemical and mass-transfer processes are possible. Some coal-fires minerals may be vectors for the transmission of toxins to humans by inhaled dust particles or by food grown in soils that contain these minerals.

Keywords: gas vents, gas-vent minerals, coal fires, coal-fire gas, fumarolic minerals.

INTRODUCTION

Gas vents associated with coal fires are circular to semicircular conduits in the earth or openings at its surface through which the by-products of coal combustion are transported. Fissures associated with such fires are elongated cracks in the earth or at its surface through which these by-products pass. Near or at the surface, these vents and fissures are often encrusted with minerals commonly assumed to form by the condensation of coal-fire gas. As the gas is exhaled at the surface it cools. Subsequently, one or more minerals may sublimate or a liquid may form, which, when cooled further, solidifies to form a mineral assemblage

---

*stracher@ega.edu

(Lapham et al., 1980; Stracher, 1995; Witzke, 1997a). These same processes may also occur at volcanic-gas vents (Stoiber and Rose, 1974), at times resulting in the formation of mineral assemblages similar to those associated with coal-fire gas. However, other complex thermochemical processes involving multiple reactions may be responsible for the gas-related origin of minerals at either type of vent.

The purpose of this paper is to delineate the processes by which mineral assemblages form at or in the vicinity of coal-fire gas vents and ground fissures as a consequence of combustion. For the sake of brevity, reference herein is made to gas vents only, since analogous processes occur at both vents and fissures. Although some such processes are in fact known to occur, others are hypothesized based on plausible-reaction paths and field evidence.

MINERAL-FORMING PROCESSES

One or more processes may be responsible for the crystallization of a mineral assemblage at a coal-fire–gas vent. The processes that occur depend on one or more combinations of gas and substrate chemistry, the presence or absence of aqueous solutions, the pressure and exhalation temperature of the gas, and the temperature of the substrate encountered by the gas during exhalation at the surface (Stracher, 1995; Stracher et al., 2005a).

Isochemical-Mineralization Processes

Condensation is the isochemical process whereby a gas undergoes a change in state during cooling and transforms directly into a liquid or a solid (Hawley, 1971, p. 231). Two condensation-related processes, discussed here, may occur during the crystallization of minerals derived from coal-fire gas. The first, sublimation, is strictly a condensation process. The second involves condensation to a liquid followed by solidification (freezing).

Sublimation

Sublimation occurs when a gas component exhaled from a coal-fire vent rapidly cools below the liquid-to-solid transformation temperature (supercools) and subsequently condenses to a solid without an intervening liquid state (Hawley, 1971, p. 832). Supercooling occurs in response to a temperature gradient between the gas and the substrate (heat sink) encountered by the gas at or near the surface. Heat sinks include rock, sediment, vegetation, the atmosphere, etc. (Stracher, 1995). During sublimation, the gas isochemically condenses to a mineral. The process is represented by the exothermic reaction

$$X(g) \xrightarrow[\text{Supercools}]{\text{Sublimates}} X(s) \qquad (1)$$

where $X(g)$ and $X(s)$ designate an element or compound in the gas and solid states, respectively. If a mineral encrusting a vent formed by sublimation, then the chemical composition of the mineral and the substrate on which it sublimated are independent of one another. In other words, no mass transfer occurs between the sublimating gas and the substrate as solidification occurs. Evidence for sublimation includes mineral encrustations with elements or compounds that are contained in the gas and euhedral crystals as well as crystal faces that do not touch, suggestive of growth in a nonrestrictive environment (Stracher et al., 2005a).

Several minerals have been reported as the sublimation products of coal-fire gas, including hydrous sulfates (Lausen, 1928), downeyite (SeO_2) (Finkelman and Mrose, 1977), orthorhombic sulfur (Stracher, 1995), coccinite (HgI_2) (Witzke, 1997b), and salammoniac (NH_4Cl) (Stracher et al., 2005a). At first glance, minerals that encrust gas vents may appear to be sublimation products of the gas, especially if the minerals appear to be powdery and the substrate that the minerals were collected from appears to be unaltered. This was the case in a recent study by Stracher et al. (2005a) of mineral assemblages collected from the Wuda Coalfield, Inner Mongolia. The gas-vent assemblages occurred on quartzofeldspathic sand and sandstone that appeared chemically unaltered in the field (Fig. 1). However, X-ray diffraction, energy-dispersive spectrometry, gas chromatography, and micro-Dumas analyses coupled with textural evidence from scanning electron microscope (SEM) images revealed that a more complex sequence of events occurred. These included condensation, hydrothermal alteration and crystallization from solution, fluctuating vent temperatures, boiling, and dehydration reactions. Consequently, it is possible that minerals originally described as sublimates may have formed by one or more complex thermochemical process.

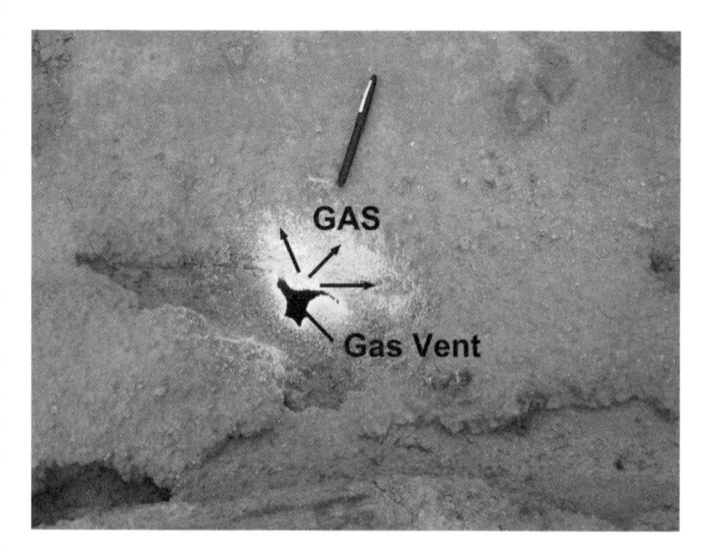

Figure 1. Alunogen, coquimbite, voltaite, and an unidentified phase comprise the white-colored mineral assemblage at a gas vent in the Wuda Coalfield, Inner Mongolia. Gas temperature immediately inside the vent was 269 °C. The underlying substrate is quartzofeldspathic sand and sandstone that appeared chemically unaltered in the field. Sample analyses and scanning electron microscope (SEM) imaging suggest otherwise. The distribution of the assemblage suggests that mineralization occurred in response to a gas moving toward the upper right in the photo. Image is from Stracher et al. (2005a), used with permission of the Mineralogical Society of America.

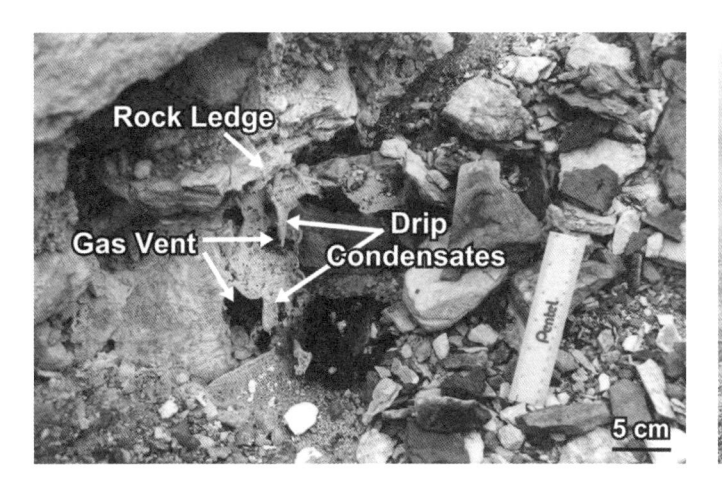

Figure 2. Gas-liquid–solidification (GLS) process suggested by unidentified "drip condensates" on a rock ledge overhanging a gas vent at the Centralia mine fire, Pennsylvania. Gas temperature immediately inside the vent was 441 °C. The drip texture suggests that a gas component exhaled from the vent condensed to a liquid that subsequently solidified.

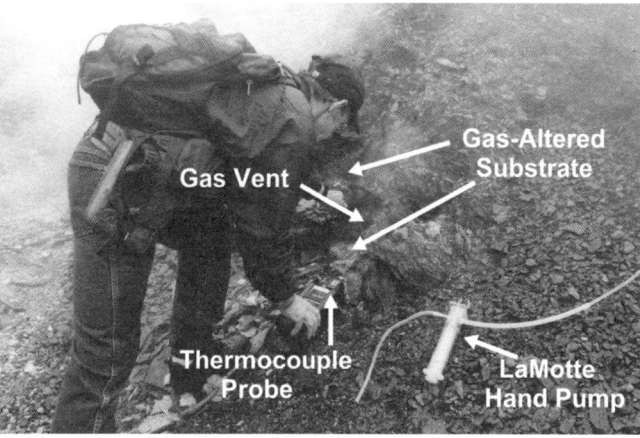

Figure 3. Possible gas-altered substrate (GAS) consisting of red clinker (light color) surrounding a coal-fire–gas vent at the Centralia mine fire, Pennsylvania. Hot gas altered the Pennsylvanian Llewellyn Formation (black shale) surrounding the vent. G.B. Stracher is measuring the vent temperature with a thermocouple probe. A LaMotte hand pump on the ground was used to collect a gas sample for analysis.

Gas-Liquid–Solidification

Gas-liquid–solidification (GLS) occurs when a gas component first condenses to a liquid, and then, as cooling continues, the liquid solidifies. Drip (stalactite) and flow (ripple) textures observed on rocks overhanging gas vents or on nearby rock facings are evidence for GLS (Fig. 2). Because they resemble dripstones and flowstones found in caves, mineral assemblages with drip and flow textures are aptly called "drip condensates" and "flow condensates," respectively. An isochemical GLS process is represented by the exothermic reaction

$$X(g) \xrightarrow{\text{Condenses}} X(l) \xrightarrow{\text{Solidifies}} X(s) \qquad (2)$$

where X(l) designates an element or compound in the liquid state.

Mass-Transfer–Mineralization Processes

Mineralization processes at coal-fire–gas vents that involve mass transfer are more complex than isochemical ones. A variety of such processes is possible, several of which are discussed next.

Gas-Altered Substrate

Gas-altered substrate (GAS) occurs when a hot-gas component(s) exhaled at a vent reacts with the substrate and alters it (Figs. 3 and 4). This is analogous to pneumatolytic alteration associated with igneous activity. Consequently,

$$X(g) \xrightarrow[\text{substrate}]{\text{Reacts with}} \text{Altered substrate} \pm Y(g) \uparrow \qquad (3)$$

where Y(g) is the composition of the remaining gas. Clinker, rock baked and oxidized by combustion, is suggestive of GAS or other substrate-altering processes, as discussed next.

Figure 4. Possible gas-altered substrate (GAS) consisting of clinker or red-baked sandstone and red-black–nonmarine shale within the Ferron Sandstone Member of the Mancos Shale, Emery Coalfield, Utah. Note the reverse fault in the Ferron sandstone and its collapse into the area where the I coal bed has completely burned, at knee level to G.B. Stracher. Image is from Stracher et al. (2005b) with modifications, used with permission of The Geological Society of America.

Gas-Liquid–Altered Substrate

Gas-liquid–altered substrate (GLAS) occurs when a liquid condensed from the gas reacts with the substrate and chemically alters it, resulting in the crystallization of minerals on the altered substrate. For example, if a liquid condensed from the gas reacts with the substrate, thereby altering the chemical composition of the liquid (and substrate) prior to mineralization, the reaction path may be

$$X(g) \xrightarrow{\text{Condenses}} X(l) \xrightarrow[\text{substrate}]{\text{Reacts with}} Y(l) + \text{Altered substrate}$$
$$Y(l) \xrightarrow[\text{Isochemical}]{\text{Solidifies}} Y(s) \qquad (4)$$

where the reaction of X(l), perhaps an acidic solution, with the substrate forms a different liquid Y(l), which then isochemically solidifies to form the mineral Y(s). On the cryptocrystalline scale of observation, the former presence of a liquid Y(l) may be revealed by vesicles observed in SEM images (Fig. 5; Stracher et al., 2005a) or mineral chemistry, as discussed below.

If Y(s) is hydrous, and it dehydrates or hydrates, perhaps due to a temperature increase or decrease, respectively, at the vent, then Equation 4 becomes

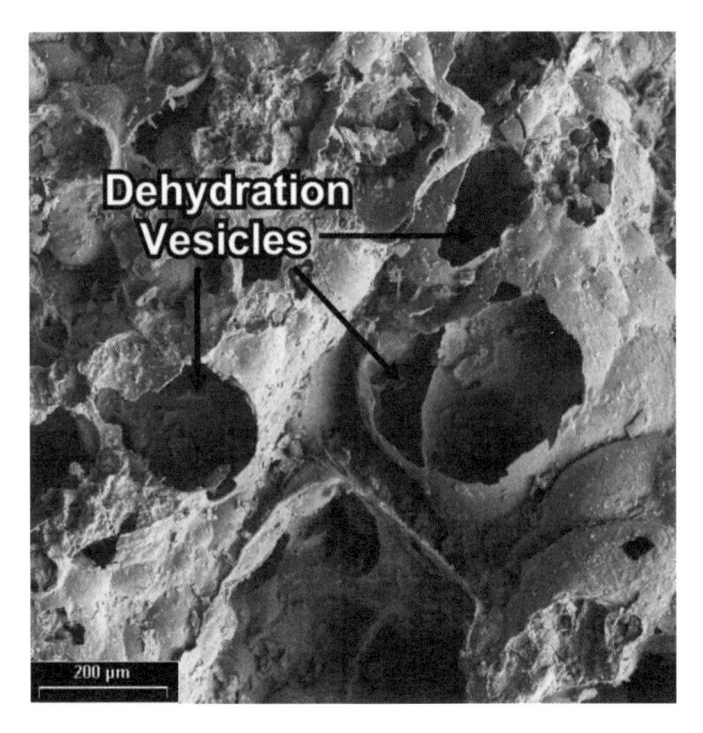

Figure 5. Gas-liquid–altered substrate (GLAS) process revealed by a cryptocrystalline mass of millosevichite and alunogen collected 1–5 cm from the outside edge of a gas vent in the Wuda Coalfield, Inner Mongolia (scanning electron microscope image). Gas temperature immediately inside the vent was 110 °C. Vesicular texture and Al ions in the mineral assemblage, collected from a quartzofeldspathic-sandstone substrate, suggest that reaction 5 had occurred, as discussed in the text. Image is from Stracher et al. (2005a) with modifications, used with permission of the Mineralogical Society of America.

$$X(g) \xrightarrow{\text{Condenses}} X(l) \xrightarrow[\text{substrate}]{\text{Reacts with}} Y(l) + \text{Altered substrate}$$
$$Y(l) \xrightarrow[\text{Isochemical}]{\text{Solidifies}} Y(s) \qquad (5)$$
$$Y(s) \xrightarrow[\text{Hydrates}]{\text{Dehydrates}} Z(s)$$

where the mineral Z(s) is the dehydration or hydration product of Y(s).

In the Wuda Coalfield of Inner Mongolia, a sample encrusting a gas vent was collected from quartzofeldspathic sandstone. The sample was vesicular (Fig. 5) and contained both alunogen, $Al_2(SO_4)_3 \cdot 17H_2O$, and millosevichite, $Al_2(SO_4)_3$. The vesicles, Al ions likely acquired from the sandstone substrate, and the simultaneous occurrence of these minerals suggest that the process given by Equation 5 occurred at the vent (Stracher et al., 2005a). The occurrence of both millosevichite and alunogen and the vesicles suggest the dehydration reaction

$$\underset{\text{Alunogen}}{Al_2(SO_4)_3 \cdot 17H_2O} \rightarrow \underset{\text{Millosevichite}}{Al_2(SO_4)_3} + 17H_2O\uparrow$$

The vesicular texture may have developed as liquid water or water vapor escaped from the assemblage. In addition, the incomplete breakdown of alunogen, evident from its occurrence with millosevichite, is the result of a temperature decrease subsequent to the increase that promoted dehydration. As suggested by Stracher et al. (2005a), the fluctuating temperatures and concomitant effects on the assemblage are accounted for by variations in combustion and atmospheric conditions and possibly removal (temperature decrease) of the assemblage from the vent during sample collecting. It is possible that the reaction path involving the vesicular assemblage found at Wuda is more complicated than that presented above. For example, millosevichite may have initially formed at the vent and, in the presence of water vapor exhaled at the surface or atmospheric moisture, hydrated to alunogen as the vent temperature fell, followed by the incomplete dehydration of this mineral as described above.

Gas-Liquid–Precipitation

Gas-liquid–precipitation (GLP) occurs when a gas first condenses to a liquid, from which a solid precipitates, followed by solidification of the remaining liquid. Described by Wagner and Ellis (1964), the first and only recorded *natural* occurrence of this process (Fig. 6) was observed by Finkelman et al. (1974) in association with burning anthracite in Forestville, Pennsylvania. According to Finkelman et al. (1974) and Lapham et al. (1980), sulfur droplets condensed from an $S_2(g)$ component in the anthracite gas. The droplets then acted as a catalyst, and absorbed Ge from the gas to form a S-Ge solution. Once the solution was saturated in Ge, germanium disulfide (GeS_2) precipitated. The process resulted in the nucleation of elongate GeS_2 crystals capped by bulb-like, amorphous sulfur that solidified from the remaining liquid-sulfur droplets, S(l), at the end of the GeS_2 crystals. The process is summarized by the reactions

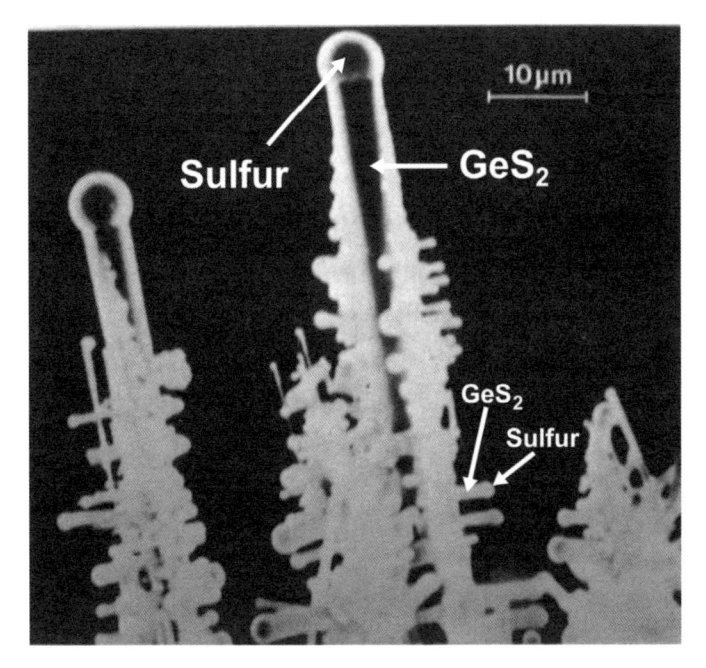

Figure 6. Gas-liquid–precipitation (GLP) process observed in a scanning electron microscope (SEM) image of elongate GeS$_2$ crystals capped by a bulb of amorphous sulfur. Note the smaller rods of GeS$_2$ capped by the small sulfur bulbs. Image is from Mineral Resources Report 78 (Lapham et al., 1980) with modifications, used with permission of the Pennsylvania Geological Survey.

$$\frac{1}{2}n_1S_2(g) \xrightarrow{\text{Condenses}} n_1S(l) \text{ (catalyst)}$$

$$\text{Ge (absorbed)} + n_1S(l) \xrightarrow{\text{Solution}} X(l) \quad (6)$$

$$\text{Ge} + 2S(l) + n_2S(l) \xrightarrow[X(l)]{\text{In solution}} GeS_2 \text{ (precipitate)} + n_2S(l) \text{ (droplets)}$$

$$n_2S(l) \xrightarrow{\text{Isochemical}} n_2S(s) \text{ (amorphous bulbous masses)}$$

where $X(l)$ designates the liquid phase containing n_1 moles of $S(l)$ and absorbed Ge, and the total number of moles of $S(l)$ in $X(l)$ is $n_1 = 2 + n_2$, where 2 is the number of moles of $S(l)$ that react with each mole of absorbed Ge to form GeS$_2$ and n_2 designates the *remaining* number of moles of $S(l)$ from $X(l)$ that condense to amorphous $S(s)$.

Gas Reaction ± Liquid-Solidification

Gas reaction ± liquid-solidification (GRLS) is a hypothetical mineral-forming process initiated by gas reactions during or after exhalation at the vent. Such reactions, however, are undocumented in the literature. GRLS consists of one or more of the steps in reactions 1 through 5. Many gas-vent assemblages attributed to sublimation may have actually formed by GRLS reaction 7 below. For example, salammoniac (NH$_4$Cl) occurs globally in association with coal fires as well as volcanic vents and burning-oil shale. It is often cited in the literature as a sublimate (Cole, 1974; Gaines et al., 1997; Lapham et al., 1980; West, 2001). Textural evidence offered in support of this includes euhedral crystals that do not share faces (Stracher et al., 2005a). Although salammoniac contains elements commonly found in the components of coal-fire gas, the salammoniac molecule is absent from gas analyses. Consequently, salammoniac may be the solid-reaction product of N-, H-, and Cl-bearing gas components. In general, the exothermic reaction for such mineralization processes is

$$\sum SGC \xrightarrow{\text{React}} X(s) \pm Z(g) \uparrow \quad (7)$$

where SGC denotes select-gas components that react to produce an element or compound $X(s)$ like salammoniac and the liquid and gas phases $Y(l)$ and $Z(g)$, respectively. This is not sublimation in the strictest sense of the definition; note that mineralization is not necessarily isochemical.

Additional plausible, although also undocumented GRLS processes include

$$\sum SGC \xrightarrow{\text{React}} X(l) \xrightarrow{\text{Solidifies}} X(s) \quad (8)$$

where the liquid-reaction product $X(l)$ solidifies to $X(s)$;

$$\sum SGC \xrightarrow{\text{React}} X(l) \xrightarrow[\text{substrate}]{\text{Reacts with}} Y(s) + \text{Altered substrate} \quad (9)$$

where $X(l)$ reacts with the substrate to form the solid $Y(s)$; and

$$\sum SGC \xrightarrow{\text{React}} \pm X(s) \pm Y(l) \pm Z(g)$$
$$\xrightarrow[\text{or each other}]{\text{React with substrate}} W(s) \pm \text{Altered substrate} \quad (10)$$

where the solid-, liquid-, and gas-reaction products $X(s)$, $Y(l)$, and $Z(g)$ react with the substrate or each other to form the solid $W(s)$.

DISCUSSION

The previous reactions exemplify mineral-forming processes at gas vents. They are not all inclusive, and additional processes are possible. For example, consider reaction 2 rewritten as

$$X(g) \xrightarrow{\text{Condenses}} X(l) \begin{array}{c} \nearrow Y_1(s) \\ \searrow Y_2(s) \end{array}$$

In this case, subsequent to solidification, two solid phases $Y_1(s)$ and $Y_2(s)$, perhaps in a drip or flow condensate, nucleated simultaneously from a liquid phase at its eutectic temperature.

The absence of a mineral assemblage at or in the vicinity of a coal-fire–gas vent does not imply that it never formed. Many assemblages consist of water-soluble–powdery materials that are easily dissolved by rainwater or eroded by wind. Consequently, if the mineralized by-products of combustion have been eroded, it may be extremely difficult to determine whether an altered substrate was formed by GAS, GLAS, or GRLS.

Determining the reactions among select-gas components of a coal fire during and after exhalation is a formidable task yet to be accomplished. As burning proceeds, gas composition fluctuates in accordance with chemical diversity in the burning coal, as well as in sediment, rock, and aqueous solutions encountered by the gas en route to the surface. Exchange reactions involving the gas, sediment, rocks, and solutions continuously change gas chemistry, thereby influencing, along with temperature and gas-component–partial pressures, the mineral assemblages that crystallize in association with the gas.

The identification of trace elements in minerals that formed as a consequence of coal combustion may reveal certain minerals that are vectors for the transmission of toxins to humans through inhalation of dust particles or even through food grown in soils that contain these minerals. The proliferation and catastrophic effects of coal fires subsequent to the industrial revolution in Europe include trace elements that promote air, soil, and water pollution (Stracher and Taylor, 2004). These pollutants have destroyed natural habitats, killed people, forced communities to relocate, and are responsible for human diseases including hyperkeratosis (arsenic poisoning), dental and skeletal fluorosis (osteosclerosis), thallium poisoning, lung cancer, and pulmonary heart disease (Johnson et al., 1997, p. 19; World Resources Institute, 1999, p. 63–67; Finkelman et al., 1999, 2001, 2002; Finkelman, 2004).

ACKNOWLEDGMENTS

The author thanks Tammy P. Taylor of Los Alamos National Laboratory, New Mexico, and Robert B. Finkelman, Emeritus, U.S. Geological Survey, Reston, Virginia, for their reviews of the manuscript. I am also grateful to Robert B. Finkelman for our discussions about the only known naturally occurring GLP process.

REFERENCES CITED

Cole, D.I., 1974, Observations on a burning cliff, *in* Proceedings of the Dorset Natural History and Archaeological Society for 1974: Dorchester, Dorset, UK, Dorset Natural History and Archaeological Society, v. 96, p. 16–19, http://www.dor-mus.demon.co.uk (accessed July 2005).

Finkelman, R.B., 2004, Potential health impacts of burning coal beds and waste banks, *in* Stracher, G.B., ed., Coal Fires Burning Around the World: A Global Catastrophe: International Journal of Coal Geology, v. 59, no. 1–2, p. 19–24.

Finkelman, R.B., and Mrose, M.E., 1977, Downeyite, the first verified natural occurrence of SeO_2: The American Mineralogist, v. 62, p. 316–320.

Finkelman, R.B., Larson, R.R., and Dwornik, E.J., 1974, Naturally occurring vapor-liquid-solid (VLS) whisker growth of germanium sulfide: Journal of Crystal Growth, v. 22, p. 159–160, doi: 10.1016/0022-0248(74)90132-8.

Finkelman, R.B., Belkin, H.E., and Zheng, B., 1999, Health impacts of domestic coal use in China: Proceedings of the National Academy of Sciences of the United States of America, v. 96, p. 3427–3431, doi: 10.1073/pnas.96.7.3427.

Finkelman, R.B., Skinner, H.C., Plumlee, G.S., and Bunnell, J.E., 2001, Medical geology: Geotimes, v. 46, no. 11, p. 21–23.

Finkelman, R.B., Orem, W., Castranova, V., Tatu, C.A., Belkin, H.E., Zheng, B., Lerch, H.E., Marharaj, S.V., and Bates, A.L., 2002, Health impacts of coal and coal use: Possible solutions: International Journal of Coal Geology, v. 50, p. 425–443, doi: 10.1016/S0166-5162(02)00125-8.

Gaines, R.V., Skinner, H.C., Foord, E.E., Mason, B., and Rosenzweig, A., 1997, Dana's New Mineralogy (8th edition): New York, John Wiley and Sons, Inc., 1819 p., http://webmineral.com/danaclass.shtml (accessed July 2005).

Hawley, G.G., 1971, editor, The Condensed Chemical Dictionary (8th edition): New York, New York, Van Nostrand Reinhold Company, 971 p.

Johnson, T.M., Liu, F., and Newfarmer, R.S., 1997, Clear Water, Blue Skies: China's Environment in the New Century: Washington, D.C., World Bank, 114 p.

Lapham, D.M., Barnes, J.H., Downey, W.F., Jr., and Finkelman, R.B., 1980, Mineralogy associated with burning anthracite deposits in eastern Pennsylvania: Harrisburg, Pennsylvania Geological Survey, Mineral Resource Report 78, 82 p.

Lausen, C., 1928, Hydrous sulphates formed under fumarolic conditions at the United Verde Mine: The American Mineralogist, v. 13, no. 6, p. 203–229.

Stoiber, R.E., and Rose, W.I., Jr., 1974, Fumarole encrustations at active Central American volcanoes: Geochimica et Cosmochimica Acta, v. 38, p. 495–516, doi: 10.1016/0016-7037(74)90037-4.

Stracher, G.B., 1995, The anthracite smokers of eastern Pennsylvania: Stability diagram by TL analysis: Mathematical Geology, v. 27, no. 4, p. 499–511, doi: 10.1007/BF02084424.

Stracher, G.B., and Taylor, T.P., 2004, Coal fires burning out of control around the world: Thermodynamic recipe for environmental catastrophe, *in* Stracher, G.B., ed., Coal Fires Burning Around the World: A Global Catastrophe: International Journal of Coal Geology, v. 59, no. 1–2, p. 7–17.

Stracher, G.B., Prakash, A., Schroeder, P., McCormack, J., Zhang, X., Van Dijk, P., and Blake, D., 2005a, New mineral occurrences and mineralization processes: Wuda coal-fire gas vents of Inner Mongolia: The American Mineralogist, v. 90, p. 1729–1739, doi: 10.2138/am.2005.1671.

Stracher, G.B., Tabet, D.E., Anderson, P.B., and Pone, J.D.N., 2005b, Utah's state rock and the Emery coalfield: Geology, mining history, and natural burning coal beds, *in* Pederson, J.L., and Dehler, C.M., eds., Interior Western United States: Geological Society of America Field Guide, v. 6, p. 199–210.

Wagner, R.S., and Ellis, W.C., 1964, Vapor-liquid-solid mechanism of single crystal growth: Applied Physics Letters, v. 4, no. 5, p. 89–90, doi: 10.1063/1.1753975.

West, I.M., 2001, Burning Beach, Burning Cliffs and the Lyme Volcano: Oil-Shale Fires; Geology of the Dorset Coast: Southampton, UK, School of Ocean and Earth Sciences, Southampton University, p. 1–16, http://www.soton.ac.uk/~imw/kimfire.htm (accessed July 2005).

Witzke, T., 1997a, A new aluminum chloride mineral from Oelsnitz near Zwickau, Saxony, Germany: Neues Jahrbuch für Mineralogie-Monatshefte, v. 7, p. 301–308.

Witzke, T., 1997b, New data on the mercury iodide mineral coccinite, HgI_2: Neues Jahrbuch für Mineralogie-Monatshefte, v. 11, p. 505–510.

World Resources Institute, 1999, 1998–1999 World Resources: A Guide to the Global Environment, Environmental Change and Human Health: New York, New York, Oxford University Press, 369 p.

MANUSCRIPT ACCEPTED BY THE SOCIETY 7 MARCH 2007

Combustion metamorphic events resulting from natural coal fires

Ellina V. Sokol*
Nina I. Volkova
Institute of Geology and Mineralogy, Siberian Branch of the Russian Academy of Sciences, Novosibirsk 630090, Russia

ABSTRACT

Fossil-fuel fires in coal-bearing and bituminous complexes and associated combustion metamorphic transformations of sedimentary protoliths have been observed adjacent to many coal and oil deposits worldwide. The geologic and topographic features governing the distribution of fossil-fuel fires are similar for the majority of combustion metamorphic complexes. There are more than 40 such complexes in Europe, Asia, North America, Australia, and New Zealand. Combustion metamorphism is a striking geologic phenomenon, and the geologic history of Tertiary and Quaternary sedimentary basins prior to the ignition of any fossil fuels they contain determines the characteristics of future combustion metamorphic rocks. The evolutionary trend of combustion metamorphic systems is: formation of a fossil fuel → ignition and combustion of the fuel → decomposition and alteration of adjacent sedimentary strata and minerals, respectively → formation of pyrometamorphic rocks → retrograde alteration. The thermal energy for high-temperature and low-pressure metamorphism is supplied by the combustion of coal, gas, oil, or bitumen.

Keywords: coal fires, bitumen fires, combustion phenomenon, geomorphic and climate controls, combustion metamorphic complexes, coal basins.

INTRODUCTION

Combustion metamorphic or pyrometamorphic (annealed, baked, melted, and slag-like) rocks, which are produced by the in situ, natural combustion of fossil fuel, are typical geological features that are observed in many coal deposits, some oil basins, and regions of bituminous sedimentation. Such rocks are known in India (Venkatesh, 1952; Chatterjee, 1988; Stracher and Taylor, 2004), Iran (McLintock, 1932), Iraq (Basi and Jassim, 1974), Tajikistan (Bogoslovsky, 1842; Ermakov, 1935; Novikov and Suprychev, 1986; Srebrodolsky, 1989; Novikov, 1993), Uzbekistan and Kyrgyzstan (Abubakirov, 1962; Zbarsky, 1963), Kazakhstan (Kalugin et al., 1991), Mongolia (Pokrovsky, 1949), China (Saidov, 1956; de Boer et al., 2001; Stracher and Taylor, 2004), Indonesia (Whitehouse and Mulyana, 2004), Israel (Bentor et al., 1963; Kolodny and Gross, 1974; Gross, 1977; Matthews and Kolodny, 1978; Matthews and Gross, 1980; Burg et al., 1991, 1999; Ron and Kolodny, 1992; Gur et al., 1995), Jordan (Khoury and Milodowski, 1992), the United States (Arnold and Anderson, 1907; Fermor, 1918; Sigsby, 1966; Bentor et al., 1981;

---

*sokol@uiggm.nsc.ru

Sokol, E.V., and Volkova, N.I., 2007, Combustion metamorphic events resulting from natural coal fires, *in* Stracher, G.B., ed., Geology of Coal Fires: Case Studies from Around the World: Geological Society of America Reviews in Engineering Geology, v. XVIII, p. 97–115, doi: 10.1130/2007.4118(07). For permission to copy, contact editing@geosociety.org. ©2007 The Geological Society of America. All rights reserved.

Hooper, 1982; Bentor, 1984; Foit et al., 1987; Cosca et al., 1989; Clark and Peacor, 1992; Heffern and Coates, 2004), Canada (Church et al., 1979; Bustin and Mathews, 1982; Goodarzi and Jerzykiewicz, 1986), Venezuela (Moticska, 1977, 1978); Colombia (Alvarez and Gomez, 1984), England (Cole, 1974), Czech Republic (Žáček et al., 2005), Russia (Eihvald, 1864; Butov and Yavorsky, 1922; Yavorsky and Radugina, 1932; Belikov, 1933; Lyakhovich, 1953; Menyailov et al., 1955; Matukhina and Van, 1965; Reverdatto, 1973; Scherbak, 1976; Tugovik, 1979), Australia (Baker, 1953; Whitworth, 1958; Rattigan, 1967; Farrand and Newton, 1985), New Zealand (Tulloch and Campbell, 1993), and Mali (Svensen et al., 2003). Coal-bed fires have been observed repeatedly in Scotland, France, Germany (Eihvald, 1864), and Morocco (Igor S. Novikov, 2002, personal commun.). Numerous works are devoted to rare and new minerals from these occurrences (Miyashiro, 1957; Schreyer and Schairer, 1961; Cosca and Peacor, 1987; Foit et al., 1987; Cosca et al., 1988, 1989; Schreyer et al., 1990; Tulloch and Campbell, 1993; Žáček et al., 2005).

Coal fires are prevalent. The area of combustion metamorphic complexes that has been studied the most extensively originated because of vast coal fires in the northern Great Plains of the United States (Foit et al., 1987; Cosca et al., 1989; Clark and Peacor, 1992; Heffern and Coates, 2004). Other well-known complexes occur in the Hatrurim Formation (The "Mottled Zone," Israel) (Gross, 1977; Burg et al., 1991, 1999; Gur et al., 1995) and the Monterey Formation (California, United States) (Arnold and Anderson, 1907; Bentor et al., 1981). Only a few quantitative data have been obtained for oil fires and related metamorphism (McLintock, 1932; Basi and Jassim, 1974; Feldman et al., 1994).

This paper reviews many published reports on natural coal fires, with special emphasis on those in Russia and central Asia that have been described in the Russian literature, and analyzes the key geological factors that control combustion metamorphic events (Table 1) and characterize combustion metamorphic process as an independent geological phenomenon.

THE COMBUSTION PHENOMENON

The primary agent of combustion metamorphism is the heat generated by the oxidation of fossil fuel. During burning, sedimentary protoliths are transformed into combustion metamorphic rocks. Combustion metamorphic temperatures may exceed 2000 °C. Heat energy and combustion products that accumulate in the area of burning are carried to adjacent combustible matter by convective heat and mass transfer, which initiates further ignition. As a consequence, the fire front may expand at a rate of 10^{-3}–10 m/s (Knunyants, 1990, 1998).

Lignite, coal, peat, oil shale, bitumen, asphaltenes, and oil commonly combust spontaneously. Self-heating is promoted by a low oxidation temperature and by the absorption of moisture by the fuel. Coal may spontaneously combust due to self-heating at temperatures of 80–140 °C. Flame combustion within the coal proceeds by diffusion or convective mechanisms and is limited by the supply of oxygen in the reaction zone. An active fire front advances mainly along the surface of a coal bed, whereas hot gases rise upward.

If heated, coal, being a complex organic matter, decomposes and releases combustible gases (Koltsov and Popov, 1978; Stracher and Taylor, 2004). Under slow heating conditions, this process has multiple stages. Initially, a distillation of gaseous products, their ignition and combustion, as well as a warm-up of the coke residue take place; only then does the coke catch fire. Chemical transformations of gaseous compounds can be expressed by the generalized formula: $(CHO) + O_2 \rightarrow CO + CO_2 + H_2 + H_2O + CH_4 + Q$ (heat of combustion). Simplified, heterogeneous reactions of solid carbon oxidation can be written as:

$$C + O_2 \rightarrow CO_2 + 409 \text{ MJ/kmol and } 2C + O_2 \rightarrow 2CO + 246 \text{ MJ/kmol}. \quad (1)$$

Coal pyrolysis and coking may be compared with the processes that occur in deep-seated zones of coal fires. Pyrolysis is a process of thermal decomposition of organic compounds, which results in the formation of products with lighter molecular weights. The process proceeds under conditions of limited access of air and water vapor at temperatures from 250 °C to 450 °C to 1000 °C. Coking is carried out during high-temperature (900–1100 °C) decomposition of fossil fuel without access to air. The main products of both processes are coke, or semi-coke, gases (H_2, H_2O, CO, CO_2, CH_4, C_2H_4, H_2S, and NH_3), benzol, pitch, and bitumen.

Fossil-fuel combustion as a specific heat source of pyrometamorphic transformations of sedimentary protolith has the following features: (1) abrupt increase in temperatures up to 1000–2000 °C in a flame zone; (2) convective heat and mass transfer that is directed mainly along the boundaries of fire beds and upward; (3) fuel pyrolysis in zones with restricted oxygen inflow; and (4) generation of combustible gases, their ignition during the interaction with an oxidizing agent, and formation of high-temperature gas vents (Cosca et al., 1989; Kalugin et al., 1991; Sokol et al., 2005). The gas regime in a zone of fuel pyrolysis is controlled mainly by the CCO buffer, for which the precise position in T-f_{O_2} (T—temperature in degrees Celsius; f_{O_2}—oxygen fugacity) coordinates can be displaced owing to redox reactions in the C-O-H system. Fossil-fuel combustion and pyrolysis result in the formation of oxygen-, sulfur-, nitrogen-, hydrogen-, and carbon-bearing gas compounds (Novikov, 1993; Stracher and Taylor, 2004; Sokol et al., 2005). Nitrogen is transported mainly as N_2, NO_x, NH_3, HCN, and CNS. Sulfur flows as H_2S (95%), CS_2 (2%–3%), SO_2 (0.4%–0.8%), and thiophene (C_4H_4S; 0.4%–0.8%). Oxygen migrates mainly as water vapor, CO, and CO_2. Hydrogen forms a wide spectrum of gas compounds, of which H_2O, H_2, and NH_3 are most active in mineral-forming processes. Hydrocarbons, including sulfur- and nitrogen-bearing ones, are toxic.

TABLE 1. MAIN CHARACTERISTICS OF REPRESENTATIVE COMBUSTION METAMORPHIC (CM) COMPLEXES

Locality	Type of fuel	Topography of region	Lithology and age of protoliths	Geological setting	Area of combustion events	Depth of CM alteration	Age of CM event; method of dating	Type of CM products	Notable or significant CM minerals and glasses	References
Tul-i-Marmar, Zoh-i-Hait, Darreh Harrachi, Naf Khaneh, southwest Iran	Hydrocarbon gases and products of bitumen and oil decomposition	Conical-shaped hills and vertical cliffs, rising abruptly to a height of 80 m. Hills are capped by CM rocks	Tertiary marls, sandstones, and mudstones, bitumen-stained limestones with vertical joints carrying bitumen	Gently dipping (3–5°) sediment horizons with vertical joints. Shattered CM rocks	The areas of CM rocks are strictly localized. Paralava spots are about 3.5 × 2.5 m in area. Breccias measure roughly 30 × 30 m with highly inclined junctions against the host sediments	The depth of CM rock extension is less than 80 m	The age of ancient fires is undetermined. There are several modern oil and gas fires in this region	Clinker, annealed marl and sandstone, ash, basic high-calcium paralava, abundant breccia, lavalike slag, recrystallized marl and limestone, sulfur, gypsum, H_2S, and SO_2	Augite, honey-yellow pyroxene (esseneite?), bytownite, wollastonite, pseudo-wollastonite, melilite, leucite, gypsum, anhydrite, glass	McLintock (1932)
The "Mottled Zone" (Hatrurim Formation), Israel and Jordan	Bitumen	Dissected mountainous terrain with height step from 200 m to 400 m. Badlands landscape is formed by low-grade carbonate CM unit. Strongly metamorphosed rocks crop out at topographical highs to form cliffs and rows of hills	The Ghareb Fm., the Taqiye Fm. of Late Cretaceous (Maastrichtian) and Early Tertiary (Paleocene) age. Bituminous chalks and marls (contain up to 30% organic matter), chert, clay, bituminous limestones and dolomites, conglomerates, phosphorites	The area of Hatrurim synclinal basin with several minor structures. CM outcrops are located in adjoining sectors of the Dead Sea Rift Valley. Mutual tectonic and contraction joints, collapsed structures. Breccias are widespread	Several outcrops with areas that exceed 1 km² up to 50 km². The total area of "Mottled Zone" is about 90 × 60 km	The total thickness of CM rocks is up to 190 m. The depth of local CM horizons exceeds 80–120 m	Major CM events took place between 2.3 and 4.0 Ma. The early event took place upon initial exposure of the protolith in the Miocene, around 16 Ma. Dating techniques are paleomagnetic, K-Ar, and $^{40}Ar/^{39}Ar$ dating, fission-track dating of apatite	Wide variety of CM rocks metamorphosed to different grades. Various types of fels and marbles; metamorphic conglomerates	About 50 high-temperature minerals, mainly Ca- and Ca-Al-silicates: spurrite, gehlenite, larnite, wollastonite, pseudo-wollastonite, anorthite, diopside, garnet. Perovskite, brownmillerite, sanidine, calcite, apatite. A remarkable feature is intensive retrograde processes (rehydration, recarbonation, and sulfatization). There are 59 minerals, including zeolites, lizardite, hydrogarnets, carbonates, gypsum, halite	Bentor et al. (1963); Gross (1977); Matthews and Kolodny (1978); Kolodny (1979); Matthews and Gross (1980); Burg et al. (1991, 1999); Khoury and Milodowski (1992); Ron and Kolodny (1992); Gur et al. (1995)

(continued)

TABLE 1. MAIN CHARACTERISTICS OF REPRESENTATIVE COMBUSTION METAMORPHIC (CM) COMPLEXES (continued)

Locality	Type of fuel	Topography of region	Lithology and age of protoliths	Geological setting	Area of combustion events	Depth of CM alteration	Age of CM event; method of dating	Type of CM products	Notable or significant CM minerals and glasses	References
Fan-Yagnob coal deposit, Ravat fire, Tajikistan	Coking coal and lignite. Seam thickness from 0.5 to 3.5 m	Cross-country, abrupt mountain relief. Canyonlike valley of Yagnob River	Jurassic shales, clays, arkosic sandstones, gritstones, siltstones, oolitic limestones, limonite	Fire zone is located at tectonic contact (thrust structure complicated by faults). Rocks are deformed intensively. Seams dip from 50° to 75°	The area of modern fire is 25 km². During the last 300 yr, fire front has moved 3.5 to 5 km	Depth of fire expansion is from 150 to 500 m	Duration of continuous combustion is not less than 2000 years	Clinker, ash, coke, basic paralava, fumarolic mineralization, bitumen, tar, naphthalene, fumes of H_2SO_4 and liquid H_2SO_4	Fayalite, sekaninaite; magnetite, hercynite, tridymite, high-silica glass; sal ammoniac, sulfur, alum	Ermakov (1935); Novikov and Suprychev (1986); Novikov (1993); Sokol et al. (2005)
Khamarin-Khural-Khid, East Mongolia	Low-rank coals; seam thickness is 0.5 to 5 m	Mountains of high relief dissected by riverbeds	Early Cretaceous sandstones, conglomerates, siliceous schists, coal-bearing siltstones	Area of combustion is bordered by a few deep cracks	The area affected by underground fire is 250 × 150 m	Subsurface	Fire was ignited in 1932–1935	Slags, annealed rocks, sal ammoniac	Sal ammoniac	Pokrovsky (1949)
Dzhergaly, Kok-Moinok, Kok-Yangak, Kyzyl-Kiy, Tash-Kumyr, Sulyuktin (Kirghizia), and Angren (Uzbekistan)	Coal seam thickness from 20–25 m up to 350 m	Dissected mountainous terrain	Jurassic sandstones, gravels, clays, siltstones	Bed structure is complicated by microflexures	The areas of CM rocks are some square km. All objects are industrial deposits of CM rocks	The depth of CM rock localization exceeds 500 m	Early Quaternary	Clinkers, annealed argillaceous rocks and gravels	Mullite, cristobalite, tridymite, cordierite, hematite, magnetite, diopside, hypersthene, garnet, wollastonite, carnegieite, high-silica glass	Abubakirov (1962); Zbarsky (1963); Skrobov (1968)
Coal deposit of the Kenderlyk Depression, east Kazakhstan	Coals, lignites	Hills and ridges of CM rocks	Triassic–Jurassic sandstones, siltstones, carbonaceous shales, siderite concretions, and petrified wood	Coal seams inclined from 50° to 60° up to overturned dip. Thrust structure is complicated by faults. CM complex defined by a fault	Two horizons of CM rocks extend continuously for 1.5 km and 2.8 km in length and to 70 m in thickness and are conformable with sedimentary host rocks	CM rocks have been traced to a depth of 300 m	According to geological data, the fire was ignited in the early Pleistocene	Clinkers, basic and ore-bearing paralavas, slags, and breccias	Diopside, feldspars, olivine, fayalite, magnetite, hematite, Ca-bearing ferrites, magnesioferrite, maghemite, cordierite, mullite, apatite, high-silica glass	Kalugin et al. (1991)

(continued)

TABLE 1. MAIN CHARACTERISTICS OF REPRESENTATIVE COMBUSTION METAMORPHIC (CM) COMPLEXES (continued)

Locality	Type of fuel	Topography of region	Lithology and age of protoliths	Geological setting	Area of combustion events	Depth of CM alteration	Age of CM event; method of dating	Type of CM products	Notable or significant CM minerals and glasses	References
Salair mountain ridge and Kemerovo district, Kuznetsk Coal Basin, southwestern Siberia, Russia	Coking, gas, and fat coals. Coal-bed methane. Seam thickness up to 10–25 m	Clinker- and breccia-controlled topography. Hills and ridges of CM rocks rise 50 to 100 m above surrounding plain	Pennsylvanian–Early Permian shales, sandstones, siltstones, siderite concretions, oolitic limonite. The rocks are intensively fractured and brecciated	The strongest fire is confined to linear zone of numerous faults. Coal seams are crumpled in abrupt folds and inclined from 70° up to subvertical dip. Zone of Mesozoic and Cenozoic tectonic activity	The area of the oldest fires is 10 × 45 km. Modern fires occupy hundreds of square meters	The depth of CM rock localization exceeds 250 m. Visible CM rock horizons reach 150 to 100 m in thickness	According to geological data, the first fire wave took place at the Jurassic–Triassic boundary. The second peak of ignitions occurred during the middle Miocene to Pliocene. Modern fires have burned continuously since 1730	Annealed sandstones, abundant breccias, clinkers, ash, coke, basic paralavas, slags; dropstone forms of melted rocks, sal ammoniac, sulfur, bitumens	Cordierite, mullite, fayalite, magnetite, native iron; high-silica glass	Butov and Yavorsky (1922); Yavorsky and Radugina (1932); Belikov (1933)
Markha River (inflow of Vilyui River), Tungus Coal Basin, east Siberia, Russia	Coals	Coal beds were ignited in steep walls of the river valley (40 m)	Pennsylvanian–Permian sandstones, siltstones, carbonaceous shales	Subhorizontal coal bed was burned out	CM rocks extend down to 20 km	Subsurface fire. The depth of CM rock localization exceeds 40 m	Historical times. The last fire was ignited in 1953	Annealed and baked sandstones, clinkers, breccias, slags; stalactites of glassy paralavas	Plagioclase, pyroxene, glass	Lyakhovich (1953)
Powder River Basin, Wyoming, USA	Subbituminous rank-coals rich in volatile matter	Coal beds were ignited on the highwall in the gully. The topography is dominated by hills, ridges, and escarpments topped by clinkers	Paleocene Fort Union and Eocene Wasatch Formations consist of mudstone, shale, siltstone, sandstone, limestone, Fe nodules	The coal beds are nearly horizontal or dip gently. Clinker forms distinctive layers that cap plateaus, hilltops, and escarpments. The clinker is highly fractured	Clinker outcrops cover more than 4100 km^2	Most ignitions were subsurface. The depth of CM horizon localization exceeds 120 m. Fires could not extend to depth because of local water table	Coal beds have burned during at least the last 4 m.y. Zircon fission-track ages of clinkers are 4.0 ± 0.7 Ma, 2.8 ± 0.6 Ma, 2.1 ± 0.6 Ma, 1.4 (±0.4)– 0.7 (±0.3) Ma, 0.5 (+0.3) to less than 0.1 Ma. Additional methods of dating are U-Th/He and paleomagnetic	Clinkers, welded breccias, basic paralavas, ashes	Dorrite, diopside, melilite, anorthite, K-Ba feldspars, magnetite, hematite-ilmenite solid solution, fayalite, sekaninaite, titanian andradite, nepheline, wollastonite, pseudo-brookite; mullite, apatite, high-silica glass	Cosca et al. (1988, 1989); Clark and Peacor (1992); Heffern and Coates (2004)
Santa Barbara, Santa Ynez Mountains, Grimes Canyon, San Marcos Ranch, Santa Maria Valley, and other localities, California, USA	Hydrocarbon fuel gases, bitumen, pyrobitumen, naphtha	Dissected mountainous terrain with height step from 250 m to 650 m. The highest hills in the basin (up to 600 m above sea level) are capped by fused CM rocks	Bituminous Monterey Formation of middle Miocene age. Mudstones, siltstones, and diatomites are dominant; shales, limestones, and dolomites are rare	The altered shales are situated almost invariably in the vicinity of oil seepages, which usually connote a fractured condition of the rock	The area of CM rocks varies from hundreds of square meters to 1.5 km^2. In Grimes Canyon, CM rocks form an almost continuous belt, 20 km long and 1–3 km wide	Annealed shales were distinguished at a depth of 320 to 350 m. They are always situated in the vicinity of oil seepages	Numerous fragments of CM rocks were found in undisturbed beds of Pleistocene age. Most of CM outcrops date from the last few hundred years	Different types of melted CM rocks, clinkers, coke, sulfur, epsomite	High-, intermediate-, and low-silica glasses. Cristobalite, tridymite, sanidine, cordierite, mullite, plagioclases, augite, garnet, hematite, corundum, wollastonite, apatite	Arnold and Anderson (1907); Bentor et al. (1981); Bentor (1984)

(continued)

TABLE 1. MAIN CHARACTERISTICS OF REPRESENTATIVE COMBUSTION METAMORPHIC (CM) COMPLEXES (*continued*)

Locality	Type of fuel	Topography of region	Lithology and age of protoliths	Geological setting	Area of combustion events	Depth of CM alteration	Age of CM event; method of dating	Type of CM products	Notable or signficant CM minerals and glasses	References
Aldridge Creek, Canadian Cordilleras, Canada	High-volatile bituminous coals (mainly vitrinite)	Coal bed was ignited on the highwall in the gully	Late Jurassic–Early Cretaceous Mist Mountain Formation consists of carbonate-bearing sandstones, siltstones, carbonaceous shales	The seam, up to about 6 m thickness, dips from 35° to 50°. CM roof rocks have fractured and collapsed locally	The length of burned zone is 1 km	The upper 3 m of coal bed is consumed to an estimated depth of 20 m	Coal seam was ignited by forest fire in 1936	Ash, coke, welded breccias, paralava, including stalactites	Cristobalite, tridymite, anorthite, diopside, glass, sulfur	Bustin and Mathews (1982)
Ravensworth, NSW, Australia	Coal	CM rocks form the tops of low, rounded hills	Tomago beds of the Main Permian coal basin. Clay shales, sandstones with siderite cement, siderite concretions and beds; limonite	Coal seams dip at 55°. Area of CM rocks is bordered with an anticline fold. CM roof rocks have fractured and collapsed	Outcrops of CM rocks are small; most are less than 15 m in diameter. The total area subjected to coal fires is about 2.5 km^2	The thickness of CM rocks is up to 30 cm	The age of ancient fires is unknown. More recent fires appeared sporadically during last several centuries	Slaglike fused and partially fused rocks, buchites	Clinohypersthene, hypersthene, cordierite, plagioclase, spinel, high-silica glass	Whitworth (1958); Rattigan (1967)
Leigh Creek, south Australia	Subbituminous coal	No data	Triassic fossiliferous shales	Roof-rock collapses structures. Local melting events connected with chimney structures and interlayer boundaries	No data	Surface and subsurface fire above the water table	Prehistoric fire	Fused and clinker ash, baked shales, clinkers, breccias	Magnetite, native iron, hematite, pyrrhotite, Ti-augite, fassaite, gehlenite, perovskite, apatite, spinel, glass	Baker (1953)

GENERAL FEATURES OF NATURAL COAL FIRES

Geological and Topographic Control of the Distribution of Fossil-Fuel Fires

Wildfires, such as grass fires and forest fires, ignite coal beds regularly. Many of these wildfires are caused by lighting strikes (Eihvald, 1864; Arnold and Anderson, 1907; Yavorsky and Radugina, 1932; Scherbak, 1976; Heffern and Coates, 2004; Stracher and Taylor, 2004). Thus, most natural coal fires have an exogenic trigger. In respect to present-day fires in Dorset, England (Cole, 1974), and Smoking Hills, Canada (Bustin and Mathews, 1982), a possible reason for ignition may be an endogenous factor (i.e., oxidation of pyrite, which accelerates the spontaneous heating of fuel).

The two necessary components for combustion are fuel and oxygen. Rugged relief and steeply dipping fuel-bearing strata provide ideal conditions for natural coal-bed fires. Coal beds usually are ignited in the steep walls of river valleys and on steep hillsides that are not protected by Quaternary sedimentation. Fuel-bearing layers that are shielded by alluvial sediments commonly are not ignited (Burg et al., 1991, 1999). When fuel-bearing strata burn, a flame front develops along steeply dipping beds, and the fire descends quickly to depth. Fires occur mostly in beds that dip from 30° up to subvertical (see Table 1). A unique example of a fire spreading within near-horizontal coal strata is in the Powder River Basin, Wyoming, United States (Heffern and Coates, 2004). When a fire starts, it first spreads along an outcrop. With time, it burns deeper into the hillside and causes the overlying rocks to subside progressively into the burned-out void. The resulting fracturing allows air to enter and gas to escape, so that the fire continues (Fig. 1). These fires extinguish themselves naturally at the point when the overburden becomes so thick that fractures from the collapse fail to reach the surface to draw in more air, or when the fire burns down to the water table in the coal.

The depth of fire spreading usually does not exceed several tens of meters because of the deficiency of oxygen at deep levels (see Table 1). However, during long-term fires, the combustion metamorphic rocks are found at depths of ~350 m (Santa Barbara, United States; Kenderlyk, Kazakhstan). The present-day Ravat coal-bed fire in Tajikistan is located at a depth of ~500 m. At this locality, the total thickness of combustion metamorphic rocks is 300 m. The maximal thickness of burned rocks is ~150 m in the Kuznetsk Coal Basin, Russia, and 80–120 m in the Hatrurim Basin, Israel.

Joints affect the scale and depth of fire spreading directly. Most combustion metamorphic localities are situated in axial areas of folds, near the margins of rift valleys, or they are controlled by thrust structures. A system of tectonic joints and cracks provides routes for air to a depth of several hundred meters and promotes underground fire expansion.

Areas of combustion metamorphic complexes are controlled by structural parameters of fuel-bearing beds, topography, and depth of erosion. The well-known Clinker beds in the western United States cover ~500,000 km^2 (Bentor et al., 1981). Commonly, ancient combustion metamorphic complexes attain an area of several thousand square kilometers (de Boer et al., 2001; Heffern and Coates, 2004), whereas present-day fires are limited in size and occupy areas that range from a few hundred square

Figure 1. Fractured clinker beds, Falcon Hills, Kuznetsk Coal Basin, southwest Siberia (2005).

Figure 2. Collapsed and fractured roof rock represented by welded pyrogenic breccia, Falcon Hills, Kuznetsk Coal Basin, southwest Siberia (2005).

meters up to several square kilometers (Bogoslovsky, 1842; Eihvald, 1864; Arnold and Anderson, 1907; Ermakov, 1935; Bustin and Mathews, 1982).

The Structure of Combustion Metamorphic Complexes

The term "burned rocks" encompasses all of the rocks that have been altered by heating from fossil-fuel fires. The degree of thermal alteration is variable, and a single spot may contain rocks that range from slightly baked to hardened lava. The completely fused and scoriaceous rocks essentially are welded and, thus, exhibit greater coherence. The baked, but unfused, rocks provide little roof support. Extensive cracking also may occur because of a loss of volatiles, principally H_2O and CO_2. The specific volume decrease of high-temperature phases results in the formation of rock-contraction cracks. Fuel-rich horizons may be consumed entirely or partially and result in the formation of numerous cavities. All of these processes lead to the appearance of collapse patterns, numerous cracks, and welded breccias that contain burned rock fragments (Fig. 2) (McLintock, 1932; Yavorsky and Radugina, 1932; Ermakov, 1935; Lyakhovich, 1953; Whitworth, 1958; Abubakirov, 1962; Rattigan, 1967; Bustin and Mathews, 1982; Cosca et al., 1989; Srebrodolsky, 1989; Burg et al., 1999). Temperatures during combustion frequently are high enough to promote partial or even total melting of sediments. In some cases, breccias are cemented by melted rocks, or paralavas (Fig. 3). Characteristic features of combustion metamorphic complexes are the subvertical extended cracks that can reach up to several meters in diameter, which are named "chimney" structures (Whitworth, 1958; Rattigan, 1967; Bentor et al., 1981; Bustin and Mathews, 1982; Cosca et al., 1989; Clark and Peacor, 1992; Heffern and Coates, 2004). Partial or total melting of roof rocks and the location of paralavas are associated closely with "chimneys," through which hot gases escape. Commonly, gas jet walls are coated with a dense crust of fine dispersed hematite.

As underground burning proceeds, the gradual collapse and cracking of roof rocks leave sinkholes that allow air to enter into fresh coal. By this mechanism, fire spreads along the bed and to adjacent horizons. Simultaneously, the roof collapse and induced inflow of cold air promote the quenching of combustion metamorphic rocks. Therefore, most clinkers and paralavas are similar to buchites (Venkatesh, 1952; Menyailov et al., 1955; Matukhina and Van, 1965; Bentor et al., 1981; Bentor, 1984; Foit et al., 1987; Cosca et al., 1988, 1989). Melt crystallization with oxygen excess causes oxidation of Fe^{2+} to Fe^{3+}. The dependence of the redox regime of cooling on paralava mineral composition has been discussed in detail elsewhere (Cosca et al., 1988, 1989; Sokol et al., 2002, 2005; Žáček et al., 2005).

Heat Transfer during Combustion Metamorphic Events

Burned rocks may overlie unmetamorphosed ones or be in lateral contact with them (McLintock, 1932; Yavorsky and Radugina, 1932; Skrobov, 1968; Basi and Jassim, 1974; Bustin and Mathews, 1982; Cosca et al., 1989; Burg et al., 1991; Kalugin et al., 1991; Heffern and Coates, 2004). In the scarp of the Taimur River (East Siberia, Russia), the unmodified coal bed directly underlies the burning coal seam, which contains heated sandstones and natural

Figure 3. Welded breccia, containing clinker fragments (A) and breccia cemented by ferriferous paralava (B), Kuznetsk Coal Basin, southwest Siberia (2005).

coke (Eihvald, 1864). In the Canadian Cordilleras (Bustin and Mathews, 1982), a fire has developed within a 6 m coal seam at a depth of 20 m. The upper 2–3 m of coal has been consumed. About half of the coal seam has been converted to residual ash and coke. The ash horizon (5–60 cm) is separated from the underlying coal by natural coke (10–50 cm), and the coal shows no evidence of heating or devolatization. The roof rocks are annealed, melted, and fractured. Some brecciated rocks were welded together by paralavas. Near the surface, combustion of gases and sulfur precipitation has taken place. The temperature of the gas emanations has been estimated to be ~1000 °C. Holes that provide air inflow to depth, which are peculiar natural "blast pipes," have been found. Similar patterns of underground fuel combustion, accompanied by coke formation, have been described previously (Arnold and Anderson, 1907; Skrobov, 1968; Church et al., 1979; Cosca et al., 1989).

Within a single horizon, the fire is propagated along the top of the fuel layer in accordance with the specificity of "burn front" motion in a solid medium. As a result, the flame can extend swiftly to depth when layers are dipping steeply. The combustion metamorphic complex in the Kenderlyk Depression (east Kazakhstan), which contains ore-bearing basic paralavas, is a typical example of this process (Kalugin et al., 1991). The horizon of combustion metamorphic rocks extends continuously down to 2800 m, attains a thickness of 70 m, and is conformable with sedimentary host rocks. However, large-scale fires are not framed by the boundaries of a single coal bed. Frequently, they spread from underlying coal horizons to overlying ones. At the Ravat fire in Tajikistan, 11 closely spaced coal beds were burned out, but the fire did not expand to five horizons below the base of the originally ignited coals.

When oxygen flows into the burn zone through fractures within roof rocks and interlayer boundaries, only the top of a coal horizon is partly subjected to pyrolysis (gasification) or oxidation, until the roof rocks collapse. This results in the formation of dense ash layers that inhibit air access to the coke residue. At depth, at higher temperatures, the devolatization (or coking) of coal proceeds, and flammable gases are generated. Rising along the cracks, the volatile combustible constituents are driven off and, when mixed with air, burn en route to the surface. Thus, combustion of the gas products of pyrolysis is the primary factor in the combustion metamorphic transformation of coal-bearing roof rocks. Such factors as local melting of metasedimentary rocks near gas jets, high-temperature annealing and melting of roof rocks, negligible thermal alteration of underlying strata, and the presence of unaffected coals under thin coke or ash layers suggest that under combustion metamorphic conditions, heat and mass transfer occur in forward and upward directions by flame or gas convection. Only a small amount of the heat is transferred downward by conduction, and it affects less than 1 m of rocks within the base of thick, burned-out coal beds.

Duration and Age of Fossil-Fuel Fires

Once ignited, an underground fire burns all of the coal that it can reach before running out of air. Complete coal or coke consumption is rare in natural fires. This is evidence of the short duration of combustion events in any local spot. Commonly, a large fire is not a one-act phenomenon. Burning spreads from one point to another and is frequently accompanied by long pauses. Tertiary surface combustion of Cretaceous bituminous sediments in Israel took place between 2.3 and 4.0 Ma and probably was preceded by a combustion metamorphic event around 16 Ma (Gur et al., 1995). According to Pliny Senior (first century A.D.), Arabian geographers (tenth century A.D.), and Chinese chronicles, the Ravat fire in Tajikistan has lasted for more than 2000 yr. During this period, the fire front has advanced ~10 km (Ermakov, 1935; Novikov, 1993).

The burning of thick coal beds generally produces higher temperatures and lasts for longer periods (Yavorsky and Radugina, 1932; Kalugin et al., 1991; Heffern and Coates, 2004). The burning of gas torches provides the highest temperature in local spots near gas vents, but this burning usually is of short duration (McLintock, 1932). According to air photography of the Aldridge Creek coal fire in the Canadian Cordilleras, the front of the underground fire has advanced at a rate of 5–13 m/yr (Bustin and Mathews, 1982). During the twentieth century, the Ravat fire has spread at a similar rate (10–17 m/yr). For bituminous sediments of the Hatrurim Basin, Burg et al. (1999) estimated that the combustion within an 8-m-diameter sphere that contained 12 wt% of fuel could last for more than 300 d, but not more than a few years. According to the same estimates, during the first days of burning, the temperature increases sharply, and then increases more gradually after that. These data suggest that high-temperature heating in near-surface burning foci can only be of short duration. The further heating of rocks is carried out by hot gases that are supplied from adjacent burning spots. Commonly, the duration of high-temperature combustion metamorphism in local sites (~10 m^2) does not exceed 0.5–2 yr (Sokol et al., 2005).

The oldest, well-grounded age of combustion metamorphic events was deduced for the Hatrurim Formation (Ron and Kolodny, 1992; Gur et al., 1995; Burg et al., 1999). The first combustion metamorphic event took place upon initial exposure of the bituminous protolith in the Miocene (around 16 Ma), after the opening of the Dead Sea Rift Valley. For this territory, the post-Miocene age for a major combustion event, established from geological data, agrees well with K-Ar and ^{40}Ar/^{39}Ar dates of 2.3–4.0 Ma and 1.2–1.7 Ma (Gur et al., 1995). In the Powder River Basin (Wyoming, United States), natural coal fires are ongoing events that has continued since the late Tertiary. Ages of clinker outcrops range from younger than 0.02 Ma to as old as 2.8 Ma. Stream gravels on high terraces contain clinker boulders that are as old as 4 Ma (Heffern and Coates, 2004). However, most combustion metamorphic complexes currently known have formed during the past 2000 yr, and many of the episodes of ignitions have been recorded in written and folk sources.

Physical Conditions of Combustion Metamorphic Events and General Features of Combustion Metamorphic Rocks

The peak temperatures deduced from combustion metamorphic mineral assemblages are estimated to be 1200–1450 °C for coal fires (Cosca et al., 1989; Kalugin et al., 1991), whereas bitumen fire temperatures may reach 1650 °C (Bentor et al., 1981). The temperature of petroleum gas torches is up to 2400 °C (Feldman et al., 1994). In intensely fractured areas where oxygen supply is abundant, the combustion focus contains high-temperature minerals and glasses, but the affected volume will be relatively small. In these localities, the transition from high-grade combustion metamorphic rocks to unmetamorphosed sediments occurs over a few meters and creates steep thermal gradients, up to 140 °C/m. Underground fossil-fuel fires usually occur at a pressure of less than 5 bar. A maximal lithostatic pressure of ~25 bar was reconstructed for the combustion metamorphic complex of the Hatrurim Formation (Burg et al., 1999).

Combustion metamorphic transformations at a moderate temperature (500–800 °C) lead to dehydration, decarbonation, and do not involve any melting. These kinds of annealed, baked, but not fused, rocks are abundant among all combustion metamorphic complexes. Melting that occurs during combustion takes place under low-pressure, dry conditions. The first liquids to form have eutectic compositions. When parent rocks are represented by pelites, the eutectic melts are similar to S-type granites. Clinkers, which are the products of the partial melting of metapelitic rocks at 900–1600 °C, are the most widespread (Bentor et al., 1981; Bentor, 1984; Sokol et al., 2005). The formation of ferriferous basic paralavas only occurs at high temperatures (1000–1450 °C) and under reducing conditions, and it involves the total melting of ferruginous sedimentary parent rocks (Foit et al., 1987; Cosca et al., 1989; Kalugin et al., 1991; Sokol et al., 2002, 2005). Compounds of Fe^{2+} (mainly, siderite decomposition products), which form under reducing conditions, essentially decrease the melting temperature of the protolith. This produces considerable volumes of iron-rich basic melts at 1100–1450 °C (Cosca et al., 1989; Kalugin et al., 1991).

The combustion metamorphism of low-silica carbonate protoliths at up to 1300 °C is characterized by progressive solid-state decarbonation and the appearance of minerals such as those that occur in Portland cement, including grossite ($CaAl_4O_7$), hatrurite (Ca_3SiO_5), periclase (MgO), and lime (CaO) (Gross, 1977; Matthews and Kolodny, 1978; Matthews and Gross, 1980; Burg et al., 1991, 1999). At the same conditions, clay marls and calcareous sandstones melt and form high-calcium basic paralavas (McLintock, 1932; Basi and Jassim, 1974). Every fire generates huge amounts of gas and water-soluble compounds, including toxic ones. Often, combustion metamorphic rocks are accompanied by fumarolic mineralization (hematite, ammonium sulfates and chlorides, sulfur, alum) and bitumens (Fig. 4) (Bogoslovsky, 1842; Eihvald, 1864; Arnold and Anderson, 1907; McLintock, 1932; Ermakov, 1935; Pokrovsky, 1949; Lyakhovich, 1953; Whitworth, 1958; Rattigan, 1967; Scherbak, 1976; Church et al., 1979; Bustin and Mathews, 1982; Novikov and Suprychev, 1986; Novikov, 1993; Srebrodolsky, 1989; Clark and Peacor, 1992; Stracher and Taylor, 2004).

Thus, fossil-fuel fires represent an unusual natural environment for the formation of minerals, comparable to sanidinite facies rocks (Table 2). The combustion metamorphic transformation of a carbonate protolith produces rocks that are characteristic of contact metamorphic facies. A good example is the high-tem-

Figure 4. (A) Modern hot-gas jets and associated fumarolic mineralization (mainly sal ammoniac and sulfur) above burning coal seams at the Kiselevsk coal-mining region. (B) Sulfur "icicles" surrounding vents on the surface of an ancient pyrogenic breccia, Kuznetsk Coal Basin, southwest Siberia (2005).

TABLE 2. MAIN CHARACTERISTICS OF COMBUSTION METAMORPHISM IN COMPARISON WITH SANIDINITE FACIES METAMORPHISM

Characteristics	Sanidinite facies metamorphism	Combustion metamorphism
Geological setting and the source of heat	Rocks occurred in the exocontact or as xenoliths in traps, dikes, necks, sills	Rocks occurred above and within the burning fossil-fuel layers
The thickness of thermal perturbation	Contact aureoles are up to 40–90 m. High-temperature zones are located at the contact with magmatic bodies. Low-temperature zones usually are absent	Thickness is up to 100 m. High-temperature rocks (>900 °C) are located in the vicinity of the burning seams or gas channels. Low- and moderate-temperature rocks (600–800 °C) predominate
Pressure (bar)	Up to some hundreds bars	Up to 25 bars
Temperature (°C)	700–1100 °C	550–1450 °C for coal fires; 550–1650 °C for bitumen fires; 550–2400 °C for oil fires
Temperature gradient	10–40 °C/m	From 25–45 °C/m up to 140 °C/m
The main direction of heat transfer	Radial heat transfer from magmatic body to host rocks	Vertical heat transfer, controlled by the direction of gas flow
The predominant mechanism of heat transfer	Conductive heat transfer with subordinate convective transfer	Gas convection with minor conductive heat transfer
The duration of metamorphic events	2–3 yr is the most realistic period of high-temperature contact metamorphism	The duration of burning in local spots (10 m²) is from 6 mo to 2 yr. The period of burning in separate areas is up to 2000 yr

perature Hatrurim Formation, where spurrite, merwinite, and rankinite, characteristic minerals of both combustion metamorphic and sanidinite facies rocks, occurred when $P_{CO_2} \leq 0.5 P_{total}$. During further decarbonation of carbonate-bearing parent rocks, the partial pressure of CO_2 decreases more, resulting in the formation of lime (CaO) and calcium aluminates (Reverdatto, 1973; Gross, 1977; Matthews and Kolodny, 1978; Kolodny, 1979; Burg et al., 1991, 1999).

The principal difference between combustion metamorphic rocks and sanidinite facies rocks is the intensive melting of metapelites, as well as other silicate protoliths. This process is completed by the formation of calcium-rich and ferrous-rich basic paralavas. Commonly, these rocks are extremely heterogeneous on a scale of tens to hundreds of micrometers and are characterized by an unusual mineral composition, including the presence of various glasses. Compared with hornfels of the sanidinite facies, the bulk melted and fused combustion metamorphic rocks represent disequilibrium; however, local equilibrium may exist within individual domains (McLintock, 1932; Foit et al., 1987; Cosca et al., 1989; Sokol et al., 2005).

The formation of a combustion metamorphic association is determined by a specific set of parameters: (1) high temperature, low pressure, and a heating duration of several months to several years in any local spot; (2) high chemical heterogeneity of the protolith; (3) active gas convection; and (4) high temperature and redox-potential gradients. These factors result in the generation of a wide variety of unique prograde combustion metamorphic minerals (mainly anhydrous oxides and silicates) that include as many as 63 species (Table 3). The mineralogy of the retrograde products of combustion metamorphic rocks has only been studied in detail for the Hatrurim Formation. Gross (1977) determined that the major retrograde processes are oxidation, recarbonation, rehydration, and sulfatization (Table 4). The retrograde process starts when the high-temperature metacarbonate combustion

TABLE 3. CRYSTALLOCHEMICAL TYPES OF PROGRADE COMBUSTION METAMORPHIC MINERALS FROM NATURAL OCCURRENCES

Types of compounds	Number of mineral species	Percentage (%)
Native elements	1	1.6
Sulfides	4	6.4
Oxides	19	30.2
Simple silicates	31	49.0
Fluor-silicates	1	1.6
Carbonate-silicates	1	1.6
Phosphate-silicates	1	1.6
Phosphates	1	1.6
Carbonates	2	3.2
Sulfates	1	1.6
Fluorides	1	1.6
Total	63	100.0

Note: All minerals are anhydrous. Calculation was carried out using the data from McLintock (1932); Yavorsky and Radugina (1932); Belikov (1933); Baker (1953); Whitworth (1958); Zbarsky (1963); Gross (1977); Kolodny (1979); Bentor et al. (1981); Bustin and Mathews (1982); Cosca and Peacor (1987); Foit et al. (1987); Cosca et al. (1988, 1989); Burg et al. (1991); Kalugin et al. (1991); Clark and Peacor (1992); Tulloch and Campbell (1993).

metamorphic rocks are still hot (≤300 °C) and accessible for weathering. In the Hatrurim Formation, the process occurred so quickly that a considerable portion of the rocks was replaced by hypergenic products. Therefore, on a geological time scale, combustion metamorphic complexes may be compared with "1-day butterflies," and the probability of finding ancient combustion metamorphic rocks is low.

Clinker, which is rock hardened by high-temperature annealing and melting processes, forms distinctive weather-resistant layers that cap plateaus, hilltops, and escarpments in the northern Great Plains of the United States (Cosca et al., 1989; Heffern and

TABLE 4. CRYSTALLOCHEMICAL TYPES OF MINERALS FROM RETROGRADE ASSEMBLAGES

Types of compounds	Number of mineral species	Percentage (%)
Native elements	1	1.7
Anhydrous oxides	1	1.7
Hydroxides and H_2O-bearing oxides	10	16.9
Anhydrous carbonates	3	5.1
Hydrocarbonates	3	5.1
Anhydrous sulfates	2	3.4
(OH)⁻- and H_2O-bearing sulfates	4	6.8
Anhydrous silicates	1	1.7
(OH)⁻- and H_2O-bearing silicates	32	54.2
Complex hydrogenous compounds	2	3.4
Total	59	100.0

Note: Calculation was carried out using the data from Gross (1977), Kolodny (1979), and Burg et al. (1991, 1999). Additional data are from McLintock (1932), Baker (1953), and Bentor et al. (1981).

Figure 6. Typical landscape of the combustion metamorphic complex of the Hatrurim Basin, Israel. (Source: Yehoshua Kolodny, The Hebrew University of Jerusalem.)

Figure 5. One of the Falcon Hills, capped by clinker (annealed and fused sandstones), Kuznetsk Coal Basin, southwest Siberia (2005).

Coates, 2004) and in the Kuznetsk Coal Basin (Siberia, Russia) (Fig. 5). Another type of resistant rock, basic paralavas, occurs as cap rock in eastern Kazakhstan (Kalugin et al., 1991) and the "Burnt Hills" of the Injana area, Iraq (Basi and Jassim, 1974). Similarly, calcium-rich basic paralavas, together with slags, breccias, and recrystallized calcareous marls, form conical-shaped hills in southwest Iran (McLintock, 1932). A badlands landscape is formed by metacarbonate combustion metamorphic units of the Hatrurim Formation (Fig. 6). High-grade metamorphic rocks crop out at topographical highs and form cliffs and hill rows with a height step from 200 m to 400 m (Burg et al., 1991).

COMBUSTION METAMORPHIC EVENTS IN THE GEOLOGICAL HISTORY OF SEDIMENTARY BASINS

Geological Review

Data on fire ages are not numerous, but they allow one to compare combustion metamorphic events with the tectonic history of coal-bearing sedimentary basins (Fig. 7). Until the late Paleogene, the foothill and intermountain troughs of various ages (from Mississippian and Permian in eastern Siberia up to the Jurassic in Tajikistan) were relatively stable. In the intracontinental lake and bog basins, upper Paleozoic to Mesozoic coal-bearing deposits were overlain by clay sediments and peat. Oligocene tectonics led to the formation of molasse units that overlapped Mesozoic sections. This sequence of events was reconstructed for the southern European part of Russia, south Siberia, the coal basins of central Asia, Mongolia, the Far East, northern and central China, and the Rocky Mountains and adjacent margins of the North American Platform. Twenty-five percent of the world's reserves of oil and gas, including Californian oil-and-gas–bearing provinces and some petroliferous provinces of Iran, were formed during the Tertiary (Kozlovsky, 1986, 1987, 1989; Dobretsov et al., 2001).

As a result of tectonic movements, older, deeply buried, and gently dipping upper Paleozoic to Mesozoic coal-bearing series were disrupted, folded, and often exposed to the surface, where they came into direct contact with the atmosphere. However, two conditions are required for fossil-fuel combustion. The rocks need to be eroded rapidly, and the time between their exposure and ignition should be short. Otherwise, weathering processes affect the potential fuel. During the prolonged oxidation, a significant part of the organic matter turns to humic acids, which are

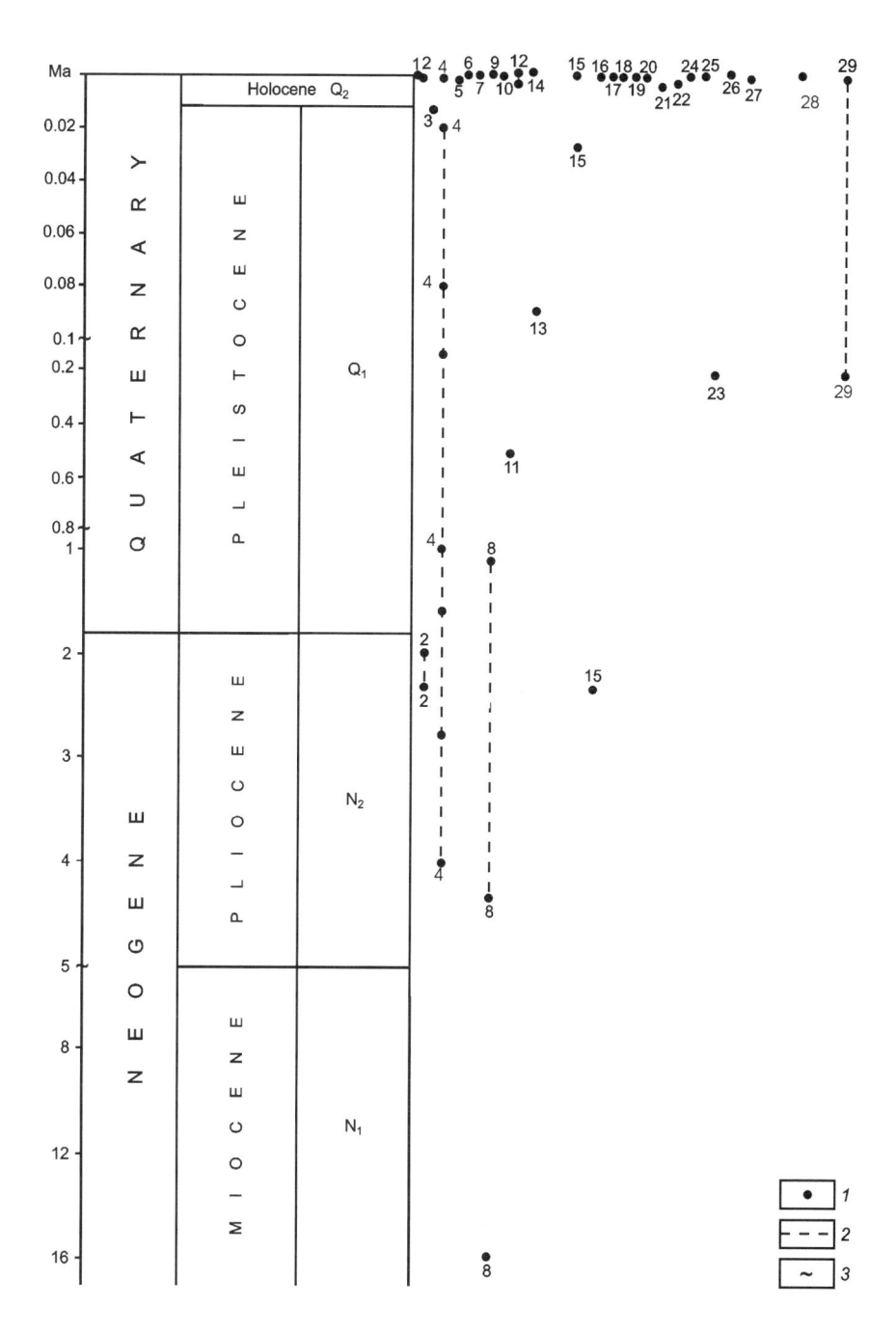

Figure 7. Ages of combustion events in Earth history. 1—Mist Mountain Formation (Canadian Cordilleras, Canada), Bustin and Mathews (1982); 2—Monterey Formation (California, USA), Arnold and Anderson (1907), Bentor et al. (1981); 3—Hat Creek (British Columbia, Canada), Church et al. (1979); 4—Powder River Basin (Wyoming, USA), Cosca et al. (1989), Heffern and Coates (2004); 5—Yangantau (Ural Mountains, Russia), Scherbak (1976); 6—Burning Cliff (Dorset, England), Cole (1974); 7—South-Scottish coal basin, UK, Eihvald (1864); 8—Hatrurim Formation (Israel, Jordan), Burg et al. (1991, 1999), Gur et al. (1995); 9—Khamarin-Khural-Khid (East Mongolia), Pokrovsky (1949); 10—Tul-i-Marmar, Zoh-i-Hait, Darreh Harrachi, Naf Khaneh (Iran), McLintock (1932); 11—Kenderlyk (Kazakhstan), Kalugin et al. (1991); 12—Ravat fire (Tajikistan), Ermakov (1935), Novikov (1993); 13—Angren (Uzbekistan), Zbarsky (1963); 14—Norilsk Plateau (north Siberia, Russia), Matukhina and Van (1965); 15—Kuznetsk Basin (southwest Siberia, Russia), Yavorsky and Radugina, (1932), Belikov (1933); 16—Jilinda River (east Siberia, Russia), Eihvald (1864); 17—Taymur River (east Siberia, Russia), Eihvald (1864); 18—Zhdanikha River and Chunya River (central Siberia, Russia), Menyailov et al. (1955); 19—Kureika River (east Siberia, Russia), Eihvald (1864); 20—Markha River (east Siberia, Russia), Lyakhovich (1953); 21—Ravensworth, (New South Wales, Australia), Whitworth (1958), Rattigan (1967); 22—Leigh Creek (Australia), Baker (1953); 23—Glenroy Valley (New Zealand), Tulloch and Campbell (1993); 24—Saarbruken (Germany), Eihvald (1864); 25—Lyons region (France), Eihvald (1864); 26—Chuya Valley (Altai Mountain, south Siberia, Russia); 27—Island Large Begichev (Taymyr Peninsula, north Siberia, Russia), Reverdatto (1973); 28—Borneo and East Kalimantan Islands (Indonesia), Whitehouse and Mulyana (2004); 29—Xinjiang (northwest China), de Boer et al. (2001). Boxes: 1—Reliably determined ages of fires (Ma); 2—time interval of discontinuous burning; 3—unconformities.

converted into low-molecular products and incombustible gases (mainly CO_2, CO, and H_2O). In this case, the released heat dissipates, and temperatures are not hot enough for ignition (Heffern and Coates, 2004). This is how weathered coals are formed; spontaneous ignitions of such coals are unknown. The oldest combustion metamorphic rocks from the Hatrurim Formation have been dated as ca. 16 Ma, which coincides with the beginning of late Miocene tectonic activity in the Dead Sea Rift Valley, when fast uplift and exposure of bituminous Mesozoic series occurred (Gur et al., 1995).

Coal Fires in Central Asia and Siberia

The sequence of geological events in the coal basins of central Asia can be illustrated by the history of mountain building and intermountain basins in Altai and Tien Shan (Fig. 8). These ranges are divided by a series of Mesozoic-Cenozoic depressions, the largest of which are the Junggar and Zaisan Basins. The Kuznetsk Basin, one of the richest coal fields in Russia, is located northeast of Altai. Formation of coal-bearing strata in central Asia and Siberia proceeded from the Pennsylvanian (Kuznetsk Basin) until the Middle Triassic–Jurassic (Zaisan and Junggar Basins) (Kamen-Kaye et al., 1988; Erofeev et al., 1991; Trifonov, 1999). The sedimentary basins evolved from shallow sea, through lagoon, to intracontinental freshwater basins, with reduced areas. The Cretaceous-Paleogene was characterized by intensive denudation of ancient mountains and the accumulation of fragmental material over coal-bearing strata. The situation changed dramatically in the Oligocene, when this territory was influenced again by the Indo-Eurasian collision. The whole Tien Shan, as well as the northeast and southwest margins of Altai, underwent orogenesis. From the end of the Neogene to the early Quaternary, areas of freshwater sedimentation were reduced sharply. The climate in the southern parts of Altai became more arid and dry, which is reflected in the appearance of red beds (Novikov, 2002). In the early Pleistocene, the peripheral regions of the Junggar and Zaisan depressions were uplifted. At this time, the present-day structure of the Zaisan Coal Basin and the Kenderlyk combustion metamorphic complex of ore paralavas were formed (Kalugin et al., 1991).

The new phase of orogenesis resulted in similar geological events in other parts of central Asia and south Siberia. In the Kuznetsk Depression, Permian to Pennsylvanian 10-km-thick coal-bearing sediments overlie Mississippian marine limestones and mudstones. This territory was downwarped at the end of the Permian (Novikov, 2002). In the Triassic, new erosion and accumulation of terrestrial volcanogenic-sedimentary series took place. At the Triassic-Jurassic boundary, at the southwestern margin of the Kuznetsk Basin, an elongate zone (90 × 10 km) that contains thick coal beds was deformed into steeply dipping folds cut by numerous faults. Further downwarping of the basin and aqueous sedimentation proceeded from the Jurassic until the Early Cretaceous. Based on the geological evidence (i.e., the considerable vertical displacement of thick burned-rock beds), Butov and Yavorsky (1922) related the first large-scale fires in the Kuznetsk Basin to a stage of Mesozoic tectonic activity. The second peak of ignitions occurred during the Pliocene, when a dry and warm climate prevailed (Yavorsky and Radugina, 1932). In historical times, small-scale fires have originated repeatedly throughout the Kuznetsk Basin; however, their consequences have been insignificant compared with ancient fires. Thus, there is reason to believe that combustion metamorphic events are synorogenic processes.

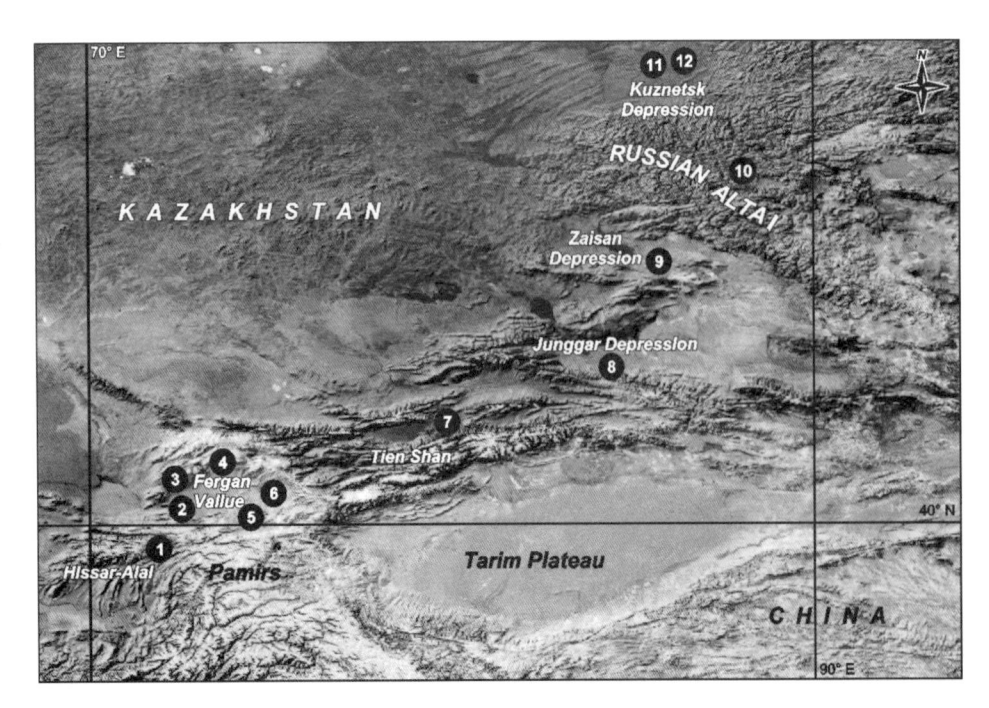

Figure 8. Locations of natural-combustion metamorphic complexes surrounding the Altai and Tien Shan Mountains, central Asia. 1—Ravat fire (Tajikistan), Ermakov (1935), Novikov, (1993); 2—Sulyuktin (Kyrgyzstan), Zbarsky (1963); 3—Angren (Uzbekistan), Abubakirov (1962), Zbarsky (1963); 4–7—Tash-Kumyr, Kyzyl-Kiy, Kok-Yangak, Dzhergaly (Kyrgyzstan), Zbarsky (1963); 8—combustion metamorphic rocks of Junggar (China), Saidov (1956); 9—Kenderlyk (Kazakhstan), Kalugin et al. (1991); 10—Chuya Valley (Altai Mountain, south Siberia, Russia); 11, 12—Kuznetsk Basin (south Siberia, Russia), Yavorsky and Radugina (1932), Belikov (1933).

The geological events that preceded underground coal fires may be reconstructed in detail for the Fan-Yagnob coal deposit in the southern Tien Shan of Tajikistan (Ermakov, 1935; Schukin, 1936; Skrobov, 1968). The Ravat fire burns in the Yagnob River valley, which separates the Zeravshan and Hissar Ranges in the Pamir-Alai (Fig. 9). Before the Pennsylvanian, the area was covered by a sea. During the Jurassic, shallow lagoons, lakes, and marshes formed in the depressions, where coal accumulation took place. A severe arid climate in the Late Jurassic led to a change from lagoon deposits to red sandstones and conglomerates. The arid climate persisted throughout the Paleogene in Turkestan and caused intensive weathering. At that time, gravel and sand series formed and giant taluses filled up mountain creeks and valleys. In the Neogene, the situation changed drastically. Mesozoic and Paleogene sedimentary rocks were deformed to form sharp, often overturned, folds. This resulted in numerous faults, fractures, and thrusts, along which vertical displacements are evident. In the Neogene, this territory was uplifted ~2.5 km and later an additional 1–5 km in the Quaternary. Today, the mountain peaks of the Zeravshan Ridge are 4500–5000 m high. Most of the Jurassic deposits in the Zeravshan region have been eroded. They are preserved only as isolated areas, one of which is the Fan-Yagnob coal deposit.

The Ravat fire burns in a canyon of the Yagnob River, which has slopes that have a total height difference of 2960 m to 1700 m. The fire is confined to a thrust zone of Cenozoic age. Jurassic coal-bearing rocks are deformed strongly and fractured. Bedding dips are 40°–50°. Thus, throughout the Pliocene-Quaternary, this region was characterized by "ideal" geological and topographical conditions for fire ignition. However, combustion has occurred only in historical times. The trigger of the Ravat fire seems to have been postponed by Quaternary glaciations. The moraine deposits, which were washed out by the Yagnob River, suggest that this valley was covered by a glacier in recent times.

In the same region, fires started much earlier in areas that were not affected by glaciation. An example of such a fire burns in the Angren coal deposit in Uzbekistan (Abubakirov, 1962; Skrobov, 1968). It is located between the Kuraminsky and Chatkalsky ranges in northeast Tien Shan (see Fig. 8). The Jurassic coal-bearing unit is overlain by Paleogene to Neogene sediments (~500 m). Quaternary sediments (100 m) overlie an eroded surface of Pliocene molasse and form five river terraces. The burned-rock zone reaches 52 m in thickness and extends to 80 m depth. Melted and slag-like rocks are located in the bottom part of the burned horizon; the middle section is occupied by glassy clinkers; and the upper part is composed of slightly annealed metapelites. Quaternary loess sediments of two low terraces have been affected by fire. The Holocene gravels of the third terrace, which were deposited directly on the burned rocks,

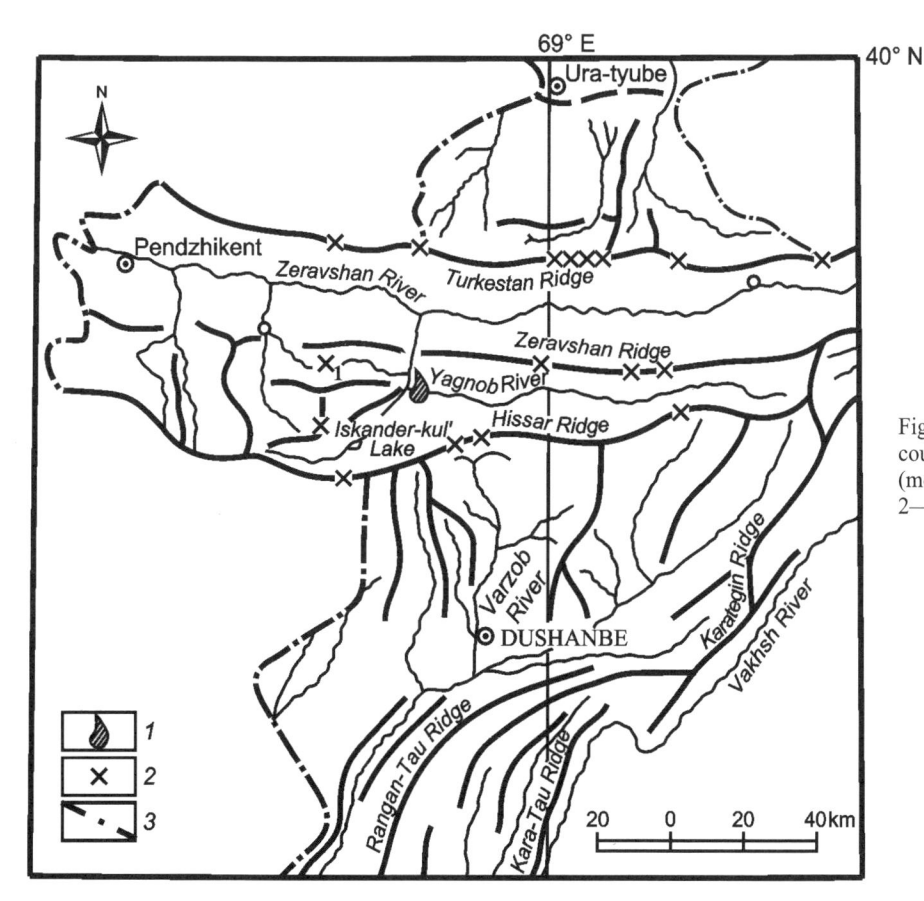

Figure 9. Location of Ravat fire in the mountain country of the Pamir-Alai, northwest Tajikistan (modified from Schukin, 1936). 1—Ravat fire; 2—ridges with passes; 3—frontier of Tajikistan.

are not baked. Hence, the ignition occurred much earlier here (i.e., in the early Quaternary) compared with the Ravat fire. The absence of moraine deposits in the lower and middle courses of the Angren River indicates that the glaciers of the Chatkalsky and Kuraminsky Ranges did not descend down to this level and did not prevent fire ignition. Coal horizons were ignited directly during development of the river valley.

The coal fire at Hat Creek, British Columbia, has been burning since before the last glacial cycle in Canada (Church et al., 1979) as a consequence of ignition due to forest or brush fires during interglacial periods. It is worth noting that all coal fires in eastern Siberia that occur in the giant Tungus Coal Basin (Fig. 10) are Holocene in age. These include fires in valley slopes of the Nizhnyaya Tunguska, Taymur, and Kureika Rivers (Eihvald, 1864); the Chunya River (inflow of the Podakamennaya Tunguska River) (Menyailov et al., 1955); and the Markha River (inflow of the Vilyui River) (Lyakhovich, 1953). Thin layers of combustion metamorphic rocks of Quaternary age also have been found in the arctic territories of Russia (the Norilsk plateau, Matukhina and Van, 1965; Big Begichev Island on the east shore of the Taimyr Peninsula, Reverdatto, 1973). Most likely, all these fires became possible during the Holocene warming periods. Thus, glaciation, one of the major factors that shaped the Earth's surface during the Quaternary, also is a major factor in young combustion metamorphic events.

CONCLUSIONS

Pyrogenic landscapes occur in fossil-fuel basins all over the world. For the development of prolonged fires, a combination of factors is necessary, including significant amounts of coal, bitumen, oil, or gas; steeply dipping beds; the occurrence of fuel beds above burning horizons; rock joints; and deep water tables.

The unique phases that occur in combustion metamorphic rocks are determined by the protolith and thermal source, i.e., the fossil fuel available for burning. The hot gaseous products of combustion rise upward and cause the ignition of overlying fuel horizons and the thermal alteration of host rocks. Heat and mass transfer in combustion metamorphic systems occurs mainly by flame (or gas flow). The convective redistribution of heat is a major factor in the thermal metamorphism of host rocks. Therefore, the zone of maximum combustion metamorphic transformation of a protolith always occurs above the burning horizon. During coal fires, the protolith-transformation temperature may reach as high as 1450 °C—for bitumen combustion it is 1650 °C, and for oil fires, 2400 °C.

At moderate temperatures (500–800 °C), combustion metamorphism mainly stimulates dehydration and decarbonation; it is not accompanied by melting. These kinds of combustion metamorphic rocks comprise the main part of combustion metamorphic complexes.

The diversity of combustion metamorphic rocks is limited and is governed mainly by protolith compositions. Clinkers and buchites derived from pelites are the most common. Basic ferriferous paralavas form from protoliths enriched in iron minerals when heated to temperatures of 1000–1300 °C. Combustion metamorphism of calcareous silica-deficient rocks occurs according to the "decarbonation series" (Bowen, 1940) and, ultimately, results in the appearance of "Portland cement" minerals. Often,

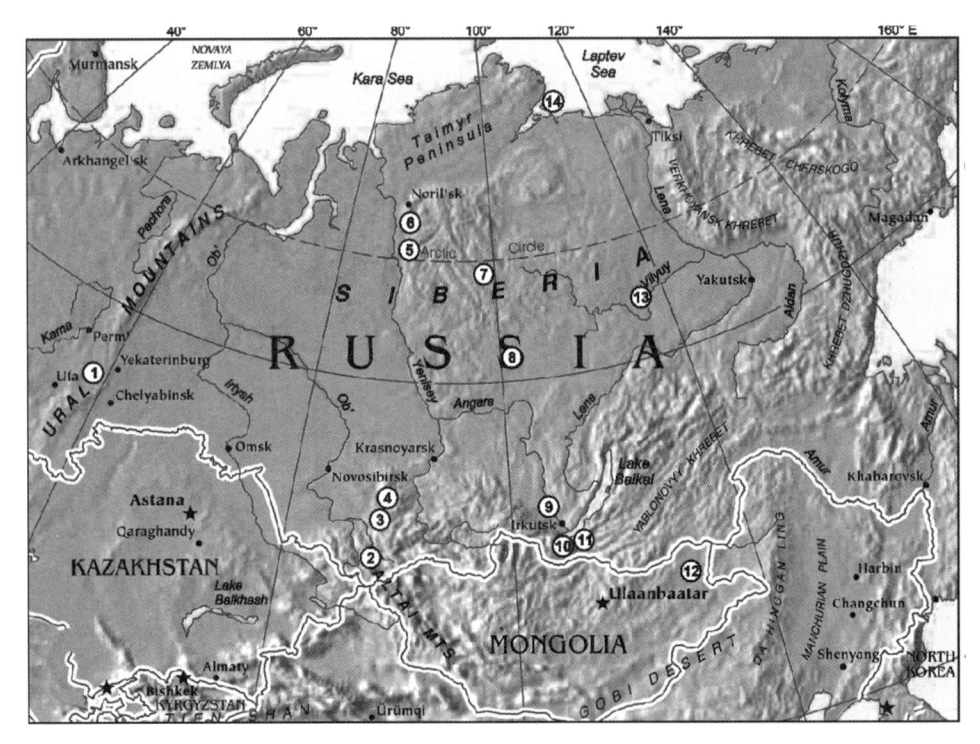

Figure 10. Known combustion metamorphic complexes in the Siberia region. 1—Yangantau (Ural Mountains), Scherbak (1976); 2—Chuya Valley (Altai Mountain); 3–4—Kuznetsk Basin (southwest Siberia), Butov and Yavorsky (1922), Yavorsky and Radugina (1932), Belikov (1933); 5—Kureika River (north Siberia), Eihvald (1864); 6—Norilsk Plateau (north Siberia), Matukhina and Van (1965); 7—Taymur River (east Siberia), Eihvald (1864); 8—Chunya River (central Siberia), Menyailov et al. (1955); 9—Cheremkhovo Coal Basin (east Siberia); 10–11—Gusinoosersky and Tankhoy coal deposits (Buryatia, east Siberia), Tugovik (1979); 12—Khamarin-Khural-Khid (East Mongolia), Pokrovsky (1949); 13—Markha River (east Siberia), Lyakhovich (1953); 14—Large Begichev Island (Laptev Sea, north Siberia, Russia), Reverdatto (1973).

combustion metamorphic rocks are accompanied by fumarolic mineralization and bitumen.

The mineral diversity of combustion metamorphic rocks is considerable. Anhydrous oxides, silicates, and aluminosilicates are dominant. At the surface, high-temperature parageneses are replaced rapidly by products of rehydration, recarbonation, sulfatization, and oxidation. Low- and medium-grade metacarbonate combustion metamorphic rocks are most intensely affected by alteration. Alternatively, clinkers, paralavas, breccias, and high-grade metacarbonate rocks are resistant to weathering. They control the topography of hills and ridges that are composed of these combustion metamorphic rocks.

The distribution of combustion metamorphic complexes in time and space is governed by the geodynamic evolution of terrains, as well as climate and geomorphologic factors. The maximum number of pyrogenic processes coincides with the main phases of tectonic activity and associated episodes of fracturing and erosion of fuel-bearing sedimentary rocks. Most commonly, Holocene fires are confined to interglacial periods. In the global carbon cycle, natural fossil-fuel fires return carbon to the atmosphere, mostly in the form of CO_2, which has been stored in rocks and withdrawn from biogeochemical circulation for millions of years. The spontaneous combustion of fossil fuels also affects the environment adversely, by releasing smoke and noxious gases, including CO, CO_2, NO_x, SO_x, H_2S, and C_nH_m.

ACKNOWLEDGMENTS

Our grateful thanks go to reviewers Yehoshua Kolodny (Picard Professor Emeritus of Geology, Institute of Earth Sciences, the Hebrew University of Jerusalem, Israel), Vladimir Reverdatto (Professor-Academician, Institute of Mineralogy and Petrography of the Siberian Branch of the Russian Academy of Sciences, Novosibirsk), and Hal Gluskoter (U.S. Geological Survey, Reston, Virginia), who radically improved the final form of this paper. The authors thank Yehoshua Kolodny for his permission to use his photo of the Hatrurim Formation landscape. We also gratefully acknowledge Glenn B. Stracher (East Georgia College, Swainsboro, Georgia, USA) and Yevgeny Vapnik (Ben-Gurion University of the Negev, Israel), who reviewed the draft and made helpful suggestions for improvement. In addition, we thank Yevgeny Vapnik and Rodney Grapes (Institut für Mineralogie, Petrologie und Geochemie, Freiburg, Germany), and the library service of the United Institute of Geology, Geophysics, and Mineralogy of the Siberian Branch of the Russian Academy of Sciences, Novosibirsk, for helping us collect information about the geology and environmental impact of fossil-fuel fires. This work was supported by the Foundation of the President of the Russian Federation for the Support of Leading Scientific Schools (program: Siberian Metamorphic School, project number NSh-4922.2006.5), the Russian Foundation for Basic Research (project number 05-05-65036), and the Siberian Division of the Russian Academy of Sciences (integration project number 105).

REFERENCES CITED

Abubakirov, I.K., 1962, Underground fire and conditions of burned rocks formation at the Angren deposit Voprosy, in Geologii Uzbekistana (Questions of Uzbekistan Geology), Issue 3: Tashkent, Uzbekistan, Tashkent Publishing House, Uzbekskoi Akademii Nauk SSR, p. 30–37.

Alvarez, R., and Gomez, C., 1984, Para-lava basalticas y metamorfitas generadas por combustion espontanea de mantos de carbon, Formation Cerrejon, Colombia: Geologia Norandina, v. 10, p. 3–9.

Arnold, R., and Anderson, R., 1907, Metamorphism by combustion of the hydrocarbons in the oil-bearing shale of California: The Journal of Geology, v. 15, p. 750–758.

Baker, G., 1953, Naturally fused coal ash from Leigh Creek, South Australia: Transaction of the Royal Society of South Australia, v. 76, p. 1–20.

Basi, M.A., and Jassim, S.Z., 1974, Baked and fused Miocene sediments from Injana area, Hemrin South, Iraq: Journal of the Geological Society of Iraq, v. VII, p. 1–14.

Belikov, B.P., 1933, Composition of some burned rocks from the Kuzbass: Trudy Petrographicheskogo Instituta Akademii Nauk SSSR, no. 4, p. 91–100.

Bentor, Y.K., 1984, Combustion metamorphic glasses: Journal of Non-Crystalline Solids, v. 67, p. 433–448, doi: 10.1016/0022-3093(84)90168-6.

Bentor, Y.K., Gross, S., and Heller, L., 1963, Some unusual minerals from the "Mottled Zone" complex, Israel: The American Mineralogist, v. 48, p. 924–930.

Bentor, Y.K., Kastner, M., Perlman, I., and Yellin, Y., 1981, Combustion metamorphism of bituminous sediments and the formation of melts of granitic and sedimentary composition: Geochimica et Cosmochimica Acta, v. 45, p. 2229–2255, doi: 10.1016/0016-7037(81)90074-0.

Bogoslovsky, F., 1842, Notes about Zeravshan River value and the surrounding mountains: Gornyi Zhurnal (Mining Journal), Part 4, book 10, p. 1–21 (in Russian).

Bowen, N.L., 1940, Progressive metamorphism of siliceous limestone and dolomite: The Journal of Geology, v. 48, p. 225–274.

Burg, A., Starinsky, A., Bartov, Y., and Kolodny, Y., 1991, Geology of the Hatrurim Formation ("Mottled Zone") in the Hatrurim Basin: Israel Journal of Earth Sciences, v. 40, p. 107–124.

Burg, A., Kolodny, Y., and Lyakhovsky, V., 1999, Hatrurim-2000: The "Mottled Zone" revisited, forty years later: Israel Journal of Earth Sciences, v. 48, p. 209–223.

Bustin, R.M., and Mathews, W.H., 1982, In situ gasification of coal, a natural example: History, petrology, and mechanics of combustion: Canadian Journal of Earth Sciences, v. 19, p. 514–523.

Butov, P.I., and Yavorsky, V.I., 1922, Materialy dlya geologii Kuznetskogo kamennougol'nogo basseyna, Yugo-zapadnaya okraina (Materials for the geology of Kuznetsk coal basin, south-west district of the basin): Petersburg, Russia, Publishing House of Ivan Fedorov, 58 p.

Church, B.N., Matheson, A., and Hora, Z.D., 1979, Combustion metamorphism in the Hat Creek area, British Columbia: Canadian Journal of Earth Sciences, v. 16, p. 1882–1887.

Clark, B.H., and Peacor, D.R., 1992, Pyrometamorphism and partial melting of shales during combustion metamorphism; mineralogical, textural, and chemical effects: Contributions to Mineralogy and Petrology, v. 112, no. 4, p. 558–568, doi: 10.1007/BF00310784.

Cole, D., 1974, A recent example of spontaneous combustion of oil shales: Geological Magazine, v. 111, p. 355–356.

Cosca, M.A., and Peacor, D.R., 1987, Chemistry and structure of esseneite, $(CaFe^{3+}AlSiO_6)$, a new pyroxene produced by pyrometamorphism: The American Mineralogist, v. 72, p. 148–156.

Cosca, M.A., Roland, R.R., and Essene, E.J., 1988, Dorrite $[Ca_2(Mg_2Fe_4^{3+})(Al_4Si_2)O_{20}]$, a new member of the aenigmatite group from a pyrometamorphic melt-rock: The American Mineralogist, v. 73, p. 1440–1448.

Cosca, M.A., Essene, E.J., Geissman, J.W., Simmons, W.B., and Coates, D.A., 1989, Pyrometamorphic rocks associated with naturally burned coal beds, Powder River Basin, Wyoming: The American Mineralogist, v. 74, p. 85–100.

de Boer, C.B., Dekkers, M.J., and van Hoof, T.A.M., 2001, Rock-magnetic properties of TRM carrying baked and molten rocks straddling burnt coal seams: Physics of the Earth and Planetary Interiors, v. 126, p. 93–108, doi: 10.1016/S0031-9201(01)00246-1.

Dobretsov, N.L., Kirdyashkin, A.G., and Kirdyashkin, A.A., 2001, Deep-level geodynamics: Novosibirsk, Russia, Publishing House of the Siberian Branch of the Russian Academy of Sciences, filial "GEO", 409 p.

Eihvald, E.I., 1864, About discoveries of coal and graphite deposits and about the underground fires on Lower Tunguska and Taymur: Gornyi Zhurnal (Mining Journal), book 3, no. 7, p. 117–153 (in Russian).

Ermakov, N.P., 1935, Pasrud-Yagnob coal deposit and burning mines of Kan-Tag mountain, in Buracheck, A.P, and Chuenko, P.P., eds., K Geologii Kamennougol'nyh Mestorozhdenyi Tadzhikistana (Geology of Tajikistan Coal Deposits): Moscow, Russia, Publishing House Akademii Nauk SSSR, p. 47–66 (in Russian).

Erofeev, V.S., Kakenov, S.G., and Nahgal, G.P., 1991, Structure of folded base of Zaisan Depression: Izvestiya Akademii Nauk Kazakhskoi S.S.R., Seria Geologicheskaya, no. 4, p. 15–22 (in Russian).

Farrand, M.G., and Newton, A.W., 1985, Pseudo-igneous phenomena in the Springfield Basin: Quarterly Geological Notes: Geological Survey of South Australia, v. 95, p. 2–7.

Feldman, V.I., Bychkov, A.M., Dikov, Y.P., and Krivtsova, V.Y., 1994, Tengezines—Glasses from the focus of oil fire: Doclady Akademii Nauk, v. 339, no. 2, p. 239–242.

Fermor, L.L., 1918, Preliminary note on the burning of coal seams at the outcrop: Transactions of the Mining, Geological, and Metallurgical Institute of India, v. 12, p. 50–63.

Foit, F.F., Hooper, R.L., and Rosenberg, P.E., 1987, An unusual pyroxene, melilite, and iron oxide mineral assemblage in a coal-fire buchite from Buffalo, Wyoming: The American Mineralogist, v. 72, p. 137–147.

Goodarzi, F., and Jerzykiewicz, T., 1986, Petrology of a burning bituminous coal seam at Coalspur, Alberta: Geological Survey of Canada Bulletin, v. 86, no. 1B, p. 421–427.

Gross, S., 1977, The mineralogy of the Hatrurim Formation, Israel: Geological Survey of Israel Bulletin, v. 70, 80 p.

Gur, D., Steinitz, G., Kolodny, Y., Starinsky, A., and McWilliams, M., 1995, $^{40}Ar/^{39}Ar$ dating of combustion metamorphism ("Mottled Zone", Israel): Chemical Geology, v. 122, p. 171–184, doi: 10.1016/0009-2541(95)00034-J.

Heffern, E.L., and Coates, D.A., 2004, Geological history of natural coal-bed fires, Powder River Basin, USA, in Stracher, G.B., ed., Coal Fires Burning Around the World: A Global Catastrophe: International Journal of Coal Geology, v. 59, no. 1–2, p. 25–47.

Hooper, R.L., 1982, Mineralogy of a Coal Burn near Kemmerer, Wyoming [Master's thesis]: Pullman, Washington, Washington State University, 86 p.

Kalugin, I.A., Tretyakov, G.A., and Bobrov, V.A., 1991, Zhelezorudnye Basal'ty v Gorelyh Porodah Vostochnogo Kazakhstana (Iron Ore Basalts in Burned Rocks of Eastern Kazakhstan): Novosibirsk, Russia, Publishing House Nauka, 80 p.

Kamen-Kaye, M., Meyerhoff, A.A., and Taner, I., 1988, Junggar Basin: A Permian to Cenozoic intermontane complex in northwestern China: Senckenbergiana Lethaea, v. 69, no. 3–4, p. 289–313.

Khoury, H.N., and Milodowski, T.E., 1992, High-temperature metamorphism and low-temperature retrograde alteration of spontaneously combusted marls, the Maqarin cement analogue, Jordan, in Proceedings of the 7th International Symposium Water-Rock Interaction, Volume 2, Park City, Utah: Rotterdam, The Netherlands, Brookfield, p. 1515–1518.

Knunyants, I.L., ed., 1990, Himicheskaja Jenciklopedija (Chemical Encyclopedia), Volume 1: Moscow, Russia, Publishing House Bol'shaja Rossijskaja Jenciklopedija, p. 300–428.

Knunyants, I.L., ed., 1998, Himicheskaja Jenciklopedija (Chemical Encyclopedia), Volume 3: Moscow, Russia, Publishing House Bol'shaja Rossijskaja Jenciklopedija, p. 594–598.

Kolodny, Y., and Gross, S., 1974, Thermal metamorphism by combustion of organic matter: Isotopic and petrological evidence: The Journal of Geology, v. 82, p. 489–506.

Koltsov, K.S., and Popov, B.G., 1978, Samovozgoranie tverdyh veshhestv i materialov i ih profilaktika (Self-Ignition of Solid Compounds and Materials and their Preventive Maintenance): Moscow, Russia, Publishing House Khimicheskaya Promyshlenost', 184 p.

Kozlovsky, E.A., ed., 1986, Gornaja Jenciklopedija (Mining Encyclopedia), Volume 2: Moscow, Russia, Publishing House Sovetskaja Jenciklopedija, 575 p.

Kozlovsky, E.A., ed., 1987, Gornaja Jenciklopedija (Mining Encyclopedia), Volume 3: Moscow, Russia, Publishing House Sovetskaja Jenciklopedija, 592 p.

Kozlovsky, E.A., ed., 1989, Gornaja Jenciklopedija (Mining Encyclopedia), Volume 4: Moscow, Russia, Publishing House Sovetskaja Jenciklopedija, 623 p.

Lyakhovich, V.V., 1953, Alteration of sedimentary rocks caused by underground coal fire: Priroda, no. 7, p. 107–110 (in Russian).

Matthews, A., and Gross, S., 1980, Petrologic evolution of the "Mottled Zone" (Hatrurim) metamorphic complex of Israel: Israel Journal of Earth Sciences, v. 29, p. 93–106.

Matthews, A., and Kolodny, Y., 1978, Oxygen isotopes fractionation in decarbonation metamorphism—The Mottled Zone event: Earth and Planetary Science Letters, v. 39, p. 179–192, doi: 10.1016/0012-821X(78)90154-1.

Matukhina, B.G., and Van, A.V., 1965, Permian burnt rocks of the Norilsk region: Geologiyai Geofizika, no. 6, p. 119–122 (in Russian).

McLintock, W.E.P., 1932, On the metamorphism produced by the combustion of hydrocarbons in the Tertiary sediments of southwest Persia: Mineralogical Magazine, v. 23, p. 207–226, doi: 10.1180/minmag.1932.023.139.01.

Menyailov, A.A., Lapin, V.V., and Lebedev A.P., 1955, Some fused sedimentary rocks from Central Siberia: Izvestiya Akademii Nauk SSSR, Seria Geologicheskaya, v. 3, p. 106–113 (in Russian).

Miyashiro, A., 1957, Cordierite–indialite relations: American Journal of Science, v. 255, p. 43–62.

Moticska, P., 1977, Generacion de magmas y autometamorfismo por combustion subterranean de carbones y de limolitas carbonosas en la Formacion Marcelina: Perija: Congreso Geologico Venezolano, v. 2, p. 663–691.

Moticska, P., 1978, Generacion de magmas y autometamorfismo por combustion subterranean de carbon y de limolitas carbonosas en la Formacion Marcelina, Perija: Boletin de Geologia, Caracas, v. 13, no. 24, p. 183–217.

Novikov, I.S., 2002, Late Paleozoic, middle Mesozoic, and late Cenozoic stages of the Altai orogeny: Russian Geology and Geophysics, v. 43, no. 5, p. 434–445.

Novikov, V.P., 1993, Organic derivatives of a coal fire on the Fan-Yagnob deposit: Izvestiya Akademii Nauk Respubliki Tajikistan, Otdelenie Geokogicheskikh Nauk, no. 4, p. 51–58 (in Russian).

Novikov, V.P., and Suprychev, V.V., 1986, Parameters of modern mineral-forming processes associated with underground coal combustion at Fan-Yagnob deposit, in Mineralogija Tadzhikistana (Mineralogy of Tajikistan): Dushanbe, Tajikistan, Publishing House of the Academy of Sciences of Tajik SSR, p. 91–104 (in Russian).

Pokrovsky, P.V., 1949, Sal ammoniac from lignite deposit Khamarin-Khural-Khid, in the Mongolian National Republic: Zapiski Vsesoyusnogo Mineralogicheskogo Obshchestva (Proceedings of Russian Mineralogical Society), part 78, no. 1, p. 8–45 (in Russian).

Rattigan, J.H., 1967, Phenomena about Burning Mountain, Wingen, New South Wales: Australian Journal of Earth Sciences, v. 30, no. 5, p. 183–184.

Reverdatto, V.V., 1973, The facies of contact metamorphism: Canberra, Australia, Australian National University of Canberra, Geological Publication no. 223, 263 p.

Ron, H., and Kolodny, Y., 1992, Paleomagnetic and rock magnetic study of combustion metamorphic rocks in Israel: Journal of Geophysical Research, v. 97, p. 6927–6939.

Saidov, M.N., 1956, Mesoneozoic terrestrial strata of Junggar Depression: Izvestiya Akademii Nauk SSSR, Seria Geologicheskaya, no. 10, p. 85–97 (in Russian).

Scherbak, V.P., 1976, Spa at the burning hill: Khimiya i Zizn, no. 1, p. 83–85 (in Russian).

Schreyer, W., and Schairer, J.F., 1961, Composition and structural states of anhydrous Mg-cordierite: A reinvestigation of the central part of the system $MgO-Al_2O_3-SiO_2$: Journal of Petrology, v. 2, p. 324–406.

Schreyer, W., Maresch, W.V., Daniels, P., and Wolfsdorff, P., 1990, Potassic cordierites: Characteristic minerals for high-temperature, very low-pressure environments: Contributions to Mineralogy and Petrology, v. 105, p. 162–172, doi: 10.1007/BF00678983.

Schukin, I.S., 1936, General essay of Tajikistan, in Schukin, I.S., ed., Materialy Tadzhiksko- Pamirskoj jekspedicii 1933 (Materials of Tajik-Pamirs Expedition in 1933): Leningrad, Russia, Publishing House of Tajik-Pamirs Expedition, Issue 23, p. 7–86 (in Russian).

Sigsby, R.J., 1966, The present general lack of "scoria" in two burning lignite areas in North Dakota: Proceedings of the North Dakota Academy of Science, v. 9, p. 7–14.

Skrobov, S.A., ed., 1968, Geologija mestorozhdenij uglja i gorjuchih slancev SSSR, t.6: Ugol'nye bassejny i mestorozhdenija Srednej Azii (Geology of Coal and Combustible Shale Deposits of the USSR, Volume 6: Coal Basins and Deposits of Central Asia): Moscow, Russia, Publishing House Nedra, 600 p.

Sokol, E.V., Sharygin, V.V., Kalugin, V.M., Volkova, N.I., and Nigmatulina, E.N., 2002, Fayalite and kirschsteinite solid solutions in melts from burned spoil-heaps, South Urals, Russia: European Journal of Mineralogy, v. 14, p. 795–807, doi: 10.1127/0935-1221/2002/0014-0795.

Sokol, E.V., Maksimova, N.V., Nigmatulina, E.N., Sharygin, V.V., and Kalugin,

V.M., 2005, Combustion Metamorphism: Novosibirsk, Russia, Publishing House of the Siberian Branch of the Russian Academy of Sciences, 286 p. (in Russian with English summary).

Srebrodolsky, B.I., 1989, Tajny Sezonnyh Mineralov (Secrets of Seasonal Minerals): Moscow, Russia, Publishing House Nauka, 144 p.

Stracher, G.B., and Taylor, T.P., 2004, Coal fires burning out of control around the world: Thermodynamic recipe for environmental catastrophe, in Stracher, G.B., ed., Coal Fires Burning Around the World: A Global Catastrophe: International Journal of Coal Geology, v. 59, no. 1–2, p. 7–17.

Svensen, H., Dysthe, D.K., Bandlien, E.H., Sacko, S., Coulibaly, H., and Planke, S., 2003, Subsurface combustion in Mali: Refutation of the active volcanism hypothesis in West Africa: Geology, v. 31, no. 7, p. 581–584, doi: 10.1130/0091-7613(2003)031<0581:SCIMRO>2.0.CO;2.

Trifonov, V.G., 1999, Neotektonika Evrazii (Neotectonics of Eurasia): Moscow, Russia, Publishing House Nauchny Mir, 252 p.

Tugovik, G.I., 1979, Prirodno-Gorelye Porody Burjatii (Natural Burned Rocks of Buryatia): Ulan-Ude, Russia, Publishing House Buryatia, 61 p.

Tulloch, A.J., and Campbell, J.K., 1993, Clinoenstatite-bearing buchites probably from combustion of hydrocarbon gases in a major thrust zone, Glenroy Valley, New Zealand: The Journal of Geology, v. 101, p. 404–412.

Venkatesh, V., 1952, Development and growth of cordierite in para-lavas: The American Mineralogist, v. 37, p. 831–848.

Whitehouse, A.E., and Mulyana, A.A.S., 2004, Coal fires in Indonesia, in Stracher, G.B., ed., Coal Fires Burning Around the World: A Global Catastrophe: International Journal of Coal Geology, v. 59, no. 1–2, p. 91–97.

Whitworth, H.F., 1958, The occurrence of some fused sedimentary rocks at Ravensworth, N.S.W.: Journal of the Royal Society of New South Wales, v. 92, p. 204–208.

Yavorsky, V.I., and Radugina, L.V., 1932, Coal-fire combustion and attendant events in Kuznetsk Basin: Gornyi Zhurnal, v. 10, p. 55–59.

Žáček, V., Skála, R., Chlupáčová, M., and Dvořák, Z., 2005, Ca-Fe^{3+}-rich Si-undersaturated buchite from Želénky, North-Bohemian brown coal basin, Czech Republic: European Journal of Mineralogy, v. 17, p. 623–633.

Zbarsky, M.I., 1963, Mineralogical and petrographical features of burned rocks from Central Asia: Zapiski Kirgizskogo Otdeleniya Vsesouznogo Mineralogicheskogo Obshchestva (Proceedings of Kirgiz Division of Russian Mineralogical Society): Frunze, Kyrgyztan, Publishing House Ilim, Issue 4, p. 53–67.

MANUSCRIPT ACCEPTED BY THE SOCIETY 7 MARCH 2007

Mineralogy and petrography of iron-rich slags and paralavas formed by spontaneous coal combustion, Rotowaro coalfield, North Island, New Zealand

M. Naze-Nancy Masalehdani*
Centre National de la Recherche Scientifique, 8110 "Processus et Bilans des Domaines Sédimentaires," Université de Lille 1, Bâtiment SN5, 59655 Villeneuve d'Ascq Cedex, France

Philippa M. Black
Huldrych W. Kobe
Geology Department, Auckland University, Private Bag 92019, Auckland, New Zealand

ABSTRACT

Pyrometamorphism of coal measures that overlie underground burnt-coal seams in the southern area of the Rotowaro coalfield, New Zealand, has produced porcellanites that enclose lenses of iron-rich magnetite and hematite-bearing slag-like rocks, which show various degrees of oxidation. Paralavas that are associated with the iron-rich lenses form stalactites around fissures and gas-escape vents and intrude the porcellanites. The slags are unusually rich in iron and contain magnetite, hematite, hercynite, titaniferous magnetite, and minor fayalite and silicate glass. Iron oxides in these rocks exhibit a variety of textures and morphologies, including dendritic, quench, exsolution, and oxidation. The paralavas contain abundant glass, which encloses feathery crystals of fayalite and orthoferrosilite, plagioclase, tridymite, cristobalite, and minor magnetite.

Phase-equilibria data indicate that the paralavas and slags were formed at temperatures in the range of 1000–1600 °C. Composition plots of local coal measure sediments, a siderite nodule, and various porcellanites, slags, and paralavas with respect to SiO_2, Fe_2O_3, and Al_2O_3 indicate that the porcellanites are iron-rich when compared with the unmetamorphosed coal measures. The slags, irrespective of their oxidation state, plot on the iron-enrichment trend shown by the porcellanites. Although the siderite nodule lies within the iron-enrichment trend, most of the slags are more iron-rich than the siderite nodule. The paralavas diverge from the iron-enrichment trend, suggesting that they formed by partial melting of the porcellanites. The Rotowaro samples represent some of the most iron-rich natural slags collected from a combustion-metamorphic environment to date.

Keywords: spontaneous-coal fire, combustion metamorphism, paralava, iron-rich slag, high-temperature minerals.

*Naze-Nancy.Masalehdani@univ-Lille1.fr

INTRODUCTION

Thermally baked and fused sediments produced by coal-seam combustion are common geological features that have been described from many localities throughout the world, including the United States (Foit et al., 1987; Cosca et al., 1989; Heffern and Coates 2004), Romania (Rădan and Rădan, 1998), Czech Republic (Tyráček, 1994), Australia (Ellyett and Fleming, 1974), New Zealand (Adams, 1978; Masalehdani, 1985), and China (Zhang, 1998; de Boer et al., 2001). Combustion metamorphic rocks, also known as pyrometamorphic rocks, contain characteristic high-temperature minerals and silicate melts with a range of unusual physical features. These features have been described by a variety of terms, including paralavas (Fermor, 1918; Sen Gupta, 1960; Essene et al., 1984), pseudo-igneous rocks (Johnson and Bucknell, 1959), scoria (Adams, 1978), buchites (Hensen and Gray, 1979; Foit et al., 1987), melted-vitrified scoriaceous rocks (Bustin and Mathews, 1982), and clinker (Lindqvist et al., 1985; Heffern and Coates, 1997). In many localities, the baked and fused sediments are enriched in iron; in a few localities, extreme iron enrichment also has led to the occurrence of iron-rich slag-like rocks (Cosca et al., 1989; Clark and Peacor, 1992).

Coal fires can occur in natural, near-surface exposures, underground and in openpit mines, and in coal-waste heaps and stockpiles. Phenomena such as lightning strikes and bush fires may start coal-seam fires directly or indirectly; however, in situ spontaneous combustion of coal is more common. In situ combustion results from heat generated by the exothermic oxidation of organic matter and frequently of pyrite in the coal. A coal's propensity to combust spontaneously depends on the physical and chemical properties of the coal, the nature and condition of the surrounding sediments (e.g., capacity to dissipate heat) (Colaizzi, 2004), and access to oxygen. In the case of buried coal seams, atmospheric oxygen needs to penetrate into the rock sequence, often to a depth of several hundred meters, for combustion to occur.

During in situ coal combustion, overburden rocks also are affected by heating, which results in changes to their strength, composition, and coherence. As baking and dehydration of sediments progress, shrinkage occurs, which leads to the opening of large lateral fissures and cavities. These openings control the paths of escaping hot gases and lead to the formation of well-defined chimneys, through which toxic hot gases, water vapor, and more volatile constituents are expelled from the underground coal combustion. The chimneys also facilitate the ingress of the oxygen that is necessary to support and advance the subsurface combustion of the coal seam, which eventually reduces it to a fused ash deposit. A reduction in volume of 40% to 60% may take place in the enclosing sediments that are being baked and/or melted; this reduces support for the overburden rock and often leads to local collapse.

In the northeastern and southeastern parts of the Rotowaro coalfield, spontaneous combustion of coal seams has metamorphosed and melted the enclosing coal measure sediments. The degree of thermal alteration that is produced by burning of the coal is variable and very much localized. This study focuses on a single outcrop that is situated 2 km southeast of Rotowaro, Waikato coal region, on the North Island of New Zealand (Fig. 1). The pyrometamorphosed coal measures include baked, partially melted, and entirely melted rocks. Metamorphism has produced mineral assemblages that indicate crystallization under a range of oxidizing conditions. The objectives of this paper are to (1) describe the field occurrence of the porcellanites, melts (paralavas), and the exceptionally iron-rich slags, (2) explain in detail the mineralogy, petrography, and chemistry, and the mode of formation of different phases in the iron-rich slags and paralavas,

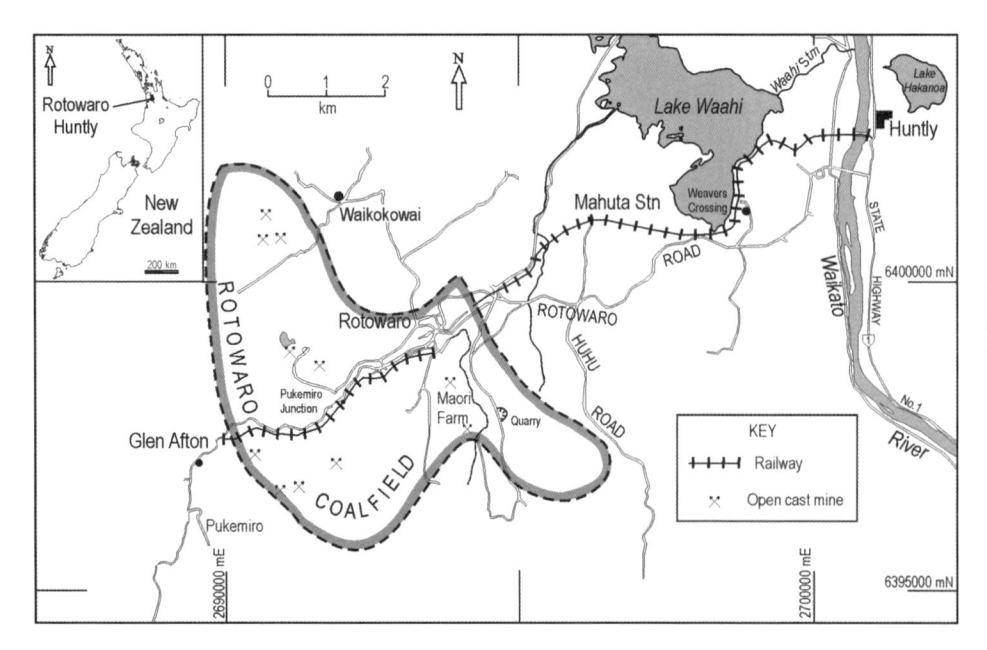

Figure 1. Location map of the Rotowaro coalfield (redrawn from New Zealand 260 Map Series, 1998, 2005), scale 1:50,000.

(3) estimate the possible temperatures of pyrometamorphism under which different minerals and glass formed from the melt, in the slag, and in the paralavas, and (4) determine the origin of the iron enrichment that produced these unusual pyrometamorphic slags and paralavas.

THE ROTOWARO COALFIELD

The Waikato coal region, North Island, New Zealand, is made up of 13 coalfields that contain coals ranging from subbituminous C to A. The Rotowaro coalfield, defined as "the coal-bearing area entered around Rotowaro, Waikokowai, Glen Afton, and Pukemiro" (Kear and Waterhouse, 1978), is situated 8 km southwest of Huntly (Fig. 1) and is the Waikato region's most productive coalfield. Underground mining began in 1915, and openpit mining commenced in 1936; ~55 million tons of coal have been removed.

The Waikato coal measures are an Upper Eocene to Oligocene sequence onto Mesozoic basement rocks. They were deposited in a north to north-northwest–trending valley system, which is ~35 km wide and 200 km long and parallels the broad structural grain of the region. In the Rotowaro coalfield, the coal measures are confined by basement topography and syndepositional faulting (Edbrooke et al., 1994). The gentle north to northwest regional dip of the coal measures (5°–10°) is locally variable because of changes in seam thickness and uneven compaction, as well as irregular pre–coal measure topography and faults (Kear and Waterhouse, 1978). The coal in the Rotowaro coalfield is entirely Eocene in age and is overlain by sediments of the Upper Eocene to Oligocene Te Kuiti Group.

Elsewhere in the Waikato coal region, the Te Kuiti Group is unconformably overlain by Miocene sediments; however, these are absent in the Rotowaro area. A regional unconformity separates the Miocene sediments from post-Miocene fluviolacustrine peats, mud, gravel, and volcanic ash deposits of the Tauranga Group. The general stratigraphic sequence of the Rotowaro area is presented in Figure 2.

The unmetamorphosed coal measure fireclays that enclose the coal seams in the Rotowaro coalfield are high in silica and alumina. Mineralogically, they are dominated by kaolin with quartz and lesser amounts of illite. Some of the sediments contain small amounts of hematite; locally, a horizon that contains nodules of siderite with minor quartz is seen immediately above the coal seams.

Three main sets of coal seams are recognized. From bottom to top, these include: (1) the Taupiri coal seams, enclosed in the lower Waikato coal measures; (2) the Kupakupa coal seams toward the base of the upper Waikato coal measures; and (3) the Renown Seam, which often is thin or absent at the top of the upper Waikato coal measures. Coal from the Rotowaro coalfield largely is subbituminous A, which is on the higher end of the range for the Waikato coal region. The properties of the coals in the Rotowaro coalfield, given on an air-dried basis, are listed in Table 1.

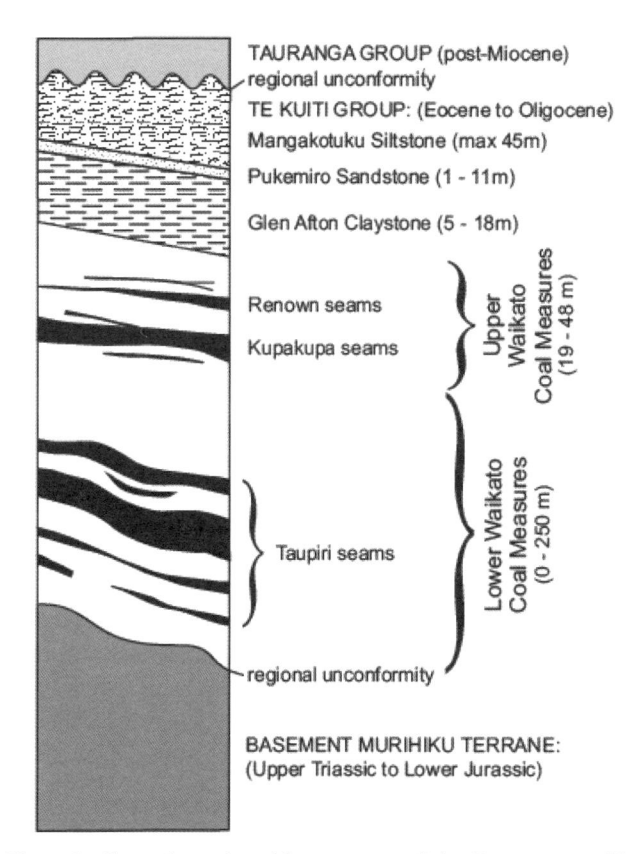

Figure 2. General stratigraphic sequence of the Rotowaro coalfield area. Data were supplied by Solid Energy New Zealand for the Callaghans Sector, Rotowaro (Edbrooke et al., 1994).

TABLE 1. PROPERTIES OF COAL IN THE ROTOWARO COALFIELD ON AN AIR-DRIED BASIS (AFTER GRAY AND DALY, 1981)

Seam	% moisture	% ash	% volatile matter	% fixed carbon	% sulfur	Specific energy (MJ/kg)
Renown	11.9	8.8	40.4	38.9	1.25	24.11
Kupakupa	13.6	2.2	39.2	45.0	0.27	24.61
Taupiri	14.3	3.8	37.2	44.7	0.27	24.19

The chemistry of representative Waikato coals and coal ashes was described by Gray and Daly (1981). Although sulfides (notably pyrite and marcasite) occur in Waikato coals, 90% of the sulfur in the coals is organic (Gray and Daly, 1981). Many of the elements found in the coal as complexes on the organic material can be removed by cation exchange or leaching of the coal and are liberated readily by combustion. The major inorganic constituents of the Rotowaro coals have been determined using the methods described in Black (1982) (Table 2).

Testing of the spontaneous combustion propensity of New Zealand coals has shown that the subbituminous coals of the

TABLE 2. MAJOR INORGANIC CONSTITUENTS OF
ROTOWARO COALS (AFTER BLACK, 1982)

Element	Cl	F	Fe	Ca	Mg	Na	K	Al
Weight %	0.04	0.01	0.4	1.2	0.1	0.1	0.05	0.07

Waikato coalfield could be classified as "extremely prone to spontaneous combustion" (Beamish et al., 2001).

COMBUSTION METAMORPHISM

The principal outcrop of pyrometamorphics is a natural scarp face that is ~25 m long and 2 m to 6 m high (Fig. 3). Although neither unmetamorphosed sediments nor the coal seam itself is exposed, it is likely that the combusted-coal seam was an extension of the major Taupiri coal seam located around the quarry.

The major portion of the exposure consists of a variety of thermally metamorphosed sediments (porcellanites) that range in color from cream to yellow to orange and red and reflect different amounts of iron enrichment and degrees of oxidation. The porcellanites, although variable in color, contain similar minerals: mullite, cristobalite, tridymite, and hematite, with residual low quartz, rutile, and zircon. Very high-temperature rocks contain cristobalite, mullite, and glass. Many of the porcellanites show disseminated black iron-rich spots that are composed mainly of hematite (Masalehdani, 1985).

In the main exposure, solid yellow porcellanites form discontinuous beds that are up to 50 cm thick, with well-developed horizontal and vertical prismatic cracks that formed as a result of shrinkage during the baking process. The complete combustion of the underlying coal led to the collapse of the overburden and created large cavities. Locally, the strata are highly disturbed with well-preserved breccias and broken rocks.

A lensoid body (2.5 m × 0.4 m) in the lower part of the outcrop, within the light-colored baked porcellanites, contains the main exposure of iron-rich, magnetite- and hematite-bearing pyrometamorphics (Fig. 4). These rocks are black-blue and magnetite-rich at the bottom and ends of the lens, whereas they are red and hematite-rich in the middle of the lens, which indicates variable oxidation conditions.

In hand specimens, melted iron-rich rock (the slag) is hard, heavy, vesicular, and dark gray to black with a dull luster. Except for its vesicularity, it is similar in appearance to a furnace slag (Fig. 5). Vesicles in the slag are less than 1 cm in diameter, with spherical or irregular shapes that result from the trapping of gas bubbles.

Melts of iron-rich silicate rock appear in irregular masses on both sides of the lens of slag and porcellanites (Fig. 6) and are referred to herein as paralava. The paralavas are dark gray to black with a ropy, lobate, rounded pod-like form, and they have dull, glazed, smooth surfaces (Fig. 7). The dark color of most paralavas indicates that they were formed in a reducing environment. "Stalactites" of the solidified paralavas were observed around vents, and tongues of paralava have penetrated the overlying porcellanites along horizontal fissures (Figs. 8A and 8B) and fill the cavities of brecciated rocks (Fig. 7C). Ropy paralava coats the surface of some

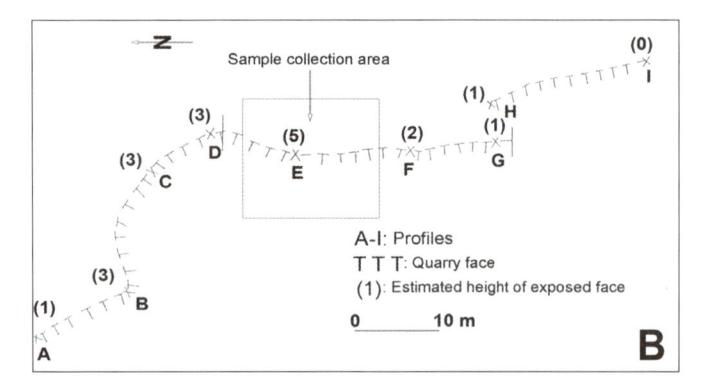

Figure 3. (A) General view of quarry face. Arrows point to the photos taken from the central part of the quarry and sample-collection locations from profile E. (B) Quarry face shown and sample-collection area (profile E).

Figure 4. A lensoid body of baked (dark blue–black) iron-rich pyrometamorphosed rock (arrow). Scale: 2.5 m × 0.4 m at the highest point (see also Fig. 6B).

Figure 5. (A–B) Cut slabs of iron-rich, slag-like material. (B) Vesicular sample S1 surrounded by reddish porcellanite.

Figure 6. (A) Shrinkage cracks along and across porcellanites (light tan–yellow) containing irregular slag-like masses of melted rock (gray-black). Lens cap is 5 cm in diameter. (B) Highly vesicular slag-like masses of iron-rich rock (arrows). Enlargement of center portion of Figure 4, profile E. Hammer is 30 cm long.

fissures (Fig. 8C). In the outcrop in profile E (Fig. 3A), the slags and paralavas occur in close vicinity to fissures, cracks, and crevices (chimneys) where most of the hot gases from underground coal combustion have escaped to the atmosphere. This indicates that the combustion gases were the source of the heat that produced the very high-temperature pyrometamorphism, rather than the combustion of the coal itself, and that the gases transported the elements that are concentrated in the slags and paralavas.

METHODS OF STUDY

For this study, samples of the pyrometamorphics were taken mainly from profile E (Figs. 3A and 3B; Tables 3 and 4). Samples of unmetamorphosed fireclays and siderite-rich nodules from openpit mines in the Rotowaro coalfield were provided by the State Coal Mines, Huntly.

X-ray powder diffraction data were obtained using a Phillips PW 1130 diffractometer. For petrographic studies of polished sections and polished thin sections, Leitz Labor Lux II transmitted-light and Reichert reflected-light microscopes were used. Quantitative electron-microprobe determinations of the mineralogy of slag and paralava samples were performed using a JEOL JXA-5A electron microprobe. Data acquisition and analysis were carried out using a "Link System" energy-dispersive analyzer and software.

Bulk-rock major-element chemical analyses were obtained using an automated Philips PW 1410 X-ray fluorescence spectrometer. Data reduction was carried out by the methods described in Parker and Willis (1977).

CHEMISTRY

Chemical analyses of representative slags, paralavas, porcellanites, and fireclays are given in Tables 5 and 6. Figure 9 shows a SiO_2-Fe_2O_3-Al_2O_3 ternary diagram of the Rotowaro slags, paralavas, and porcellanites plotted with unmetamorphosed Rotowaro sediments. The various fields show little overlap. The trends indicate that the porcellanites are iron-enriched with respect to the coal measures. Although the paralavas show an amount of iron enrichment that is similar to the porcellanites, their trend diverges toward enrichment in SiO_2 with decreasing Fe_2O_3. The slags show much more iron-enrichment and lie on the extension of the porcellanite field toward the Fe_2O_3 corner of the ternary diagram. The Rotowaro examples of slags represent some of the most iron-rich natural slags yet observed in a combustion-metamorphic environment.

The variations in chemistry of the main rock groups also are reflected in their petrography. The iron-rich slags are composed dominantly of magnetite, accompanied by other iron-rich oxides and fayalite. The iron oxides show a range of textural features indicating that complex oxidation and exsolution effects took place as the slag cooled. The paralavas differ from the slags and other pyrometamorphics in that they contain abundant glass-enclosing, quench-textured silicate minerals, which include plagioclase, tridymite, and feathery crystals of fayalite and orthoferrosilite, with only minor amounts of magnetite.

PETROGRAPHY AND MINERALOGY OF THE IRON-RICH SLAGS

The iron-rich slags have assemblages that are dominated by magnetite, hematite, and spinels. These minerals occur in many different textural associations with each other, and they reflect the original crystallization of the slags at high temperatures and under a range of oxidation conditions followed by periods of re-equilibration of the primary phases as the rocks cooled (Fig. 10). In addition to oxides, minor amounts of fayalite and silicate glass are found in the slags. Iron silicate (fayalite) in slags is an accessory phase. It appears as fine-grained, dark-gray inclusions in magnetite hosts and/or the matrix (Fig. 11C). The principal oxide associations observed in the slags are magnetite-hematite-spinel and hematite–Ti-spinel assemblages.

Magnetite-Hematite-Spinel Assemblages

- Relatively coarse skeletal and fine dendritic intergrowths of magnetite-hematite-spinel occur in a microcrystalline silicate mineral or silicate-rich glass matrix. Typically, the skeletal crystals are elongated and consist of a simple set of cross-arms at right angles that correspond to the cubic crystallographic axes; it appears that growth was initiated from the interior surface and progressed toward the arms and arrowheads. The outlines of these skeletal crystal-face boundaries are often incomplete (Fig. 10).

- Exsolved fine spinel lamellae occur parallel to crystallographic planes in the host magnetite. The spinel lamellae were identified in polished sections by their isotropic nature and gray color. Fine exsolved and tabular hematite intergrowths in magnetite appear white-gray and aniso-

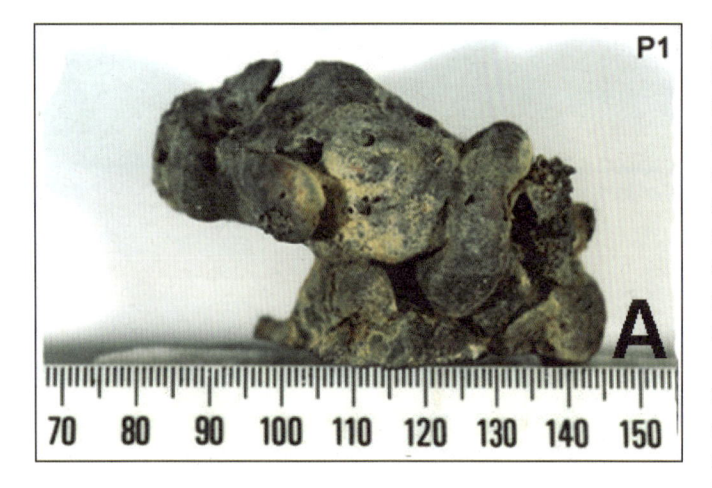

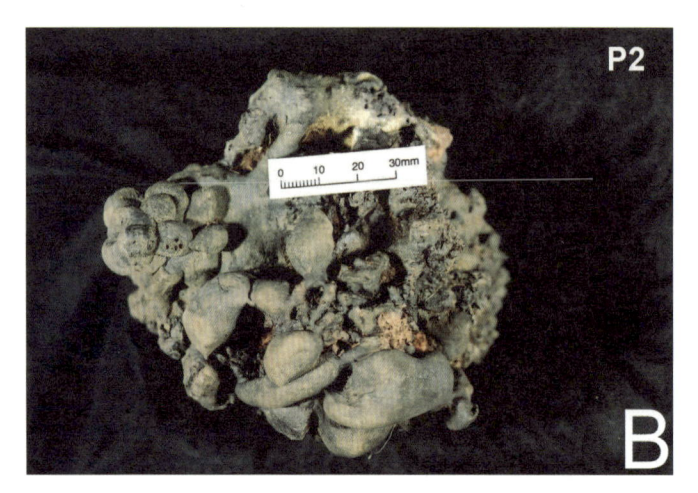

Figure 7. (A–B) Cut slabs of representative paralava samples (P1 and P2) from burnt-coal seam, Rotowaro coalfield. Note gray-black color, ropy, lobate, and rounded pod-like shape, and dull glazed surfaces of the paralavas. (C) Filling of spaces by paralava (gray-black) in brecciated porcellanite (yellow) in sample P4. Scale is in centimeters.

tropic and are highly reflective. Exsolved intergrowths of the fine hematite and Al-spinel lamellae in the host magnetite commonly occur in the central portion of the slag lens (Fig. 11). Qualitative scanning traverses across a representative crystal (Fig. 12) showed a good negative correlation between Al and Fe. The concentration of Al toward the center of the crystal could be evidence of crystallization at very high temperatures. The exsolved fine spinel lamellae in host magnetite are aluminous spinels. Unfortunately, because of the very size of the spinel, it has not been possible to obtain an electron-microprobe analysis of the individual spinel lamellae.

- Fine intergrowths of magnetite-hematite-spinel with myrmekitic textures are present as interstitial crystals in a silicate matrix. The grain boundaries of these crystals are rounded, so that in thin sections, the texture resembles a finely woven fabric. Although it is difficult to observe the details of the exsolution textures of these assemblages, they are considered to be the same as those described previously. These dendrites often form "coronas" around

Figure 8. Profile E. (A) "Stalactites" (arrows) of solidified paralava in a cavity (note the open vertical crack). (B) Paralava (gray-black) associated with and/or flowing over porcellanite. Hammer is 40 cm long. (C) Ropy paralava coating the surface of some fissures in profile E, at the study area. Lens cap is 5 cm in diameter.

TABLE 3. SAMPLE DESCRIPTIONS OF REPRESENTATIVE SLAGS AND PARALAVAS
AND THE MINERALS IDENTIFIED

Sample number	Descriptions and minerals
E1 (slag)	Numerous circular vesicles <1 cm in diameter, dark gray to black, dull luster, heavy and hard. Contains magnetite, hematite, Al-spinel, Ti-spinel ± fayalite ± glass.
RO1 (slag)	Vesicular, dark gray to black, dull luster, heavy and hard, dark gray, surrounded by reddish porcellanite. Contains magnetite, hematite, Al-spinel, Ti-spinel, ± fayalite ± glass ± unknown Fe-Al-Ti-Si-oxide.
RO2 (slag)	Circular vesicles <1 cm in diameter, rough surface, dark gray to black, dull luster. Contains magnetite, hematite, Al-spinel, Ti-spinel, ± fayalite ± glass.
P1 (paralava)	Dark gray to black in color, with a ropy, lobate, rounded pod-like form, dull glazed smooth surface, hollow and light. Contains fayalite, orthoferrosilite, anorthite, tridymite, cristobalite, glass ± magnetite.
P2 (paralava)	Gray color, with a ropy, lobate round or ellipsoidal shape, dull glazed luster, smooth surface, hollow and light. Contains fayalite, orthoferrosilite, anorthite, tridymite, cristobalite, glass ± magnetite.

TABLE 4. METHODS OF MINERAL IDENTIFICATION IN SLAGS AND PARALAVAS

Identified minerals	XRD*	OP[†]	EMP[§]
Magnetite	x	x	x
Hematite	x	x	x
Al-spinel	x	x	x
Ti-spinel	x	x	x
Fayalite	x	x	x
Orthoferrosilite	x	x	x
Anorthite	x	x	x
Tridymite	x	x	
Cristobalite	x	x	
Glass		x	x
Fe-Al-Ti-Si-oxide		x	x

*XRD—X-ray diffraction.
[†]OP—optical microscope.
[§]EMP—electron microprobe.

TABLE 5. CHEMICAL ANALYSES OF REPRESENTATIVE SLAG (E1) AND PARALAVA (P1 AND P2) SAMPLES

Composition (wt%)	(E1)	(P1)	(P2)
SiO_2	32.48	55.24	54.92
TiO_2	0.63	0.86	0.77
Al_2O_3	12.80	17.2	16.78
Fe_2O_3*	46.13	21.39	21.49
MnO	0.54	0.01	0.01
MgO	1.42	1.7	1.89
CaO	0.42	0.2	0.22
Na_2O	0.28	0.03	0.07
K_2O	0.25	1.39	1.83
P_2O_5	0.20	0.19	0.28
Cl	N.D.[†]	0.017	0.029
SO_3	N.D.	0.027	0.029
H_2O	2.06	N.D.	N.D.
LOI[§]	1.61	N.D.	N.D.
Total	98.82	98.254	98.318

*Fe_2O_3 refers to total iron.
[†]N.D.—no data.
[§]LOI—loss on ignition.

TABLE 6. CHEMICAL ANALYSES OF REPRESENTATIVE SIDERITE-NODULE (SQ8), FIRECLAYS (F1 AND F2), AND PORCELLANITES (E10 AND B2)

Composition (wt%)	(SQ8)	(F1)	(F2)	(E10)	(B2)
SiO_2	9.16	62.52	62.60	43.12	52.44
TiO_2	0.25	1.26	1.25	0.79	1.11
Al_2O_3	5.08	26.08	26.26	19.68	24.87
Fe_2O_3*	62.99	3.99	3.98	29.15	16.25
MnO	1.58	0.01	0.02	0.48	0.30
MgO	0.45	0.69	0.70	1.14	1.01
CaO	0.82	0.05	0.06	0.11	0.06
Na_2O	N.D.[†]	0.02	0.04	0.00	0.14
K_2O	0.29	1.73	1.72	0.19	0.18
P_2O_5	0.25	0.18	0.70	0.16	0.11
H_2O	0.60	0.18	0.18	2.77	2.13
LOI[§]	17.83	3.19	3.19	2.85	2.13
Total	100.00	99.89	100.70	100.42	100.73

*Fe_2O_3 refers to total iron.
[†]N.D.—no data.
[§]LOI—loss on ignition.

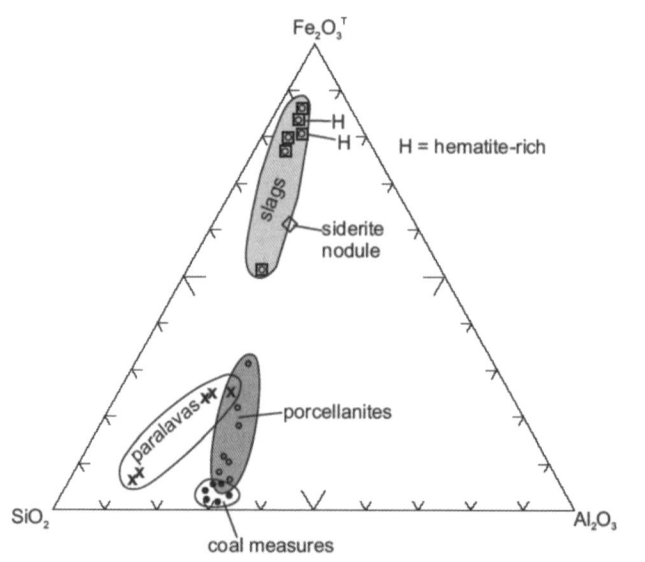

Figure 9. Plot of SiO_2, Fe_2O_3, and Al_2O_3 of various Rotowaro coalfield pyrometamorphics and coal measure sediments.

Figure 10. Reflected-light micrograph, sample E1 (×22 objective lens). Skeletal crystals of exsolved magnetite-hematite-spinel assemblages (white) in silicate matrix (gray) of the slag. The black areas are vesicles.

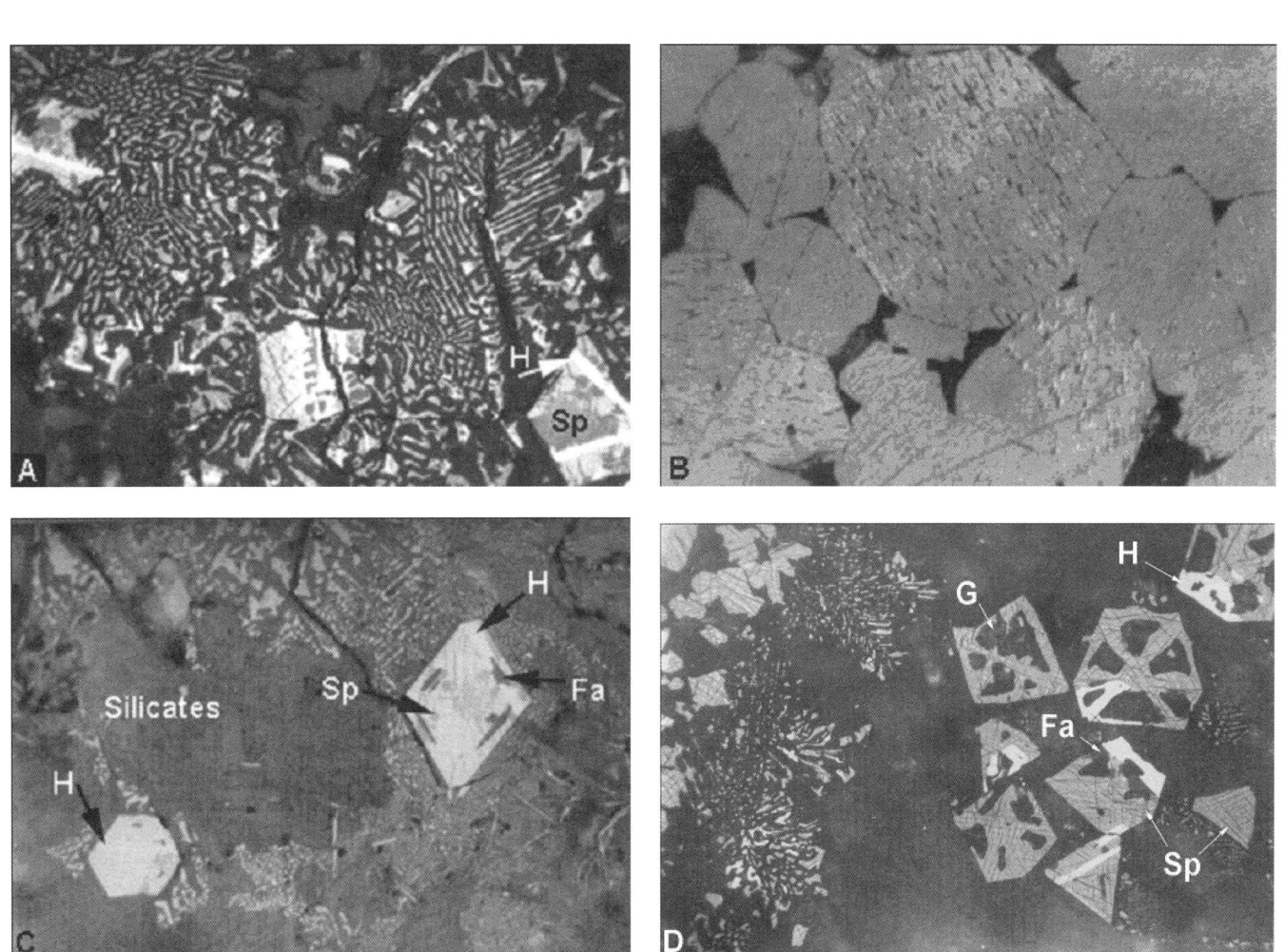

Figure 11. (A) Reflected-light micrograph, slag sample E1 (×45 objective lens) showing exsolved magnetite-hematite-spinel assemblages (white) in granophyric-like intergrowth relation with silicate minerals (gray). H—hematite; Sp—spinel. (B) Reflected-light micrograph, slag sample RO2 (×110 objective lens) showing crystals of magnetite with finely exsolved hematite (white) and spinel (gray), arranged in a mosaic texture. The black areas between the crystals are silicates. (C) Reflected-light micrograph, slag sample E1 (×45 objective lens) showing euhedral magnetite with finely exsolved spinel (Sp) lamellae (light gray) and tabular intergrowths of hematite (H) (white) in silicate (cristobalite, fayalite, and glass) matrix (dark gray). The dark-gray inclusions in these crystals are fayalite (Fa). (D) Reflected-light micrograph, slag sample E1 (×45 objective lens) showing euhedral magnetite skeletal crystals with finely exsolved spinel (Sp) lamellae (light gray) and tabular intergrown hematite (H) (white) in a silicate matrix (dark gray). The areas between the internal dendrite within these crystals are filled with fayalite (Fa) and/or glass (G).

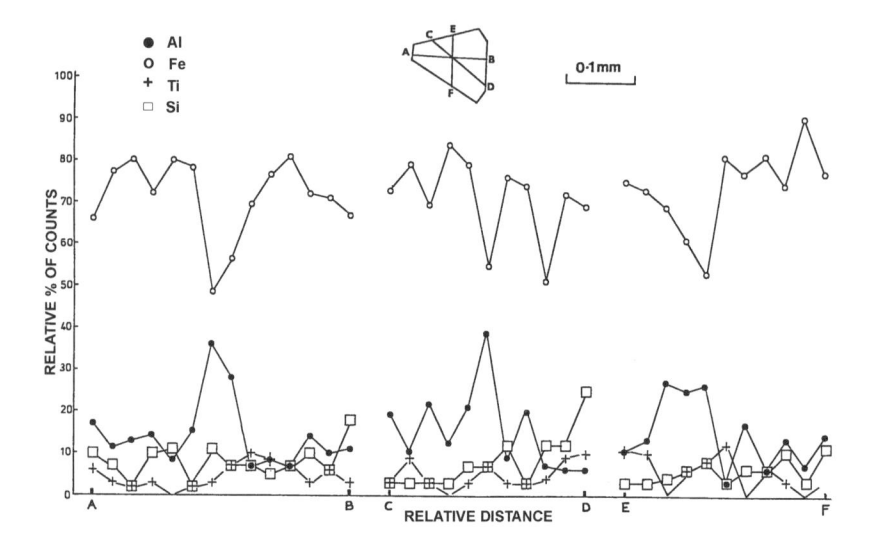

Figure 12. Plot of the qualitative electron-microprobe analyses of representative exsolved magnetite-hematite-spinel assemblages in the slag (sample E1).

coarse central solid crystals of the same phases; in some cases, they form subparallel acicular branches in the "corona" (Fig. 11A).

- Euhedral crystals (up to 0.4 mm in diameter) of an exsolved magnetite-hematite-spinel assemblage occur as aggregates of fine equant crystals that are grouped together and that have a mosaic texture (Fig. 11B), or they have hexagonal outlines enclosed in a silicate matrix (Fig. 11C). The interior and interstices between the hollow hexagonal dendritic crystals are filled with silicates and/or glass (Fig. 11D).

Hematite and Ti-Spinel Assemblages

Ti-spinel is found in a relatively homogeneous composition, which contains 9.6% to 11.8% TiO_2 and less than 4% Al_2O_3, and it occurs with hematite as anhedral "mosaic" textured aggregates that are concentrated at the rims of the vesicles in the slag sample E1 (Fig. 13). Minerals in this assemblage lack the exsolution features that are observed in the magnetite-hematite-spinel assemblages.

Late Oxidation of Magnetite to Hematite and Zoning Effects

Oxidation of magnetite to hematite occurs as relatively broad lamellae and often extends from one edge of the magnetite crystal to the other (Fig. 11). This particular oxidation process apparently took place while the combustion of the coal seam was in progress, as evidenced by similar occurrences that were observed in the "furnace slags of a central heating system, from places where the draft has been especially intermittent" (Ramdohr, 1982, p. 45).

Some large euhedral magnetite crystals (up to 0.4 mm in diameter) show a characteristic zoning that is expressed most obviously by color variation; the innermost part of the magnetite crystal is gray-brown, whereas the margins are whitish-blue (Fig. 14). These optical variations indicate the presence of hematite at the margins of the magnetite crystals. This incipient "martitization" most likely resulted from oxidation of magnetite to hematite during the cooling process at temperatures of at least 600 °C (Haggerty, 1976).

PETROGRAPHY AND MINERALOGY OF THE PARALAVAS

The paralavas are distinct in that they contain more glass and an abundance of silicate minerals, including orthoferrosilite, fayalite, tridymite, and cristobalite, occasional high-temperature feldspar, and only minor amounts of magnetite. All minerals in the paralavas show ubiquitous feathery quench textures.

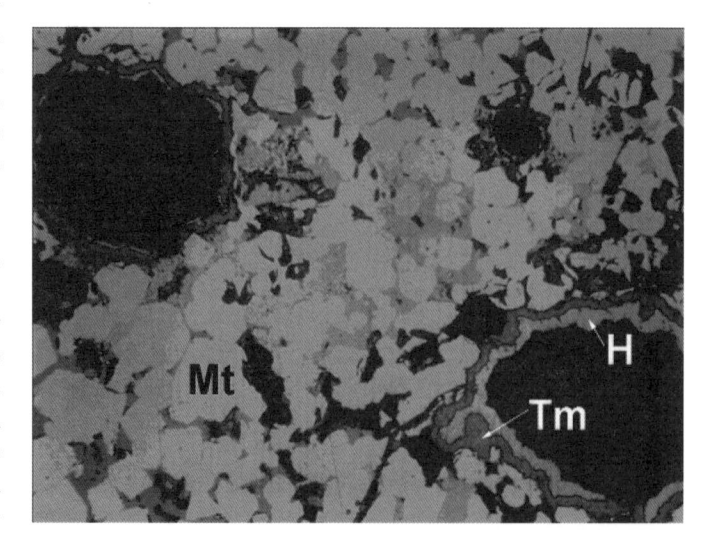

Figure 13. Reflected-light micrograph, sample E1 (×45 objective lens) showing hematite (H) and titaniferous magnetite (Tm) crystals concentrated around the vesicles (black). Matrix (dark gray) consists of tridymite, fayalite, and glass. Black spots are cavities. Mt—magnetite.

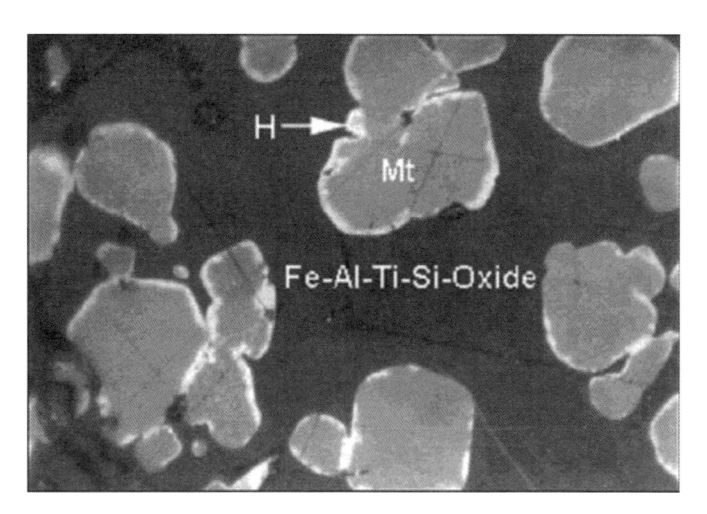

Figure 14. Reflected-light micrograph, sample RO1 (×110 objective lens plane polarized light) showing magnetite (Mt) crystals with rims of hematite (H). Matrix consists of unknown Fe-Al-Ti-Si oxide. Black spots are dirt on the surface of the sample.

Iron Silicates

In thin sections, fayalite appears as pale yellow crystals with high relief and high birefringence. Orthoferrosilite is colorless to pale yellow, with nearly parallel extinction (0°–8°) and 2Vα values, as measured by the universal stage method, ranging from 82° to 108° (average, 90°). Skeletal crystals of fayalite and orthoferrosilite, which vary in size from a few microns to phenocrysts of up to 0.3 mm in diameter, occur in all paralavas. These crystals may be embayed at their extremities or cellular (filled with glass) throughout much of their length and elongated parallel to the a-axis, or they form as large hopper-shaped crystals, generally with the gross outline of the common dome and prismatic forms of olivine (Fig. 15).

Electron-microprobe analyses of fayalite and orthoferrosilite have been recalculated to give an empirical structural formula, calculated on the basis of four oxygen atoms for fayalite and three oxygen atoms for orthoferrosilite, of $(Fe_{1.91}Mn_{0.03}Mg_{0.06})(Al_{0.07}Si_{0.92})O_4$ and $(Fe_{0.80}Mn_{0.01}Mg_{0.15}Ca_{0.01})(Al_{0.13}Si_{0.90})O_3$, respectively.

Feldspar, Silica Minerals, Opaques, and Glass

High-calcium feldspar crystals occur as colorless low-relief and birefringence laths or as tabular crystals intergrown with orthoferrosilite and fayalite and enclosed by reddish-brown glass. The tabular plagioclase crystals, when viewed in sections cut perpendicular to their a-axes, show a characteristic hollow, rectangular "belt-buckle" form (Fig. 15A).

Electron-microprobe analyses reveal that the plagioclase contains notable, although variable, amounts of iron (1.9%–3.97% total iron calculated as Fe_2O_3). Feldspars in melted sediments associated with coal seams have been reported by Phemister (1942), Bishop (1965), Bentor and Kastner (1976), Bustin and Mathews (1982), Mason and Davis (1983), Essene et al. (1984), Cosca et al. (1989), and Grapes (2004).

Cristobalite and tridymite were identified by their X-ray powder diffraction data. In thin section, tridymite occurs as acicular (up to 0.3-mm-long) or "fan"-shaped arrays, which contain fine feather-like orthoferrosilite (Fig. 15D). Cristobalite in paralavas possibly crystallized from the residual silicate melt in the temperature range of 1500 °C to 1600 °C, whereas tridymite, in the same specimens, could have been formed directly from the melt at lower temperatures (1500–1400 °C) or by conversion of cristobalite during the cooling process according to Schairer and Bowen (1947) (Fig. 16).

Magnetite in paralavas forms fine feathery inclusions intergrown on the rims and/or the concave embayments of lobate crystals of fayalite and orthoferrosilite. Glass occurs in the interstices between the silicates (Fig. 15).

The glass enclosing the high-temperature silicate and oxide minerals is high in silica (60 wt%) and contains variable amounts of iron (≤7 wt%). Notably, all glasses contain substantial amounts of K_2O (≤4 wt%) and very low amounts of Na_2O (≤1 wt%) (Table 7).

DISCUSSION OF CONDITIONS OF FORMATION OF SLAGS AND PARALAVAS

It is clear that the paralavas and slags from Rotowaro coalfield in New Zealand formed from high-temperature melting at atmospheric pressure, but there is no direct way to determine the melting temperatures or the crystallization temperatures of the various mineral assemblages. Occurrences of mineral assemblages in these rocks and phase-equilibria studies from the literature do allow us to make estimates of temperatures for the formation of the slags and paralavas at the study site.

The concentration of Al toward the center of the magnetite crystals in the slags is good evidence of crystallization at very high temperatures. Phase relations in the system Fe-Al-O at temperatures ≥1000 °C were studied by Richards and White (1954), Hoffmann and Fisher (1956), Muan and Gee (1956), and Atlas and Sumida (1958). They observed that at these temperatures and atmospheric pressure, complete solid solution between the spinels, magnetite (Fe_2O_3), and hercynite ($FeAl_2O_4$) occurred. Exsolution between Fe_3O_4 and $FeAl_2O_4$ is reported to begin at temperatures $(T) \leq 1000$ °C at a given composition of 55 mol% hercynite (Turnock and Eugster, 1962).

The concentration of homogeneous crystals of hematite and titaniferous magnetite around the vesicles indicates that these mineral assemblages crystallized later than the assemblages in the slag itself, and under more oxidizing conditions than those that controlled the magnetite-hercynite exsolved assemblages found in the interior of the slag samples. It is also possible that the titaniferous magnetite and hematite rapidly quenched around the vesicles and remained unexsolved. Similar occurrences of unexsolved Ti-spinel have been described from extrusive or plutonic

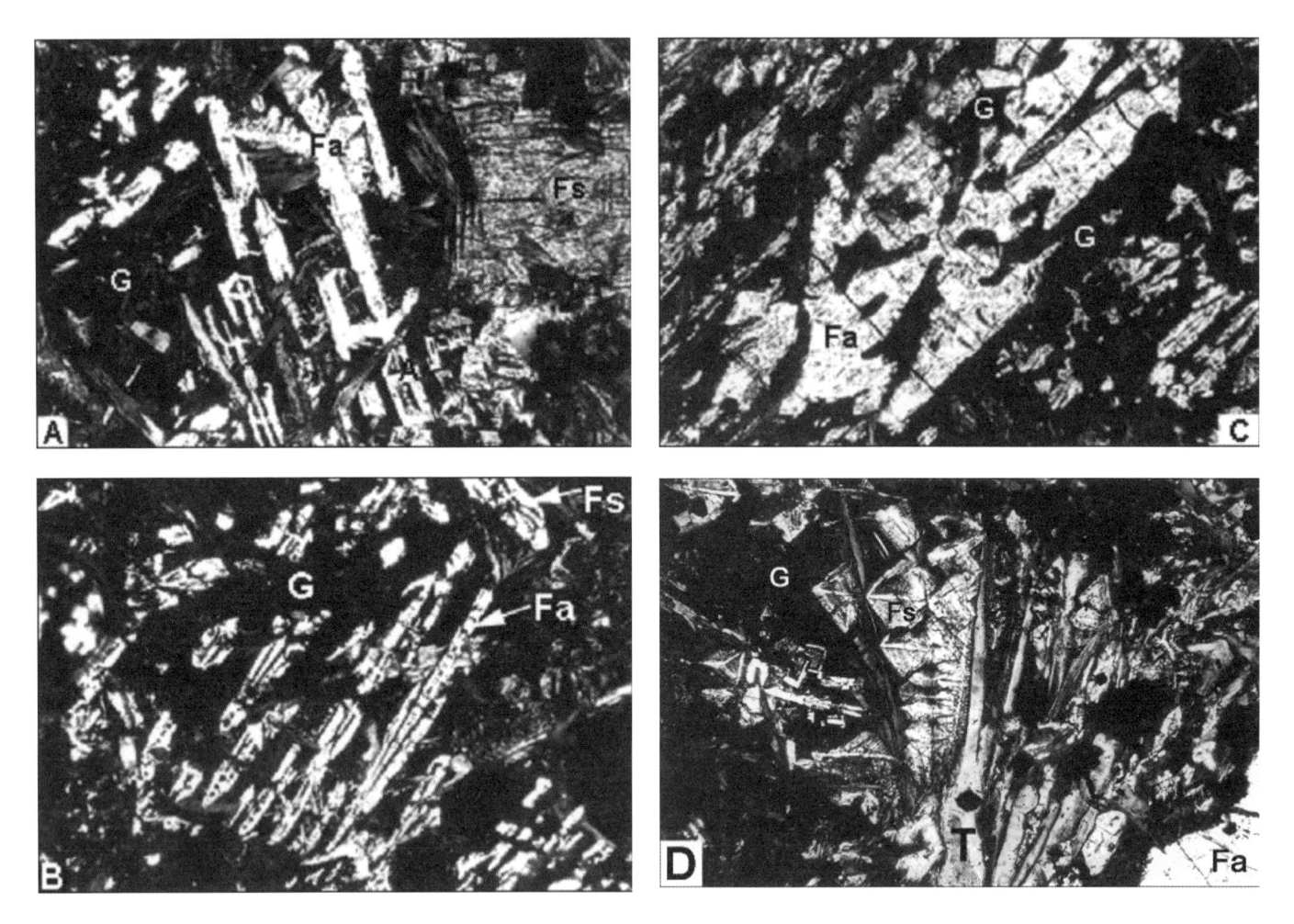

Figure 15. (A) Transmitted-light micrograph of paralava sample P1 showing typical branching dendritic texture of fayalite (Fa) that contains an overgrowth of skeletal anorthite. The gray phase at right top corner is orthoferrosilite (Fs) with inclusions of glass (G) (black). Scale: ×220, objective lens plane polarized light (PPL). (B) Transmitted-light micrograph of paralava sample P1 (×220 objective lens PPL) showing elongated skeletal crystals of fayalite and orthoferrosilite in a glass matrix (×220 objective lens PPL). (C) Transmitted-light micrograph of paralava sample P1 (×80 objective lens PPL) showing large skeletal fayalite; the black material in and around the crystals is glass. (D) Transmitted-light micrograph of paralava sample P1 (×220 objective lens PPL) showing dendrites of orthoferrosilite concentrated between acicular tridymite (T) crystals. The white phase at right bottom corner is fayalite.

rocks that were brought rapidly from the high-temperature zones to Earth's surface (Ramdohr, 1982, p. 924).

The crystallization of the titaniferous magnetite in basalt is a function of temperature and partial pressure of oxygen (pO_2) (Haggerty, 1976). To determine the pO_2 during development of this mineral within the slag sample E1, the composition of Ti-magnetite obtained from the electron-microprobe analyses was plotted in the system $FeO-TiO_2-Fe_2O_3$ at 1200 °C (Fig. 17), assuming the temperature of the crystallization was the same. From this figure, it seems likely that the crystallization of titaniferous magnetite occurred under high-pO_2 conditions, as also evidenced by the presence of coexisting hematite.

The appearances of silicate minerals in the paralavas also provide some evidence of temperatures higher than 1000 °C. Investigations by Smith (1971) showed that orthoferrosilite ($FeSiO_3$) crystallizes from fayalite and silica at $T \geq 900$ °C and a pressure of 1 bar. Higher pressures (12–73 kbar) increase the stability of $FeSiO_3$ at high temperatures (620–1270 °C), as found by Akimoto et al. (1964). Bohlen et al. (1980a) determined a temperature and pressure that was lower (700 °C and 8 kbar) than estimated previously for the coexistence of the fayalite-orthoferrosilite assemblage.

The coexistence of orthoferrosilite and fayalite and their dependence on temperature and pressure have been the subject of extensive investigation since the initial work of Bowen and Schairer (1932). Later, they studied the ternary system, $MgO-FeO-SiO_2$, under atmospheric pressure and high temperature (≥ 1200 °C) and found that pure or substantially pure $FeSiO_3$ dissociates into fayalite and silica under such conditions (Bowen and Schairer, 1935).

TABLE 7. CHEMICAL ANALYSES OF GLASS IN
REPRESENTATIVE PARALAVA SAMPLES
(P1 AND P2)

Composition (wt%)	(P1)	(P2)
SiO_2	60.75	49.13
TiO_2	0.28	0.30
Al_2O_3	19.11	17.89
FeO*	5.81	6.76
MnO	0.40	0.35
MgO	0.00	0.39
CaO	8.38	9.90
Na_2O	0.00	0.60
K_2O	3.94	2.94
P_2O_5	0.09	0.25
Total	98.76	88.51

*FeO refers to total iron.

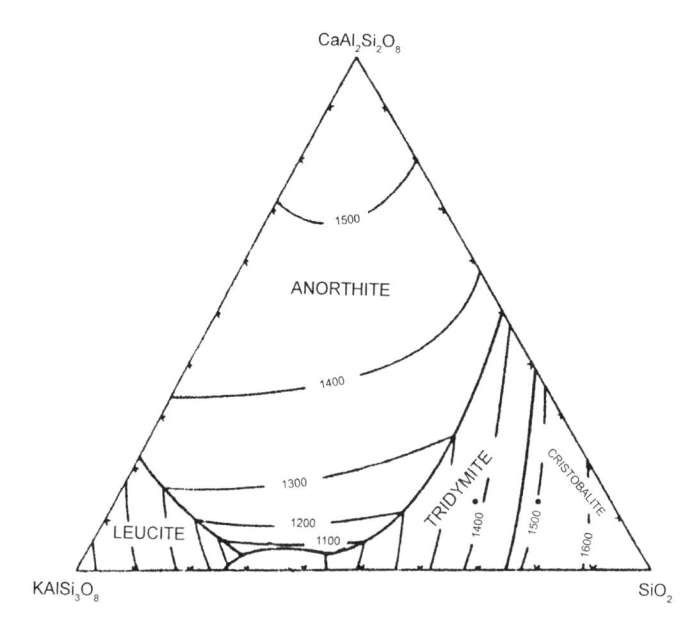

Figure 16. Plot of composition of glass (filled circles) in paralavas (Table 7) projected onto the liquidus of the system $CaAl_2Si_2O_8$-$KAlSi_3O_8$-SiO_2 (redrawn from Schairer and Bowen, 1947).

In the Rotowaro paralava specimens, there is no textural evidence to support the hypothesis that orthoferrosilite ($FeSiO_3$) crystallized after fayalite (Fe_2SiO_4). Simultaneous crystallization of orthoferrosilite and fayalite may have taken place from the reaction of iron and silica present in the paralava melt at atmospheric pressure and temperature ≤1200 °C (Fig. 18). The reaction could have been stabilized by partitioning of the other components, such as Mg, Mn, and Al, into the fayalite and orthoferrosilite structures as described by Bohlen et al. (1980b). Rapid quenching of these minerals might have inhibited their decomposition.

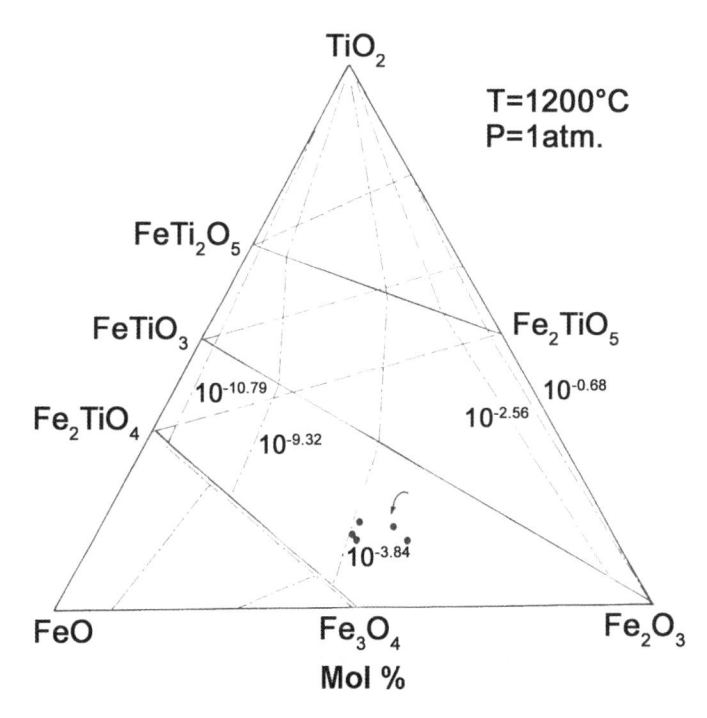

Figure 17. Plot of composition of titaniferous magnetite in slag (sample EI) from electron-microprobe analyses (redrawn from Webster and Bright, 1961). Ulvöspinel—Fe_2TiO_4; pseudobrookite series—$FeTiO_5$–Fe_2TiO_5; magnetite—Fe_3O_4; hematite—Fe_2O_3; ilmenite—$FeTiO_3$; oxidation-reaction lines—dashed; solid solution lines—straight.

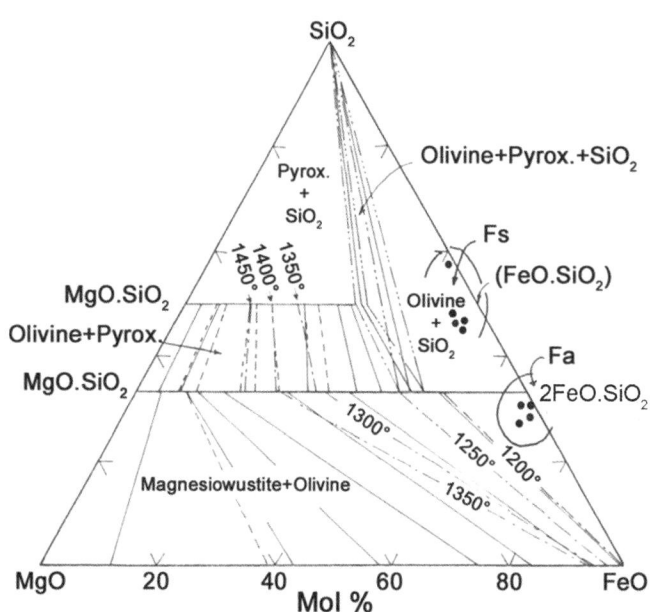

Figure 18. Plot of electron-microprobe compositions of fayalite (Fa) and orthoferrosilite (Fs) in paralava sample P1 (redrawn from Nafziger and Muan, 1967). Pyrox.—pyroxene. Degrees given in Celsius.

The occurrences of (tridymite + anorthite + orthoferrosilite + fayalite) assemblages in the paralavas indicate very high temperatures, certainly above 1000 °C. Crystallization of anorthite from a silicate melt probably occurs at ≤1470 °C (Morse, 1980) (Fig. 16). The general convergence of the data indicates formation of the slags and paralavas in the temperature range of 1000 °C to 1600 °C.

SUMMARY AND CONCLUSIONS

Thermal metamorphism of coal measures immediately overlying burnt-coal seams in the Rotowaro coalfield, North Island, New Zealand, produced unusual iron-rich slags that contain the crystalline phases: magnetite, hematite, hercynite, titaniferous magnetite, and accessory fayalite. Silicate melts (paralavas) associated with the slags contain abundant glass, fayalite, orthoferrosilite, anorthite, cristobalite, and tridymite and minor magnetite. Evidence from known equilibrium relationships between these phases indicates that the slags and paralavas were formed at ultrahigh temperatures in the range of 1000 °C to 1600 °C and at atmospheric pressure.

The iron oxides in the slags show variations in texture and composition. In the interior of the slags, the magnetite crystals show well-developed exsolution and oxidation textures, whereas near the surface around the vesicles, hematite and titaniferous magnetite coexist, and these minerals do not show any oxidation or exsolution textures. The observed variations in the iron oxides indicate differences in composition, pO_2 conditions, time of crystallization, and cooling rate between the area immediately adjacent to the vesicles and the central portion of the slag. The occurrence of high-Ti minerals near the surface around the vesicles indicates that the Ti must have been transported by hot gases. It is suggested that as a result of the early high-temperature crystallization of high-Al phases in the slag, the relative proportion of Ti increased in the residual system. It may have been transported by the hot gases escaping from the slag melt to be precipitated near the surface around the vesicles as titaniferous magnetite.

Exceptionally iron-rich slag-like rocks are unusual, and the Rotowaro examples represent some of the most iron-rich natural slags presently known from a combustion-metamorphic environment. The plotted trends of Rotowaro slag and porcellanite chemistries indicate that some iron mobilization process occurred during the pyrometamorphism, as also was suggested by Clark and Peacor (1992) in their study of similar rocks from the Powder River Basin in Wyoming. The source of iron required to produce these iron-rich slags is an enigma because the coal measures themselves are poor in iron. Nevertheless, the slags, irrespective of their oxidation state, lie on a continuation of the iron-enrichment trend observed in the porcellanites, which suggests that iron was concentrated by the coal-combustion process.

During coal combustion at atmospheric pressures and ultrahigh (>1000 °C) temperatures, hot gases escaped into the open through surficial fractures and cracks. The combination of the escaping gases with atmospheric oxygen resulted in sustained high temperatures, which caused intensive melting of the protolith sediments in an environment similar to that of a blast furnace. Melting of the sediments also may have been affected by fluxes of alkali elements, which are known to be a complex in the coal (Table 2) and which are liberated by its combustion. Evidence for this is seen in the high content of K_2O in the paralava glasses (Table 7). The hot gases that were vented to the surface through open fissures also may have leached out elements from the enclosing sediments and transported them to be deposited in the melt and pyrometamorphics.

ACKNOWLEDGMENTS

The authors gratefully acknowledge Glenn Stracher of East Georgia College, Swainsboro, Georgia, who encouraged us to contribute this article, as well as Jeffrey Swope, M.J. Hibbard, and R. McDowell for reviewing this paper. We wish to acknowledge staff of the Geology Department, Auckland University, who provided help with the electron microprobe and X-ray fluorescence analyses. S. Maquiné de Souza and L. Cotterall provided assistance with figure preparations. The first author would like to express very special thanks to Jean-Luc Potdevin of the Geology Department, University of Lille, and to Daniel Milhau for their help and support.

REFERENCES CITED

Adams, J., 1978, A melt rock near Murchison, New Zealand: Basalt or fused sandstone?: New Zealand Journal of Geology and Geophysics, v. 21, no. 6, p. 743–745.

Akimoto, S., Fujisawa, H., and Katsura, T., 1964, Synthesis of $FeSiO_3$ pyroxene (ferrosilite) at high pressures: Proceedings of the Japan Academy, v. 40, p. 272–275.

Atlas, L.M., and Sumida, W.K., 1958, Solidus, subsolidus, and subdissociation phase, equilibria in the system Fe-Al-O: Journal of the American Ceramic Society, v. 41, p. 150–160, doi: 10.1111/j.1151-2916.1958.tb13532.x.

Beamish, B.B., Barakat, M.A., and St. George, J.D., 2001, Spontaneous-combustion propensity of New Zealand coals under adiabatic conditions: International Journal of Coal Geology, v. 45, p. 217–224, doi: 10.1016/S0166-5162(00)00034-3.

Bentor, Y.K., and Kastner, M., 1976, Combustion metamorphism in Southern California: Science, v. 193, p. 486–488, doi: 10.1126/science.193.4252.486.

Bishop, D.G., 1965, The geology of the Clinton District, South Otago: Transactions of the Royal Society of New Zealand, Geological Series, v. 2, p. 205–230.

Black, P.M., 1982, Petrology and inorganic chemistry of South Island lignites, New Zealand: Melbourne, Australia, Geological Society of Australia, Coal Group Symposium, p. 131–144.

Bohlen, S.R., Essene, E.J., and Boettcher, A.L., 1980a, Reinvestigation and application of olivine-quartz-orthopyroxene barometry: Earth and Planetary Science Letters, v. 47, p. 1–10, doi: 10.1016/0012-821X(80)90098-9.

Bohlen, S.R., Boettcher, W.A., Dollase, W.A., and Essene, E.J., 1980b, The effect of manganese on olivine-quartz-orthopyroxene stability: Earth and Planetary Science Letters, v. 47, p. 11–20, doi: 10.1016/0012-821X(80)90099-0.

Bowen, N.L., and Schairer, J.F., 1932, The system $FeO-SiO_2$: American Journal of Science, v. 24, p. 177–213.

Bowen, N.L., and Schairer, J.F., 1935, The system $MgO-FeO-SiO_2$: American Journal of Science, v. 29, p. 151–217.

Bustin, R.M., and Mathews, W.H., 1982, In situ gasification of coal, a natural example: History, petrology and mechanics of combustion: Canadian Journal of Earth Sciences, v. 19, p. 514–523.

Clark, B.H., and Peacor, D.R., 1992, Pyrometamorphism and partial melting of shales during combustion metamorphism: Mineralogical, textural, and chemical effects: Contributions to Mineralogy and Petrology, v. 112, p. 558–568, doi: 10.1007/BF00310784.

Colaizzi, G.J., 2004, Prevention, control and/or extinguishment of coal-seam fires using cellular grout, in Stracher, G.B., ed., Coal Fires Burning Around the Word: A Global Catastrophe: International Journal of Coal Geology, v. 59, no. 1–2, p. 75–81.

Cosca, M.A., Essene, E.J., Geissman, J.W., Simmons, W.B., and Coates, D.A., 1989, Pyrometamorphic rocks associated with naturally burned coal beds, Powder River Basin, Wyoming: The American Mineralogist, v. 74, p. 85–100.

de Boer, C.B., Dekkers, M.J., and van Hoof, T.A.M., 2001, Rock-magnetic properties of TRM carrying baked and molten rocks straddling burnt coal seams: Physics of the Earth and Planetary Interiors, v. 126, p. 93–108, doi: 10.1016/S0031-9201(01)00246-1.

Edbrooke, S.W., Sykes, R., and Pocknall, D.T., 1994, Geology of the Waikato Coal Region, New Zealand: Lower Hutt, New Zealand, Institute of Geological and Nuclear Sciences Monograph 6, 236 p. (plus index and 8 map sheets).

Ellyett, C.D., and Fleming, A.W., 1974, Thermal infrared imagery of the burning mountain coal fire: Remote Sensing of Environment, v. 11, p. 221–229.

Essene, E.J., Coates, D.A., Geissman, J.W., and Simmons, W.B., 1984, Paralavas formed by burning of coal beds: Eos (Transactions, American Geophysical Union), v. 65, no. 16, p. 17.

Fermor, L.L., 1918, Preliminary note on the burning of coal seams at the outcrop: Transactions of the Mining, Geological, and Metallurgical Institute of India, v. 12, p. 50–63.

Foit, F.F., Jr., Hooper, R.L., and Rosenberg, P.E., 1987, An unusual pyroxene, melilite, and iron oxide mineral assemblage in a coal-fire buchite from Buffalo, Wyoming: The American Mineralogist, v. 72, p. 137–147.

Grapes, R., 2004, Marl to basalt: A case of extreme pyrometamorphism, in Annual Report of the Institute for Mineralogy, Petrology and Geochemistry: Freiburg, Germany, Albert-Ludwigs-Universität, p. 26–27.

Gray, V.R., and Daly, T.A., 1981, Chemical properties and composition of Waikato coals: New Zealand Journal of Science, v. 24, p. 179–202.

Haggerty, S.E., 1976, Opaque mineral oxides in terrestrial igneous rocks, in Rumble, D., ed., Oxide Minerals: Mineralogical Society of America Short Course Notes, v. 3, p. Hg101–Hg300.

Heffern, E.L., and Coates, D.A., 1997, Clinker—Its occurrence, uses, and effects on coal mining in the Powder River Basin, in Jones, R.W., and Harris R.E., eds., Proceedings of the 32nd Annual Forum on the Geology of Industrial Minerals: Wyoming State Geological Survey Public Information Circular 38, p. 151–165.

Heffern, E.L., and Coates, D.A., 2004, Geological history of natural coal-bed fires, Powder River basin, USA, in Stracher, G.B., ed., Coal Fires Burning Around the Word: A Global Catastrophe: International Journal of Coal Geology, v. 59, no. 1–2, p. 25–47.

Hensen, B.J., and Gray, D.R., 1979, Clinohypersthene and hypersthene from a coal buchite near Ravensworth, New South Wales, Australia: The American Mineralogist, v. 64, no. 1–2, p. 131–135.

Hoffmann, A., and Fisher, W.A., 1956, Bildung des spinels $FeOAl_2O_3$ und seiner Mischkristalle mit Fe_3O_4 und Al_2O_3: Zeitschrift für Physikalische Chemie, Neve Folge, v. 7, p. 80–90.

Johnson, W., and Bucknell, M.J., 1959, Pseudo-igneous rocks in the Triassic succession of the Springfield Basin, Gordon-Cradock district: Transactions of the Royal Society of South Australia, p. 245–257.

Kear, D., and Waterhouse, B.C., 1978, Waikato Coalfields: Wellington, New Zealand, Rotowaro Coalfield, 1:15,840 scale map: Lower Hutt, New Zealand, Dept. of Scientific and Industrial Research, Geology and Geophysics, 24 p.

Lindqvist, J.K., Hatherton, T., and Mumme, T.C., 1985, Magnetic anomalies resulting from baked sediments over burnt coal seams in southern New Zealand: New Zealand Journal of Geology and Geophysics, v. 28, p. 404–412.

Masalehdani, M.N., 1985, Thermal Metamorphism Associated with Burnt Coal Seams, Rotowaro Coalfield, Huntly, New Zealand (M.Sc. thesis): Auckland, Auckland University, New Zealand, 102 p.

Mason, D.R., and Davis, R., 1983, Geology and mineral chemistry of fused shale within overburden strata of Liddel seam in the Bowman's Creek Area, Hunter Valley, NSW: Proceedings Abstracts 17th Newcastle Symposium on Advances in the Study of the Sydney Basin, April 29–May 1: University of Newcastle, Newcastle, Australia, 4 p.

Morse, S.A., 1980, Basalts and Phase Diagrams: An Introduction to the Quantitative Use of Phase Diagrams in Igneous Petrology: New York, Springer, 493 p.

Muan, A., and Gee, C.L., 1956, Phase equilibrium studies in the system iron oxide–γAl_2O_3 in air at 1 atmosphere O_2 pressure: Journal of the American Ceramic Society, v. 39, p. 207–214, doi: 10.1111/j.1151-2916.1956.tb15647.x.

Nafziger, R.H., and Muan, A., 1967, System MgO-FeO-SiO_2: The American Mineralogist, v. 52, no. 9–10, p. 1377.

New Zealand 260 Map Series, 1998, Map 260–S14-Hamilton: Wellington, Land Information New Zealand, scale 1:50,000.

New Zealand 260 Map Series, 2005, Map 260–S13-Huntly: Wellington, Land Information New Zealand, scale 1:50,000.

Parker, R.J., and Willis, J.P., 1977, Computer programs SORT, REORD, and MW for major element XRF data processing: Computers & Geosciences, v. 3, no. 1, p. 115–171, doi: 10.1016/0098-3004(77)90037-1.

Phemister, J., 1942, Note on fused spent shale from a retort at Pumpherston, Midlothian, Transactions of the Geological Society of Glasgow, no. 18 (XVIII), p. 238–246.

Rădan, S.C., and Rădan, M., 1998, Rock magnetism and paleomagnetism of porcellanites/clinkers from the western Dacic Basin (Romania): Geologica Carpathica, v. 49, p. 209–211.

Ramdohr, P., 1982, The Ore Minerals and Their Intergrowths (2nd ed.): London, Pergamon Press, 1250 p.

Richards, R.G., and White, J., 1954, Phase relations of iron-oxide containing spinels. Volume 1: Relations in the system Fe-Al-O: Transactions of the British Ceramic Society, v. 53, p. 233–270.

Schairer, H., and Bowen, N.L., 1947, The system leucite-anorthite-silica: Geological Society of Finland Bulletin, v. 20, p. 67–87.

Sen Gupta, S., 1960, Petrography of the para-lavas of the eastern part of Jharia coalfield: Quarterly Journal of the Geological Mining and Metallurgical Society of India, v. 23, p. 79–101.

Smith, D., 1971, Stability of the assemblage iron-rich orthopyroxene-olivine-quartz: American Journal of Science, v. 271, p. 370–383.

Turnock, A.C., and Eugster, H.P., 1962, Fe-Al oxides: Phase relationships below 1000°C: Journal of Petrology, v. 3, no. 3, p. 533–565.

Tyráček, J., 1994, Stratigraphical interpretation of the paleomagnetic measurements of the porcellanites of the Most Basin, Czech Republic: Vestnik Českého Geologického Ústavu, v. 69, no. 2, p. 83–87.

Webster, A.H., and Bright, N.F.H., 1961, The system iron-titanium-oxygen at 1200°C and oxygen partial pressure between 1 atmosphere and 2×10^{-14} atmospheres: Journal of the American Ceramic Society, v. 44, p. 110–116, doi: 10.1111/j.1151-2916.1961.tb13723.x.

Zhang, X., 1998, Coal Fires in Northwest China: Detection, Monitoring, and Prediction using Remote Sensing Data [Ph.D. thesis]: Delft, The Netherlands, Delft University of Technology, 135 p.

MANUSCRIPT ACCEPTED BY THE SOCIETY 7 MARCH 2007

Paralavas in a combustion metamorphic complex: Hatrurim Basin, Israel

Yevgeny Vapnik*
Department of Geological and Environmental Sciences, Ben-Gurion University of the Negev, P.O.B. 653, Beer-Sheva 84105, Israel

Victor V. Sharygin
Ella V. Sokol
Institute of Geology and Mineralogy, Siberian Branch of the Russian Academy of Sciences, pr. Koptuyga 3, Novosibirsk 630090, Russia

Reginald Shagam
Department of Geological and Environmental Sciences, Ben-Gurion University of the Negev, P.O.B. 653, Beer-Sheva 84105, Israel

ABSTRACT

Unusual dike-like bodies and lenses of paralava, up to 1–5 m long and 10 cm thick, are found in the "olive" unit of the combustion metamorphic complex in the Hatrurim Basin, Israel. High-temperature rocks of the "olive" unit are composed of anorthite and clinopyroxene (diopside-hedenbergite-esseneite) and are connected with pipe-like explosion structures. The paralavas are cryptocrystalline rocks that exhibit vesicular and fluidal textures. The main mineral assemblage in these rocks, identified by electron-microprobe analysis and X-ray diffraction, consists of basic plagioclase + Fe-Ti oxides + clinopyroxene + K-feldspar + tridymite ± apatite. The silica content of the paralavas is similar to that of basalt, whereas the high calcium content suggests similarity to anorthosite. The occurrence of glass in the paralava is evidence for melting. The glasses are compositionally similar to rhyolites and more acidic melts. Melting temperatures were at least 1100 °C. The presence of pipe-like explosion structures, the occurrence of melted rocks, and various geological relationships throughout the Hatrurim Basin provide evidence for a new hypothesis about the genesis of this pyrometamorphic complex. The combustion metamorphic rocks in the Hatrurim Basin in Israel formed as a result of repeated ignition of hydrocarbon gases. The setting we envisage has many geological features typical of mud-volcano provinces. The occurrence of paralavas is restricted to the areas of gas ignition.

Keywords: combustion metamorphism, Hatrurim Formation, Dead Sea Transform Valley, bituminous strata, melting processes, paralava, gas explosions, gas ignition.

---

*vapnik@bgu.ac.il

INTRODUCTION

Combustion metamorphic complexes form during the long-term natural combustion of sedimentary rocks that are rich in coal seams and other types of organic matter. The natural combustion usually occurs as coal deposits are uplifted during orogeny. Commonly, combustion temperatures are high enough (900–1650 °C) to produce paralavas by partial or total melting of the sedimentary strata. The melts produced vary in composition from mafic to felsic (Bentor et al., 1981; Foit et al., 1987; Cosca et al., 1989; Kalugin et al., 1991). The formation of paralavas is determined by two main factors: the composition of the protolith and the temperature of combustion. Commonly, the composition of paralavas corresponds to eutectic or cotectic melts. However, at the extreme temperatures of combustion, complete and/or non-equilibrium melting of protolith may occur.

Optimal conditions for melting processes may be achieved during combustion metamorphism in coal-bearing strata composed of pelitic rocks, such as shaly sandstones, argillites, mudstones, and shales, and iron-rich rocks, such as siderites, chlorite-illite segregations, and limonite concretions. Under reducing conditions, the Fe^{2+}-bearing components significantly decrease the melting temperature of sedimentary rock, making partial melting possible at temperatures as low as 1000 °C to 1100 °C. Iron-rich basic paralavas commonly connected with coal fires are known in combustion metamorphic complexes in the United States, Canada, Australia, Russia, and Kazakhstan (Yavorsky and Radugina, 1932; Whitworth, 1958; Foit et al., 1987; Cosca et al., 1989; Kalugin et al., 1991; Sokol et al., 2005). The spatial and genetic relationships between basic iron-rich paralavas and chimney structures, which control the movement of combustible hot gases, have been documented for a high-temperature (up to 1450 °C) coal fire in the Powder River Basin, Wyoming, USA (Cosca et al., 1989). A reconstruction of the melting process of sedimentary protoliths indicates that the melts commonly form under reducing conditions corresponding to oxygen fugacity between wüstite-magnetite and quartz-magnetite-fayalite (QMF) buffers.

Melting occurs in a different way when oil- and gas-bearing silicate-carbonate sedimentary strata are subjected to combustion metamorphism. The minimum temperature of the pseudowollastonite-tridymite-anorthite eutectic in the $CaO-Al_2O_3-SiO_2$ system is 1170 °C. The displacement into higher calcium compositions is followed by a significant increase in the eutectic temperature (Schairer, 1942). Such a situation makes the appearance of a melt phase in a silicate-carbonate protolith at a temperature of less than 1200 °C problematic for combustion metamorphism. Large-scale, high-temperature (1000–1650 °C) combustion metamorphism events can be observed in the oil-bearing Monterey Formation (California, USA), where melting of the metapelitic protolith proceeds from the appearance of melt of eutectic composition to bulk melting of the calcium-rich protolith (Bentor et al., 1981). The partial, relatively low-temperature melting ($T \geq 1000$ °C) of metapelitic (diatomite) substrate is a local phenomenon; melt volume is less than 5%. The melt composition corresponds to dry eutectic melt similar to S-granites but with a strongly increased K_2O/Na_2O ratio. The dry acid melt has little chance to flow and commonly is quenched to glass in its place of generation. These rocks are identified easily, and the presence of glasses clearly indicates that partial melting has occurred. With an increase in temperature (up to 1650 °C), larger volumes of rock may be subject to melting. The melts become progressively enriched in MgO and CaO, and the high-temperature melt phase gradually attains the bulk composition of the dehydrated and decarbonated sedimentary protolith (Bentor et al., 1981). The low-silica and high-calcium paralavas, which are formed under high-temperature combustion metamorphism conditions, do not have magmatic analogs. Such rocks have been found in other regions where combustion metamorphism was promoted by the combustion of bitumen, oil, and gases. A lake of high-calcium silicate melt was formed at the site of oil-reservoir burning at the Tengiz deposit, Kazakhstan, where the temperature of burning oil was estimated to be as high as 2400 °C (Feldman et al., 1994). When the temperature is lower ($T \leq 1200$ °C), melting of the silicate-carbonate protolith has been observed only in the vicinity of cryptoexplosion (diatreme) structures that are associated with gas vents and/or flammable gas emissions. High-calcium paralavas composed of calcium silicates, and rarely, glass, are known from Iranian and Iraqi oil fields (McLintock, 1932; Basi and Jassim, 1974).

The subject of the present study is a rare sequence of high-temperature, low-pressure (sanidinite facies), calcareous silica-deficient combustion metamorphic rocks in Israel. These rocks are known as the Hatrurim Formation or the "Mottled zone" complex. The combustion origin of these rocks has been established by petrological, mineralogical, and isotopic studies conducted during the past 40 yr by Israeli geologists (Bentor et al., 1963; Gross et al., 1967; Kolodny and Gross, 1974; Gross, 1977, 1984; Matthews and Kolodny, 1978; Matthews and Gross, 1980; Burg et al., 1991; Gur et al., 1995). During surface combustion metamorphism events, the protolith of the pyrometamorphic rocks close in composition to chalk and marl underwent decarbonation reactions and chemical reactions between coexisting carbonate and silicate matter. The low total pressure (≤25 bar) of the combustion metamorphism of the Hatrurim Formation allowed the decarbonation reactions to occur at much lower temperatures than is common for regional or even contact metamorphic terrains. These reactions, which are endothermic, consumed a significant fraction of the heat provided by the combustion reactions, inhibiting any large-scale melting. Additionally, because of the very high content of calcium in the protolith (up to 41–45 wt% of CaO) (Bogoch et al., 1999), the generated melts behaved like metallurgical slag; with the rapid decrease in temperature, the low viscosity of the melt favored quick crystallization without quenching to glass. Some indirect indicators of local melting have been observed in previous studies. For example, coarse-grained, high-temperature assemblages (Gross, 1977;

Matthews and Gross, 1980) were found in fine-grained rocks. The occurrence of pseudowollastonite in coarse-grained schorlomite (Ca$_3$(Fe^{3+},Ti)$_2$[SiO$_4$]$_3$)-wollastonite-gehlenite rocks clearly suggests that they crystallized from melt at temperatures higher than 1100 °C. Moreover, the intergrowth of skeletal nagelschmidtite (Ca$_7$[SiO$_4$]$_3$[PO$_4$]$_2$) and schorlomite in these rocks also indicates crystallization from melt (Gross, 1977, her plate XIV). Similar rocks, with Ti-garnets, have been reported in combustion metamorphic complexes in the United States and interpreted as a special type of paralava (Bentor et al., 1981; Cosca et al., 1988, 1989). The fluidal texture reported in some hornfelses of the Hatrurim Basin is evidence of plastic behavior (Burg et al., 1991). At low total pressure (<25 bar) and temperatures of less than 1000 °C, the plastic behavior of the rock may be explained only by the presence of an interstitial melt phase.

This paper emphasizes the description of the mineralogy, petrography, and genesis of unusual calcium-rich basic paralavas that form dike-like bodies in the "olive" rock unit (hereafter olive unit) in the Hatrurim Basin.

GEOLOGICAL SETTING

The Hatrurim Basin

This unusual rock complex is situated near the Hatrurim crossroads on the Arad–Dead Sea road (Figs. 1 and 2). Here, rocks of the Hatrurim Formation have developed an irregular, blotchy pattern of colors (red, brown, black, green, and yellow), strong veining, and loss of primary sedimentary features. The Hatrurim Basin is a relatively small open syncline of rectangular outcrop adjoining several open anticlines on the southwestern flanks of the Dead Sea (Fig. 2); it is a component of the Syrian arc fold system, which is affected by the Dead Sea rift, the major structural feature of the region. The Campanian Mishash Formation (Fig. 3), which consists of cherts, bituminous chalky marls, and phosphorite, is at the base of the stratigraphic section in the basin. In nonaltered sedimentary sections in the vicinity of the Hatrurim Basin and in some parts of the Hatrurim Basin, the Mishash Formation is overlain by the Maastrichtian Ghareb Formation, which consists of a soft, slightly phosphatic, light-gray chalky marl. This marl is commonly accompanied by gypsum-bearing marls. In some cases, the rocks of the Ghareb Formation are highly bituminous. Bituminous matter (up to 20–25 wt%) is mostly sapropelic type II kerogen that was deposited in a highly organic-rich environment. The Paleocene Taqiye Formation consists of dark-gray to green, highly bituminous, gypsiferous clays, shales, and limestones, with scattered limonite concretions (Bentor and Vroman, 1960). Miocene (Hazeva Formation) and younger continental clastics overlie the older units (Burg et al., 1991; Gur et al., 1995). The Hatrurim Formation usually has been considered as a metamorphic correlative of the Ghareb and Taqiye Formations (Fig. 3).

Four mapping units were distinguished in the Hatrurim Basin by Burg et al. (1991). Three rock types describe the high-temperature transformations of calcium-rich protolith: (1) gehlenite-bearing marble, (2) calcite-spurrite and larnite-gehlenite-spurrite marble, and (3) bedded larnite rocks. There is a subordinate volume of high-grade metamorphic rocks compared to the huge amount of low-temperature hydrothermal rocks, which are represented by a range of rocks composed mainly of hydro-calcium silicates and sulfates.

A fourth unit occurs as a thin cover, in some places represented by a hard resistant crust, especially near the crests of low hills (Figs. 4A and 4B), and it is distributed randomly over the high- and low-temperature rocks described previously. Three

Figure 1. Outcrops of the "Mottled zone" complex in Israel and Jordan (after Burg et al., 1991).

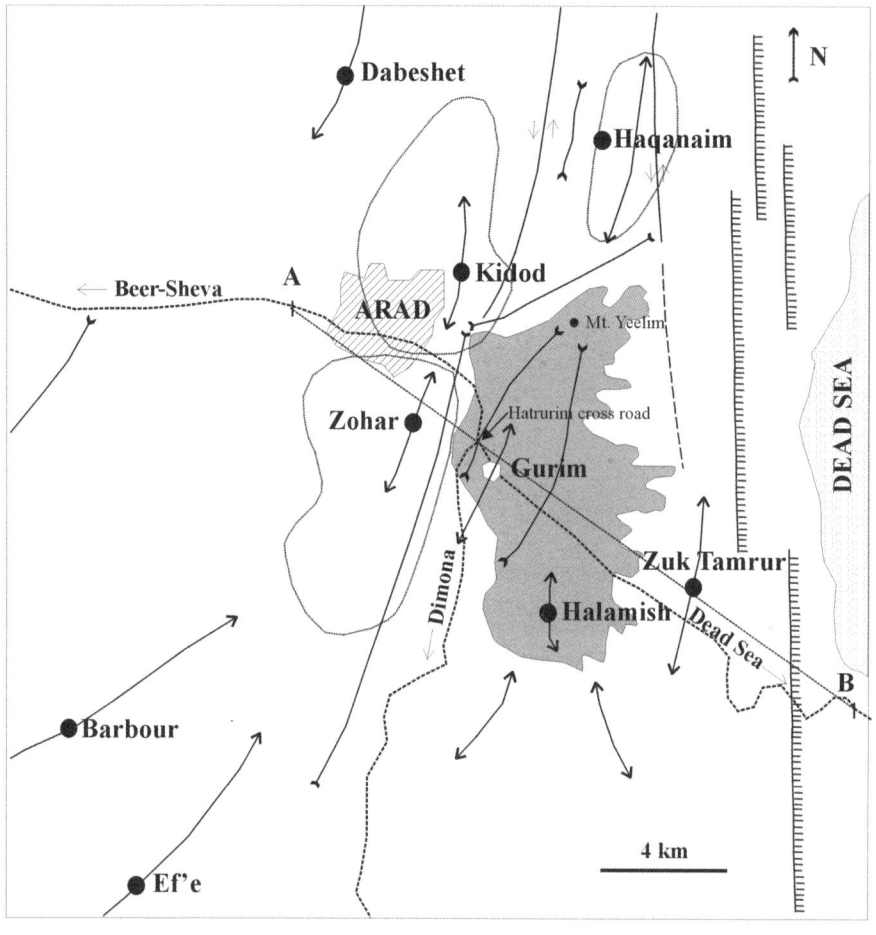

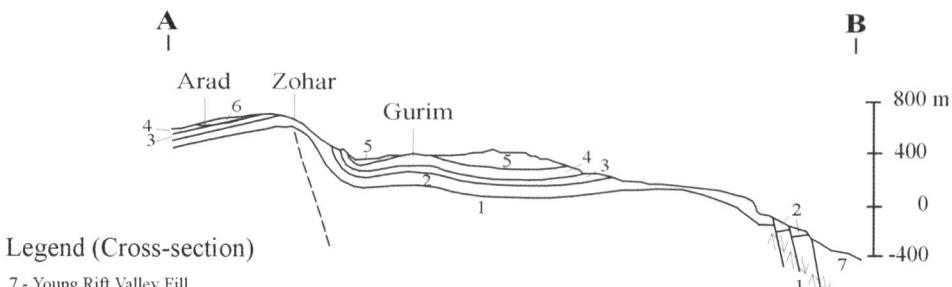

Figure 2. The tectonic setting and structural framework of the Hatrurim Basin. The eastern margin abuts the Jordan Rift; the western limit is defined by folds of the Syrian Arc. The map is modified after Gilat (1998b) and Gvirtzman and Stanislavsky (2000). The cross section is after Bentor (1966).

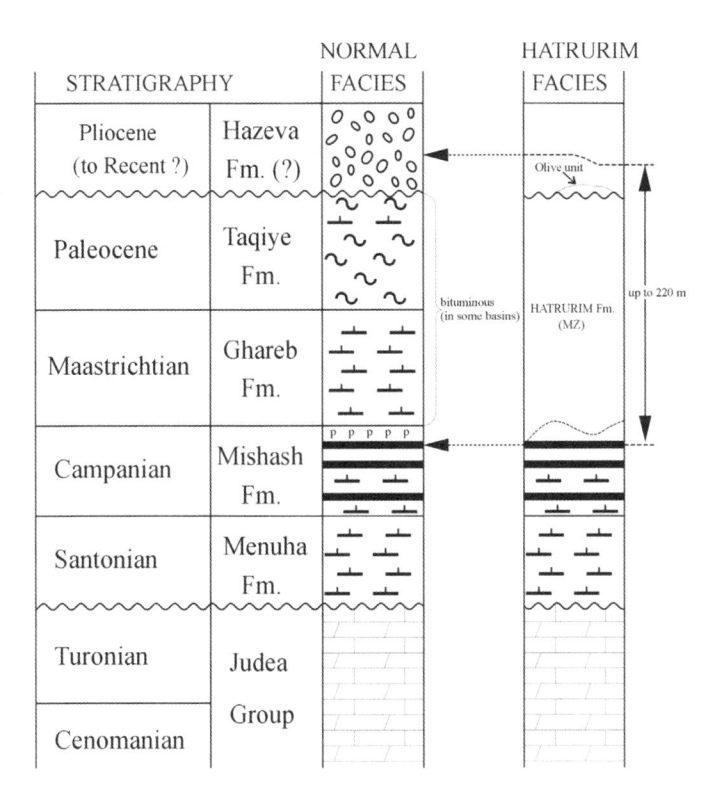

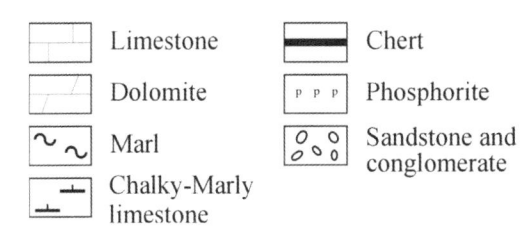

Figure 3. Stratigraphic position of the Hatrurim Formation (after Gross, 1977).

mineral assemblages have been recognized in this unit: anorthite-diopside, anorthite-diopside-calcite, and calcite-zeolite-hydrogarnet (each assemblage has subordinate phases listed in the following paragraphs and tables). These rocks have a dark greenish-gray color, and they are termed the "olive" unit. Of the aforementioned three assemblages, the anorthite-diopside assemblage characterizes a hornfels, the dominant resistant rock of the olive unit considered to represent high-grade metamorphism. The anorthite-diopside-calcite and calcite-zeolite-hydrogarnet assemblages make up the porous olive rocks that were formed as a by-product of hydrothermal alteration of hornfels and the fragments of sedimentary rocks.

The original material for the olive unit is problematic. Burg et al. (1991) suggested that it is a metamorphic equivalent of the lower part of the Taqiye Formation, whereas Shahar (1993) assumed that the olive unit may be related to a Miocene terrestrial sediment, the Hazeva Formation, especially to its clay component. It is clear that the protolith for the olive unit was different from the calcium-rich protolith for the other units and that it was the most enriched in clay components.

General Characteristics of the "Mottled Zone"

The "Mottled zone" (the Hatrurim Formation) has come to be recognized as an example of low-pressure, high-temperature sanidinite and pyroxene-hornfels facies metamorphism. Thermal energy was supplied by the near-surface oxidation and combustion of hydrocarbon compounds accumulated in a sedimentary facies. The Hatrurim Formation is exposed in at least 11 separate outcrops in Israel and on the western bank of Jordan (Fig. 1). Most outcrops of "Mottled Zone" rocks occur in the western vicinity of the Dead Sea Rift Valley. The Hatrurim Basin is the largest "Mottled zone" area in Israel (Fig. 2).

The origin of the Hatrurim Formation has been controversial for many decades. The following hypotheses have been suggested: contact metamorphism (Tristram, 1865), volcanism (Hull, 1886), hydrothermal alteration (Blanckenhorn, 1912), diagenesis (Picard, 1931), late diagenesis (Avnimelech, 1964), and pronounced subsurface combustion of the Senonian bituminous marls and chalks (Lees, 1928; Bentor and Vroman, 1960). Hydrothermal alteration related to tectonically introduced, overheated fluids that were rich in hydrocarbons was proposed by Gilat (1998a). The generally accepted hypothesis for the origin of the Hatrurim Formation involves internal combustion of bitumen. It has been suggested (Bentor et al., 1963; Gross et al., 1967; Kolodny and Gross, 1974; Gross, 1977, 1984; Matthews and Kolodny, 1978; Matthews and Gross, 1980; Burg et al., 1991; Gur et al., 1995) that during Tertiary surface combustion, the bituminous chalks and marls of the Ghareb and Taqiye Formations were converted into unusual high-temperature, low-pressure assemblages of calc-silicates, calc-aluminosilicates, and calc-aluminates. On the basis of mineral assemblages, the temperature is believed to have been in the range of 550 °C to 900 °C. The high-temperature processes occurred extremely close to the surface (where oxygen was available), in many local foci, and were of a relatively short duration (Matthews and Kolodny, 1978; Burg et al., 1991). It has been suggested that the combustion metamorphism events occurred at least twice. Available K-Ar and $^{40}Ar/^{39}Ar$ dating indicates major combustion metamorphism events in the Miocene at ca. 16 Ma and in the Pliocene-Pleistocene at 4.0–1.5 Ma (Gur et al., 1995).

The Setting of Paralavas within the Olive Unit in the Hatrurim Basin

The Hatrurim Basin is made up of numerous conical hills; the base of the formation is composed of intensively brecciated metasedimentary rocks with relicts of unaltered sedimentary protolith, and it is topped by separate blocks and lenses of high-grade combustion metamorphic rocks. The intensively brecciated metasedimentary zones are roughly horizontal and show no trace of primary layering. In some cases, these layers are very hard; however, in many cases, they are strongly altered and weathered with the appearance of a hydrogarnet + calcite + zeolite + gypsum mineral assemblage (Fig. 4F). The base of this problematic sequence invariably is underlain by phosphates, bituminous carbonates, and cherts of the Mishash Formation.

Some of the hills in the central and northern parts of the basin are topped by the rocks of the olive unit. There are ~25 individual outcrop areas (ranging in size from a few tens of square meters to several hundreds of square meters) of the olive unit. All outcrops are located at or near the crests of hills, ~60–120 m above the valleys (Fig. 4A). Several of the larger outcrops may be traced from crests 20–40 m downslope to lower elevations. In most cases, on closer inspection, olive rocks are a porous grayish-yellow fine-grained aggregate composed of calcite, zeolites, and hydrogarnet. Subordinate minerals include aragonite, barite, serpentine, andradite-grossular garnet, and apatite (Gross, 1977). Another rock type represented by porous dusky-yellow rocks is also widely distributed. The main minerals in such rocks are clinopyroxene and calcite, with minor plagioclase and zeolites. Some of these rocks look homogeneous to the naked eye; however, they show a fine clastic structure at higher magnification. In several cases, outcrops of hornfels occur in the uppermost part of the olive unit. These are dark, dense, massive, hard, fine- to micrograined rocks (Fig. 4B) and are composed mainly of plagioclase and clinopyroxene with subsidiary K-feldspar and apatite. Heating has completely obliterated the original structure of their protolith.

Erosion-resistant breccias, composed of angular clasts of hornfels and the products of their retrograde alteration, form several low (up to 1 m high) collar-like rings of agglomerate with diameters between 2.5 and 15 m (Fig. 4B). The breccias have an oval or circular trace in outcrop, and they narrow quickly with depth, making them similar to typical explosion pipes (Vapnik and Sokol, 2006). The hard, welded breccias, completely composed of large fragments of hornfelses (fragment sizes up to 0.5–0.7 m), occur only in the upper part of the agglomeratic collar encircling the pipe. To a depth of 1 m, the hornfels have undergone intensive replacement by an aggregate composed of calcite, zeolite, and, in some cases, hydrogarnet with relicts of diopside. At a depth of 2–4 m, the breccias are completely represented by fragments of retrograde units composed of calcite, zeolite, and hydrogarnet, whereas the main part of the breccias has been replaced by cement made up mainly of the same minerals.

In one outcrop, erosion-resistant hornfels forms two sickle-shaped bodies (Fig. 4C). The sickle-shaped bodies also consist of brecciated rock containing angular clasts of hornfels. Some of these clasts in the breccias have undergone plastic deformation (Fig. 4E). The diameter of the sickle-shaped bodies is up to 5.0–7.0 m, the height of the wall is up to 1 m, and the thickness is 0.5–1.5 m. Usually, there is a clear depression in the central part of the pipes. This central part, located within the agglomeratic collar of hornfels breccias and the subvertical neck, commonly is composed of fine-grained clastic matter with fragments of retrograde rocks of olive color. There also are different proportions of fragments of the underlying sedimentary rocks, such as Turonian dolomite, Campanian phosphorite, and chert. All of this fine-grained clastic matter, together with larger fragments, is cemented firmly in a matrix of retrograde mineral assemblages (Fig. 4D).

Three bodies of cemented breccias, laid down on an uneven surface, occur in the same area as the olive unit. Fragments of two of the breccia bodies consist mainly of rocks of the olive unit. Fragments of the third body mainly are strongly brecciated cobbles, pebbles, and boulders of chert of the Mishash Formation, with minor fragments of spurrite-larnite-brownmillerite rock, and, in addition, these fragments may contain rocks of the olive unit. The chert fragments are up to 30 cm in diameter. All fragments are cemented by a retrograde matrix that also includes quartz grains.

The occurrence of ribbons and lenses of dense black and red rocks (paralavas), commonly with fluidal and vesicular textures, is restricted completely to the olive unit (Figs. 4G and 4H). In some cases, the vesicular texture, normally of microscopic scale, can be seen clearly in hand samples. Lenses and ribbons, up to 5 m long and 10 cm thick, are enclosed by hornfels and hydrogarnet-zeolite-calcite rocks. In many cases, retrograde zeolite-bearing rocks crosscut ribbon-like bodies, dividing them into several lenses (Fig. 4G).

ANALYTICAL TECHNIQUES

Several samples of paralavas and their host rocks, namely anhydrous hornfels and retrograde products, from the typical outcrops of the olive unit near the Hatrurim crossroads of the Arad–Dead Sea region were studied. X-ray diffraction (XRD) analysis was used for identification of the crystalline components in amounts greater than 5 wt% of sample using standard methodology and a DRON-3 powder diffractometer ($Cu_{K\alpha}$-radiation, accelerating potential 30–40 kV, beam current 30 mA).

X-ray fluorescence analysis of major components was performed using a VRA-20 R X-ray detector. Rock samples and a standard sample (essexite gabbro, #SGD-1) were prepared as fused disks. The detection limit for most components (SiO_2, Al_2O_3, Fe_2O_3 total, MnO, CaO, K_2O, P_2O_5) was 0.02 wt%; for Na_2O, it was 0.2 wt%, and for MgO, it was 0.1 wt%. The concentrations were calculated by the method of main parameters (Afonin et al., 1984). Losses on ignition (LOI) were determined by weighing samples, and sulfur, FeO, and Fe_2O_3 were determined by chemical methods. Determination of H_2O and CO_2 was done using a chromatograph LX-8 MD if concentrations were less than 1.5 wt%; for higher concentrations, determinations were made by the gravimetric method and titration, respectively. Fluorine was analyzed by the photometric method with a precision of 0.003 wt%.

Trace-element contents were obtained using precise X-ray fluorescence analysis with synchrotron radiation (XRF-SR). The synchrotron data were collected with EDS—energy-dispersion

Figure 4. The outcrops and rocks of the Hatrurim Basin. (A) Panoramic view of the north part of the basin. Several hills are topped by gray-weathering rocks of the olive rock unit. (B) Typical outcrop of hornfelses. (C) Sickle-shaped hornfels body. (D) Explosion breccia from pipe crater, made up of fragments of the olive rock unit and scaly fragments of chert. (E) Explosion breccia composed of plastically deformed fragments of hornfelses. (F) Breccia composed of the fragments of spurrite-larnite-gehlenite rocks enclosed in the hydrocarbonate-silicate-sulfate matrix. (G) Paralava lens from a hill near the Hatrurim crossroads, sample 5200. (H) Paralava vein, sample Y-10–1, west slope of Mt. Yeelim.

spectrum (Phedorin et al., 2000). Two values of excitation energy were used: 22 kV for analysis of V, Cr, Co, Ni, Cu, Zn, Ga, As, Rb, Sr, Y, Zr, and Nb, and 42 kV for analysis of Cd, Sn, Sb, I, Cs, Ba, La, and Ce. The calculation of element concentrations was performed on the basis of an external standard. The detection limits for trace elements were as follows (in ppm): V = 30, Cr = 30, Mn = 30, Fe = 30, Ni = 30, Cu = 5, Zn = 5, Ga = 2, As = 0.8, Se = 0.5, Br = 0.5, Rb = 0.5, Sr = 3, Zr = 5, Nb = 0.5, Cd = 0.2, Sn = 0.5, Y = 5, La = 5, Ce = 5, I = 0.5, Th = 1, and U = 1.

Quantitative analyses of minerals were performed at the Institute of Mineralogy and Petrography (Novosibirsk) using a CAMEBAX electron microprobe. The operating conditions were optimized with an accelerating potential of 20 kV and a beam current of 15–30 nA, counting time of 10 s, and beam diameter of 2–3 μm. This regime was used for most analyses of anhydrous silicate and oxide compounds. For beam-sensitive materials, such as serpentine, zeolites, hydrogarnets, and apatite, a beam current of 10 nA and a beam diameter of up to 5 or 10 μm were used. Natural and synthetic minerals of similar composition served as standards. Analyses were performed at a take-off angle of 40° with 10 s counts. The Ca and P concentrations were corrected for overlap of the $Ca_{K\beta1}$ and $P_{K\alpha}$ peaks. Similar corrections also were provided for Ti-Ba, Cr-Mn, and Si-Sr pairs. The standard procedure of "final correction" indicated that the estimated precision was better than 2% for all major elements and ~5% for minor elements. The detection limit was calculated using 2σ criterion (at a confidence level of 95%).

TABLE 1. MAJOR- AND TRACE-ELEMENT COMPOSITION OF PARALAVAS AND HOST ROCKS

Sample number:	5200[1]	5200[2]	5201[1]	5201[2]	5201[3]	5206[1]	5207[1]	5207[3]	5208[1]
Oxides (wt%)									
SiO_2	48.41	34.67	46.81	34.03	43.36	47.63	46.07	44.43	49.35
TiO_2	0.82	0.62	0.87	0.55	0.80	0.84	0.82	0.77	0.85
Al_2O_3	24.82	16.08	24.82	15.18	21.37	24.12	22.75	21.39	25.11
Fe_2O_3	3.56	5.77	6.35	4.55	3.73	6.67	3.39	3.21	7.22
FeO	2.05	0.07	0.29	0.07	2.26	0.29	2.51	2.51	N.D.
MnO	0.03	0.04	0.03	0.06	0.03	0.03	0.03	0.03	0.15
MgO	2.60	3.59	2.78	5.00	3.86	2.78	3.62	4.04	1.96
CaO	14.02	17.97	14.08	27.40	18.56	13.33	16.64	18.20	11.73
Na_2O	0.75	2.70	0.93	0.35	1.63	0.78	0.67	0.91	0.46
K_2O	1.63	1.97	1.42	1.27	1.28	2.28	1.75	1.34	2.60
P_2O_5	0.56	0.69	0.57	0.74	0.72	0.53	0.63	0.69	0.57
H_2O^+	0.00	5.89	0.88	2.60	1.37	0.21	0.55	1.01	0.00
H_2O^-	0.16	5.55	0.11	0.97	0.27	0.16	0.14	0.24	N.D.
S_{tot}	0.03	0.03	0.00	0.58	0.02	0.02	0.00	0.03	N.D.
CO_2	0.20	4.47	0.00	6.70	0.81	0.41	0.41	0.81	N.D.
F	0.01	0.04	0.00	0.00	0.06	0.00	0.04	0.06	N.D.
Total	99.65	100.15	99.94	100.05	100.13	100.08	100.02	99.66	100.00
Trace elements (ppm)									
V	112	110	87	78	115	66	98	96	354
Cr	228	116	215	88	220	221	228	172	200
Ni	225	175	205	124	234	171	212	165	225
Cu	38	20	18	25	32	33	26	25	5.3
Zn	255	76	185	231	294	224	188	225	401
Cd	0.2	0.8	0.5	1.6	0.4	0.5	0.2	0.7	0.9
Sn	0.6	1.3	0.8	2.2	1.0	3.5	0.3	2.0	4.1
Sb	0.2	0.9	0.5	0.7	0.6	1.0	1.7	0.8	0.9
Ga	17	16	19	14	18	15	22	18	25
As	5.5	1.8	2.5	5.5	2.7	6.9	5.5	2.5	1.9
Rb	94	64	54	35	29	73	57	32	81
Cs	6.2	0.5	3.4	0.4	1.2	9.0	4.0	0.8	7.3
Ba	770	1547	428	1183	1054	387	838	1304	194
Sr	564	1280	610	921	601	475	615	615	499
Zr	123	88	118	87	117	119	114	110	131
Nb	23	14	20	14	17	22	20	20	24
Y	38	37	34	37	46	33	40	41	46
La	50	40	36	38	47	47	49	54	51
Ce	78	66	57	59	77	76	77	83	85
I	1.4	6.1	0.9	4.0	1.9	0.9	1.8	2.7	0.9

Note: 1—paralava; 2—calcite-zeolite-hydrogarnet rock; 3—hornfels. N.D.—not determined.

ANALYTICAL RESULTS

Whole-Rock Compositions and Trace-Element Geochemistry of Paralavas and Their Host Rocks of the Olive Unit

The major- and trace-element analyses of paralavas and host rocks are shown in Table 1. The major components of the paralavas and their host hornfels are rather similar. The main difference is the low content of structural water and CO_2 in paralavas. They also contain somewhat lower amounts of calcium and higher amounts of Al_2O_3. According to the $K_2O + Na_2O$ versus SiO_2 classification diagram, the paralavas are similar to basalt, whereas their high calcium content suggests a similarity to anorthosite. Paralavas and hornfels differ markedly from retrograde rocks. Retrograde rocks differ by their high water and CO_2 content and have the least amounts of silica and aluminum among all compared rock units. All rocks are conspicuously enriched in V, Cr, Ni, and Zn. The similar set of trace elements (particularly the high content of V) is typical for oil-bearing sediments and oils of the Middle East basins (Basi and Jassim, 1974) and phosphorites (Nathan et al., 1979). Compared to hornfels and hydrogarnet-zeolite-calcite rocks, the paralavas are enriched in Cr, Ni, Rb, and Cs and are depleted in Ba, Sr, and I.

Petrography of Paralavas and Host Rocks

Paralavas collected by the authors consist mainly of plagioclase, clinopyroxene, K-feldspar, and Fe-Ti oxides. In contrast to calcite-zeolite-hydrogarnet rocks and hornfels, they are enriched in Fe-Ti oxides. Two kinds of paralavas can be distinguished by their association with Fe-Ti oxides. The first type is more widespread and contains hematite and ilmenite (samples 5201 and 5208). The second type occurs rarely and contains magnetite and ilmenite (sample 5200). Minor phases in the paralavas include apatite, pyrrhotite, fayalite, and low-calcium pyroxene. Tridymite was identified in all paralava samples by XRD (Table 2).

The paralavas are characterized by cryptocrystalline fluidal texture with alternating light and black bands. The black bands clearly are enriched in Fe-Ti oxides. The samples have abundant vesicles and lensoid relict fragments (Figs. 5A and 5B) of host rocks (up to 1.5 cm long), which are oriented according to the fluidal texture. Calcite, hematite, and zeolites fill some cavities. Between the host-rock relicts and magnetite-rich paralava, there usually are several zones with different mineral compositions, color, and crystallinity. In some cases, smaller relicts of host rock are replaced completely in paralavas with the formation of relict structure; this has been noted in other mafic paralavas (Kalugin et al., 1991; Sokol et al., 2005).

The paralava samples 5200 and 5208 contain numerous thin vesicles (up to 1–2 mm in diameter) with oval cross sections. The walls of such ovoids are zoned and enriched in oxides and silicates of iron. Larger, branched channels have oval cross sections (up to 5–7 mm), filled mainly by fine-grained hematite in hydrogarnet-zeolite-calcite rocks of the olive unit (Figs. 5C and 5D). The structure and phase composition of the walls of the ovoids permit one to recognize these structures as small gas channels that were formed by the flow of heated gases, whereas the zoning of assemblages reflects a variation in temperature and oxygen fugacity across the rock surrounding the gas-flow channels.

Hornfelses are fine- to medium-grained rocks and consist of plagioclase, clinopyroxene, K-feldspar, and minor apatite. They contain uniformly distributed, but few (3–5 vol%), isolated closed vesicles (≤0.2 mm diameter). In some cases, the vesicles are filled by calcite and zeolites, and, rarely, by fine-grained hematite. The similarity of paralavas to hornfelses is evidenced by the presence of clinopyroxene and plagioclase ± K-feldspar ± apatite, whereas the principal petrographic differences are the fluidal texture, the appearance of numerous gas channels and vesicles, the abundance of Fe-Ti oxides, and the presence of fayalite and tridymite in the paralavas.

Retrograde hydrogarnet-zeolite-calcite rocks form cryptocrystalline to finely crystalline aggregates and contain relicts of clinopyroxene and plagioclase (affected by corrosion) with

TABLE 2. MINERALOGY OF ROCKS OF THE OLIVE UNIT

Sample	Rock species	Cpx	Pl	Hem	Mag	Kfs	Zeol	Hgr	Cal	Trd	Minor phases (<5 vol%) (petrographic data)
5200 host	Olive rock	+++	++	−	−	−	++++	+++	++++	−	Ap
5200	Paralava	+++	++++	−	+++	++	+	−	+	++	Ilm, Fa, Ap, Po, Serp, Opx
5201 host1	Olive rock	+++	++	−	−	−	++++	+++	++++	−	Ap
5201 host2	Hornfels	++++	++++	+	−	++	+	−	+	−	Ap
5201	Paralava	+++	++++	++	−	++	+	−	+	++	Ilm, Ap
5208	Paralava	+++	++++	+++	−	++	+	−	+	++	Ilm, Ap, Whit, Opx

Note: Data were obtained by X-ray diffraction (XRD) and petrographic analyses. Cpx—clinopyroxene; Pl—plagioclase; Hem—hematite; Hgr—hydrogarnet and other water-bearing calcium silicates; Mag—magnetite or magnesioferrite; Kfs—K-feldspar, Zeol—zeolite (mainly, phillipsite); Cal—calcite; Trd—tridymite; Ap—apatite, Ilm—ilmenite, Fa—fayalite; Po—pyrrhotite; Serp—Fe-rich serpentine (greenalite); Whit—a mineral of the whitlockite group; Opx—low-calcium pyroxene. Maghemite is absent. ++++ = very abundant; +++ = abundant; ++ = minor; + = very minor; − = absent. Olive rock—retrograde calcite-zeolite-hydrogarnet rock.

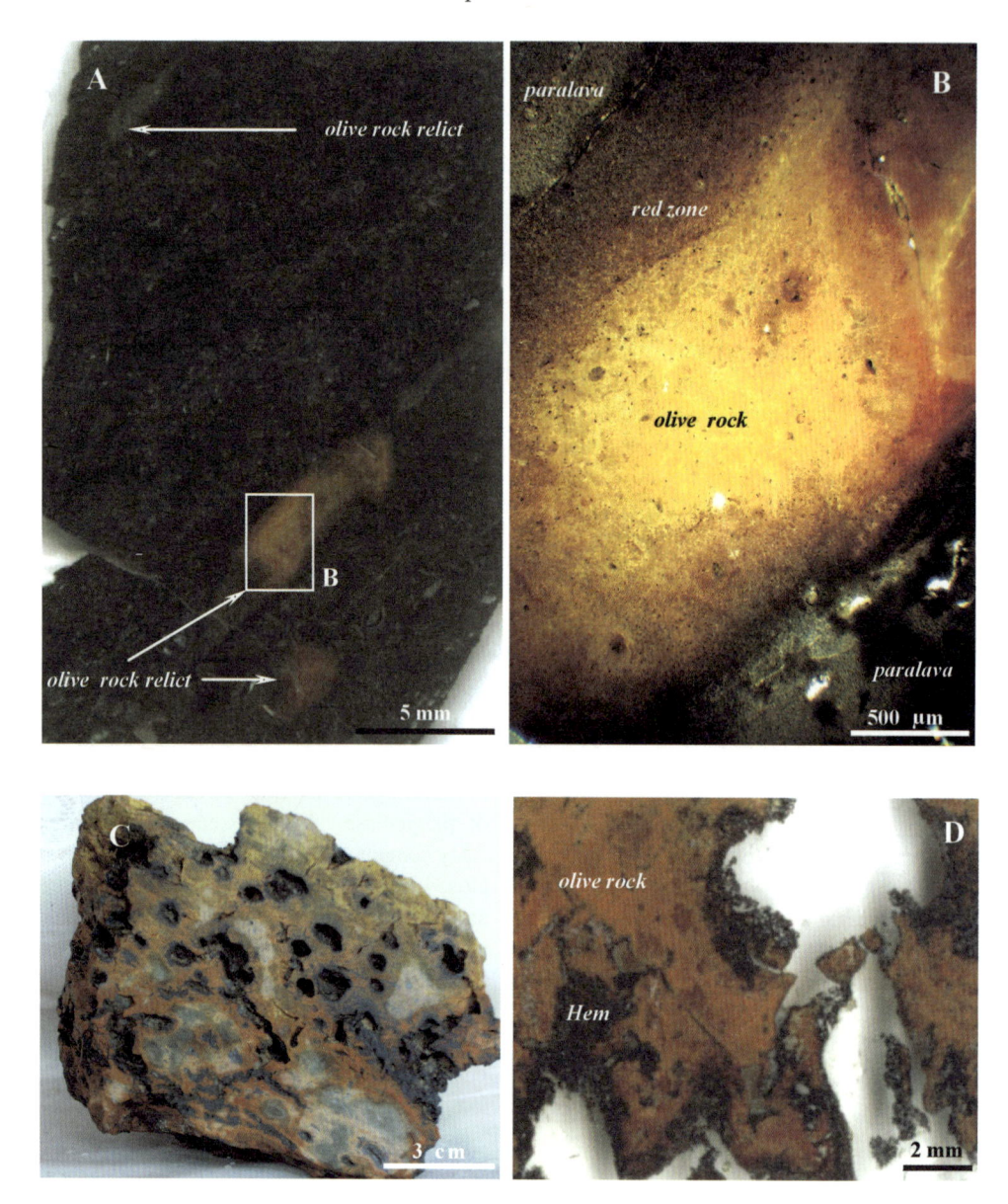

Figure 5. (A–B) Relicts of the olive rocks in magnetite-rich paralava, sample 5200. (C–D) Melting and crystallization around gas channels in the olive rock unit. (C) Hydrogarnet-calcite-zeolite rock (former hornfels) from the olive unit totally reworked by high-temperature gas vents, sample 5210. (D) Hematite (Hem)-enriched zone around cavities in the olive rock (same sample).

minor apatite and serpentine. Calcite and zeolites (mainly phillipsite) usually fill cavities or form thin veinlets.

Mineral Composition of Paralavas

In general, paralavas contained mineral grains that usually did not exceed 5 µm diameter. Thus, it was impossible to analyze individual minerals with the electron microprobe. Only some grains (10–20 µm) of Fe-Ti oxides in these fine-grained rocks were suitable for microprobe analysis (Table 3). However, mineral grains from "ovoid zones" in the paralavas were larger and permitted easier analysis of the main minerals. Two paralava samples (5200 and 5208; Fig. 6) were investigated in detail.

Assemblages in the Ovoids of Magnetite-Rich Paralavas

Sample 5200, a magnetite-bearing paralava, has vesicular walls coated with two types of zonal assemblages that differed in mineral composition: clinopyroxene-magnetite-plagioclase and fayalite-clinopyroxene-magnetite-plagioclase (Figs. 6A–6C). The first mineral assemblage is found around open vesicles (0.5–1.0 mm diameter). The replacement of mafic zones by leucocratic zones is common for these ovoids. The mafic zones are enriched in green clinopyroxene, magnetite, ilmenite, apatite, and, less commonly, low-calcium pyroxene. Leucocratic zones are composed mainly of plagioclase ± K-feldspar. In some cases, the outermost zone is made up of a coarse-grained magnetite-clinopyroxene assemblage. The open space, seen commonly in

TABLE 3. REPRESENTATIVE COMPOSITIONS OF ORE MINERALS FROM GROUNDMASS OF PARALAVAS

Sample number:	5200		5201			5208		
Mineral:	Mag	Ilm	Ilm	Hem	Hem	Ilm	Hem	Hem
Oxides (wt%)								
TiO_2	13.49	48.46	40.22	7.68	8.16	44.62	12.25	6.91
Cr_2O_3	0.73	0.08	0.21	0.21	0.36	0.10	0.17	0.70
V_2O_3	0.52	0.23	N.D.	0.11	N.D.	0.00	0.00	0.49
Al_2O_3	1.77	0.06	2.79	0.87	1.64	1.31	1.15	2.42
FeO_t	76.53	47.80	53.89	79.82	78.48	47.48	72.14	78.72
MnO	0.14	0.19	0.00	0.19	0.17	0.06	0.22	0.07
MgO	1.31	2.28	1.26	1.85	1.85	4.12	5.65	1.72
CaO	0.19	0.08	0.00	0.15	0.22	0.32	0.22	0.27
NiO	N.D.	N.D.	N.D.	N.D.	N.D.	0.01	0.24	0.36
ZnO	0.48	0.00	N.D.	0.22	N.D.	0.02	0.21	0.04
Total	95.16	99.18	98.37	91.10	90.88	98.03	92.25	91.70
Fe_2O_3	39.69	9.54	22.19	85.34	83.23	16.42	79.83	84.49
FeO	40.82	39.22	33.92	3.03	3.59	32.70	0.31	2.69
Total	99.14	100.13	100.59	99.65	99.22	99.67	100.25	100.16
Atomic proportions								
Ti	0.381	0.907	0.750	0.150	0.159	0.828	0.230	0.134
Cr	0.022	0.002	0.004	0.004	0.007	0.002	0.003	0.014
V	0.016	0.005	N.D.	0.002	N.D.	0.000	0.000	0.010
Al	0.078	0.002	0.082	0.027	0.050	0.038	0.034	0.073
Fe^{3+}	1.122	0.179	0.414	1.667	1.624	0.305	1.502	1.635
Fe^{2+}	1.282	0.816	0.704	0.066	0.078	0.675	0.007	0.058
Mn	0.004	0.004	0.000	0.004	0.004	0.001	0.005	0.001
Mg	0.073	0.085	0.047	0.072	0.072	0.151	0.211	0.066
Ca	0.008	0.002	0.000	0.004	0.006	0.008	0.006	0.007
Ni	N.D.	N.D.	N.D.	N.D.	N.D.	0.000	0.005	0.008
Zn	0.013	0.000	N.D.	0.004	N.D.	0.000	0.004	0.001

Note: Magnetite (Mag) was calculated on the basis of three cations and four oxygen; ilmenite (Ilm) and hematite (Hem) were calculated on the basis of two cations and three oxygen. Iron for Fe-Ti-oxides is divided into FeO and Fe_2O_3 according to stoichiometry. N.D.—not determined.

the center of the vesicles, may be filled with zeolites. This type of assemblage is identical in its mineral composition to paralava, except for the appearance of low-calcium pyroxene.

The fayalite-clinopyroxene-magnetite-plagioclase association is located in zones, up to 3 mm wide, around open narrow channels, and it has minerals that differ noticeably from the host rock (Figs. 6A–6C). The central zone (in the vicinity of an open vesicle) is relatively coarse-grained and is composed mainly of fayalite and magnetite with minor basic plagioclase, K-feldspar, and pyrrhotite (Fig. 6B). Fe-rich serpentine (greenalite) appears to be the alteration product of fayalite. The intermediate zone contains poikilitic fayalite and magnetite with a small amount of green clinopyroxene and apatite dispersed in a basic plagioclase matrix (Fig. 6C). The outermost zone is a fine-grained feldspar matrix with minor grains of clinopyroxene, magnetite, pyrrhotite, and apatite. Fayalite is rare here and occurs as rounded inclusions in clinopyroxene. This is a border zone to the host paralava, with which it forms a transitional contact marked by decreasing grain size.

Although the phase ratio differs in successive zones of fayalite- and clinopyroxene-magnetite–rich ovoid assemblages, the chemistry of minerals is almost identical (Table 4). The presence of fayalite was supported by XRD and optical analysis (refractive index nZ of ~1.880). It has the following composition: $Fa_{81.7-82.2}Fo_{17.3-17.5}La_{0.4-0.6}$ with minor P_2O_5 content. Greenalite shows admixture of K_2O, Al_2O_3, and P_2O_5. Magnetite is significantly enriched in TiO_2 (4–32 mol% of the ulvöspinel) and V_2O_3 (up to 1.4 wt%). Other components, namely Cr_2O_3, Al_2O_3, MgO, ZnO, and NiO, are minor. There are no major differences between magnetites from different types of ovoids and magnetites in the groundmass of paralava. Plagioclase is represented by basic plagioclase ($An_{79-95}Ab_{5-19}Or_{0.5-11}$), whereas Na-enriched compositions are more common for fayalite-free assemblages. K-feldspar corresponds to orthoclase ($Or_{90.5-90.8}Ab_{1.5-4.9}An_{4.7-7.7}$) with minor BaO content (up to 0.3 wt%). Clinopyroxene has a composition ($Di_{31-65}Hd_{1-45}Fs_{12-27}Ess_{5-10}Acm_{0.5-1}$) that corresponds to ferroaugite. These clinopyroxenes from fayalite-bearing and magnetite-pyroxene ovoid assemblages differ slightly in their MgO and FeO content. However, they differ from clinopyroxenes of hydrogarnet-zeolite-calcite rocks and hornfels by their low CaO, Al_2O_3, TiO_2, and Cr_2O_3 content and high SiO_2 content. Low-calcium pyroxene ($En_{53.5}Fs_{43.2}Wo_{3.3}$) contains negligible Al_2O_3 (0.8 wt%); thus, it does not correspond to pigeonite. It may be classified as orthopyroxene or as a monoclinic analog in the clinoenstatite-clinoferrosilite group. The small grain sizes and

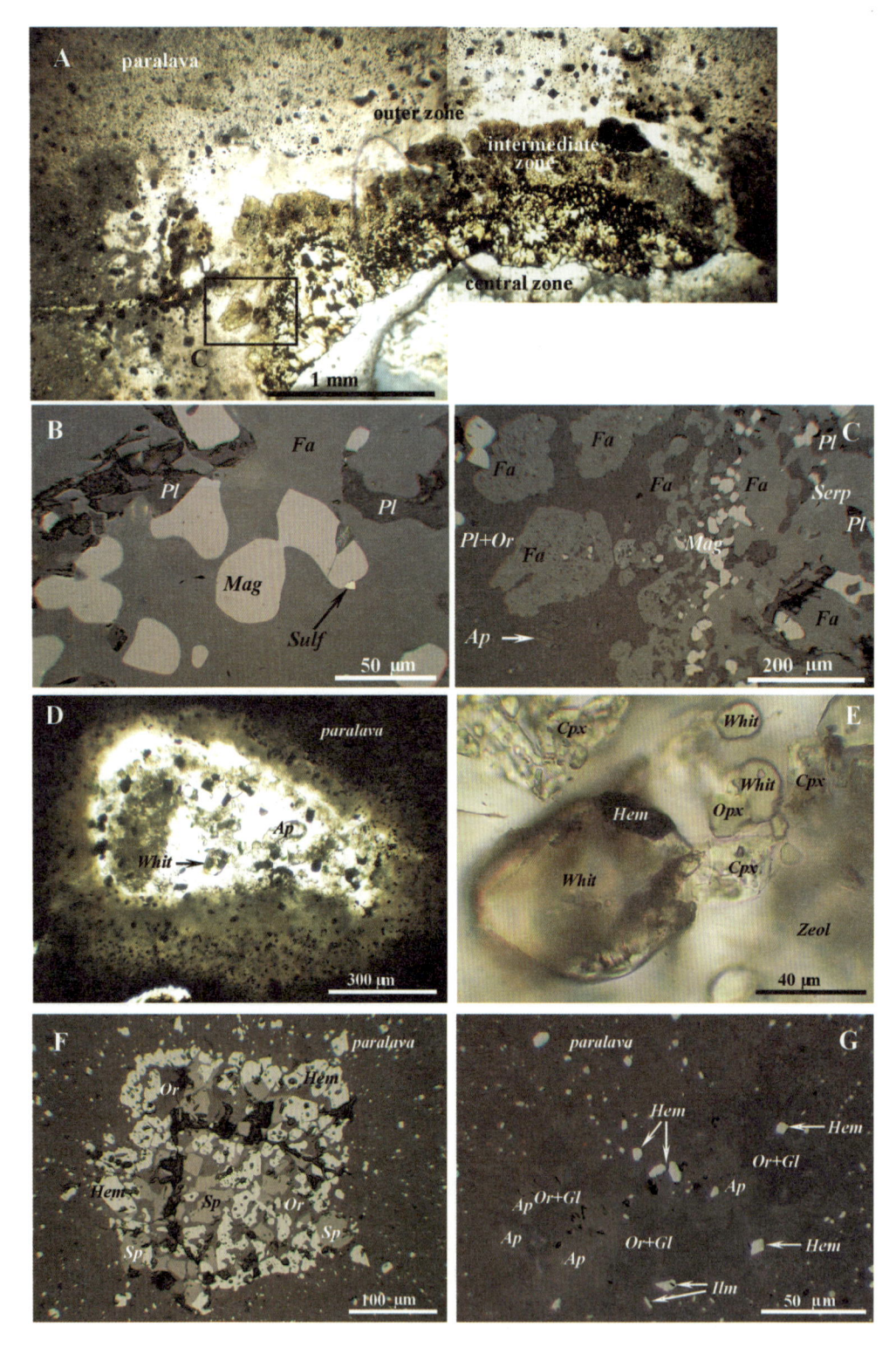

Figure 6. (A–C) Large zoned fayalite-bearing ovoid around empty vesicle, sample 5200. (A) General image, transmitted light. Core zone is composed of fayalite + magnetite + plagioclase; intermediate zone is fayalite + plagioclase + magnetite + orthoclase + apatite + clinopyroxene; outer zone is clinopyroxene + plagioclase + magnetite + orthoclase + apatite. (B) Magnetite and sulfide (pyrrhotite) in fayalite within the core zone, reflected light. (C) Poikilitic fayalite within the intermediate zone, reflected light. (D–G) Ovoids in hematite-rich paralava, sample 5208. Numbers in parentheses are ovoids listed in Tables 5 and 6. (D–E) Large ovoid containing whitlockite-like phase, apatite, pyroxenes, and hematite. (D) General view of ovoid (2), transmitted light. (E) Euhedra of whitlockite and pyroxenes in zeolite matrix filling the cavity, transmitted light. (F) Fe-Ti oxide–rich assemblages, ovoid (13), reflected light. (G) Leucocratic ovoid (9) containing apatite, K-feldspar, and glass, reflected light. Abbreviations: Fa—fayalite; Cpx—clinopyroxene (ferroaugite); Pl—plagioclase; Mag—titanomagnetite; Or—orthoclase; Ap—apatite; Sulf—pyrrhotite; Serp—Fe-rich serpentine (greenalite); Opx—orthopyroxene (or low-calcium clinopyroxene?); Ilm—ilmenite; Hem—hematite; Sp—magnesioferrite spinel; Whit—mineral of the whitlockite group; Zeol—zeolites (mainly phillipsite); Gl—glass.

TABLE 4. REPRESENTATIVE COMPOSITIONS OF MINERALS FROM OVOID ZONES IN MAGNETITE-RICH PARALAVA, SAMPLE 5200

	Fayalite-rich ovoid assemblages																					Pyroxene-magnetite–rich ovoid assemblages							
	Core zone						Intermediate zone						Border zone							Ovoid 1				Ovoid 4		
Mineral:	Fa	Srp	Mag	An	Or		Fa	Mag	An				Fa	Mag	Cpx	Ap	An	Or		Mag	Cpx	Ap	An		Mag	Opx
Oxides																										
SiO$_2$	30.87	31.73	0.00	45.69	60.29		31.11	0.00	44.99				30.97	0.00	48.80	0.72	45.06	64.69		0.00	49.84	0.12	45.29		0.00	51.52
TiO$_2$	0.14	0.10	10.29	0.12	0.07		0.10	10.96	0.07				0.08	12.25	0.28	0.08	0.06	0.16		8.49	0.43	0.07	0.03		8.41	0.12
Cr$_2$O$_3$	0.00	0.00	0.16	0.00	0.00		0.00	0.47	0.00				0.00	0.71	0.01	0.00	0.00	0.00		0.36	0.02	0.00	0.00		0.29	0.02
V$_2$O$_3$	N.D.	N.D.	0.76	N.D.	N.D.		N.D.	0.45	N.D.				0.00	0.53	N.D.	N.D.	N.D.	N.D.		0.91	N.D.	N.D.	N.D.		0.58	N.D.
Al$_2$O$_3$	0.00	0.61	1.74	33.03	16.73		0.00	1.95	34.44				0.00	2.85	1.51	0.00	34.04	16.97		2.06	1.32	0.00	33.66		1.95	0.76
FeO$_t$	61.40	56.31	81.44	2.01	8.12		61.12	80.59	0.95				60.80	77.87	25.41	0.61	1.53	1.33		81.92	18.84	0.52	0.90		81.79	26.75
MnO	0.17	0.17	0.02	0.00	0.00		0.16	0.02	0.00				0.12	0.05	0.05	0.00	0.06	0.00		0.05	0.18	0.04	0.00		0.05	0.32
MgO	7.05	6.10	0.23	0.05	0.46		7.35	0.42	0.04				7.64	0.29	6.98	0.06	0.05	0.34		0.87	10.37	0.14	0.07		1.08	18.79
CaO	0.26	0.32	0.04	18.31	1.26		0.34	0.15	18.84				0.25	0.14	17.05	53.22	18.46	0.87		0.10	18.71	55.69	18.66		0.29	1.51
BaO	0.00	0.00	N.D.	0.05	0.01		0.00	N.D.	0.01				0.00	N.D.	0.00	0.00	0.02	0.32		N.D.	N.D.	N.D.	N.D.		N.D.	N.D.
Na$_2$O	0.00	0.07	N.D.	0.73	0.14		0.00	N.D.	0.50				0.00	N.D.	0.11	0.08	0.69	0.50		N.D.	0.11	0.21	0.79		N.D.	0.06
K$_2$O	0.00	0.19	N.D.	0.11	12.51		0.00	N.D.	0.09				0.00	N.D.	0.00	0.00	0.24	14.01		N.D.	0.00	0.08	0.22		N.D.	0.00
P$_2$O$_5$	0.19	0.18	N.D.	0.00	0.00		0.16	N.D.	0.00				0.20	N.D.	0.00	39.90	0.00	0.00		N.D.	0.00	41.44	0.00		N.D.	0.00
ZnO	N.D.	N.D.	0.32	N.D.	N.D.		N.D.	0.24	N.D.				N.D.	0.29	N.D.	N.D.	N.D.	N.D.		0.37	N.D.	N.D.	N.D.		0.42	N.D.
Total	100.07	95.77	95.00	100.10	99.59		100.33	95.24	99.92				100.05	94.98	100.20	94.98	100.15	99.18		95.14	99.81	98.95	99.61		94.86	99.84
Fe$_2$O$_3$			45.97					44.64						40.18	1.83					49.40	2.20				50.08	1.18
FeO			40.08					40.42						41.71	23.76					37.47	16.86				36.72	25.69
Total	100.07	95.77	99.61	100.10	99.59		100.33	99.71	99.92				100.05	99.00	100.38	94.98*	100.15	99.18		100.09†	100.03	98.95§	99.61		100.02#	99.96
Atomic proportions																										
Si	0.991	1.768	0.000	2.123	2.885		0.993	0.000	2.086				0.990	0.000	1.933	0.121	2.091	3.014		0.000	1.929	0.020	2.108		0.000	1.965
Ti	0.003	0.004	0.293	0.004	0.003		0.002	0.310	0.002				0.002	0.348	0.008	0.004	0.002	0.006		0.239	0.012	0.004	0.001		0.237	0.003
Cr	0.000	0.000	0.005	0.000	0.000		0.000	0.014	0.000				0.000	0.021	0.000	0.000	0.000	0.000		0.011	0.000	0.000	0.000		0.009	0.001
Al	0.000	0.040	0.078	1.809	0.943		0.000	0.087	1.882				0.000	0.127	0.071	0.000	1.862	0.932		0.091	0.060	0.000	1.847		0.086	0.034
Fe^{3+}		1.309						1.265						1.141	0.055					1.393	0.064				1.413	0.034
Fe^{2+}	1.648	2.624	1.268	0.078	0.325		1.632	1.273	0.037				1.626	1.316	0.787	0.088	0.059	0.052		1.174	0.546	0.071	0.035		1.152	0.819
Mg	0.337	0.507	0.013	0.003	0.003		0.350	0.024	0.002				0.364	0.016	0.412	0.016	0.004	0.023		0.049	0.598	0.034	0.005		0.060	1.068
Ca	0.009	0.019	0.002	0.912	0.065		0.012	0.006	0.936				0.009	0.006	0.724	9.869	0.918	0.043		0.004	0.776	9.750	0.931		0.012	0.062
Na	0.000	0.008	N.D.	0.066	0.013		0.000	N.D.	0.045				0.000	N.D.	0.008	0.026	0.062	0.045		N.D.	0.008	0.067	0.071		N.D.	0.005
K	0.000	0.013	N.D.	0.006	0.764		0.000	N.D.	0.005				0.000	N.D.	0.000	0.000	0.014	0.833		N.D.	0.000	0.017	0.013		N.D.	0.000
P	0.005	0.008	N.D.	0.000	0.000		0.004	N.D.	0.000				0.005	N.D.	0.000	5.812	0.000	0.000		N.D.	0.000	5.732	0.000		N.D.	0.000

Note: Iron is apportioned into FeO and Fe$_2$O$_3$ according to stoichiometry. Fa—fayalite; Mag—titanian magnetite; Srp—Fe-rich serpentine (greenalite); An—anorthite; Or—orthoclase; Cpx—Fe-rich clinopyroxene; Opx—low-calcium pyroxene; Ap—apatite. Totals include: *Cl—0.04 wt% (0.010); †NiO—0.16 wt% (0.005); §SrO—0.19 wt% (0.018), Y$_2$O$_3$—0.15 wt% (0.013), La$_2$O$_3$—0.13 wt% (0.008), Ce$_2$O$_3$—0.16 wt% (0.010), SO$_3$ and Cl <0.01 wt% (<0.003); and #NiO—0.14 wt% (0.004). Fayalite was calculated on the basis of four oxygen; magnetite was calculated on the basis of three cations and four oxygen; greenalite was calculated on the basis of five cations; anorthite and orthoclase were calculated on the basis of eight oxygen; clinopyroxene and orthopyroxene were calculated on the basis of six oxygen; apatite was calculated on the basis of ten cations in the Ca position. N.D.—not determined.

their rarity in paralava (<5 vol%) prevent identification by routine XRD measurements. Pyroxenes of the clinoenstatite-clinoferrosilite group are extremely rare in terrestrial rocks; only two occurrences have been confirmed in combustion metamorphic complexes. These are the burned rocks of problematic genesis in the Glenroy Valley, New Zealand (Tulloch and Campbell, 1993), and the so-called "black blocks" from coal spoil-heaps in the Chelyabinsk Coal Basin, Russia (Chesnokov and Shcherbakova, 1991). Thus, it is more likely that the calcium-poor pyroxene of this study is an ordinary orthorhombic species. Apatite (likely fluorapatite) is virtually free of SO_3 and Cl. Ilmenite is a rare mineral in ovoid assemblages and groundmass of magnetite-rich paralavas. It usually forms very thin blades that are not suitable for microprobe analysis. We analyzed only one blade in association with titaniferous magnetite in the groundmass (sample 5200, Table 3). This ilmenite contained 7.8 and 16.3 mol% of geikielite and hematite, respectively.

Assemblages in the Ovoids of Hematite-Rich Paralavas

Sample 5208 is a hematite-rich paralava. Three types of ovoid assemblages are located around and within the vesicles; they are approximately similar in their mineralogical associations but differ strongly with respect to the ratios of essential minerals (Figs. 6D–6G).

The first type of mineral assemblage, confined to vesicles (Figs. 6D, 6E), is the most common. Low-calcium pyroxene, clinopyroxene, apatite, a mineral of the whitlockite group, magnesioferrite, hematite, and ilmenite usually occur in the walls of vesicles. Zeolites or K-feldspar commonly fill the intergranular space in the vesicles. The paralava around such vesicles commonly contains a zone enriched in feldspars and depleted in mafic minerals.

The second type of ovoid assemblage forms large, rounded ovoids (up to 1 mm long) and is rich in Fe-oxides (Figs. 6F, 6G). Hematite and magnesioferrite are the principal minerals; K-feldspar, low-calcium pyroxene, clinopyroxene, plagioclase, and apatite are minor phases. The interstices between these minerals usually are filled by zeolites.

The third type of ovoid assemblage forms small, leucocratic isolated ovoids (100–150 μm) in the paralava. K-feldspar, apatite, and, in some cases, low-calcium pyroxene, are principal phases. Some of these isolated ovoids contain colorless and slightly devitrified glass that usually occupies the interstices between grains of K-feldspar and other minerals (Fig. 6G). The abundance of glass does not exceed 5–10 vol%.

Hematites from paralava groundmass and ovoids are strongly enriched in Ti, Mg, and Al (Tables 3 and 5). Magnesioferrite is homogeneous; its composition is uniform within a given assemblage, but it varies strongly from one association to another. Spinel mole fraction in magnesioferrite varies from 13 to 50 mol%. The more aluminous compositions are more common for magnesioferrite from the first type of ovoid assemblage. The content of Fe_3O_4 ranges from 6 to 16 mol%, and NiO, ZnO, and Cr_2O_3 are the minor components. Ilmenite occurs in a few ovoid assemblages and forms very minute blades, not suitable for microprobe analysis. However, ilmenite from the groundmass in hematite-rich paralavas is larger, and electron microprobe data indicate a hematite mole fraction in the range of 27–37 mol%. Other components, such as Al_2O_3, MnO, and MgO, are minor.

In contrast to magnetite-bearing paralava, clinopyroxene from sample 5208 is enriched in MgO, CaO, Al_2O_3, Na_2O, and TiO_2 and depleted in FeO_{tot}. Its composition corresponds to $Di_{58-80}En_{6-10}Ts_{15-35}Acm_{2-3}$. The charge balance calculation shows that most iron is in the Fe^{3+} state (8–27 mol% of essencite $CaFeSiAlO_6$). Clinopyroxene from hematite-rich paralava also differs strongly from clinopyroxene of hydrogarnet-zeolite-calcite rocks and hornfels because it has a higher Mg number (75–89) and a higher essencite content. Low-calcium pyroxene also is Mg-rich ($En_{92-94}Fs_{4-6}Wo_{1-3}$), corresponding to enstatite or clinoenstatite.

Plagioclase is bytownite-anorthite ($An_{79-91}Ab_{9-19}Or_{0.5-15}$) with high FeO_{tot} (1–1.9 wt%). K-feldspar corresponds to Na-bearing orthoclase ($Or_{82-91}Ab_{9-14}An_{1-4}$) with high FeO_{tot} (0.5–2.1 wt%) and minor SrO and BaO content (up to 0.15 wt%).

Two phosphates occur in the assemblages of hematite-rich paralavas: colorless apatite and a yellowish mineral. These two phosphates may coexist within a particular ovoid (Fig. 6D, Table 5). Considering the major components, the yellowish phosphate seems to belong to the whitlockite group (Griffin et al., 1972). This phase is enriched in SO_3. Unfortunately, XRD identification of both phosphates was not available because of the small grain sizes. Their Raman spectra did not show any pronounced peaks in the region of $\nu = 1000–1100$ cm^{-1} (characteristic of the $[CO_3]^{2-}$ group), as well as in the regions of $\nu = 1500–1600$ cm^{-1} and $\nu = 3200–3700$ cm^{-1} (characteristic for OH-vibrations). According to these data, both minerals are virtually free of OH- and CO_3-groups. Therefore, the first mineral may be identified as fluorapatite, and the second phase seems similar to merrillite-(Na), $Ca_{18}Na_2Mg_2(PO_4)_{14}$, an extraterrestrial mineral that has not been approved by the International Mineralogical Association.

Glasses were found in only two leucocratic ovoid assemblages. Glass compositions are similar to rhyolite-like and more acidic melts. In general, the glass composition may be expressed as $Qz_{35-50}Or_{35-55}Ab_{7-15}$. Glasses from particular assemblages also differ slightly in the amount of normative corundum (Table 6). Inasmuch as totals of analyses are ~99–100 wt%, it may be suggested that glasses contain a minimal quantity of water.

Mineral Compositions of Hydrogarnet-Zeolite-Calcite Rocks and Hornfelses

Representative analyses of the main minerals from hydrogarnet-zeolite-calcite olive rocks and hornfelses are given in Table 7. Clinopyroxene and plagioclase are the significant phases of retrograde olive rocks; however, their abundance is strongly subordinate to calcite, zeolites, and hydrogarnet. Both phases were analyzed in host rocks and host-rock relicts in paralavas. Clinopyroxenes from hornfels, their relicts in paralavas,

TABLE 5. REPRESENTATIVE ANALYSES OF MINERALS FROM SOME OVOID ZONES IN HEMATITE-RICH PARALAVA, SAMPLE 5208

Mineral:	Opx	An	Kfs	Hem	Whit	Ap	Cpx	Cpx	Spl	Cpx	Cpx	An	Spl	Hem	Opx	An	Kfs	Hem	Kfs	Spl	Hem
	\multicolumn{6}{c}{Ovoid 2}	\multicolumn{3}{c}{Ovoid 5}	\multicolumn{5}{c}{Ovoid 11}	\multicolumn{4}{c}{Ovoid 12}	\multicolumn{3}{c}{Ovoid 13}
Oxides																					
SiO$_2$	56.58	44.99	63.32	0.00	0.04	0.06	51.34	44.28	0.00	47.68	47.51	46.37	0.00	0.00	55.54	45.54	63.57	0.00	62.97	0.00	0.00
TiO$_2$	0.21	0.06	0.21	9.75	0.06	0.04	0.68	2.25	0.89	0.97	1.39	0.09	1.20	10.96	0.11	0.06	0.13	10.96	0.10	0.22	5.18
Cr$_2$O$_3$	0.03	0.00	0.00	0.32	0.00	0.00	0.00	0.01	0.70	0.00	0.00	0.00	1.52	0.69	0.00	0.00	0.00	0.69	0.00	1.01	0.41
Al$_2$O$_3$	3.37	33.16	18.30	1.08	0.00	0.00	2.98	8.75	14.48	7.91	5.63	33.16	15.95	1.83	4.25	33.35	18.44	1.83	18.81	29.27	1.69
FeO$_t$	2.75	1.79	0.75	76.98	0.13	0.03	3.92	7.45	55.68	5.04	6.26	1.07	53.50	72.51	3.46	0.95	0.56	72.51	1.35	41.35	81.30
MnO	0.41	0.03	0.01	0.25	0.64	0.08	0.13	0.18	1.06	0.19	0.19	0.01	0.83	0.07	0.25	0.00	0.03	0.07	0.01	0.37	0.07
MgO	35.12	0.09	0.01	2.09	4.25	0.47	16.89	12.77	16.78	14.59	15.09	0.05	15.66	4.42	35.61	0.12	0.01	4.42	0.05	18.37	2.34
NiO	N.D.	N.D.	N.D.	0.27	N.D.	N.D.	N.D.	N.D.	1.02	N.D.	N.D.	N.D.	1.07	0.31	N.D.	N.D.	N.D.	0.31	N.D.	1.68	0.20
ZnO	N.D.	N.D.	N.D.	0.23	N.D.	N.D.	N.D.	N.D.	3.49	N.D.	N.D.	N.D.	4.27	0.16	N.D.	N.D.	N.D.	0.16	N.D.	3.77	0.05
CaO	1.18	18.51	0.82	0.33	46.30	55.62	23.34	23.22	0.11	22.50	23.56	17.06	0.22	0.38	0.38	17.16	0.33	0.38	0.87	0.07	0.01
Na$_2$O	0.17	0.99	1.43	N.D.	1.75	0.12	0.38	0.48	N.D.	0.57	0.15	1.66	N.D.	N.D.	0.04	1.32	1.23	N.D.	1.63	N.D.	N.D.
K$_2$O	0.00	0.11	14.54	N.D.	0.08	0.03	0.00	0.00	N.D.	0.00	0.00	0.07	N.D.	N.D.	0.00	0.84	14.91	N.D.	14.00	N.D.	N.D.
P$_2$O$_5$	0.00	0.00	0.00	N.D.	44.87	41.59	0.00	0.00	N.D.	0.00	0.00	0.00	N.D.	N.D.	0.00	0.00	0.00	N.D.	0.00	N.D.	N.D.
Total	99.82	99.73	99.52*	91.37†	100.37§	98.35#	99.66	99.39	94.18	99.46	99.78	99.54	94.21	91.48††	99.64	99.34	99.21	91.48††	99.78	96.04	91.32§§
Fe$_2$O$_3$	0.25			81.22			4.23	8.20	58.32	5.36	6.83		54.24	79.49	1.47			79.49		42.88	90.16
FeO	2.53			3.90			0.11	0.07	3.20	0.22	0.12		4.69	0.98	2.14			0.98		2.76	0.17
Total	99.84	99.73	99.52	99.51	100.37	98.35	100.09	100.21	100.03	99.99	100.46	99.54	99.64	99.44	99.78	99.34	99.21	99.44	99.78	100.34	100.35
Atomic proportions																					
Si	1.925	2.102	2.952	0.000	0.014	0.010	1.873	1.647	0.000	1.749	1.750	2.152	0.000	0.000	1.894	2.127	2.965	0.000	2.929	0.000	0.000
Ti	0.005	0.002	0.008	0.190	0.016	0.002	0.019	0.063	0.021	0.027	0.039	0.003	0.029	0.154	0.003	0.002	0.004	0.208	0.003	0.005	0.100
Al	0.135	1.826	1.006	0.033	0.000	0.000	0.128	0.384	0.543	0.342	0.244	1.814	0.600	0.032	0.171	1.836	1.014	0.055	1.031	1.016	0.051
Fe^{3+}	0.006			1.580			0.116	0.230	1.397	0.418	0.189		1.304	1.652	0.038			1.512		0.951	1.740
Fe^{2+}	0.072	0.070	0.029	0.084	0.039	0.004	0.003	0.002	0.085	0.007	0.004	0.042	0.125	0.002	0.061	0.037	0.022	0.021	0.053	0.068	0.004
Mn	0.012	0.001	0.000	0.006	0.192	0.011	0.004	0.006	0.028	0.006	0.006	0.001	0.022	0.007	0.007	0.000	0.001	0.001	0.000	0.009	0.002
Mg	1.785	0.006	0.001	0.081	2.244	0.115	0.918	0.708	0.796	0.797	0.828	0.003	0.745	0.126	1.810	0.008	0.000	0.167	0.003	0.800	0.089
Ca	0.043	0.927	0.041	0.009	17.588	9.800	0.912	0.926	0.004	0.884	0.930	0.848	0.008	0.011	0.014	0.859	0.017	0.010	0.043	0.002	0.000
Na	0.012	0.089	0.129	N.D.	1.202	0.038	0.027	0.034	N.D.	0.041	0.010	0.149	N.D.	N.D.	0.003	0.120	0.111	N.D.	0.147	N.D.	N.D.
K	0.000	0.006	0.865	N.D.	0.036	0.006	0.000	0.000	N.D.	0.000	0.000	0.004	N.D.	N.D.	0.000	0.050	0.887	N.D.	0.831	N.D.	N.D.
P	0.000	0.000	0.000	N.D.	13.456	5.785	0.000	0.000	N.D.	0.000	0.000	0.000	N.D.	N.D.	0.000	0.000	0.000	N.D.	0.000	N.D.	N.D.

Note: Iron for pyroxenes and Fe-Ti oxides is apportioned into FeO and Fe$_2$O$_3$ according to stoichiometry. Whit—a mineral of the whitlockite group (merrilite-[Na]?), Hem—Ti-rich hematite, Spl—Zn-Ni-rich magnesioferrite. Other symbols, see Table 4. Totals include: *SrO—0.13 wt% (0.007 wt%); †SrO—0.07 wt% (0.003); ‡V$_2$O$_3$—0.05 wt% (0.003), Y$_2$O$_3$—0.16 wt% (0.033), La$_2$O$_3$—0.06 wt% (0.008), Ce$_2$O$_3$—0.02 wt% (0.003), SO$_3$—1.95 wt% (0.518); #SrO—0.19 wt% (0.018), Y$_2$O$_3$—0.04 wt% (0.003), La$_2$O$_3$—0.03 wt% (0.002), Ce$_2$O$_3$—0.02 wt% (0.001), SO$_3$—0.02 wt% (0.001), Cl—0.02 wt% (0.006); **V$_2$O$_3$—0.03 wt% (0.001); ††V$_2$O$_3$—0.16 wt% (0.003); §§V$_2$O$_3$—0.08 wt% (0.002). Hematite was calculated on the basis of two cations and three oxygen; magnesioferrite was calculated in the same way as shown in Table 4. N.D.—not determined.

TABLE 6. REPRESENTATIVE ANALYSES (WT%) OF PHASES FROM LEUCOCRATIC OVOID ZONES
IN HEMATITE-RICH PARALAVA, SAMPLE 5208

	Ovoid 9							Ovoid 10			
Mineral:	Ap	Kfs	Kfs	Gl	Gl	Gl	Gl	Kfs	Gl	Gl	Gl
SiO_2	0.06	64.17	64.23	77.70	78.46	77.85	77.46	64.58	75.87	76.35	75.93
TiO_2	0.01	0.17	0.13	0.38	0.34	0.44	0.23	0.19	0.47	0.43	0.70
Al_2O_3	0.00	18.65	18.49	11.87	12.17	11.70	11.63	18.57	11.18	11.11	11.05
FeO	0.05	0.45	0.53	0.16	0.18	0.19	0.20	0.42	0.33	0.42	0.55
MnO	0.09	N.D.	0.03	N.D.	N.D.	N.D.	0.02	N.D.	N.D.	N.D.	N.D.
MgO	0.74	0.00	0.01	0.18	0.15	0.40	0.20	0.01	0.32	0.09	0.20
CaO	55.46	0.34	0.30	0.40	0.39	0.36	0.45	0.17	0.78	0.65	0.56
BaO	N.D.	0.15	0.12	0.06	0.03	0.03	N.D.	0.09	0.00	0.02	0.02
Na_2O	0.16	1.05	1.31	1.23	1.02	0.94	1.06	0.96	0.72	0.70	0.66
K_2O	0.07	14.66	14.45	7.18	6.46	6.87	6.55	14.69	9.13	8.55	8.94
P_2O_5	41.46	0.00	0.00	0.08	0.03	0.07	0.15	0.00	0.38	0.17	0.15
Total	98.73*	99.63	99.58	99.24	99.22	98.84	97.96	99.69	99.17	98.49	98.74

Note: Cr_2O_3 and Cl are below detection limits. Ap—apatite; Kfs—K-feldspar; Gl—glass. N.D.—not determined.
*Total includes: SrO—0.17 wt%; Y_2O_3—0.05 wt%; La_2O_3—0.10 wt%; Ce_2O_3—0.10 wt%; SO_3—0.21 wt%.

and hydrogarnet-zeolite-calcite rocks have rather similar compositions. Their distinguishing feature is the abundance of esseneite, and they are classified as aluminous diopside and hedenbergite. The most esseneite-rich compositions are common for hydrogarnet-zeolite-calcite rocks: $Ess_{36-51}Di_{34-44}Hd_{2-26}TiTs_{2-7}Acm_{1-5}$. Clinopyroxenes from hornfels and from the relics of olive rocks in paralavas are richer in hedenbergite: $Ess_{21-39}Di_{30-55}Hd_{10-27}TiTs_{4-7}CaTs_{0-13}Acm_{1-2}$ and $Ess_{22-25}Di_{24-25}Hd_{48-50}TiTs_{1-2}Acm_{1-2}$, respectively. Plagioclase analyses are rather calcic. Most of them correspond to anorthite ($An_{92-99}Ab_{1-6}Or_{0-1}$) or bytownite ($An_{85-87}Ab_5Or_{9-11}$), or, in some cases, high FeO_{tot} (0.6–4.2 wt%). The composition of K-feldspar is $Or_{92}Ab_5An_3$. Apatite is common in retrograde rocks of the olive unit. Apatite composition is similar to analytical data presented by Gross (1977). Analyzed apatite does not contain SO_3 and Cl; fluorine was not analyzed. Gross (1977) described this apatite as fluorapatite. Carbonate is an ordinary calcite with a low (<0.03 wt%) SrO content. The oxide contents of zeolites range widely. According to XRD, phillipsite is the major zeolite mineral in these rocks.

DISCUSSION

Whole-Rock Compositions of Hornfelses and Paralavas

The major-element data suggest that hornfelses and paralavas are high-temperature products of metamorphism of the same protolith. The bulk chemical composition of the protolith was similar to a marl with a high clay component (Table 1). The absence of noteworthy differences in the petrochemical compositions of hornfelses and paralavas suggests that the melting processes were controlled mainly by local variation in heat source rather than protolith composition. In most cases of coal fires, it seems that differences in the composition of the protolith determine the different compositions of paralavas (Yavorsky and Radugina, 1932; Whitworth, 1958; Kalugin et al., 1991; Sokol et al., 2002, 2005).

Preliminary Estimation of Temperature and Oxygen Fugacity

Previously, the mineral assemblages of the Hatrurim Formation were interpreted to indicate maximum temperatures of combustion metamorphism of 1000 °C to 1100 °C (Gross, 1977); paralava indicates that temperatures were much higher locally. Unfortunately, we encountered numerous problems with direct thermodynamic calculations of temperature and oxygen fugacity during paralava formation. Only preliminary, qualitative results are suggested here. The normative components of paralava plotted on different phase diagrams, such as An + Wo + FeO, Di + An + Wo, and Qz + Or + An (Schairer, 1942), indicate temperatures of between 1220 °C and 1500 °C. The melting temperature of natural systems might be significantly lower because of impurities, i.e., freezing-point depression. However, the ambiguity of temperature estimation here, unlike with endogenic melts, is due to the fact that the liquids do not flow away. Consequently, the melts may be heated to superliquidus temperatures (Cosca et al., 1989). Glass compositions from the Hatrurim paralavas indicate that the temperature of melt quenching was between 950 °C and 1150 °C (Schairer, 1950). The coexistence of fayalite, magnetite, and tridymite in the zonal mineral assemblages around vesicular walls in the Hatrurim paralava (sample 5200) suggests that subsequent melt crystallization and its quenching took place under redox conditions near the QMF buffer. The widespread occurrence of hematite and esseneite in hematite-bearing paralava requires much more oxidizing conditions around the hematite-magnetite buffer (Cosca et al., 1989).

Origin of Paralavas in the Hatrurim Basin

It is commonly accepted that high-temperature gas flows rising through the sedimentary sequence are the main agent of combustion metamorphism. This approach was proved unambiguously using physico-chemical analysis of the burning process (Knunyants, 1998) and is supported by geological observations

TABLE 7. REPRESENTATIVE ANALYSES OF MINERALS IN CALCITE-ZEOLITE-HYDROGARNET ROCKS AND HORNFELS

Sample:	5200(h)							5200(r)					5203				H-5-93		
Mineral:	Cpx	Cpx	An	Ap	Hgr	Cal	Zeol	Zeol	Cpx	An	Ap	Zeol	Cpx	Cpx	An	Cpx	Cpx	An	Kfs
Oxide (wt%)																			
SiO$_2$	37.48	40.92	43.30	2.14	31.72	0.00	43.99	39.14	43.52	44.60	1.27	43.48	39.31	45.18	43.83	42.75	40.84	43.83	58.45
TiO$_2$	2.46	2.58	0.02	0.00	2.23	0.00	0.07	0.08	1.48	0.19	0.06	0.02	2.32	1.42	0.05	2.29	1.51	0.00	0.06
Cr$_2$O$_3$	0.04	0.10	0.00	0.00	0.16	0.00	0.00	0.00	0.17	0.00	0.00	0.00	0.18	0.07	0.00	0.13	0.17	0.00	0.00
Al$_2$O$_3$	10.84	6.83	34.19	0.00	7.10	0.00	25.06	28.57	4.30	32.94	0.00	24.40	10.11	8.02	34.52	8.51	11.79	35.44	23.17
FeO$_t$	18.68	17.41	0.71	0.27	13.34	0.01	0.09	0.18	22.82	2.20	0.24	0.21	18.29	10.58	0.85	15.59	15.88	0.64	0.26
MnO	0.12	0.17	0.03	0.01	0.13	0.02	0.01	0.02	0.16	0.01	0.02	0.03	0.12	0.02	0.00	0.13	0.09	0.00	0.00
MgO	6.09	7.67	0.13	0.11	5.21	0.00	0.02	0.09	4.09	0.11	0.04	0.02	5.92	9.84	0.02	6.50	5.24	0.00	0.01
CaO	23.28	23.08	19.29	54.88	30.34	55.13	5.33	10.26	22.89	18.54	55.52	7.83	23.05	24.54	19.89	23.54	23.45	19.82	0.59
SrO	0.00	0.00	0.19	0.11	0.00	0.03	0.12	0.27	0.00	0.00	0.17	0.00	0.00	0.00	N.D.	0.00	0.00	N.D.	N.D.
Na$_2$O	0.35	0.49	0.43	0.06	0.52	0.03	2.54	1.53	0.18	0.37	0.12	0.35	0.19	0.07	0.15	0.18	0.15	0.24	0.66
K$_2$O	0.00	0.00	0.20	0.01	0.05	0.05	3.61	4.30	0.00	0.62	0.03	6.03	0.00	0.00	0.03	0.00	0.00	0.04	16.56
P$_2$O$_5$	0.00	0.00	0.00	39.41	0.00	0.00	0.00	0.00	0.00	0.00	40.16	0.00	0.00	0.00	0.00	0.00	0.00	0.00	0.00
Total	99.33	99.24	98.49	97.00	90.79	55.27	80.83	84.44	99.60	99.57	98.15	82.35	99.49	99.74	99.34	99.62	99.12	100.00	99.76
Fe$_2$O$_3$	15.63	13.14			14.82				7.82				11.44	5.13		5.70	6.51		
FeO	4.62	5.58							15.78				7.99	5.97		10.46	10.02		
Total	100.90	100.56	98.49	97.00	92.28	55.27	80.83	84.44	100.39	99.57	98.15*	82.35	100.64	100.26	99.34	100.19	99.77	100.00	99.76
Atomic proportions																			
Si	1.462	1.592	2.047	0.360	2.685	0.000	4.786	4.300	1.741	2.095	0.210	4.815	1.535	1.709	2.051	1.660	1.592	2.035	2.750
Ti	0.072	0.076	0.001	0.000	0.142	0.000	0.006	0.007	0.045	0.007	0.002	0.002	0.068	0.040	0.002	0.067	0.044	0.000	0.002
Al	0.498	0.313	1.905	0.000	0.708	0.000	3.214	3.700	0.203	1.823	0.000	3.185	0.465	0.358	1.904	0.389	0.542	1.939	1.285
Fe^{3+}	0.459	0.385			0.944				0.235				0.336	0.146		0.166	0.191		
Fe^{2+}	0.151	0.182	0.028	0.038	0.657	0.000	0.008	0.017	0.528	0.086	0.033	0.019	0.261	0.189	0.033	0.340	0.327	0.025	0.010
Mg	0.354	0.445	0.009	0.028	2.752	0.000	0.003	0.015	0.244	0.007	0.010	0.003	0.345	0.555	0.001	0.376	0.304	0.000	0.000
Ca	0.973	0.962	0.977	9.901	2.752	0.997	0.621	1.208	0.981	0.933	9.862	0.929	0.965	0.995	0.997	0.979	0.980	0.986	0.030
Na	0.026	0.037	0.040	0.020	0.085	0.001	0.536	0.326	0.014	0.034	0.039	0.075	0.015	0.005	0.014	0.014	0.011	0.022	0.061
K	0.000	0.000	0.012	0.002	0.005	0.002	0.501	0.603	0.000	0.037	0.006	0.852	0.000	0.000	0.002	0.000	0.000	0.002	0.994
P	0.000	0.000	0.000	5.613	0.000	0.000	0.000	0.000	0.000	0.000	5.631	0.000	0.000	0.000	0.000	0.000	0.000	0.000	0.000

Note: 5200(h)—calcite-zeolite-hydrogarnet olive rock; 5200(r)—relic of calcite-hydrogarnet rock in paralava; 5203 and H-5-93—hornfels. Iron for clinopyroxenes is divided into FeO and Fe$_2$O$_3$ according to stoichiometry. Cpx—clinopyroxene; Kfs—K-feldspar; An—anorthite; Ap—apatite; Hgr—hydrogarnet; Zeol—zeolite; Cal—calcite. Hydrogarnet was calculated on the basis of eight cations; calcite was calculated on the basis of one cation; zeolite was calculated on the basis of (Si + Al) = 8; other mineral phases were calculated in the same way as shown in Tables 4 and 5.
*Total includes: Y$_2$O$_3$—0.10 wt% (0.009); La$_2$O$_3$—0.18 wt% (0.011); Ce$_2$O$_3$—0.19 wt% (0.012); SO$_3$ and Cl <0.03 wt% (<0.010).

on numerous combustion metamorphic complexes worldwide (McLintock, 1932; Yavorsky and Radugina, 1932; Cosca et al., 1989; Kalugin et al., 1991; Heffern and Coates, 2004; Stracher and Taylor, 2004; Sokol et al., 2005). The areas of highest-temperature metamorphism, the local melting phenomena, and paralavas are located on the migration pathways of overheated gases from a deep zone of underground fire.

Nevertheless, internal combustion of organic matter over a large area within an undisturbed section of bituminous rocks of the Ghareb and Taqiye Formations, down to depths of 80–120 m, is considered by many investigators to be the main cause for combustion metamorphic complexes in the Hatrurim Formation (Bentor et al., 1963; Gross et al., 1967; Kolodny and Gross, 1974; Gross, 1977, 1984; Matthews and Kolodny, 1978; Matthews and Gross, 1980; Burg et al., 1991; Gur et al., 1995). This hypothesis for the origin of the Hatrurim Formation does not answer the following questions:

- What mechanism was responsible for rock heating and eventual ignition in numerous local foci?

- How did high-temperature oxidation of organic matter and sequential bitumen burning occur in the subhorizontal undisturbed sedimentary sequence at a depth of 80–120 m?

- What type of fuel was burned in the combustion metamorphic zone at a distance of more than 100 m above the nearest bitumen-rich sedimentary rocks?

- What caused the fracturing, brecciation, and recrystallization of the rock sequence and extensive hydrothermal alteration (Fig. 4F) with the unusual mineralization (including Cr, Au, U, Th, Ba, V, rare earth elements) (Gilat, 1994)?

The new geological observations presented here and the results of the study of the olive rock unit and its related paralavas are the basis of a new hypothesis for the genesis of the Hatrurim Basin. This hypothesis proposes the burning of hydrocarbon gas ejections.

Of the ~20 outcrops of the olive rock unit in the Hatrurim Basin, most outcrops contain explosion pipes surrounded by agglomeratic collars made of metasedimentary breccias. The character of all of these outcrops is similar. High-temperature rocks, such as hornfelses and paralavas, are distributed locally but are associated closely with pipe-like explosion bodies. The explosion pipes are always located within the complex of breccias that are composed of sedimentary and metasedimentary rocks, which in some cases, have been subjected to intensive hydrothermal alteration. The contacts between explosion pipes and host breccias reflect a lack of baking, whereas in all cases, the agglomeratic collar of explosion breccias shows evidence of baking.

In the Hatrurim Basin, combustion metamorphic rocks have a spotty local distribution (Burg et al., 1991); the same is valid for the olive rock unit. Burg et al. (1991) also showed that in numerous cases, the burning was focused along fractures that served as conduit channels. Nevertheless, the Hatrurim Basin is still interpreted as a strongly eroded combustion metamorphic complex, where combustion occurred over a large area in the same way as for areas subject to coal burning. This interpretation does not agree with the basin's geomorphology and geology. For example, combustion of dispersed organic matter as the cause of combustion metamorphism does not explain the intensive erosion of hard, weathering-resistant hornfels and the simultaneous preservation of soft low-temperature rocks that form the main landscape elements of the basin. It is evident that in other areas of combustion metamorphism, the layers of rocks that were subjected to high-temperature metamorphism determined the relief (Kalugin et al., 1991; Heffern and Coates, 2004; Sokol et al., 2005). A unique example of a fire spreading within a near-horizontal coal stratum (near-horizontal stratum resembles the structure in the Hatrurim Basin) has been documented for the Powder River Basin, Wyoming, USA (Heffern and Coates, 2004). The table-like hills that are armored by the resistant clinker layers, produced in response to combustion, are the dominant landform.

The existence of numerous kinds of breccias, which are developed on all hypsometric and stratigraphic levels, is not explained readily by the combustion of organic matter. The spotty distribution of combustion metamorphic rocks between soft, weakly consolidated, and brecciated sedimentary and metasedimentary rocks—in many cases, subjected to intensive hydrothermal alteration (Gilat, 1998a)—is impossible to explain by the uniform burning of bitumen over a vast area. The extensive distribution of hydrothermally altered rocks in the Hatrurim Basin also is in contradiction to observations on combustion metamorphic complexes. All of them have been classified as anhydrous mineral-forming systems. Hydrothermal alteration of high-temperature combustion metamorphic rocks has not been reported elsewhere (Sokol et al., 2005).

With regard to geological characteristics and rock sequences, the Hatrurim Basin differs significantly from combustion metamorphic complexes that formed as a result of solid fuel, in particular, coal combustion. However, it does resemble combustion metamorphic complexes in southwest Iran (McLintock, 1932). The combustion metamorphic complex Tul-i-Marmar is a conical-shaped hill that rises rather abruptly to a height of ~70 m. On the slopes of the hill, there is a local outcrop (an area of ~30 × 30 m) of "a curious mass of breccia...with highly inclined junction against the adjacent sediments" (McLintock, 1932, p. 208). The breccia consists of angular blocks of sedimentary rocks (marl, commonly hard and clinkery, limestone, sandstone, and grit), paralavas, lava-like slags, as well as fused and, in some cases, almost totally melted marls. The significant feature of this complex is "the strictly localized nature of metamorphism" (McLintock, 1932, p. 208) that occurred only on the tops of hills, whereas combustion metamorphic rocks are surrounded by undisturbed and unaltered sedimentary rocks. McLintock (1932) analyzed the geology in the Tul-i-Marmar region and compared it with gas ejections in the Persian Gulf. He concluded that the explosive force of the gas formed vent-like structures and shattered the rocks surrounding the orifice. Subsequently, the gases ignited and resulted in local fusion, recrystallization, and metamorphism of country rocks. Thus, this specific case of combustion metamorphism accompanied by explosions and rock brecciation is related completely to hydrocarbon gas emission, explosion, and ignition.

In summary, explosion pipes in the Hatrurim Basin, the upper part of which is represented by combustion metamorphic rocks and paralavas, and the combustion metamorphic features in Iran have similar characteristics: (1) geomorphological features of the areas; (2) details of the geological structure of the hills; (3) the localization of combustion metamorphic rocks within the complex of massive fragmented baked breccias located at the tops of hills; and (4) the combination of unaltered sedimentary rocks, low-temperature hydrothermal and combustion materials within the blocks of fragmented breccias at lower hypsometric levels (Fig. 4F).

Thus, the development of the explosion pipes and breccia complexes in the Hatrurim Basin has to be related to episodic breakthrough of hydrocarbon gases from the deep sedimentary sequence. These gas breakthroughs were accompanied by explosions that formed numerous layers of breccia at different levels. The discharge of sedimentary material and its ejection onto the surface by the pressurized gas emission led to the formation of explosive agglomeratic collars; in numerous cases, these were accompanied by gas ignition. These events were interrelated and were realized only in the vicinity of the surface. The surface character of the burning determined that the high-temperature combustion metamorphic rocks, represented by hornfelses and paralavas, have a restricted distribution and are always located in the vicinity of gas explosion craters.

Gas emissions and their time of ignition are of particular interest. For vast oil and gas provinces in the Middle East and Caspian Sea regions, the emissions and subsequent spontaneous ignition of gas were common phenomena during the geological past and in historical times (McLintock, 1932; Kholodov, 2002). Such events were never considered in the territory of Israel. Nevertheless, the geological sequence and the scope of sedimentary rocks in the Negev Desert are similar to those in the Middle East oil and gas provinces (Hall et al., 2005). Gvirtzman and Stanislavsky (2000) provided a detailed analysis of the diagenetic evolution of this sedimentary sequence, hydrocarbon maturation, pathways of gas and oil migration, and hydrocarbon accumulation in the Dead Sea rift and its western vicinity. According to their data, the Hatrurim Basin is a rectangular basinal structure, ~8 km long and 5 km wide, with several anticlinal neighbors on the southwestern flanks of the Dead Sea (Fig. 2). The region also is of interest because of the occurrence of noncommercial asphalts and light and heavy oils that occur from the surface down to 4000 m. These anticlines are mostly dry (Dabeshet, Barbour, Ef'e, and Halamish), whereas in others (Zohar, Kidod, and Haqanaim), three small gas fields have been found along the western border of the Hatrurim Basin. Noneconomic pockets of oil and gas also have been found in the Gurim anticline (a minor anticline in the Hatrurim Basin, Fig. 2). The one subcommercial oil field (Zuk-Tamrur, Fig. 2) also is situated here (Gardosh et al., 1996). Limestones and sandstones of the Middle Jurassic Zohar Formation, at a depth of 1000–1250 m, are the reservoir rocks for the gas fields; these are overlain and sealed by impermeable Jurassic shales of the Kidod Formation.

The migration and accumulation of hydrocarbons in the deep layers of the sedimentary sequence in the Hatrurim Basin, bordered by the rift valley, occurred differently than in the region located further to the west. In this eastern region, there are no impermeable Jurassic shales of the Kidod Formation under the rocks of the Hatrurim Basin (Gvirtzman and Stanislavsky, 2000). This likely is the cause of the discharge of hydrocarbon gas flows over the last 3 m.y. and in the recent geological past (20–30 k.y.) (Gardosh et al., 1996).

It is well known that there is a strong flow pattern of water and hydrocarbons through thick marine limestones and sandstones enriched in dispersed organic matter. Often, diagenetic reactions are accompanied by the appearance of anomalous gas pressures in the sedimentary column, resulting in the extrusion of highly mineralized mud that is enriched in clay and marl (Kholodov, 2002). The gas emissions, mainly of methane composition, are extruded with the mud. The extrusion of mud occurred simultaneously with the gas emissions, mainly of methane composition. Asphalts and oil seepages accompany mud lakes and springs in the zones of discharge. This phenomenon is manifested in the Middle East and Caspian Sea regions, classic provinces for mud volcanoes. Emissions of burning gas from the tops of mud volcano cones, composed of brecciated rocks known as mud breccias, are the culmination of a mud volcano event. Commonly, the fragments in mud breccias are composed of angular pieces and pebbles of combustion metamorphic rocks that formed during the previous gas ejection and ignition. The number of horizons that are filled with breccias containing the fragments of combustion metamorphic rocks is determined by the prevalence of strong explosions of hydrocarbon gases in consolidated rocks within the crater and the quantity of ignited gas extruded. These cones seem to be very similar to hills within the Hatrurim Basin that consist of numerous types of breccias.

CONCLUSIONS

The Hatrurim Basin in Israel is bordered directly by the Dead Sea Transform Valley on the east and by a series of anticlines hosting noneconomic gas domes on the west. High-temperature combustion metamorphic rocks, hornfels and paralavas, are distributed locally, whereas low-temperature hydrothermal rocks compose the major part of the complex. Hornfels and paralavas are related to the olive rock unit. The nature of the olive rock unit has been problematic, and its sedimentary protolith has not been defined unambiguously (Burg et al., 1991; Shahar, 1993). Several explosion pipes, consisting of explosion breccias and fragments of rocks of different petrographic composition and nature (metasedimentary, combustion metamorphic, and hydrothermally altered rocks), have been identified in the olive unit. Hard, high-temperature combustion metamorphic rocks (hornfelses) are located exclusively in the explosion agglomeratic collars around explosion pipes. The paralavas are represented by small lens- and dike-like bodies and are most commonly associated with hornfels. The paralavas are basic high-calcium igneous rocks, yet they are also similar to anorthosite. The paralavas are the products of the total melting of a metasedimentary protolith localized in zones of the highest-temperature combustion metamorphism. Several geomorphological, geological, and mineralogical observations, as well as the character of the hydrothermal alteration and the geochemical particularities of mineralization, suggest that the Hatrurim Basin has characteristics commonly associated with mud volcanoes accompanied by episodic burning of gas ejections.

ACKNOWLEDGMENTS

The authors thank L.N. Pospelova for help with electron-microprobe analyses of minerals. We are grateful to N.V. Maksimova and A.V. Dariin for XRF-SR analyses of paralavas and olive rocks of the Hatrurim Basin. We also thank V.S. Pavlyuchenko for XRD data on rocks and minerals. The comments and suggestions of Prof. E.J. Essene were extremely helpful for improving the English as well as the mineralogical sections of the paper. The constructive and thoughtful comments of Dr. A. Gilat, Prof. G.G. Lepezin, Prof. M. Roden, and an anonymous reviewer greatly improved the early version of this manuscript. We are thankful to Dr. A. Burg and Prof. Y. Kolodny for their valuable input and field discussions with us in the Hatrurim Basin. We thank Prof. V.V. Reverdatto for his input regarding contact metamorphism, Prof. V.N. Sharapov for discussing diatreme formation mechanisms with us, and Dr. I.S. Novikov for information about regional geology and geomorphology. We also are much obliged to Prof. G.B. Stracher for his support and editorial handling of our manuscript. This work was supported, in part, by the Russian Foundation for Basic Research (grant 05-05-65036) and a grant to E.V. Sokol, and grants awarded by the President of the Russian Federation ("Siberian Scientific School" No. HSH-4922.2006.5) for the support of young Russian scientists and leading scientific schools. R. Shagam acknowledges financial support from research funds supplied by Ben-Gurion University of the Negev.

REFERENCES CITED

Afonin, V.P., Gunicheva, T.N., and Piskunova, L.F., 1984, Rengenofluorescentniy silicatniy analiz (X-ray fluorescence silicate analysis): Novosibirsk, Nauka, 227 p. (in Russian).

Avnimelech, M., 1964, Remarks on the occurrence of unusual high-temperature minerals in the so-called "Mottled zone" complex of Israel: Israel Journal of Earth Sciences, v. 13, no. 3–4, p. 102–110.

Basi, M.A., and Jassim, S.Z., 1974, Backed and fused Miocene sediments from Injana area, Hemrin South, Iraq: Journal of the Geological Society of Iraq, v. VII, p. 1–14.

Bentor, Y.K., 1966, The Clays of Israel. Guide-book to the Excursions of the International Clay Conference: Jerusalem, Israel Program for Scientific Translations, 121 p.

Bentor, Y.K., and Vroman, A., 1960, The geological map of Israel on a 1:100,000 scale, in Series A—The Negev, Sheet 16: Mount Sedom (with Explanatory Text): Jerusalem, Israel, Government Printer, 117 p.

Bentor, Y.K., Gross, S., and Heller, L., 1963, High-temperature minerals in non-metamorphosed sediments in Israel: Nature, v. 199, p. 478–479, doi: 10.1038/199478a0.

Bentor, Y.K., Kastner, M., Perlman, I., and Yellin, Y., 1981, Combustion metamorphism of bituminous sediments and the formation of melts of granitic and sedimentary composition: Geochimica et Cosmochimica Acta, v. 45, no. 11, p. 2229–2255, doi: 10.1016/0016-7037(81)90074-0.

Blanckenhorn, M., 1912, Naturwissenschaftliche Studien am Toten Meer und im Jordantal: Berlin, Friendlander & Sohn, 478 p.

Bogoch, R., Gilat, A., Yoffe, O., and Ehrlich, S., 1999, Rare earth trace element distributions in the Mottled zone complex, Israel: Israel Journal of Earth Sciences, v. 48, no. 3–4, p. 225–234.

Burg, A., Starinsky, A., Bartov, Y., and Kolodny, Y., 1991, Geology of the Hatrurim Formation ("Mottled zone") in the Hatrurim Basin: Israel Journal of Earth Sciences, v. 40, no. 1–4, p. 107–124.

Chesnokov, B.V., and Shcherbakova, E.P., 1991, The mineralogy of burned heaps in the Chelyabinsk coal basin: Moscow, Publishing House of Nauka, 152 p. (in Russian).

Cosca, M.A., Rouse, R.R., and Essene, E.J., 1988, Dorrite [$Ca_2(Mg_2Fe_4^{3+})(Al_4Si_2)O_{20}$], a new member of the aenigmatite group from a pyrometamorphic melt-rock: The American Mineralogist, v. 73, no. 11–12, p. 1440–1448.

Cosca, M.A., Essene, E.J., Geissman, J.W., Simmons, W.B., and Coates, D.A., 1989, Pyrometamorphic rocks associated with naturally burned coal beds, Powder River Basin, Wyoming: The American Mineralogist, v. 74, no. 1–2, p. 85–100.

Feldman, V.I., Bychkov, A.M., Dikov, Yu.P., and Krivtsova, V.Ya., 1994, Tengizites—Glasses from center of oil fire: Doklady of the Russian Academy of Sciences, v. 339, no. 2, p. 239–242.

Foit, F.F., Hooper, R.L., and Rosenberg, P.E., 1987, An unusual pyroxene, melilite, and iron oxide mineral assemblage in a coal-fire buchite from Buffalo, Wyoming: The American Mineralogist, v. 72, no. 1–2, p. 137–147.

Gardosh, M., Kashai, E., Salhov, S., Shulman, H., and Tannenbaum, E., 1996, Hydrocarbon explosion in the southern Dead Sea area, in Niemi, T.N., Ben-Avraham, Z., and Gat, L., eds., The Dead Sea: The Lake and Its Setting: Oxford Monographs in Geology and Geophysics 36, p. 57–72.

Gilat, A., 1994, Tectonics and Associated Mineralization Activity, Southern Judea, Israel [Ph.D. thesis]: Jerusalem, Ministry of Energy and Infrastructure, Geological Survey of Israel, Mineral and Energy Resources Division, 280 p. (in Hebrew).

Gilat, A., 1998a, Hydrothermal activity and hydro-explosions as a cause of natural combustion and pyrolysis of bituminous rocks: The case of Pliocene metamorphism in Israel (Hatrurim Formation): Geological Survey of Israel: Current Research, v. 11, p. 96–102.

Gilat, A., 1998b, Strike-slip movements along the Qana'im fault zone, Judean Desert: Geological Survey of Israel: Current Research, v. 11, p. 33–41.

Griffin, W.L., Åmli, R., and Heier, K.S., 1972, Whitlockite and apatite from lunar rock 14310 and from Ödegården, Norway: Earth and Planetary Science Letters, v. 15, p. 53–58, doi: 10.1016/0012-821X(72)90028-3.

Gross, S., 1977, The Mineralogy of the Hatrurim Formation, Israel: Geological Survey of Israel Bulletin 70, 80 p.

Gross, S., 1984, Occurrence of ye'elimite and ellestadite in an unusual cobble from the "pseudo-conglomerate" of the Hatrurim Basin, Israel: Geological Survey of Israel: Current Research, v. 1983–84, p. 1–4.

Gross, S., Mazor, E., Sass, E., and Zak, I., 1967, The "Mottled zone" complex of Nahal Ayalon (central Israel): Israel Journal of Earth Sciences, v. 16, no. 2, p. 84–96.

Gur, D., Steinitz, G., Kolodny, Y., Starinsky, A., and McWilliams, M., 1995, $^{40}Ar/^{39}Ar$ dating of combustion metamorphism ("Mottled zone", Israel): Chemical Geology, v. 122, no. 1–4, p. 171–184, doi: 10.1016/0009-2541(95)00034-J.

Gvirtzman, H., and Stanislavsky, E., 2000, Palaeohydrology of hydrocarbon maturation, migration and accumulation in the Dead Sea Rift: Basin Research, v. 12, no. 1, p. 79–93.

Hall, J.K., Krasheninnikov, V.A., Hirsch, F., Benjamini, Ch., and Flexer, A., eds., 2005, Geological Framework of the Levant, Volume II: The Levantine Basin and Israel: Jerusalem, Israel, Historical Productions-Hall, 826 p.

Heffern, E.L., and Coates, D.A., 2004, Geological history of natural coal-bed fires, Powder River Basin, USA: International Journal of Coal Geology, v. 59, no. 1–2, p. 25–47.

Hull, E., 1886, The Survey of Western Palestine: Memoir on the Physical Geology and Geography of Arabia Petraea, Palestine and Adjoining Districts: London, Committee of the Palestine Exploration Fund, 154 p.

Kalugin, I.A., Tretyakov, G.A., and Bobrov, V.A., 1991, Iron Ore Basalts in Burned Rocks of Eastern Kazakhstan: Novosibirsk, Publishing House of Nauka, 80 p. (in Russian).

Kholodov, V.N., 2002, Mud volcanoes: Regularity of location and genesis: Litologiya i Poleznii Iskopaemii, no. 3, p. 227–241 (in Russian).

Knunyants, I.L., ed., 1998, Chemical Encyclopedia, Volume 3: Moscow, Russia, P.H. Bol'shaja Rossijskaja Jenciklopedija, p. 594–598 (in Russian).

Kolodny, Y., and Gross, S., 1974, Thermal metamorphism by combustion of organic matter: Isotopic and petrological evidence: The Journal of Geology, v. 82, no. 4, p. 489–506.

Lees, G.M., 1928, The chert beds of Palestine: Proceedings of Geological Association of London, v. 39, p. 445–462.

Matthews, A., and Gross, S., 1980, Petrologic evolution of the "Mottled zone" (Hatrurim), metamorphic complex of Israel: Israel Journal of Earth Sciences, v. 29, no. 1–2, p. 93–106.

Matthews, A., and Kolodny, Y., 1978, Oxygen isotope fractionation in decarbonation metamorphism: The Mottled zone event: Earth and Planetary Science Letters, v. 39, p. 179–192, doi: 10.1016/0012-821X(78)90154-1.

McLintock, W.F.P., 1932, On the metamorphism produced by the combustion of hydrocarbons in the Tertiary sediments of south-west Persia: Mineralogical Magazine, v. 23, p. 207–227, doi: 10.1180/minmag.1932.023.139.01.

Nathan, Y., Shiloni, Y., Roded, R., Gal, Y., and Deutsch, Y., 1979, The Geochemistry of the Northern and Central Negev Phosphorites (Southern Israel): Geological Survey of Israel Bulletin 73, 41 p.

Phedorin, M.A., Bobrov, V.A., Chebykin, E.P., Goldberg, E.L., Melgunov, M.S., Filippova, S.V., and Zolotarev, K.V., 2000, Comparison of synchrotron radiation X-ray fluorescence with conventional techniques for the analysis of sedimentary samples: The Journal of Geostandards and Geoanalysis, v. 24, no. 2, p. 205–216.

Picard, L., 1931, Geological Research in the Judean Desert: Jerusalem, Goldberg Press, 108 p.

Schairer, J.F., 1942, The system $CaO-FeO-Al_2O_3-SiO_2$. I. Results of quenching experiments on five joins: Journal of the American Ceramic Society, v. 25, p. 241–274, doi: 10.1111/j.1151-2916.1942.tb14292.x.

Schairer, J.F., 1950, The alkali-feldspar join in the system $NaAlSiO_4-KAlSiO_4-SiO_2$: The Journal of Geology, v. 58, p. 512–517.

Shahar, J., 1993, Comment on "Geology of the Hatrurim Formation ('Mottled zone') in the Hatrurim Basin," by A. Burg, A. Starinsky, Y. Bartov, and Y. Kolodny (Letter to Editor): Israel Journal of Earth Sciences, v. 42, no. 2, p. 97.

Sokol, E.V., Sharygin, V.V., Kalugin, V.M., Volkova, N.I., and Nigmatulina, E.N., 2002, Fayalite and kirschsteinite solid solutions in melts from burned spoil-heaps, South Urals, Russia: European Journal of Mineralogy, v. 14, no. 4, p. 795–807, doi: 10.1127/0935-1221/2002/0014-0795.

Sokol, E.V., Maksimova, N.V., Nigmatulina, E.N., Sharygin, V.V., and Kalugin, V.M., 2005, Combustion metamorphism: Novosibirsk, Publishing House of Siberian Branch of Russian Academy of Sciences, 284 p. (in Russian).

Stracher, G.B., and Taylor, T.P., 2004, Coal fires burning out of control around the world: Thermodynamic recipe for environmental catastrophe, in Stracher, G.B., ed., Coal Fires Burning Around the World: A Global Catastrophe: International Journal of Coal Geology, v. 59, no. 1–2, p. 7–17.

Tristram, H.B., 1865, The Land of Israel: A Journal of Travels in Palestine, Undertaken with Special Reference to Its Physical Character: London, The Society for the Promotion of Christian Knowledge, 651 p.

Tulloch, A.J., and Campbell, J.K., 1993, Clinoenstatite-bearing buchites possibly from combustion of hydrocarbon gases in a major thrust zone: Glenroy Valley, New Zealand: The Journal of Geology, v. 101, no. 3, p. 404–412.

Vapnik, Y., and Sokol, E., 2006, Explosion Breccias and Diatremes as Key Structures in the Formation of the Hatrurim Formation [abs.]: Bet-Shean, Israel Geological Society, p. 131.

Whitworth, H.F., 1958, The occurrence of some fused sedimentary rocks at Ravensworth, N.S.W.: Journal and Proceedings of the Royal Society of New South Wales, v. 92, part 4, p. 204–208.

Yavorsky, V.I., and Radugina, L.V., 1932, Coal-fire combustion and attendant events in Kuznetsk Basin: Gornyi Zhurnal, v. 10, no. 10, p. 55–59.

MANUSCRIPT ACCEPTED BY THE SOCIETY 7 MARCH 2007

Geochronology of clinker and implications for evolution of the Powder River Basin landscape, Wyoming and Montana

Edward L. Heffern*
U.S. Bureau of Land Management, 5353 Yellowstone Road, Cheyenne, Wyoming 82009, USA

Peter W. Reiners
Department of Geosciences, University of Arizona, Tucson, Arizona 85721, USA

Charles W. Naeser
U.S. Geological Survey, MS 926A, Sunrise Valley Drive, Reston, Virginia 20192-0002, USA

Donald A. Coates
Consultant, P.O. Box 1726, Bodega Bay, California 94923, USA

ABSTRACT

In the Powder River Basin of southeast Montana and northeast Wyoming, coal beds exposed by regional erosion have burned naturally from as early as the Pliocene to the present. Layers of reddish clinker, formed by baking, welding, and melting of sediments above burned coal beds, cover over 4000 km^2 and cap ridges and escarpments throughout the dissected landscape of the Powder River Basin. Fission-track (ZFT) and (U-Th)/He (ZHe) ages of zircon grains from baked sandstones in clinker provide new insights about rates of regional erosion as well as episodic advance of coal fires into hillsides. Older, resistant clinker layers up to 60 m thick, formed by the burning of thick coal beds, cap summits and broad benches. Younger clinker rims, from thinner coals, form ledges on valley sides. ZHe ages of clinker, mainly from the Wyodak-Anderson coal zone of the Fort Union Formation in the Rochelle Hills east of Wright, Wyoming, and from the Wyodak-Anderson and Knobloch coal zones in the Tongue River valley near Ashland and Birney, Montana, range from 1.1 Ma to 10 ka. These dates generally agree with ZFT ages of clinker analyzed in the early 1980s, but they are a more precise record of ancient coal fires in the region. Our data indicate 0.2–0.4 km of vertical erosion in the past 1 m.y. Spatial-temporal patterns of clinker ages may prove to be useful in deciphering the patterns of fluvial incision and basin excavation in the Powder River Basin during the late Cenozoic and in weighing the relative importance of uplift, variations in climate, and base-level change.

Keywords: coal-bed fires, clinker, geochronology, geomorphology, Powder River Basin.

*ed_heffern@blm.gov

Heffern, E.L., Reiners, P.W., Naeser, C.W., and Coates, D.A., 2007, Geochronology of clinker and implications for evolution of the Powder River Basin landscape, Wyoming and Montana, *in* Stracher, G.B., ed., Geology of Coal Fires: Case Studies from Around the World: Geological Society of America Reviews in Engineering Geology, v. XVIII, p. 155–175, doi: 10.1130/2007.4118(10). For permission to copy, contact editing@geosociety.org. ©2007 The Geological Society of America. All rights reserved.

INTRODUCTION

In the western United States, coal beds have burned naturally in several basins in Wyoming, Montana, North Dakota, Colorado, Utah, New Mexico, and Arizona (Sigsby, 1966; Hoffman, 1996). The most extensive burning has been in the Powder River Basin of northeast Wyoming and southeast Montana (Rogers, 1918; Coates and Heffern, 2000). The Powder River Basin covers ~56,000 km², an area about four times larger than the state of Connecticut. About 4100 km² (7%) of the Powder River Basin is covered by outcrops of clinker—baked, welded, and melted rocks formed by the natural burning of coal beds. These outcrops record the natural burning of tens of billions of tons of coal in the geologic past (Heffern and Coates, 2004).

More than one-third of U.S. coal comes from the Powder River Basin. Seventeen coal mines, including some of the largest on Earth, produced 430 million short tons (390 million metric tons) of coal there in 2005 using surface-mining methods (Energy Information Administration, 2006). Beginning in the 1990s, biogenic methane originating in the coal beds has been extracted at increasing rates. The coal is subbituminous and is found in the Upper Paleocene Tongue River Member of the Fort Union Formation and Lower Eocene Wasatch Formation.

The heat generated from in-situ burning of coal beds alters the strata overlying the coal, much as firing alters bricks. Clinker, hardened by heating, forms distinctive erosion-resistant reddish layers that cap plateaus, hilltops, and escarpments. The highly fractured nature of the clinker allows rainfall and snowmelt to infiltrate rather than run off the surface and erode the outcrop (Coates, 1991). Clinker generally erodes more slowly than surrounding unbaked rocks in this environment, leaving the resistant clinker standing in relief. Clinker fragments are abundant in landslides and talus deposits below clinker cap rock.

Clinker-controlled topography dominates the exposed areas of the Upper Paleocene Tongue River Member of the Fort Union Formation in the northern, eastern, and western parts of the Powder River Basin (Fig. 1). This topography extends north to the Yellowstone River near Miles City, Montana (Heffern et al., 1993), and south along the west flank of the Powder River Basin from the Wolf Mountains of Montana past the towns of Sheridan and Buffalo, Wyoming. It extends south along the east flank of the basin past Broadus, Montana, and into northeast Wyoming. Clinker in the Tongue River Member covers over 400 km² in the Rochelle Hills east of the towns of Gillette and Wright, Wyoming (Heffern and Coates, 2004). Clinker also caps hills and buttes in the overlying Eocene Wasatch Formation, from east of Sheridan and Buffalo to Gillette and Wright (Heffern and Coates, 1997).

The landscapes of the northern and southern parts of the Powder River Basin are different in several ways. Where the Tongue River Member is exposed in the northern part of the basin, thick sandstone beds form cliffs and ledges. The less-consolidated exposures of the Wasatch Formation in the southern part of the basin have gentler topography. However, relief throughout the basin is largely controlled by the distribution of clinker (Heffern and Coates, 2000). It is literally a landscape formed by fire—albeit many separate fires in many different places over time scales of 10^3–10^6 yr. Most parts of the southern Powder River Basin are characterized by broad rolling hills, and flat-topped buttes capped by clinker, with relief typically less than 100 m. However, in the Rochelle Hills (Fig. 1), natural burning of the Wyodak-Anderson coal zone has left an eastward-facing escarpment 100–200 m high that is capped by a 20–50-m-thick layer of reddish clinker (Fig. 2). In the central and northern parts of the basin, the major drainages of the Powder River, Tongue River, and Rosebud Creek, and their dominantly ephemeral tributaries, form an intricate trellis pattern. Here, generally flat-lying clinker beds create flights of terraces that ascend from stream level to clinker-capped plateaus or ridges, with a total relief of ~200–400 m.

The semiarid climate, low-rank coal rich in volatile matter, common range fires, and regional erosion in the Powder River Basin provide ideal conditions for natural coal-bed fires. Several coal beds 20 m or more thick, as well as many thinner beds, have burned over large areas of the Powder River Basin, firing the overlying rock to a brick-like hardness and melting the rock in places. As erosion lowers the general land surface, coal is exposed in stream beds, gullies, and hillsides. In some places, the coal is degassed and oxidized by the time it is exposed to the air, and it does not combust spontaneously. However, where fresh coal is exposed to the air by rapid erosion or removal of overburden along stream banks and gullies, in the headwalls of landslides, in roadcuts, or in mines, spontaneous combustion is common, especially when heat of oxidation is abetted by heat of wetting. The openpit coal mines, for example, experience the greatest numbers of spontaneous fires in the spring when humidity increases due to melting snow and seasonal precipitation. When Lewis and Clark explored this region two centuries ago, they reported numerous coal-bed fires and correctly noted the relation between clinker and coal beds in the "burnt hills" of the northern Great Plains (Thwaites, 1969). Clark named one river the Redstone River because of the many reddish cobbles of clinker in its streambed where it joined the Yellowstone River. That river was later renamed the Powder River because the smell of the ever-burning coal fires evoked the smell of burning gunpowder.

The numerous thick coal beds that originally accumulated in vast peat swamps of the Paleocene intermontane foreland basin have provided the setting for a natural geomorphic experiment with the ability to reveal a detailed record of fluvial incision and landscape evolution in the Powder River Basin. This record has the potential to constrain the timing, rate, and ultimate tectonic and/or climatic driving forces behind late Cenozoic exhumation of the Powder River Basin and other Rocky Mountain foreland basins in the region. As coals in the basin are exhumed by fluvial incision and scarp retreat and burn naturally, they leave clinker as a record. The age of the clinker can be dated with roughly 10% or better precision using zircon (U-Th)/He (ZHe) chronometry, and to a lesser precision using zircon fission-track (ZFT) dating. In several cases, Pliocene through Recent ZHe

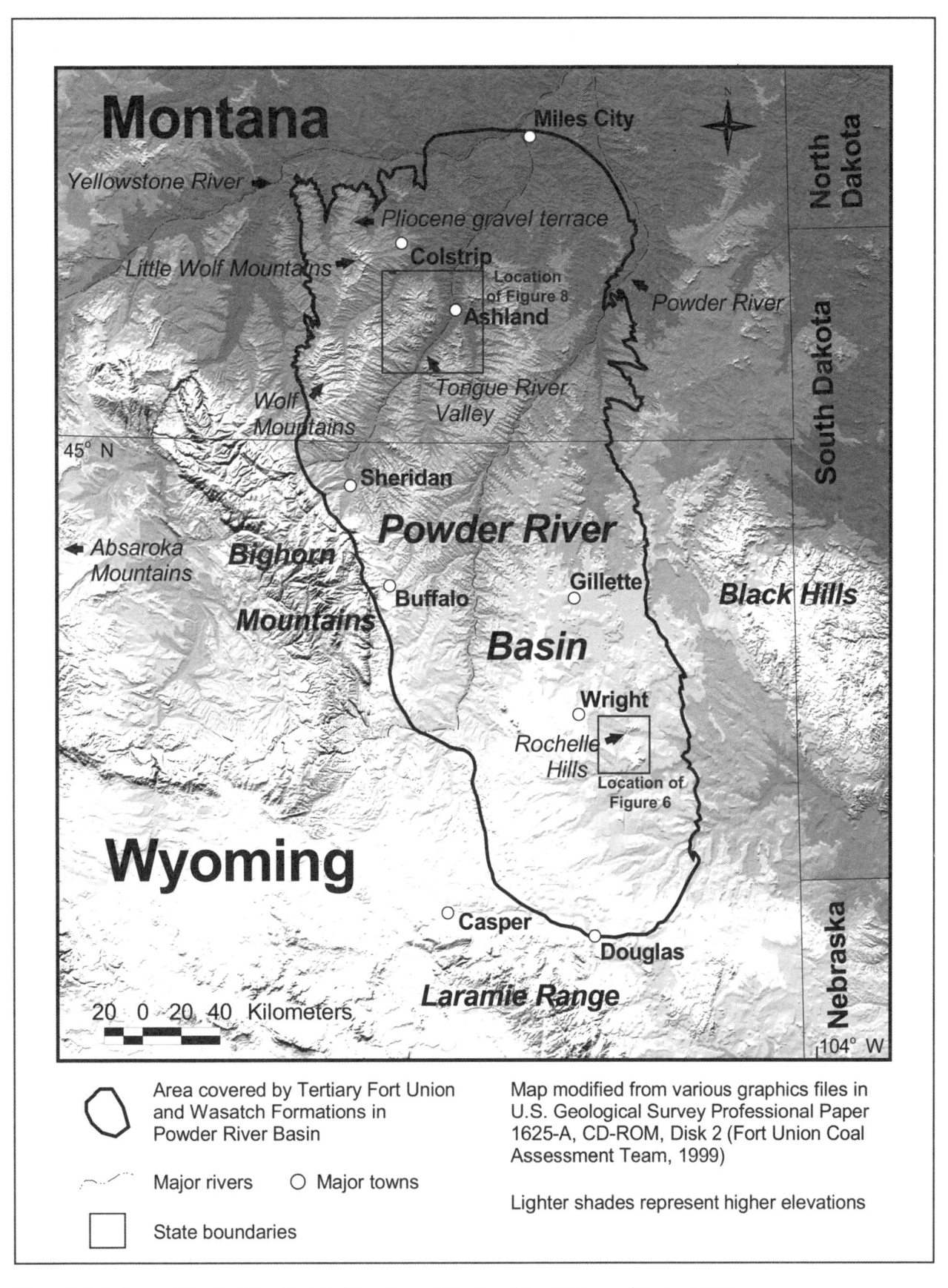

Figure 1. Relief map of Powder River Basin in Wyoming and Montana (after Heffern and Coates, 2004).

and ZFT clinker ages show reproducible and systematic spatial-temporal patterns with respect to fluvial networks and other topographic features across the Powder River Basin.

COAL-BED FIRES

Coal-bed fires are typically ignited by either spontaneous combustion or wildfires. Spontaneous combustion of Powder River Basin coals is largely due to exothermic reactions associated with oxidation of organic matter, and it is aided by the high volatile content of the low-grade coal (Lyman and Volkmer, 2001; Coates and Heffern, 2000; Heffern and Coates, 2004). In dry years, lightning- and human-induced wildfires can ignite hundreds of coal fires by direct burning of coal outcrops or of trees and bushes rooted in fresh coal seams (Heffern and Coates, 2004). Aggressive control measures have extinguished many coal-bed fires in the Powder River Basin, but in the prehistoric past, wildfires probably were a continuous source of natural ignition.

Hundreds of natural coal-bed fires are currently burning in the Powder River Basin (Fig. 3). The coal beds in the Powder River Basin are nearly horizontal, and they commonly dip less than one degree except for the western and southern edges of the basin, where dips are steeper. When a fire starts, it first spreads along the horizontal outcrop because this is the path with the most oxygen. This pattern of advance occurs in coal fires in active mines as well as natural outcrops, and it is documented by the many shallow bands of clinker marking the position of coal beds along hillsides. With time, the fire burns deeper into the hillside. A 10-m-thick coal bed may burn down to form an ash layer less than 1 m thick, causing the overlying rocks to subside into the burned-out void. The resulting fracturing allows air to enter and gas to escape, feeding the fire. The removal of the coal bed disrupts and fractures the overlying rock enough that air access and venting allow the fire to burn laterally back into the hill beneath a considerable thickness of overburden—commonly as much as two to three times the original thickness of the coal bed (Heffern and Coates, 2004). The fire may smolder for long periods but sometimes erupts into a roaring inferno that melts rock.

Two major controls determine the depth to which a coal fire can burn beneath overburden and the distance it can penetrate beneath a slope. First, the fire may encounter groundwater and be unable to progress. Ultimately, an underground coal fire burns all the coal that it can before running out of air or fuel. As a coal bed burns back into a hillside and the fire encounters water-saturated coal, further spread is constrained, and the fire dies for lack of air and fuel. Along the contact between clinker and unburned coal in the subsurface (the "burn front" or "burn line"), the top part of a coal bed may be burned away and replaced by clinker that has subsided into the void, while the lower, saturated part of the coal bed may still be present below the water table. These patterns can be seen in many cross sections drawn across the coal-clinker contact in mine plans on public file with the state agencies in Wyoming and Montana delegated to enforce the Surface Mining Control and Reclamation Act of 1977.

Second, the fire may penetrate beneath a slope so far that collapse of overburden deprives it of air. The fire goes out naturally where the overburden becomes so thick that fractures from the collapse fail to reach the surface to draw in more air. The distance that the fire penetrates depends on the thickness of the coal bed and the lithology and fracture patterns in the overburden that affect the ability of the collapsed rock to allow ventilation. Burning of a thicker coal bed causes the overburden to collapse to a greater degree, which allows the fire to burn farther and deeper into the hillside. Other factors, such as surface runoff filling fractures with clay and sealing the fire from its source of oxygen, as well as splitting, faulting, and truncation of coal beds in the subsurface, may also extinguish a coal fire.

Over time, regional erosion lowers the ground surface and water table relative to coal beds and dries out the coal near the surface, which is then susceptible to burning as it is exposed to air. Present-day burn fronts are, at the deepest, 60 m below the surface, and in most cases, less than 30 m below the surface. Systematic age trends across exposed clinker-dominated landscapes (e.g., Coates and Naeser, 1984; Reiners and Heffern, 2002) strongly suggest that coal burns within a relatively restricted range of distances, horizontally and vertically, from exposure in any given location at a given time. Because coal burning and clinker formation are restricted to the uppermost tens of meters in the subsurface, and because coal in the Powder River Basin does not reside within this zone for very long, in a geologic sense, before burning, spatial patterns of clinker ages provide estimates of the timing and rates of fluvial incision and scarp retreat throughout the basin. They can be used to map the larger-scale exhumation of the basin from at least Pliocene to Holocene time.

CLINKER

The term "clinker" encompasses all the rocks that have been altered by heating from a coal-bed fire. Thus, it includes a variety of thermally metamorphosed or melted lithologies resulting from the burning of coal (Sarnecki, 1991; Papp, 1998; Coates and Heffern, 2000). These variations are partly due to different types of source rocks in the Upper Cretaceous and Lower Tertiary strata of the American West but also to variations in the dynamics of burning, ventilation, and proximity of the source rocks to the coal fire. The most distinguishing feature of clinker in the landscape is its red color. In hand specimen, clinker may display little obvious alteration besides reddening (especially for sandstones), may have a pronounced ceramic-like texture from more intense heating of shale or siltstone (porcellanite), or may be melted to form black paralava and appear nearly identical to pahoehoe. Such melting requires temperatures greater than 1300 °C (Cosca et al., 1989). Some portions of clinker, especially baked shale, may be dark gray, black, or green, because of heating under reducing conditions. Porcellanite was worked by early Native Americans for tools and blades, and it is found in archaeological sites throughout the Powder River Basin and

Figure 2. Clinker of Wyodak-Anderson coal zone capping Rochelle Hills escarpment, Wyoming.

Figure 3. Active coal-bed fire along bank of Tongue River north of Sheridan, Wyoming.

beyond. Porcellanite Folsom and Goshen points as old as 11 ka have been found (Clark, 1985; Fredlund, 1976).

The basal part of a clinker exposure is characterized by a thin zone, usually less than 1 m thick, of gray or tan ash and greenish glass from burned coal. Vesicular and shattered paralava may lie directly above the ash. This is overlain by shattered layers of porcellanite and baked shale, as well as lenses or thick sections of sandstone that show little metamorphism other than reddening. In most cases, collapse of overlying strata has formed a chaotic jumble of intensely fractured and faulted blocks. Chimneys that allowed ventilation during burning are typically found every few tens of meters laterally in a clinker unit. Rocks in these chimneys usually have evidence of intense heating and melting, including flow structures along voids and fractures. The trace of the end of clinker exposures on a plateau or hillside—the burn line—typically marks the beginning of unburned coal that is too far, horizontally and vertically, from the surface to have burned. The burn front on the eastern margin of the Powder River Basin in Wyoming marks the eastern boundary of some of the world's largest surface coal mines.

The structure of a typical clinker hillside (Fig. 4) shows progressive collapse of overburden as the coal-bed fire advances back into the hill. As the support of underlying coal is removed, the overburden subsides. Some subsidence takes the form of large, highly fractured slump blocks separated by tension cracks that allow air to flow into the burning area and direct combustion gases upward from the burn zone. Other rock subsides as a mass that has been shattered by dehydration and shrinkage of clays during heating but otherwise exhibits little deformation. Most of the heat is transferred upward by convection, leaving strata that are more than 0.5–1 m below the original base of the coal unaltered. The tension cracks are generally filled with welded breccia and paralava, indicating that they were vents where burning of coal gas created intense heating. These areas, which are harder than the surrounding clinker, are left by erosion to stand as resistant chimneys on the land surface, recording the location of fissures where the temperature was high enough to weld and melt the rock.

Published studies (Heffern et al., 1993; Heffern and Coates, 1997; Coates and Heffern, 2000) show that the Powder River Basin in Wyoming and Montana contains ~4100 km^2 of clinker outcrops: ~2800 km^2 in Montana and 1300 km^2 in Wyoming. These clinker outcrops are only remnants of a much larger volume of clinker that has been removed by erosion. As erosion deepened and widened valleys and lowered the water table, successively lower coal beds were dewatered, exposed, and burned back into the hillsides. The clinker that was formed has been eroded and carried downstream into the tributaries of the Missouri River, along with other rock and soil eroded during dissection of the landscape. Fragments of clinker are found in the gravel terraces and streambed alluvium of this major river system. The clinker remaining in its original position appears to be only a small fraction of, perhaps an order of magnitude less than, the total volume of clinker that has been produced over time (Heffern and Coates, 2004).

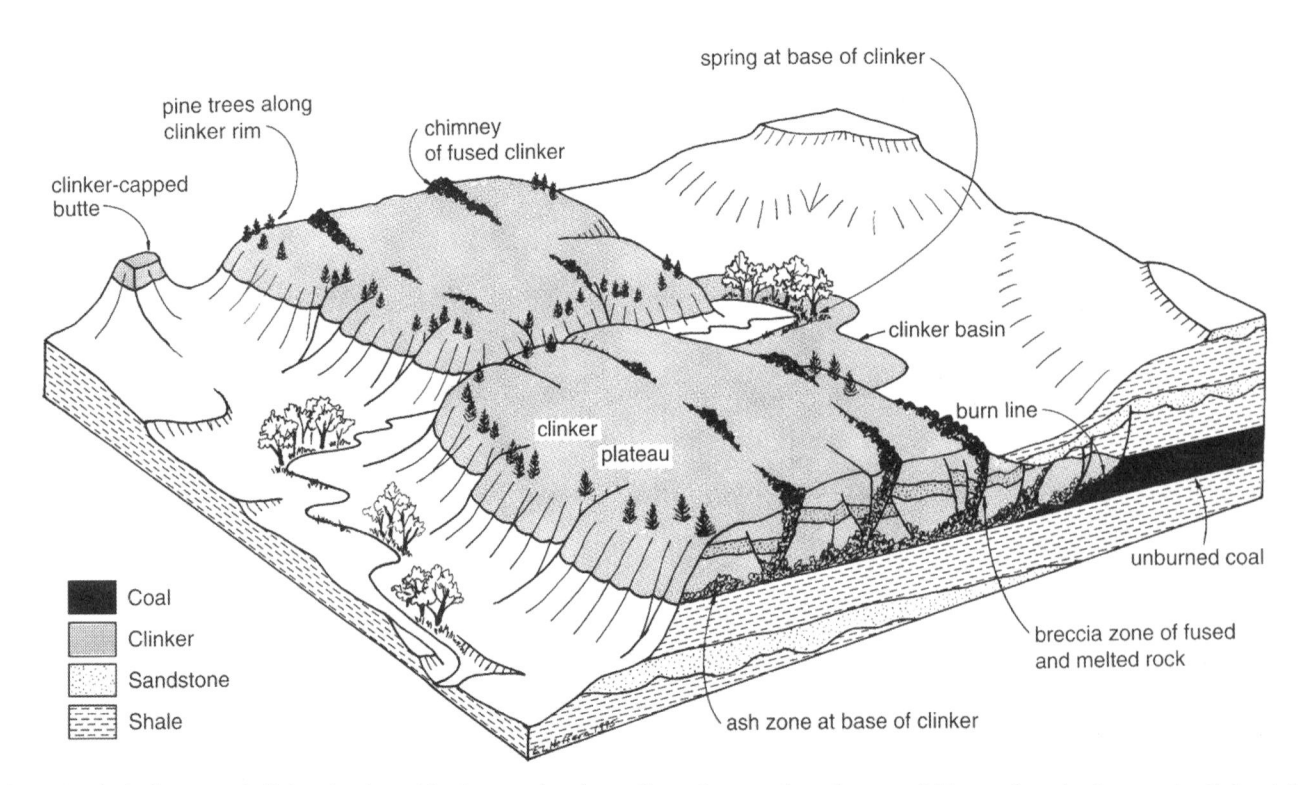

Figure 4. Block diagram of clinker-dominated landscape, showing collapse features, breccia zones (chimneys), and ash zones in clinker (after Heffern and Coates, 1997).

Most of the clinker in the Fort Union Formation was formed by burning of three regionally extensive coal zones in the Tongue River Member—the Wyodak-Anderson zone in both Wyoming and Montana and the Knobloch and Rosebud-Robinson zones in Montana (Heffern et al., 1993; Flores and Bader, 1999). Other less-extensive coal zones, such as the Wall, Pawnee, Terret, and Dominy (Bass, 1932; Matson and Blumer, 1973), have locally produced large areas of clinker in Montana. In the overlying Wasatch Formation, mostly in Wyoming, extensive clinker was produced by the burning of the Lake De Smet (Healy-Ucross) and Ulm coal zones in the western part of the basin near Sheridan and Buffalo, as well as the Felix coal zone near Gillette and Wright. The stratigraphic column in Figure 5 (from Flores, 2004) shows the relative position of these coal zones.

METHODS FOR DATING CLINKER

Clinker has been dated by two means: (1) fission-track dating of detrital zircon grains in baked sandstones; and (2) (U-Th)/He dating of detrital zircon grains in baked sandstones. A third method—paleomagnetic orientations of magnetite, hematite, and goethite in paralava and baked sediments—provides additional indirect age constraints. These dates can lead to a better understanding of landscape evolution and incision of fluvial networks into the sedimentary fill of the Powder River Basin, and they can constrain the patterns and rates by which coal beds were exposed and burned.

Zircon Fission-Track (ZFT) Dating

Zircon ($ZrSiO_4$) grains are widespread in sandstones of the Fort Union and Wasatch Formations, and they contain ppm-level uranium. Over time, the radioactive uranium decays in two different modes, alpha decay and spontaneous fission, and each mode has a different half-life. During spontaneous fission, particles smash through the lattice of the zircon crystal, leaving a track of weakened and disturbed bonds between atoms in the lattice. These fission tracks can be revealed by etching and counted under a microscope. The ratio between the density of fission tracks and

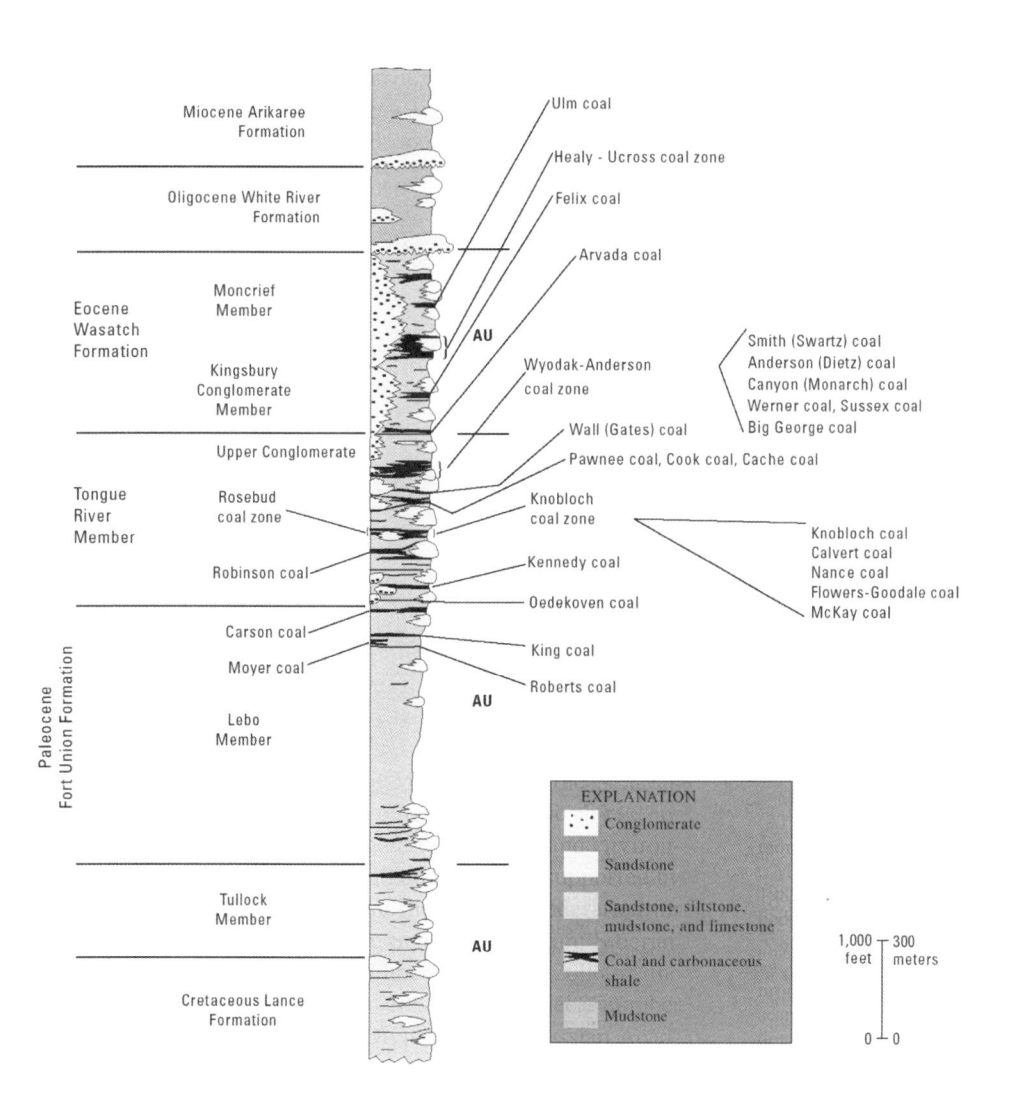

Figure 5. Stratigraphic column of coal zones in Fort Union and Wasatch Formation assessment units (AU), Powder River Basin (from Flores, 2004).

uranium concentration in a crystal provides an estimate of how long the tracks have been accumulating (Naeser, 1979). When that crystal is heated to a sufficiently high temperature, fission tracks anneal and disappear, resetting the clock. Nearly all of the sandstones that have been baked into clinker contain a small proportion of detrital zircon. Because these zircon grains are heated and annealed during the formation of clinker, it is possible to determine when the underlying coal bed burned. Once the fire is over, the fission-track clock starts again. On time scales characteristic of coal fires in the shallow subsurface (e.g., 10^1–10^2 yr), zircons will completely anneal fission tracks, fully resetting the ZFT age, at temperatures of ~375–450 °C (Figure 14 *in* Reiners, 2005). A limiting factor in determining fission-track ages of Powder River Basin clinker younger than ~0.1 Ma is the low uranium content in most zircons, which results in very low fission-track densities. Consequently, the statistical uncertainty of the determined age is high, whereas in older clinker, ages are better constrained because more fission tracks are present. Another limitation on both ZFT and ZHe age measurements is that the zircon crystals in a sample may not be big or numerous enough to date. Some clinker samples collected from the Tongue River Member in Montana did not have enough zircon crystals to date, but almost all samples collected from the uppermost Tongue River Member and the Wasatch Formation in Wyoming had a sufficient number.

U-Th/He (ZHe) Dating of Zircons

Uranium (U) and thorium (Th) in detrital zircons experience alpha decay, producing helium (^{4}He). As in the case of zircon fission-track dating, zircon (U-Th)/He dating is a thermochronometer because He diffuses out of crystals at elevated temperatures. The diffusivity of He in zircon has been studied by step-heating diffusion experiments and comparisons of zircon (U-Th)/He ages with ages of other thermochronometers with known thermal sensitivities (e.g., Reiners et al., 2004; Reiners, 2005). For normal crystal sizes and under slow-cooling conditions (dT/dt ~ 10 °C/m.y.) typical of many exhuming orogens, ZHe has an effective closure temperature (Dodson, 1973) of ~170–190 °C. On time scales characteristic of coal fires in the shallow subsurface (e.g., 10^1–10^2 yr), zircons will completely degas all He, fully resetting the ZHe age, at temperatures of ~375–450 °C (Reiners, 2005). Zircon He ages typically have a precision of ~9% (two standard deviations of replicate analyses on typical zircons such as those in the frequently used geochronologic standard of the Fish Canyon Tuff from southwest Colorado).

While fires in coal beds more than ~30 cm thick easily reset clinker zircon, there is little to no concern that surface wildfires reset zircons more than ~1 cm below the rock surface (Mitchell and Reiners, 2003). Although clinker ages are typically quite young (10 ka to 4 Ma), no corrections for U-series disequilibria are required because ages represent resetting, not formation, events. The accuracy of ZHe ages determined by conventional methods can be affected by strong and systematic intracrystalline zonation of U-Th in single zircons (Farley et al., 1996; Reiners et al., 2004; Hourigan et al., 2005).

Because these samples are detrital and single clinker samples likely contain grains from a wide range of sources, multiple single-grain analyses should not yield consistently inaccurate ages, though individual fliers in each sample may be due to U-Th zonation. For this reason, we measured multiple single-grain ages on most samples. Laser-ablation depth profiling to check core-to-rim U-Th gradients in about ten grains was also performed on zircons from one sample (CLK5) because of systematic age discrepancies with ZFT ages from this sample. No significant or systematic zonation was found in these grains.

Paleomagnetic Orientations of Clinker Outcrops

Clinker is more magnetic than unbaked overburden (Hasbrouck and Hadsell, 1978). Changes in mineralogy in the bulk of the clinker that result in increased magnetism are not yet well understood. In the chimneys, however, the reason for increased magnetism is evident. Paralava, common in chimneys, is marked by the enrichment of iron, much of it in the form of magnetite and ilmenite (Cosca et al., 1989). Formation of these magnetic minerals, which contain iron in the ferrous (Fe^{2+}) oxidation state, apparently results from the presence of carbon monoxide driven off the burning coal face below (Rogers, 1918) and a restricted oxygen supply in the chimney. As these newly formed minerals cool through the Curie temperature, they become ferromagnetic, imparting a strong and coherent magnetic field to the rock that reflects the orientation of Earth's magnetism at that time. This can be used to constrain the age of clinker. Reversed polarity in clinker indicates ages older than ca. 778 ka (Fullerton et al., 2004a, 2004b), and more-precise constraints are possible if magnetic orientation is combined with other geologic constraints.

RESULTS

Overview

The natural burning of coal beds, and therefore the formation of clinker, is a process that has been ongoing since the sedimentary fill of the Powder River Basin began to erode in the late Cenozoic. Radioisotopic ages of clinker outcrops range from 10 ka to as old as 2.9 Ma, showing a history of progressive burning through geologic time. Stream gravels on high terraces contain clinker boulders dated as old as 3.8 Ma, indicating even earlier burning in the region. Relative ages of some clinker have been constrained paleomagnetically (Jones et al., 1984), but numerical dating precise enough for geomorphic applications requires a technique such as thermochronometry that records the thermal effects of coal fires and is insensitive to protolith or detrital ages.

Zircon Fission-Track (ZFT) Ages

In the late 1970s and early 1980s, Naeser measured zircon fission-track ages of 39 baked sandstone samples from the Powder River Basin and northeast Montana, focusing mostly on the Rochelle Hills on the eastern side of the basin in Wyoming, and on the region around Ashland, Montana, in the north (Fig. 1). The samples were collected by Donald Coates and Ed Heffern, with assistance from Jason Whiteman of the Northern Cheyenne Tribe. Most of the results were published in a series of abstracts and two U.S. Geological Survey maps and were discussed in subsequent papers (Coates and Naeser, 1984; Heffern et al., 1983, 1993; Coates and Heffern, 2000; Heffern and Coates, 2004). Although these fission-track ages have large uncertainties (typical 2σ of 30%–60%), in many cases they show remarkably systematic trends with respect to geomorphic features. For this report, we have recalculated the original age determinations to the latest zeta calibration value (Hurford and Green, 1983) and rounded the ages to the nearest 0.1 Ma (Tables 1 and 2). The ZFT ages in this article are recalculated numbers, but the references are to previous publications with the original age determinations. Table 1 provides age and location data for the 39 outcrops used in this study; Table 2 provides supporting data used to calculate the ZFT ages. In the subsequent tables, "Wyodak" denotes the Wyodak-Anderson coal zone in Wyoming, and "Anderson" denotes the Wyodak-Anderson coal zone in Montana.

(U-Th)/He (ZHe) Ages

Over the past five years, we collected clinker samples from 27 locations in eastern Wyoming and Montana and analyzed zircon grains from the baked sandstones to obtain their ZHe ages (Tables 3 and 4). All sites except one were from the Powder River Basin. Some clinker samples collected from additional locations had zircon grains that were too small to date, especially in the northern part of the basin. Table 3 provides age and location data for the 27 ZHe sites; Table 4 provides supporting data used to calculate the ZHe ages, including individual analyses where a weighted mean of several measurements was used to derive a single age for a given location. Some ZHe samples were collected at or near where samples were collected for ZFT analysis; in all such cases, these ZHe ages are within the error range of ZFT ages from the same locations (Reiners and Heffern, 2002) but are generally more precise than the ZFT ages.

DISCUSSION

Interpretation of Results

The clinker ages determined in Wyoming and Montana are best understood in the context of the geologic and topographic setting of the region. This section interprets the sample ages from the tables in the results section.

Wyoming

In addition to the 19 ZFT analyses conducted by Naeser from clinker samples in the Wyoming part of the Powder River Basin in the late 1970s and early 1980s, Reiners and Heffern collected clinker samples for ZHe analysis from 15 locations in the Wyoming Powder River Basin during 2002. Most of the samples were collected from the Rochelle Hills, with a few in the Felix clinker near Wright, and a few in the Lake De Smet area near Buffalo. Figure 6 shows both ZFT and ZHe ages for clinker in the Rochelle Hills. Viewing this figure, the reader can easily compare clinker formation ages generated by each method in nearby locations.

Rochelle Hills

The Rochelle Hills, a major east-facing escarpment on the eastern flank of the Powder River Basin in Wyoming (Fig. 2), are capped by a 20- to 50-m-thick layer of clinker that marks the prehistoric burning of the Wyodak-Anderson coal zone. This is the same thick coal zone that is being mined immediately west of the band of clinker (Fig. 6). The beds dip westward 10–15 m/km, and the landscape drains to the east. Consequently, the plateaus and ridges of clinker that extend eastward from the burn line rise higher toward the east above the streams and above the local water table.

Fission-track ages of zircon grains in clinker from 14 sites on the ridge north of Little Thunder Creek in the Rochelle Hills and two sites on the ridge south of Little Thunder Creek were determined by Naeser (Coates and Naeser, 1984). These ZFT ages (Fig. 6) decrease from 0.8 ± 0.5 Ma on the eastern edge of the escarpment to less than 0.09 Ma near the burn front on the west, with average uncertainties of 55%. Along the burn front north of Little Thunder Creek, the name Burning Coal Draw indicates an area where an active coal fire was mentioned by military expeditions in the 1870s, which continued to burn until the U.S. Bureau of Mines extinguished it in 1951 (Russell and Smith, 1951). West of the burn front, coal is currently mined from the Wyodak-Anderson zone at the Jacobs Ranch, Black Thunder–North Rochelle, and North Antelope–Rochelle operations. The age progression shows variations, due to (1) different areas burning at different times in different tributaries of Little Thunder Creek, and (2) a wide margin of error in age determinations; however, the overall pattern suggests a progressive westward migration of the escarpment with time.

For the ZHe study described in this article, the authors (Reiners and Heffern) collected and analyzed samples of the Wyodak-Anderson clinker from 12 locations in the Rochelle Hills. Here, ZHe ages ranged from 0.615 ± 0.035 to 0.010 ± 0.001 Ma, systematically decreasing from east to west toward the present-day burn front (Fig. 6). These ZHe and ZFT ages suggest progression of the burn front to the west at a rate of ~10–14 mm/yr and, referenced to the elevation of the base of the clinker unit where these ages are found, vertical denudation at a rate of ~0.2–0.3 mm/yr (Fig. 7). For a given elevation, the ZFT ages tend to be slightly older. If the escarpment has maintained the same width and relief

TABLE 1. ZIRCON FISSION-TRACK AGES OF CLINKER IN THE POWDER RIVER BASIN

Sample number	Age (Ma)	Coal zone	Legal description	Latitude* (°N)	Longitude* (°W)	Elevation (m)	Comments
Wyoming							
Rochelle Hills							
77C-1	0.4 ± 0.2	Wyodak	SENW Sec. 22, T43N, R69W	43.6878	105.1329	1488	N of Little Thunder Creek
77C-18	0.7 ± 0.3	Wyodak	NESW Sec. 13, T43N, R69W	43.6996	105.0923	1530	N of Little Thunder Creek
77C-19	0.7 ± 0.3	Wyodak	SWNW Sec. 13, T43N, R69W	43.7050	105.0965	1516	N of Little Thunder Creek
77C-20	0.8 ± 0.5	Wyodak	SWSW Sec. 13, T43N, R69W	43.6964	105.1003	1524	N of Little Thunder Creek
77C-21	0.3 ± 0.2	Wyodak	NENW Sec. 14, T43N, R69W	43.7073	105.1159	1522	N of Little Thunder Creek
77C-22	0.4 ± 0.2	Wyodak	NENW Sec. 15, T43N, R69W	43.7082	105.1323	1515	N of Little Thunder Creek
77C-23	0.3 ± 0.3	Wyodak	NWNW Sec. 16, T43N, R69W	43.7064	105.1598	1506	N of Little Thunder Creek
77C-24	0.3 ± 0.2	Wyodak	NENW Sec. 29, T43N, R69W	43.6797	105.1751	1458	N of Little Thunder Creek
77C-25	<0.09	Wyodak	SESW Sec. 7, T43N, R69W	43.7115	105.1930	1476	N of Little Thunder Creek
77C-26	0.5 ± 0.3	Wyodak	NWNE Sec. 28, T43N, R69W	43.6792	105.1466	1493	N of Little Thunder Creek
77C-27	0.5 ± 0.3	Wyodak	NENW Sec. 23, T43N, R69W	43.6923	105.1152	1515	N of Little Thunder Creek
81G-10	<0.5	Wyodak	NESW Sec. 11, T43N, R69W	43.7146	105.1131	1524	S of HA Creek
81G-11	0.5 ± 0.4	Wyodak	NWNW Sec. 14, T43N, R69W	43.7074	105.1207	1518	N of Little Thunder Creek
81G-14	<0.6	Wyodak	NENE Sec. 8, T43N, R69W	43.7203	105.1648	1482	S of HA Creek
81G-15A	<0.5	Wyodak	SWNW Sec. 18, T42N, R68W	43.6169	105.0784	1555	S of Little Thunder Creek
CPSE-3, 4b, 5, 6, 9[†]	0.7 ± 0.4	Wyodak	NESE Sec. 6, T41N, R68W	43.5568	105.0647	1595	N of Keyton Creek
Felix Coal Zone							
79F-2	0.9 ± 0.6	Felix	SESE Sec. 10, T44N, R71W	43.7962	105.3674	1546	W of Hilight Road
79F-A	<0.3	Felix	NENE Sec. 28, T45N, R72W	43.8524	105.5068	1485	NW of Wright
79F-C	<0.3	Felix	NWSW Sec. 1, T44N, R72W	43.8182	105.4591	1500	N of Wright
Montana							
Little Wolf Mountains							
CP-32	2.2 ± 0.8	Anderson?	NESE Sec. 8, T1N, R39E	45.8492	106.8980	1430	Summit—Little Wolf Mts.
CP-33	2.9 ± 0.9	Anderson?	SENW Sec. 17, T1N, R39E	45.8422	106.9115	1448	Summit—Little Wolf Mts.
Gravel Terrace							
80M-2	3.8 ± 1.2	Unknown	SESE Sec. 6, T3N, R39E	46.0384	106.9189	1143	Gravel terrace SW of Forsyth
Tongue River Valley							
CP-1	0.8 ± 0.7	Anderson	NWSE Sec. 32, T5S, R42E	45.3559	106.5706	1235	NW of Birney
CP-2	0.8 ± 0.4	Anderson	SENW Sec. 32, T5S, R42E	45.3604	106.5767	1226	NW of Birney
CP-3	1.0 ± 0.4	Anderson	NWNW Sec. 29, T4S, R43E	45.4634	106.4570	1220	N of Birney Day School
CP-4	1.5 ± 0.5	Anderson	NENE Sec. 25, T4S, R42E	45.4697	106.4830	1210	NW of Birney Day School
CP-5	1.0 ± 0.4	Anderson	SESW Sec. 13, T4S, R42E	45.4862	106.4936	1226	NW of Birney Day School
CP-6	<0.2	Knobloch	SESW Sec. 33, T4S, R43E	45.4422	106.4336	951	N of Birney Day School
CP-8	1.1 ± 0.4	Anderson	SESW Sec. 29, T5S, R43E	45.3656	106.4508	1241	Summit—Browns Mtn.
CPE-30	0.6 ± 0.6	Knobloch	SESE Sec. 22, T2S, R44E	45.6420	106.2752	966	N of Ashland
83M-1	0.8 ± 0.2	Anderson	NENE Sec. 29, T2S, R43E	45.6385	106.4431	1300	N of Stebbins Creek
83M-2	1.1 ± 0.3	Anderson	NWNW Sec. 9, T2S, R43E	45.6849	106.4362	1293	Summit—Garfield Peak
83M-3	1.0 ± 0.3	Anderson	NWSE Sec. 4, T3S, R43E	45.6039	106.4255	1280	S of Stebbins Creek
83M-4	0.9 ± 0.3	Anderson	NESW Sec. 34, T3S, R43E	45.5309	106.4096	1268	N of Kelty Creek
83M-5	1.3 ± 0.3	Anderson	NENW Sec. 16, T4S, R43E	45.4953	106.4315	1250	S of Kelty Creek
83M-6	0.4 ± 0.3	Knobloch	SWNE Sec. 9, T3S, R44E	45.5936	106.3036	939	Roadcut W of Ashland
83M-7	0.2 ± 0.1	Knobloch	NWSE Sec. 8, T2S, R44E	45.6757	106.3207	957	S of Reservation Creek
83M-8	<0.4	Knobloch	NWSW Sec. 28, T1S, R44E	45.7221	106.3130	960	N of Lay Creek
Northeast Montana							
79M-11	<0.06	Unknown	NE Sec. 8, T27N, R56E	48.1100	104.5100	600	Gorge S of Missouri River, W of Highway 16 bridge

*Latitude and longitude in North American Datum 27.
[†]Average of five samples from same location.

through this time, then the lateral retreat of the escarpment would occur at the same rate as the progression of the burn front.

Three of the ZHe measurements (Reiners and Heffern, 2002) resampled three of Coates and Naeser's (1984) localities along the ridge north of Little Thunder Creek and obtained more accurate ZHe dates, all of which fell within the uncertainties of the earlier ZFT dates. These results corroborated the fission-track dating and more precisely confirmed the trend of younger ages toward the west. The largest discrepancy between the ZFT and ZHe ages was ~150 k.y. (~25%) on the eastern end of the ridge north of Little Thunder Creek. Multiple replicate analyses from several places on the east edge of the Rochelle Hills escarpment, detailed screening for U-Th zonation in these samples by laser-ablation–inductively coupled plasma–mass spectrometry (LA-ICP-MS), and careful extraction of crystals from interior slabs of rock to avoid potential wildfire effects suggest that our ca. 0.54 Ma ZHe age for the eastern edge is accurate. In addition, Jones et al. (1984) reported nor-

TABLE 2. ZIRCON FISSION-TRACK DATA AND AGES FROM CLINKER IN THE POWDER RIVER BASIN

Sample number	No. of grains	Fission-track density ($\times 10^5$ t/cm^2)	Fission tracks counted	Induced-track density ($\times 10^6$ t/cm^2)	Induced tracks counted	χ^2	Dosimeter density ($\times 10^5$ t/cm^2)	Dosimeter tracks counted	Age (Ma ± 2σ)
CPSE-3	6	1.54	20	11.4	736	P	2.03	2140	0.9 ± 0.4
CPSE-4B	6	1.44	20	12.7	883	P	2.01	2140	0.7 ± 0.3
CPSE-5	6	0.81	12	8.34	618	P	1.97	2140	0.6 ± 0.4
CPSE-6	6	1.34	18	11.3	757	P	1.95	2140	0.7 ± 0.4
CPSE-9	6	1.80	30	15.7	1310	P	1.92	2140	0.7 ± 0.3
77C-1	6	1.33	14	21.1	1115	P	2.13	3242	0.4 ± 0.2
77C-18	6	1.86	19	17.7	905	P	2.13	3242	0.7 ± 0.3
77C-19	6	1.47	23	13.8	1082	P	2.11	3242	0.7 ± 0.3
77C-20	6	1.59	13	13.3	541	P	2.09	3242	0.8 ± 0.5
77C-21	6	0.678	9	13.1	872	P	2.07	3242	0.3 ± 0.2
77C-22	6	1.14	12	21.1	1110	P	2.05	3242	0.4 ± 0.2
77C-23	6	1.14	7	22.0	672	P	2.03	3242	0.3 ± 0.3
77C-24	4	0.639	5	13.9	661	P	2.03	3242	0.3 ± 0.2
77C-25	3	<0.20	0	26.2	1113	F	2.01	3242	<0.09
77C-26	7	1.01	12	13.7	816	P	1.99	3242	0.5 ± 0.3
77C-27	6	1.51	17	18.0	1009	P	1.97	3242	0.5 ± 0.3
CP-1	2	2.59	6	19.5	226	P	1.98	2226	0.8 ± 0.7
CP-2	6	1.32	19	10.7	771	P	1.98	2226	0.8 ± 0.4
CP-3	6	2.19	26	14.4	857	P	1.98	2226	1.0 ± 0.4
CP-4	6	3.46	40	14.9	864	P	1.98	2226	1.5 ± 0.5
CP-5	6	2.95	30	18.8	957	P	1.98	2226	1.0 ± 0.4
CP-6	6	1.57	4	14.9	1893	P	1.98	2226	<0.2
CP-8	6	1.96	39	11.3	1121	P	1.98	2226	1.1 ± 0.4
79F-2	7	1.40	11	10.2	400	P	2.05	2204	0.9 ± 0.6
79F-A	6	0.12	1	7.15	298	P	2.03	2204	<0.3
79F-C	6	0.240	2	11.9	494	P	2.03	2204	<0.3
79G-2	6	0.524	8	9.60	733	P	1.96	2206	0.3 ± 0.2
79M-7	6	<0.100	0	10.3	551	P	1.96	2206	<0.2
79M-11	6	<0.04	0	13.1	1607	P	1.96	2206	<0.06
80M-2	5	0.882	49	12.1	336	P	1.61	2180	3.8 ± 1.2
81G-10	7	0.270	2	7.86	291	P	1.91	2143	<0.5
81G-11	6	0.488	7	6.44	462	P	1.91	2143	0.5 ± 0.4
81G-14	6	0.568	5	11.6	508	P	1.91	2143	<0.6
81G-15A	6	0.309	3	8.62	419	P	1.91	2143	<0.5
CPE-30	8	0.609	11	6.91	624	P	1.94	2179	0.6 ± 0.6
CP-32	6	4.32	30	12.2	425	P	1.94	2179	2.2 ± 0.8
CP-33	6	4.41	51	9.31	539	P	1.94	2179	2.9 ± 0.9
81-1F	6	0.060	1	4.85	404	P	1.94	2179	<0.3
81-1C	6	0.080	1	10.4	648	P	1.94	2179	<0.2
83M-1	9	2.08	55	16.2	526	P	1.92	2694	0.8 ± 0.2
83M-2	9	2.28	58	13.0	1657	P	1.92	2694	1.1 ± 0.3
83M-4	10	2.03	31	14.3	1091	P	1.92	2694	0.9 ± 0.3
83M-5	9	2.92	67	13.6	1555	P	1.92	2694	1.3 ± 0.3
83M-6	10	0.563	6	8.49	452	P	1.92	2694	0.4 ± 0.3
83M-7	10	0.531	15	13.0	1830	P	1.92	2694	0.2 ± 0.1
83M-8	7	0.756	7	18.2	842	P	1.92	2694	<0.4

Note: Zeta: (SRM 962) = 319.6 (zircon external detector); DF = laboratory number; χ^2 = pass (P) or fail (F) chi-square test at 5% (external detector runs). Central age was used if sample failed chi-square test (Green et al., 1989; Galbraith and Laslett, 1993).

mal paleomagnetic orientations of clinker on the east end of this ridge, suggesting that the clinker at the east edge of the escarpment formed more recently than the magnetic reversal at 0.778 Ma (Fullerton et al., 2004a).

Felix

Three clinker samples from the Felix coal zone near Wright, Wyoming, had ZFT ages of 0.9 ± 0.6 Ma or younger. The Felix coal lies ~100 m above the Wyodak-Anderson coal, and its outcrop/burn front is 10–15 km west of the Wyodak-Anderson outcrop/burn front. The one ZHe clinker sample from the Felix coal zone, southwest of Wright, had a very young age of 0.024 ± 0.001 Ma.

West Side of Powder River Basin

On the west side of the basin near Buffalo, Wyoming, clinker beds from the Healy-Ucross (Lake De Smet) coal zone in the Wasatch Formation at 1402 and 1439 m above sea level yield ZHe ages of 0.019 ± 0.001 Ma and 0.117 ± 0.007 Ma, respectively, suggesting denudation rates of ~0.4 mm/yr.

TABLE 3. (U-Th)/He AGES OF CLINKER IN THE POWDER RIVER BASIN

Sample number	Age* (Ma)	No. of samples	Coal zone	Legal description	Latitude[†] (°N)	Longitude[†] (°W)	Elevation (m)	Comments
Wyoming								
Rochelle Hills								
SPL1zb, CLK1c	0.205 ± 0.012	2	Wyodak	NENW Sec. 35, T42N, R69W	43.5792	105.1159	1518	Piney Canyon Road
SPL2zA,B	0.482 ± 0.027	2	Wyodak	NWNE Sec. 28, T43N, R69W	43.6796	105.1502	1484	Ridge N of Little Thunder Creek
SPL3zA,B	0.198 ± 0.011	2	Wyodak	NENW Sec. 29, T43N, R69W	43.6797	105.1758	1460	Ridge N of Little Thunder Creek
CLK5a,c,d	0.536 ± 0.025	3	Wyodak	NESW Sec. 13, T43N, R69W	43.7002	105.0924	1527	Ridge N of Little Thunder Creek
CLK6b	0.127 ± 0.010	1	Wyodak	NWNW Sec. 6, T42N, R69W	43.6499	105.2008	1445	Quarry E of Black Thunder Mine
CLK9a,b	0.013 ± 0.002	2	Wyodak	NESE Sec. 34, T48N, R71W	44.0911	105.3652	1417	Quarry pit E of Belle Ayr Mine
PRB11zA,B	0.010 ± 0.001	2	Wyodak	NWSE Sec. 32, T42N, R69W	43.5684	105.1681	1497	S of Piney Canyon Road
PRB12zA,B	0.120 ± 0.008	2	Wyodak	SWSE Sec. 34, T42N, R69W	43.5667	105.1299	1538	100 m E of Red Spring
PRB15zA,B	0.216 ± 0.012	2	Wyodak	NESE Sec. 12, T41N, R69W	43.5397	105.0856	1563	Keyton Canyon
PRB17zA,B	0.615 ± 0.035	2	Wyodak	SWNE Sec. 5, T41N, R68W	43.5586	105.0502	1608	N of Keyton Creek
PRB18zA,B	0.550 ± 0.031	2	Wyodak	SWNE Sec. 5, T41N, R68W	43.5572	105.0504	1598	N of Keyton Creek
PRB19zA,B	0.502 ± 0.028	2	Wyodak	SWNE Sec. 6, T41N, R68W	43.5574	105.0681	1602	N of Keyton Creek
Felix coal zone								
PRB20zA,B	0.024 ± 0.001	2	Felix	NWNE Sec. 27, T43N, R72W	43.6777	105.4892	1540	SW of Wright
Lake De Smet								
CLK7c,d	0.019 ± 0.001	2	Lake De Smet	SWSW Sec. 31, T53N, R82W	44.5166	106.7879	1415	Road W of Lake De Smet dam
CLK8b,c	0.117 ± 0.007	2	Lake De Smet	NENW Sec. 31, T51N, R80W	44.3535	106.5319	1436	Hilltop NE of McNeese Draw
Montana								
Tongue River valley								
03NPRB1z A,B	0.013 ± 0.001	2	Anderson	NESE Sec. 10, T9S, R40E	45.0584	106.8087	1063	Railroad cut N of Decker mine
03NPRB3z EH1,2,3,4	0.652 ± 0.026	4	Anderson	SWSE Sec. 16, T2S, R45E	45.6568	106.1819	1295	Butte N of Cook Mountain
03NPRB4zB	0.110 ± 0.009	1	F	NWNE Sec. 21, T2S, R45E	45.6533	106.1799	1213	Butte N of Cook Mountain
03NPRB5z A,B	0.155 ± 0.009	2	E	SENW Sec. 21, T2S, R45E	45.6480	106.1822	1122	Butte N of Cook Mountain
03NPRB6zA	0.287 ± 0.023	1	Sawyer	SENW Sec. 8, T3S, R45E	45.5894	106.2034	1000	Quarry pit SW of Cook Mountain
03NPRB14zA,B	0.056 ± 0.004	2	Knobloch	NESE Sec. 28, T3S, R44E	45.5445	106.2970	912	Where Bridge Creek joins Tongue River
03NPRB16zA	1.07 ± 0.086	1	Anderson	SWNE Sec. 15, T2S, R42E	45.6687	106.5317	1305	Headwaters of Miller Creek
03NPRB18zA, B,F	0.105 ± 0.005	3	Anderson	SWSW Sec. 27, T2S, R42E	45.6327	106.5403	1267	Highway 212 S of Badger Peak
03NPRB19zA, B,C,D	0.981 ± 0.040	4	Anderson	NENE Sec. 29, T2S, R43E	45.6397	106.4442	1309	Ridge S of Garfield Peak
03NPRB22zA, C,D	0.603 ± 0.028	3	Knobloch	NENE Sec. 16, T1S, R44E	45.7591	106.2971	960	N of Roe and Cooper Creek
03NPRB25zB	0.152 ± 0.012	1	Knobloch	NWNW Sec. 18, T3S, R45E	45.5788	106.2283	928	Hwy 212 E of Ashland
Northeast Montana								
638-2z A,B	0.144 ± 0.014	2	Unknown	NWSW Sec. 16, T28N, R56E	48.177	104.498	655	Pit E of Hwy 16, N of Culbertson

*Weighted mean calculated for multiple measurements from same site; ± error is two standard deviations (internal errors only).
[†]Latitude and longitude in North American Datum 27.

TABLE 4. (U-Th)/He DATA AND AGES FOR ZIRCONS FROM POWDER RIVER BASIN CLINKER

Sample number	⁴He (fmol)	U (ng)	Th (ng)	Ft*	Mass (µg)	Half-width (µm)	U (ppm)	Th (ppm)	Age (Ma)	± (2σ)
03NPRB1zA	0.175	2.756	0.749	0.775	6.13	43	450	122	0.014	0.002
03NPRB1zB	0.624	8.219	3.371	1†	8.29	51	992	407	0.013	0.001
Weighted mean									**0.013**	**0.001**
03NPRB3z_EH1	0.756	0.307	0.149	0.669	1.98	28	155	75.5	0.612	0.049
03NPRB3z_EH2	2.84	1.066	0.332	0.715	2.75	34	388	121	0.644	0.052
03NPRB3z_EH3	1.29	0.460	0.167	0.705	2.04	35	225	82.0	0.679	0.054
03NPRB3z_EH4	1.28	0.483	0.100	0.685	1.71	32	283	58.6	0.682	0.055
Weighted mean									**0.652**	**0.026**
03NPRB4zB	0.188	0.466	0.078	0.652	1.49	30	313	52.6	0.110	0.009
03NPRB5zA	0.261	0.584	0.167	0.564	0.82	21	713	203	0.138	0.011
03NPRB5zB	0.079	0.144	0.048	0.509	0.50	19	288	95.9	0.186	0.015
Weighted mean									**0.155**	**0.009**
03NPRB6zA	0.613	0.607	0.158	0.616	1.22	25	498	130	0.287	0.023
03NPRB14zA	0.509	2.283	0.683	0.715	3.00	34	760	227	0.054	0.004
03NPRB14zB	0.055	0.195	0.124	0.673	1.59	31	122	78.8	0.068	0.010
Weighted mean									**0.056**	**0.004**
03NPRB16zA	1.35	0.248	0.299	0.732	3.09	40	80	96.8	1.07	0.086
03NPRB18zF	0.509	1.223	1.190	0.703	2.30	34	531	517	0.089	0.007
03NPRB18zA	0.577	1.115	0.663	0.717	3.30	36	338	201	0.117	0.009
03NPRB18zB	0.996	2.032	0.703	0.710	4.16	32	488	169	0.119	0.009
Weighted mean									**0.105**	**0.005**
03NPRB19zC	6.63	1.649	0.614	0.699	4.10	33	402	150	0.981	0.079
03NPRB19zD	7.63	1.962	0.529	0.660	3.15	27	623	168	1.03	0.082
03NPRB19zA	2.37	0.493	0.285	0.689	4.07	28	121	70.0	1.14	0.091
03NPRB19zB	1.93	0.559	0.148	0.704	3.03	33	185	48.8	0.857	0.069
Weighted mean									**0.981**	**0.040**
03NPRB22zA	4.24	1.615	0.599	0.754	5.76	40	280	104	0.595	0.048
03NPRB22zC	1.58	0.617	0.262	0.703	2.55	32	242	103	0.613	0.049
03NPRB22zD	1.37	0.582	0.169	0.678	1.65	30	352	102	0.602	0.048
Weighted mean									**0.603**	**0.028**
03NPRB25zB	0.172	0.376	0.218	0.491	0.48	18	784	454	0.152	0.012
638-2zA	0.094	0.204	0.070	0.576	0.65	23	316	108	0.138	0.015
638-2zB	0.051	0.081	0.028	0.549	0.48	22	169	58.6	0.198	0.048
Weighted mean									**0.144**	**0.014**
PRB20zA	0.132	1.34	0.71	0.69	2.73	31	491	261	0.0236	0.0019
PRB20zB	0.193	2.10	0.50	0.662	2.28	27	920	220	0.0245	0.0020
Weighted mean									**0.0240**	**0.0010**
PRB18zA	1.80	0.80	0.42	0.699	3.27	31	245	128	0.530	0.042
PRB18zB	1.75	0.75	0.39	0.679	2.12	32	352	183	0.572	0.046
Weighted mean									**0.550**	**0.031**
PRB17zA	5.75	2.52	0.39	0.709	3.33	33	757	118	0.576	0.046
PRB17zB	1.77	0.63	0.30	0.701	3.28	32	192	92.6	0.667	0.053
Weighted mean									**0.615**	**0.035**
PRB19zA	3.07	1.47	0.52	0.713	3.43	34	429	151	0.501	0.040
PRB19zB	1.90	1.01	0.37	0.643	1.82	26	553	204	0.502	0.040
Weighted mean									**0.502**	**0.028**
PRB11zA	0.455	8.50	1.25	0.796	1.25	46	682	101	0.0121	0.0010
PRB11zB	0.080	1.72	1.05	0.804	1.27	51	136	83.0	0.0094	0.0010
Weighted mean									**0.010**	**0.0010**
PRB12zA	0.976	1.77	0.62	0.763	6.97	40	254	89.4	0.124	0.010
PRB12zB	0.817	1.56	0.69	0.755	5.34	42	291	129	0.117	0.010
Weighted mean									**0.120**	**0.0080**

(*continued*)

TABLE 4. (U-Th)/He DATA AND AGES FOR ZIRCONS FROM POWDER RIVER BASIN CLINKER (continued)

Sample number	⁴He (fmol)	U (ng)	Th (ng)	Ft*	Mass (μg)	Half-width (μm)	U (ppm)	Th (ppm)	Age (Ma)	± (2σ)
PRB15zA	0.828	0.93	0.22	0.753	5.32	40	174	40.6	0.209	0.017
PRB15zB	1.27	1.35	0.91	0.673	2.31	30	584	395	0.224	0.018
Weighted mean									**0.216**	**0.012**
SPL1zb	1.34	1.48	0.259	0.785	8.84	46	167	29.3	0.207	0.017
CLK1c	0.939	1.09	0.553	0.717	4.84	32	220	115	0.203	0.016
Weighted mean									**0.205**	**0.012**
CLK5a	0.692	0.318	0.212	0.681	2.45	31	130	86.6	0.513	0.041
CLK5c	4.86	2.19	0.956	0.703	3.42	32	629	287	0.539	0.043
CLK5d	2.31	0.996	0.540	0.703	2.80	35	346	189	0.559	0.045
Weighted mean									**0.536**	**0.025**
SPL2zA	3.08	1.46	0.695	0.726	3.93	39	371	177	0.486	0.039
SPL2zB	2.63	1.24	0.495	0.755	6.14	32	202	80.6	0.478	0.038
Weighted mean									**0.482**	**0.027**
SPL3zA	1.20	1.74	0.641	0.621	1.44	24	1210	445	0.191	0.015
SPL3zB	0.428	0.575	0.192	0.625	1.29	26	446	149	0.206	0.016
Weighted mean									**0.198**	**0.011**
CLK6b	3.10	6.24	1.47	0.686	15.3	30	401	142	0.127	0.010
CLK7c	0.221	2.14	0.761	0.782	8.84	45	236	178	0.021	0.002
CLK7d	0.182	2.37	0.771	0.781	8.71	45	265	125	0.017	0.001
Weighted mean									**0.019**	**0.001**
CLK8b	0.290	0.624	0.360	0.711	3.23	35	187	117	0.109	0.009
CLK8c	0.385	0.774	0.349	0.669	2.08	29	364	168	0.127	0.010
Weighted mean									**0.117**	**0.007**
CLK9a	0.060	0.690	0.460	0.731	4.31	37	154	108	0.020	0.004
CLK9b	0.215	4.28	0.178	0.786	7.51	49	557	41.0	0.012	0.002
Weighted mean									**0.013**	**0.002**
638-2zA	0.0945	0.204	0.070	0.576	0.646	23	316	108	0.138	0.015
638-2zB	0.0513	0.081	0.028	0.549	0.480	22	169	58.6	0.198	0.048
Weighted mean									**0.144**	**0.014**
Incompletely reset zircons from thin clinker units										
638-1zA	2458	0.663	0.477	0.684	1.56	32	424	305	819	65
638-1zB	3328	1.326	0.153	0.743	3.15	41	421	48.6	586	47
PRB13zA	8.76	1.07	0.45	0.724	3.83	36	281	118	1.90	0.15
PRB13zB	1.75	1.77	0.54	0.682	2.77	29	637	194	0.252	0.020
PRB14zA	23.9	0.37	0.36	0.723	3.83	37	95.6	95.2	13.5	1.08
PRB14zB	2.22	0.44	0.20	0.695	2.56	33	174	78.5	1.21	0.097

Note: Uncertainties are based on internal errors only.
*Ft is the alpha-ejection correction (see Farley et al., 1996; Reiners, 2005).
†Ft of unity is used for this sample because crystal was completely coated in >17 μm thickness of clinker matrix.

Montana

In the late 1970s and early 1980s, Naeser analyzed ZFT ages of clinker samples collected from 20 locations in eastern Montana, 16 of which were in the Tongue River valley, a dissected drainage system carved through the northern Powder River Basin. Eleven of the Tongue River valley samples were from clinker of the Wyodak-Anderson coal zone, and five were from clinker of the Knobloch coal zone. In addition, ZFT ages were determined for two clinker samples from the Little Wolf Mountains, one from detrital clinker in a gravel terrace, and one from clinker in northeast Montana. Recently, Reiners obtained ZHe ages from 11 clinker samples collected in 2003 from the Wyodak-Anderson, Knobloch, and intermediate coal zones in the Tongue River valley. Figure 8 shows clinker distribution and ZFT and ZHe ages in a part of the Tongue River valley near Ashland, Montana. Reiners also analyzed one additional sample of clinkered glacial till from northeast Montana.

Little Wolf Mountains

The oldest in-place clinker samples dated thus far, with ZFT ages of 2.9 ± 0.9 Ma and 2.2 ± 0.8 Ma (Coates and Heffern, 2000; Heffern and Coates, 2004), come from the summits of the Little Wolf Mountains west of Colstrip, Montana (Fig. 1; Table 1) in the northern Powder River Basin. This narrow, isolated range stands 300–400 m above the surrounding landscape and is capped by a layer of clinker as much as 30 m thick.

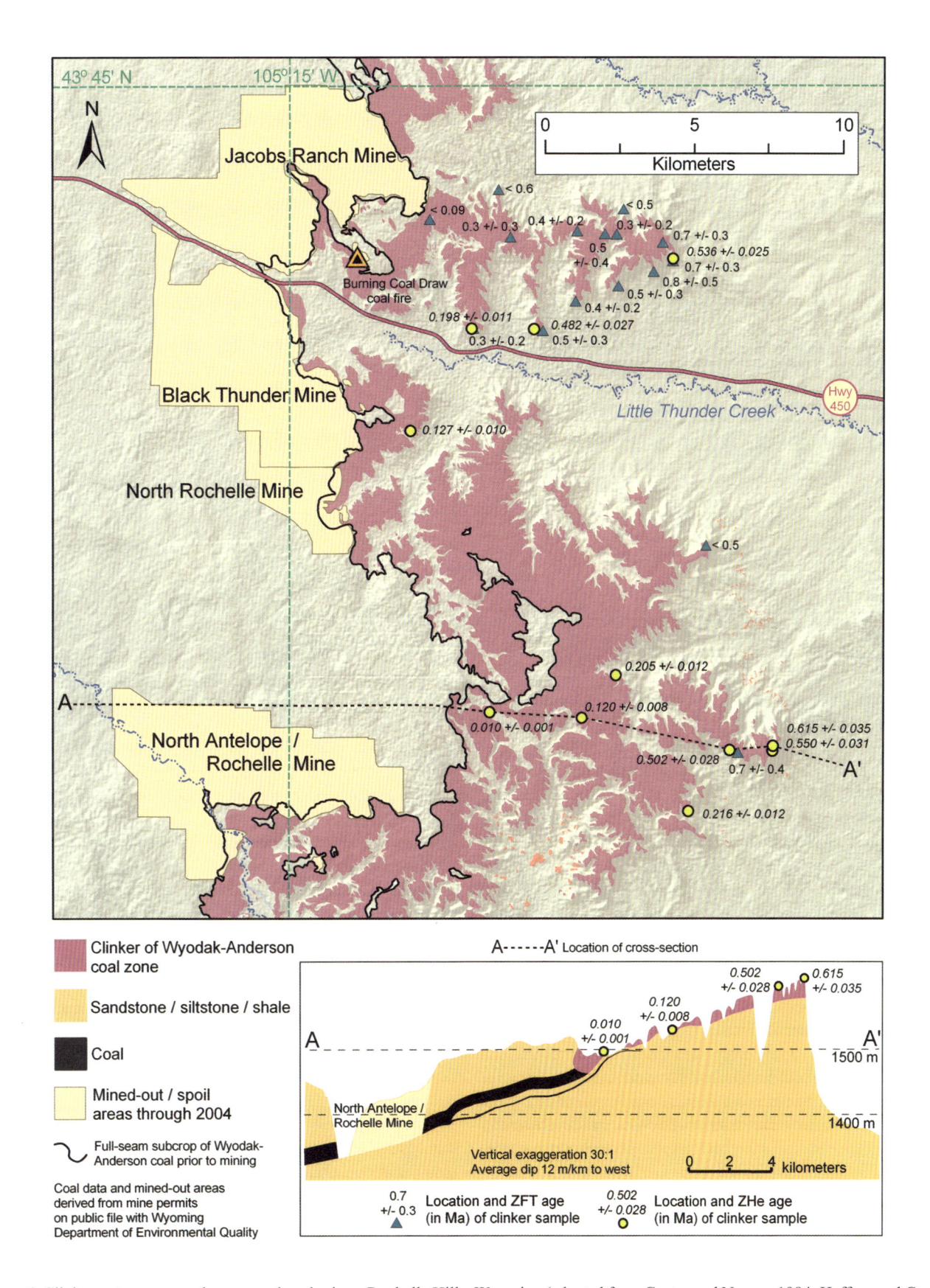

Figure 6. Clinker outcrops, sample ages, and coal mines, Rochelle Hills, Wyoming (adapted from Coates and Naeser, 1984; Heffern and Coates, 2000, 2004). ZFT—zircon fission-track age; ZHe—zircon (U-Th)/He age.

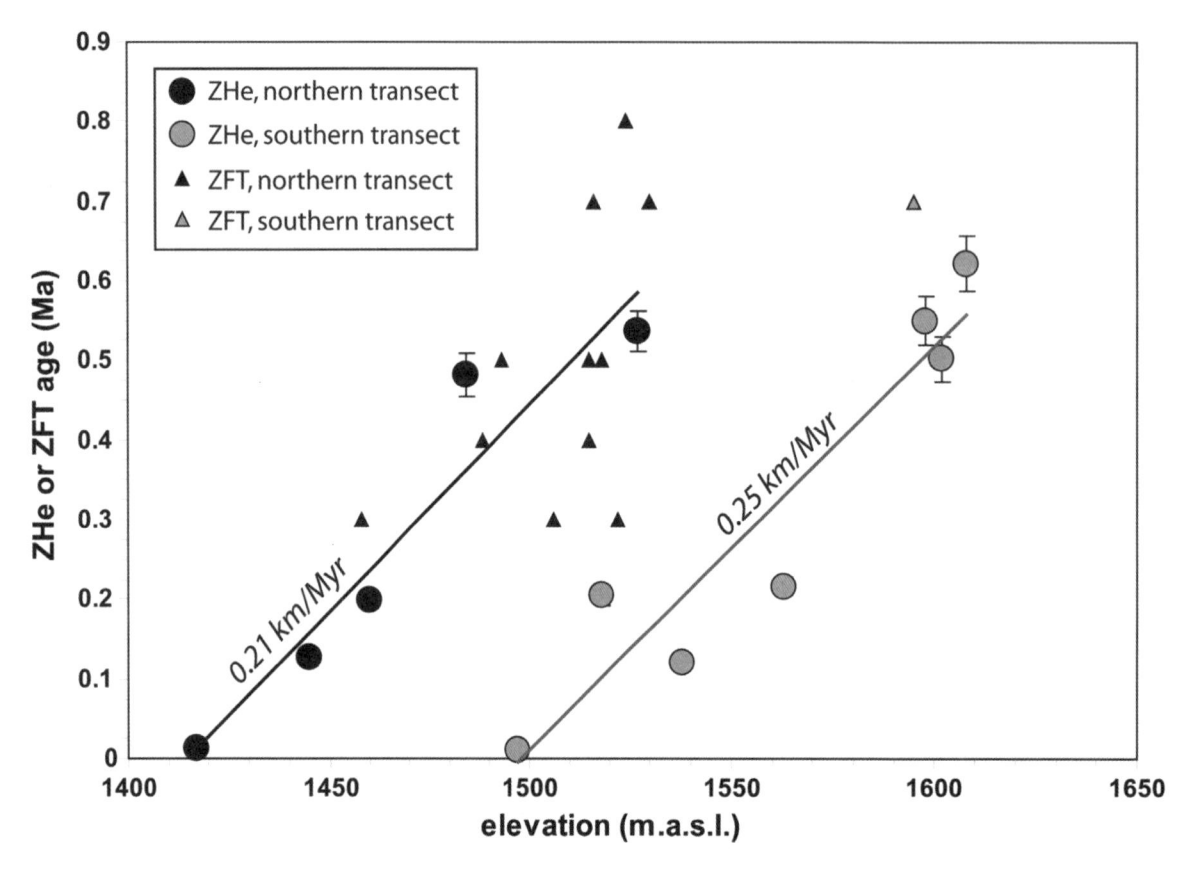

Figure 7. Zircon (U-Th)/He (ZHe) and zircon fission-track (ZFT) ages versus sample elevation in Wyodak-Anderson clinker, Rochelle Hills, Wyoming (Fig. 6). Circles—ZHe; triangles—ZFT; black symbols—northern transect; gray symbols—southern transect. Error bars are 2σ. Solid lines are regressions through ZHe data for each transect, yielding apparent vertical erosion rates of 0.21 ± 0.01 km/m.y. and 0.25 ± 0.01 km/m.y. for the northern and southern transects, respectively (a.s.l.—above sea level).

Gravel Terrace

A clinker boulder at the base of a gravel deposit on a strath terrace 360 m above the Yellowstone River, is even older, with a ZFT age of 3.8 ± 1.2 Ma (Coates and Heffern, 2000; Heffern and Coates, 2004). The site is 28 km southwest of Forsyth, Montana, and 300 m below and 22 km north of the summits of the Little Wolf Mountains (Fig. 1; Table 1). This age indicates that coal beds were exposed and subjected to burning there as early as the Pliocene. Boulders of clinker were eroded from their place of origin and deposited in a valley bottom in stream gravel. The high percentage of basaltic, porphyritic, and rhyolitic cobbles, in addition to the locally derived clinker boulders, indicates that the stream depositing these gravels originated to the southwest in the Absaroka Mountains of north-central Wyoming (Colton et al., 1996; Fig. 1). Later, this gravel-filled valley bottom was isolated as a terrace when the stream cut down to a lower level.

Tongue River Valley

In the northern Powder River Basin near Ashland, Montana, a layer of clinker as much as 60 m thick caps a plateau that forms the rim of the Tongue River valley, 300–400 m above present river level (Fig. 8). This clinker results from ancient burning of the Wyodak-Anderson coal zone. Eleven ZFT samples of this clinker range in age from 1.5 ± 0.5 Ma to 0.8 ± 0.7 Ma. The stratigraphically lower clinker of the Knobloch coal zone is less than 100 m above present river level. Five ZFT samples of the Knobloch clinker range in age from 0.6 ± 0.6 Ma to less than 0.2 Ma (Coates and Heffern, 2000; Heffern and Coates, 2004).

In 2003, we collected clinker samples for ZHe dating from 25 locations in the Tongue River valley. Samples from 11 of these locations yielded zircon grains suitable for dating. Of these 11 locations, four were from the Wyodak-Anderson clinker, three were from the Knobloch clinker, and three were from clinker of intermediate coal beds between the Knobloch and Wyodak-Anderson clinker, near Ashland, Montana; one was from the Wyodak-Anderson clinker near the Montana-Wyoming state line. These ZHe ages ranged from 0.013 ± 0.001 Ma to 1.07 ± 0.086 Ma (Fig. 8).

Strata in the Tongue River valley dip gently to the south, causing the clinker benches to descend southward to river level. However, the Tongue River flows north, intersecting and eroding stratigraphically lower coal beds. No coal has burned below the level of the river, which controls the water table. The Knobloch coal is exposed in the bed of the Tongue River near Birney Day

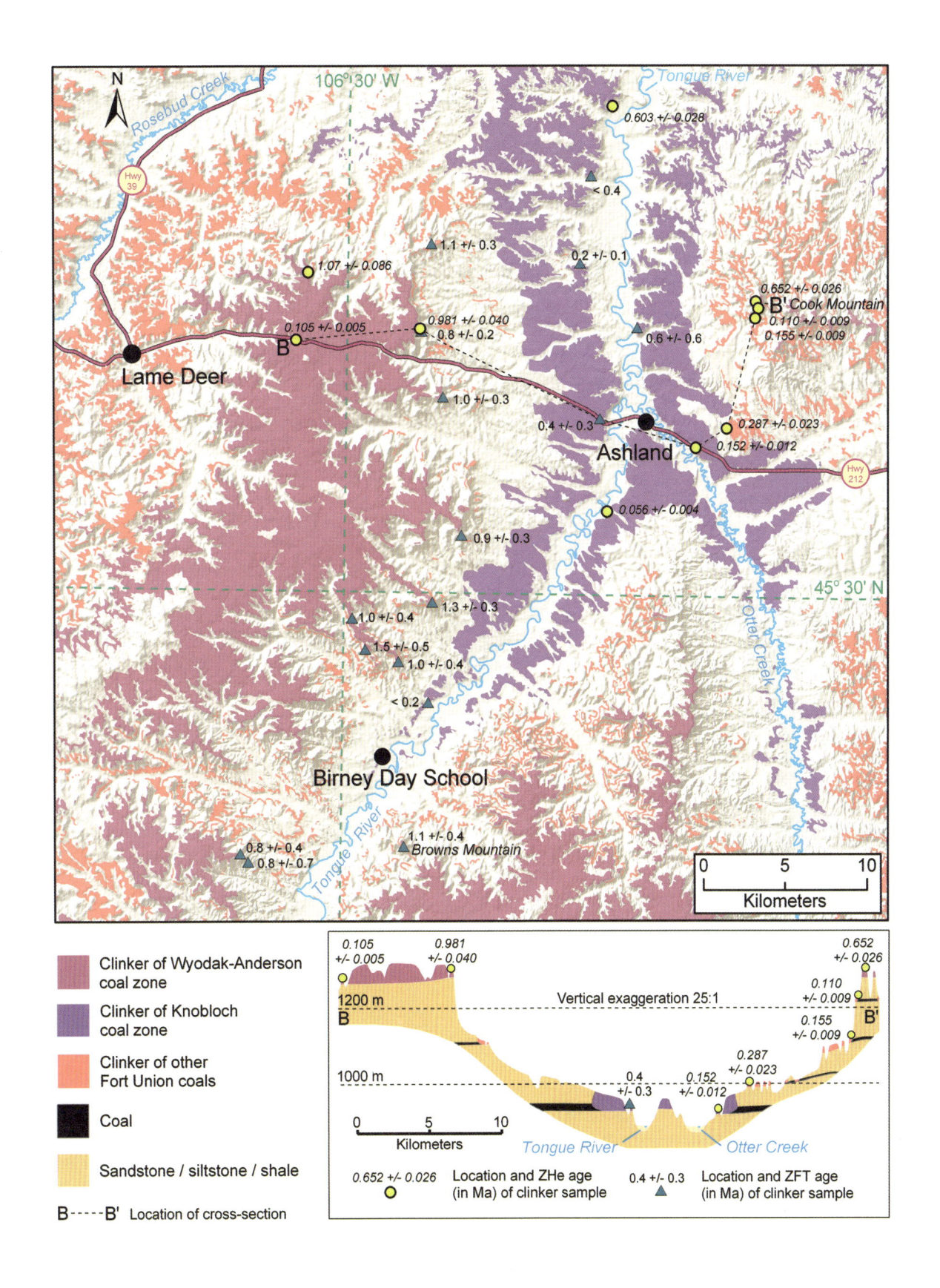

Figure 8. Clinker outcrops and sample ages, Tongue River valley, Montana (adapted from Bass, 1932; Coates and Heffern, 2000; Heffern et al., 1993; Heffern and Coates, 2004; Matson and Blumer, 1973; Vuke et al., 2001a, 2001b). ZFT—zircon fission-track age; ZHe—zircon (U-Th)/He age.

School (Fig. 8), and the coal rises above river level northward, where it has burned to produce clinker as the water table has dropped below the coal.

The ZFT and ZHe ages (Fig. 8) record downcutting of the Tongue River during the Pleistocene. The clinker ages indicate that the Wyodak-Anderson coal was first exposed by the prehistoric Tongue River and burned progressively downdip. As the Tongue River continued to cut down through the strata of the Fort Union Formation, successively lower coal beds were exposed and have burned to produce clinker, while backwasting along valley sides and incision of side drainages have exposed more of the Wyodak-Anderson coal farther away from the river. More recently, the Knobloch coal zone (Fig. 5), 300 m below the Wyodak-Anderson clinker, was exposed and has burned progressively downdip to produce broad benches of clinker near present river level.

West of Ashland, Montana, two ZHe samples on the edge of the Wyodak-Anderson clinker plateau have ages of 1.07 ± 0.086 and 0.981 ± 0.040 Ma, while one sample in the interior of the plateau has an age of 0.105 ± 0.005 Ma. The ZHe and ZFT ages in this area suggest that the plateau capped by the Wyodak-Anderson clinker ~300 m above and 15 km west of the Tongue River has an age of 0.8–1.5 Ma for a large distance around its perimeter, but that clinker at the heads of deep (~8 km) lateral incisions into the plateau has ages at least as young as 0.1 Ma. This pattern of ages would suggest a lateral scarp retreat rate of ~8–10 mm/yr, slightly less than that observed in the Rochelle Hills to the south. The Wyodak-Anderson (Garfield) clinker that caps a butte north of Cook Mountain on the east side of the Tongue River valley (Bass, 1932) is considerably younger (0.652 ± 0.026 Ma) than the Anderson clinker promontories west of the river; the reason is unclear. The Wyodak-Anderson clinker that caps Browns Mountain east of the Tongue River near Birney Day School has a ZFT age of 1.1 ± 0.4 Ma. The ZHe age of a Wyodak-Anderson clinker sample taken from an outcrop 30 km north of Sheridan is very young (ca. 0.013 ± 0.001 Ma). Its location, 70 km south of and 200 m lower in elevation than the other samples, is where the Wyodak-Anderson coal zone descends to the level of the Tongue River at the Decker mine just north of the Montana-Wyoming state line. This relationship shows that the Tongue River has cut down to the elevation of the Wyodak-Anderson coal zone at that location only recently and that the river exposed this coal zone much earlier farther north, which is both downstream and updip.

Three-hundred meters below the Wyodak-Anderson clinker, downstream (northward) ZHe age progressions in the Knobloch clinker in the Tongue River valley show ages of 0.056 ± 0.004, 0.152 ± 0.012, and 0.603 ± 0.028 Ma, with distances between the first and second and second and third samples of ~5.0 and 20 km, respectively (Fig. 8). The ZFT age of the Knobloch clinker near Birney Day School is relatively young, less than 0.2 Ma, but north of Ashland, it is as old as 0.6 Ma because the river cut down and exposed the Knobloch zone much earlier to the north. These dates show that the north-flowing Tongue River cut down through the south-dipping Knobloch coal zone rapidly enough to allow the intersection of the river and the coal zone to migrate southward at a rate of ~40–50 mm/yr. This is consistent with the pattern found for the Wyodak-Anderson coal zone. The ZHe and ZFT ages of the Wyodak-Anderson clinker compared with the Knobloch clinker 300 m below indicate fluvial incision rates of ~0.3 mm/yr.

East of the Tongue River near Ashland, three clinker rims crop out at intermediate elevations between the 30-m-thick Wyodak-Anderson clinker on a butte north of Cook Mountain (dated at 0.652 ± 0.026 Ma) and the thick Knobloch clinker to the west (dated at 0.152 ± 0.012 Ma), 360 m below the base of the Wyodak-Anderson near present-day river level (Fig. 8). Clinker of the F coal (Bass, 1932), on a steep hillside 70 m below the base of the Wyodak-Anderson clinker, was dated at 0.110 ± 0.009 Ma. Clinker of the E coal on the same slope, 160 m below the base of the Wyodak-Anderson, was dated at 0.155 ± 0.009 Ma. These young ages are from thinner coal beds that created narrow rims of clinker on the hillside and were only able to burn a short distance beneath the slope before being extinguished. The clinker bodies produced by these thinner coals are less massive and less hardened by heating and therefore less resistant to erosion. The thick resistant clinker that caps the butte likely protected the underlying coal beds from rapid erosion and resulted in those beds burning a little bit at a time. The clinker of the Sawyer coal, which forms a broader bench 280 m below the base of the Wyodak-Anderson, has a somewhat older age of 0.287 ± 0.023 Ma.

Northeast Montana

Naeser obtained a ZFT age of less than 0.06 Ma for a sample of clinker derived from glacial till and collected from the wall of a postglacial gorge of the Missouri River south of Culbertson, Montana (Table 1). Reiners obtained a ZHe age of 0.144 ± 0.014 Ma (Table 3) on a sample of clinkered glacial till collected by Wayne Van Voast of the Montana Bureau of Mines and Geology from a gravel pit north of Culbertson. This ZHe age indicates baking of Illinoian or pre-Illinoian till after the Illinoian glacier (Fullerton et al., 2004a, 2004b) receded from the area.

Regional Considerations

Regional patterns of clinker ages enable us to make geomorphic interpretations and inferences. While an individual clinker age represents the burn front of a particular coal bed at a specific point in time, the spatial patterns of clinker and clinker ages reflect the longer-term evolution of the landscape and exhumation of the Powder River Basin. Three possible controls on late Cenozoic fluvial erosion rates in the Powder River Basin region are: (1) base-level change of the Yellowstone-Missouri and/or Mississippi River systems (e.g., Zaprowski et al., 2001); (2) long-wavelength tectonic uplift (e.g., Leonard, 2002; McMillan et al., 2002); and (3) climate change that enhanced fluvial erosion rates and influenced groundwater levels (Molnar and England, 1990; Dethier, 2001; Molnar, 2001; Zhang et al., 2001). Because clinker is found throughout the Powder River Basin at a range of elevations, ZHe and ZFT dating offer a promising means to distinguish among several of these processes. We note that the limitation of burning to tens of meters below the surface does not limit the available age record of landscape changes to only the tens of meters nearest the surface because the scale of topographic relief in the Powder River Basin is hundreds of meters. This relief preserves in situ clinker in cap rock as old as 2.9 Ma, where, locally, vertical erosion rates have been relatively slow (e.g., <0.01 mm/yr, for an ~30-m-thick clinker of this age), though regional-scale erosion rates, as determined from age variations over greater distances, may be significantly faster.

Large-scale, regional fluctuations of water-table elevation resulting from tectonic or climatic factors may play some role in suppressing or enhancing coal fires close to the surface, where the coal aquifer becomes unconfined. Water-table fluctuations by themselves are probably only a minor factor in large-scale age variations because adequate ventilation by near-surface oxygen is the most direct control on coal burning and clinker formation (Lyman and Volkmer, 2001; Heffern and Coates, 2004).

Climate changes that trigger continental or alpine glacial-interglacial cycles may (a) cause variations in water or sediment flux through the basins, (b) change large-scale drainage patterns, and (c) influence exposure of coal beds and clinker formation. Our preliminary data set of replicated ZHe ages from 26 separate clinker outcrops over a large part of the Powder River Basin and one clinker outcrop from northeast Montana shows no clear relation with either the oxygen isotope record (Shackleton and Pisias, 1985; Shackleton et al., 1990, 1995), or glacial cycles in the Northern Plains or nearby Laramide ranges (Phillips et al., 1997; Chadwick et al., 1997; Fullerton et al., 2004a, 2004b); however, our data set is too limited and the resolution too coarse to compare with oxygen isotope stages.

Continental glaciations in the northern plains of Montana occurred ca. 15–30 ka (late Wisconsin), and 130–190 ka (Illinoian), with interglacial intervals in between (Fullerton et al., 2004a, 2004b; David Fullerton, 2005, personal commun.). Remnants of the Archer till suggest two other glacial intervals between 640 and 660 ka, and again between 710 and 740 ka. Deposition of the Archer till occurred during oxygen isotope stages 16 and 18, before emplacement of the Lava Creek B tephra from the Yellowstone caldera at 639 ± 2 ka (Fullerton et al., 2004a, 2004b) but after the Matuyama-Brunhes paleomagnetic reversal at 778 ka (Fullerton et al., 2004a, 2004b). An interglacial interval of nearly 500 k.y. followed deposition of the Archer till, and a long interglacial interval preceded deposition of the Archer till. At times during the Wisconsin and Illinoian glaciations, the Yellowstone River was dammed northeast of Glendive, Montana, by the advance of the continental ice lobe (Fullerton et al., 2004a, 2004b), which may have limited downcutting and exposure of coal beds when glacial lakes were present. In contrast, one could speculate that the lowering of sea level or base levels during glacial stages could have increased erosion in drainages not dammed by glacial lobes—such as the Cheyenne River drainage in the southern Powder River Basin—leading to greater exposure of coal beds and more coal fires. Coal beds were exposed in northeast Montana as a result of incision of meltwater streams and catastrophic drainage of glacial lakes and later burned to produce clinker. Clinker ages can potentially provide bracketing ages for glacial events or deposits but cannot date the actual glacial event (David Fullerton, 2005, personal commun.).

CONCLUSIONS

The most important control on clinker age patterns across the landscape is the progress of erosion that brings coal above the water table and allows air from the surface to ventilate the coal. Each natural coal fire starts at a surface exposure but may continue to burn at depth if the coal is ventilated and not saturated. At a given site, the clinker closest to the burn front (that furthest back beneath the slope) is generally younger than the clinker closest to the outcrop on the hillside. Regional clinker age patterns and observations of modern coal fires indicate that most coal in the Powder River Basin burns, forming clinker, as it is exhumed above depths shallower than several tens of meters. Thus, clinker ages provide estimates of the timing of local exhumation, and, through regional age patterns, rates of regional erosion—both vertical and lateral—due to fluvial incision.

The limited number of dates collected so far prevents us from making detailed analyses of erosion rates; however, some general patterns have become evident. Coal beds in the Powder River Basin have been exposed and have burned naturally since at least the early Pliocene. Burning progresses in a systematic pattern as erosion removes the sedimentary fill of the basin, exposing stratigraphically lower coal beds, and rivers carve valleys upstream. The burning of a thick coal bed generally produces higher temperatures over a longer time than burning of a thin bed, so that the clinker produced is thicker, harder, and more resistant. The sheets of clinker produced by major coal beds are more durable and extensive than clinker produced by thin coal beds, and the oldest clinker preserved in place is the thick clinker capping the higher divides. Thin clinker rims along valley sides are generally young. In addition, detrital clinker in gravel terraces and alluvial fans, derived from clinker bodies that have eroded away in the distant past, can in some cases provide older dates of burning than in situ clinker outcrops. In at least one case, where the gently tilted exposure of the Wyodak-Anderson coal in the Rochelle Hills has produced a laterally eroding clinker plateau, depth versus age relationships allow us to estimate a vertical erosion rate of ~0.2–0.3 mm/yr for that area.

The 27 ZHe ages and 39 ZFT ages collected from clinker in northeast Wyoming and eastern Montana provide a glimpse of the geologic history of the region over the past several million years. In limited areas, these ages have allowed us to derive rough estimates of the rates of vertical and lateral erosion by some of the rivers and streams that carved into the Paleogene strata of the Powder River Basin. Many more clinker outcrops and detrital boulders over great areas in the region have not been isotopically dated. Even in the areas where we have derived preliminary rate estimates, a greater density of data points would refine our analyses. The new ZHe data, as well as the earlier ZFT data, serve mainly to show that radioisotopic dating provides reproducible ages of clinker formation. These ages reveal systematic spatial patterns that make sense in terms of expectations about landform evolution in a basin undergoing exhumation.

ACKNOWLEDGMENTS

We wish to thank William Reiners of the University of Wyoming for helping us to collect clinker samples from the Rochelle Hills during November 2002, Jason Whiteman of the Northern Cheyenne Tribe and Paul F. Gore for helping us collect samples from the Tongue River valley during August 2003, and the late Wayne Van Voast of the Montana Bureau of Mines and Geology for providing the sample of clinkered till from northeast Montana. We appreciate the constructive reviews of our draft manuscript by Gretchen Hoffman of the New Mexico Bureau of Geology and Mineral Resources, John Garver of Union College, and Timothy Rohrbacher, Art Schultz, and Jack McGeehin of the U.S. Geological Survey. The input from Roger Colton and David Fullerton of the U.S. Geological Survey regarding glaciation and landscape evolution in eastern Montana was valuable. Catherine Riihimaki of Bryn Mawr College also provided help in interpreting clinker ages in relation to geomorphic evolution of the Powder River Basin. Thanks also go to Ken and Pam Kania of Ashland, Montana, for their hospitality, to Stefan Nicolescu for analytical assistance with ZHe dating, and to Larry Neasloney and Mike Londe of the U.S. Bureau of Land Management for explaining the mysteries of geographic information systems and global positioning systems.

REFERENCES CITED

Bass, N.W., 1932, The Ashland Coal Field, Rosebud, Powder River, and Custer Counties, Montana: U.S. Geological Survey Bulletin 831-B, 105 p.

Chadwick, O.A., Hall, R.D., and Phillips, F.M., 1997, Chronology of Pleistocene glacial advances in the central Rocky Mountains: Geological Society of America Bulletin, v. 109, p. 1443–1452, doi: 10.1130/0016-7606(1997)109<1443:COPGAI>2.3.CO;2.

Clark, G.R., 1985, The distribution and procurement of lithic raw materials of coal burn origin in eastern Montana: Archaeology in Montana, v. 26, no. 1, p. 36–43.

Coates, D.A., 1991, Clinker, in Morrison, R.B., ed., Quaternary Nonglacial Geology, Conterminous U.S.: Boulder, Colorado, Geological Society of America, Geology of North America, vol. K-2, p. 448.

Coates, D.A., and Heffern, E.L., 2000, Origin and geomorphology of clinker in the Powder River Basin, Wyoming and Montana, in Miller, R., ed., Coal Bed Methane and Tertiary Geology of the Powder River Basin: Casper, Wyoming, Wyoming Geological Association, 50th Annual Field Conference Guidebook, p. 211–229.

Coates, D.A., and Naeser, C.W., 1984, Map Showing Fission-Track Ages of Clinker in the Rochelle Hills, Southern Campbell and Weston Counties, Wyoming: U.S. Geological Survey Miscellaneous Investigations Map I-1462, scale 1:50,000.

Colton, R.B., Ellis, M.S., Coates, D.A., Heffern, E.L., Bierbach, P.R., Klockenbrink, J.L., and Grout, M.A., 1996, Photogeologic and Reconnaissance Geologic Map of the Griffin Coulee and Griffin Coulee SW Quadrangles, Rosebud and Treasure Counties, Montana: U.S. Geological Survey Miscellaneous Field Studies Map MF-2302, scale 1:24,000.

Cosca, M.A., Essene, E.J., Geissman, J.W., Simmons, W.B., and Coates, D.A., 1989, Pyrometamorphic rocks associated with naturally burned coal seams, Powder River Basin, Wyoming: The American Mineralogist, v. 74, p. 85–100.

Dethier, D.P., 2001, Pleistocene incision rates in the western United States calibrated using Lava Creek B tephra: Geology, v. 29, p. 783–786, doi: 10.1130/0091-7613(2001)029<0783:PIRITW>2.0.CO;2.

Dodson, M.H., 1973, Closure temperature in cooling geochronological and petrological systems: Contributions to Mineralogy and Petrology, v. 40, p. 259–274, doi: 10.1007/BF00373790.

Energy Information Administration, 2006, Annual Coal Report 2005: U.S. Department of Energy Report No. DOE/EIA-0584(2005), 78 p.

Farley, K.A., Wolf, R.A., and Silver, L.T., 1996, The effects of long alpha-stopping distances on (U-Th)/He ages: Geochimica et Cosmochimica Acta, v. 60, p. 4223–4229, doi: 10.1016/S0016-7037(96)00193-7.

Flores, R.M., 2004, Coalbed methane in the Powder River Basin, Wyoming and Montana: An assessment of the Tertiary–Upper Cretaceous coalbed methane total petroleum system, in Powder River Basin Province Assessment Team, eds., Total Petroleum System and Assessment of Coalbed Gas in the Powder River Basin Province, Wyoming and Montana: U.S. Geological Survey Digital Data Series DDS-69-C, Chapter 2, 62 p.

Flores, R.M., and Bader, L.R., 1999, Fort Union coal in the Powder River Basin, Wyoming and Montana: A synthesis, in Fort Union Coal Assessment Team, eds., 1999 Resource Assessment of Selected Tertiary Coal Beds and Zones in the Northern Rocky Mountains and Great Plains Region: U.S. Geological Survey Professional Paper 1625-A, Chap. PS, CD-ROM, 75 p.

Fredlund, D.E., 1976, Fort Union porcellanite and fused glass: Distinctive lithic materials of coal burn origin on the northern Great Plains: Plains Anthropologist, v. 21, p. 207–211.

Fullerton, D.S., Colton, R.B., and Bush, C.A., 2004a, Limits of mountain and continental glaciations east of the Continental Divide in northern Montana and north-western North Dakota, U.S.A., in Ehlers, J., and Gibbard, P.L., eds., Quaternary Glaciations—Extent and Chronology, Part II: Amsterdam, Netherlands, Elsevier, p. 131–150.

Fullerton, D.S., Colton, R.B., Bush, C.A., and Straub, A.W., 2004b, Map Showing Spatial and Temporal Relations of Mountain and Continental Glaciations on the Northern Plains, Primarily in Northern Montana and Northwestern North Dakota: U.S. Geological Survey Scientific Investigations Map 2843, scale 1:1,000,000, 36 p. brochure.

Galbraith, R.F., and Laslett, G.M., 1993, Statistical models for mixed fission track ages: Nuclear Tracks and Radiation Measurements, v. 17, p. 197–206.

Green, P.F., Duddy, I.R., Gleadow, A.J.W., and Lovering, J.F., 1989, Apatite fission-track analysis as a paleotemperature indicator for hydrocarbon exploration, in Naeser, N.D., and McCulloh, T.H., eds., Thermal History of Sedimentary Basins—Methods and Case Histories: New York, Springer-Verlag, p. 181–195.

Hasbrouck, W.P., and Hadsell, F.A., 1978, Geophysical techniques for coal exploration and development, in Hodgson, H.E., ed., Proceedings of Second Symposium on the Geology of Rocky Mountain Coal: Colorado Geological Survey, Resource Series no. 4, p. 187–218.

Heffern, E.L., and Coates, D.A., 1997, Clinker—Its occurrence, uses, and effects on coal mining in the Powder River Basin, in Jones, R.W., and Harris, R.E., eds., Proceedings of the 32nd Annual Forum on the Geology of Industrial Minerals: Wyoming State Geological Survey, Public Information Circular no. 38, p. 151–165.

Heffern, E.L., and Coates, D.A., 2000, Hydrology and ecology of clinker in the Powder River Basin, Wyoming and Montana, in Miller, R., ed., Coal Bed Methane and Tertiary Geology of the Powder River Basin: Casper, Wyoming, Wyoming Geological Association, 50th Annual Field Conference Guidebook, p. 231–252.

Heffern, E.L., and Coates, D.A., 2004, Geologic history of natural coal-bed fires, Powder River basin, USA, in Stracher, G.B., ed., Coal Fires Burning Around the World: A Global Catastrophe: International Journal of Coal Geology, v. 59, no. 1–2, p. 25–47.

Heffern, E.L., Coates, D.A., and Naeser, C.W., 1983, Distribution and age of clinker in the northern Powder River Basin, Montana [abs.]: American Association of Petroleum Geologists Bulletin, v. 67, no. 8, p. 1342.

Heffern, E.L., Coates, D.A., Whiteman, J., and Ellis, M.S., 1993, Geologic Map Showing Distribution of Clinker in the Tertiary Fort Union and Wasatch Formations, Northern Powder River Basin, Montana: U.S. Geological Survey Coal Investigations Map C-142, scale 1:175,000.

Hoffman, G.K., 1996, Natural clinker—the red dog of aggregates in the Southwest, in Austin, G.S., Hoffman, G.K., Barker, J.M., Zidek, J., and Gilson,

N., eds., Proceedings of the 31st Forum on the Geology of Industrial Minerals—The Borderland Forum: New Mexico Bureau of Mines and Mineral Resources Bulletin no. 154, p. 187–195.

Hourigan, J.K., Reiners, P.W., and Brandon, M.T., 2005, U-Th zonation-dependent alpha-ejection correction in (U-Th)/He chronometry: Geochimica et Cosmochimica Acta, v. 69, p. 3349–3365, doi: 10.1016/j.gca.2005.01.024.

Hurford, A.J., and Green, P.F., 1983, The zeta age calibration of fission track dating: Isotope Geoscience, v. 1, p. 285–317.

Jones, A.H., Geissman, J.W., and Coates, D.A., 1984, Clinker deposits, Powder River Basin, Wyoming and Montana: A new source of high-fidelity paleomagnetic data for the Quaternary: Geophysical Research Letters, v. 11, no. 12, p. 1231–1234.

Leonard, E.M., 2002, Geomorphic and tectonic forcing of late Cenozoic warping of the Colorado piedmont: Geology, v. 30, p. 595–598, doi: 10.1130/0091-7613(2002)030<0595:GATFOL>2.0.CO;2.

Lyman, R.M., and Volkmer, J.E., 2001, Pyrophoricity (spontaneous combustion) of Powder River Basin coals—Considerations for coal-bed methane development: Wyoming State Geological Survey, Wyoming Geo-Notes no. 69, p. 18–22.

Matson, R.E., and Blumer, J.W., 1973, Quality and Reserves of Strippable Coal, Selected Deposits, Southeastern Montana: Montana Bureau of Mines and Geology Bulletin 91, 135 p.

McMillan, M.E., Angevine, C.L., and Heller, P.L., 2002, Postdepositional tilt of the Miocene-Pliocene Ogallala Group on the western Great Plains: Evidence of late Cenozoic uplift of the Rocky Mountains: Geology, v. 30, p. 63–66, doi: 10.1130/0091-7613(2002)030<0063:PTOTMP>2.0.CO;2.

Mitchell, S.G., and Reiners, P.W., 2003, Influence of wildfires on apatite and zircon (U-Th)/He ages: Geology, v. 31, p. 1025–1028, doi: 10.1130/G19758.1.

Molnar, P., 2001, Ornate change, flooding in arid environments, and erosion rates: Geology, v. 29, p. 1071–1074, doi: 10.1130/0091-7613(2001)029<1071:CCFIAE>2.0.CO;2.

Molnar, P., and England, P., 1990, Late Cenozoic uplift of mountain-ranges and global climate change—Chicken or egg: Nature, v. 346, p. 29–34, doi: 10.1038/346029a0.

Naeser, C.W., 1979, Fission-track dating and geologic annealing of fission tracks, in Jager, E., and Hunziker, J.C., eds., Lectures in Isotope Geology: Berlin, Springer-Verlag, p. 154–169.

Papp, A.R., 1998, Coal burns, in Papp, A.R., Hower, J.C., and Peters, D.C., eds., Atlas of Coal Geology, Volume 1: American Association of Petroleum Geologists, Studies in Geology no. 45, CD-ROM, 7 p.

Phillips, F.M., Zreda, M.G., Gosse, J.C., Klein, J., Evenson, E.B., Hall, R.D., Chadwick, O.A., and Sharma, P., 1997, Cosmogenic ^{36}Cl and ^{10}Be ages of Quaternary glacial and fluvial deposits of the Wind River Range, Wyoming: Geological Society of America Bulletin, v. 109, p. 1453–1463, doi: 10.1130/0016-7606(1997)109<1453:CCABAO>2.3.CO;2.

Reiners, P.W., 2005, Zircon (U-Th)/He thermochronometry: Reviews in Mineralogy and Geochemistry, v. 58, p. 151–179, doi: 10.2138/rmg.2005.58.6.

Reiners, P.W., and Heffern, E.L., 2002, Pleistocene exhumation rates of Wyoming intermontane basins from (U-Th)/He dating of clinker: Geological Society of America Abstracts with Programs v. 34, no. 6, abstract 144-4.

Reiners, P.W., Spell, T.L., Nicolescu, S., and Zanetti, K.A., 2004, Zircon (U-Th)/He thermochronometry: He diffusion and comparisons with ^{40}Ar/^{39}Ar dating: Geochimica et Cosmochimica Acta, v. 68, p. 1857–1887, doi: 10.1016/j.gca.2003.10.021.

Rogers, G.S., 1918, Baked Shale and Slag Formed by the Burning of Coal Beds: U.S. Geological Survey Professional Paper 108-A, p. 1–10.

Russell, H.W., and Smith, J.H., 1951, Final report—Control of Coal Crop Fire in Little Thunder Basin, Hilight, Campbell County, Wyoming: U.S. Bureau of Mines, unpublished report, 38 p.

Sarnecki, J.C., 1991, Formation of clinker and its effects on locating and limiting coal resources, in Peters, D.C., ed., Geology in Coal Resource Utilization: Fairfax, Virginia, Tech Books, p. 29–35.

Shackleton, N.J., and Pisias, N.G., 1985, Atmospheric carbon dioxide, orbital forcing, and climate, in Sundquist, E.T., and Broeker, W.S., eds., The Carbon Cycle and Atmospheric CO_2: Natural Variations Archean to Present: American Geophysical Union, Geophysical Monograph 32, p. 412–417.

Shackleton, N.J., Berger, A., and Peltier, W.R., 1990, An alternative astronomical calibration of the lower Pleistocene timescale based on ODP Site 677: Transactions of the Royal Society of Edinburgh, Earth Sciences, v. 81, p. 251–261.

Shackleton, N.J., Hall, M.A., and Pate, D., 1995, Pliocene stable isotope stratigraphy of Site 846, in Pisias, N.G., Janacek, L.A., Palmer-Julson, A., and Van Andel, T.H., eds., Proceedings of the Ocean Drilling Program, Scientific Results, Volume 138: College Station, Texas, Ocean Drilling Program, p. 337–355.

Sigsby, R.J., 1966, "Scoria" of North Dakota [Ph.D. thesis]: Grand Forks, University of North Dakota, 218 p.

Thwaites, R.G., ed., 1969, Original Journals of the Lewis and Clark Expedition, 1804–1806: New York, Arno Press, vol. 1, p. 315, and vol. 5, p. 310.

Vuke, S.M., Heffern, E.L., Bergantino, R.N., and Colton, R.B., 2001a, Geologic Map of the Birney 30′ × 60′ Quadrangle, Eastern Montana: Montana Bureau of Mines and Geology Open-File Map 431, scale 1:100,000.

Vuke, S.M., Heffern, E.L., Bergantino, R.N., and Colton, R.B., 2001b, Geologic Map of the Lame Deer 30′ × 60′ Quadrangle, Eastern Montana: Montana Bureau of Mines and Geology Open-File Map 428, scale 1:100,000.

Zaprowski, B.J., Evenson, E.B., Pazzaglia, F.J., and Epstein, J.B., 2001, Knickzone propagation in the Black Hills and northern High Plains: A different perspective on the late Cenozoic exhumation of the Laramide Rocky Mountains: Geology, v. 29, p. 547–550, doi: 10.1130/0091-7613(2001)029<0547:KPITBH>2.0.CO;2.

Zhang, P.Z., Molnar, P., and Downs, W.R., 2001, Increased sedimentation rates and grain sizes 2–4 Myr ago due to the influence of climate change on erosion rates: Nature, v. 410, p. 891–897, doi:10.1038/35073504.

MANUSCRIPT ACCEPTED BY THE SOCIETY 7 MARCH 2007

Possible sources of magnetic anomalies over thermally metamorphosed carbonate rocks of the Mottled Zone in Israel

Boris Khesin*
Shimon Feinstein
Sophia Itkis
Department of Geological and Environmental Sciences, Ben Gurion University of the Negev, P.O.B. 653, Beer-Sheva 84105, Israel

ABSTRACT

Considerable natural remanent magnetization of the calcareous rocks in the Mottled Zone, Israel, is commonly related to surface combustion metamorphism. The vector sum of inductive and remanent magnetization (effective magnetization) was determined based on a study of the magnetic properties of rocks and interpretation of T (the modulus of geomagnetic field vector) and ΔZ (the increment of vertical component of geomagnetic field) anomalies. Ground multiscale magnetic measurements indicate that the high-grade metamorphic rocks of the Mottled Zone, which have a spotty distribution, are characterized by extensive variations in magnetic fields and susceptibility, with a median value of ~200 × 10^{-5} SI. This magnetic pattern is similar to other areas of combustion metamorphism. Low-temperature hydrothermal rocks have a relatively homogeneous magnetic susceptibility with the same median value.

Relatively homogeneous and moderate magnetization also was observed in some outcrops of the Mishash Formation, which underlies the Mottled Zone rocks. High-grade metamorphism and locally varying magnetization may be due to the burning of gases. The local aeromagnetic maxima observed within the Hatrurim Basin and quantitative interpretation of some magnetic anomalies suggest magnetic sources with relatively homogeneous and stable magnetization and greater area and depth. The formation of such bodies requires a regional source for magnetization processes (e.g., gas flow from depth along faults). A more detailed study, including a helicopter survey and special ground and laboratory analyses, must be performed for a complete characterization of the complex magnetic system.

Keywords: magnetic anomalies, metamorphosed rocks, Mottled Zone, diverse magnetic sources, bitumen burning, gas flow.

*khesin@bgu.ac.il

INTRODUCTION

Metamorphosed rocks of the Hatrurim Formation are distinguished from other Phanerozoic sedimentary rocks of Israel not only by their exotic mineralogy (Gross, 1977), but also by their significant magnetization (Ron and Kolodny, 1992; Khesin and Itkis, 2002; Khesin and Feinstein, 2005; Khesin et al., 2005). These rocks crop out in several areas (Bentor and Vroman, 1960; Bartov, 1990; Sneh et al., 1998) and often are described as Mottled Zone rocks (e.g., Matthews and Gross, 1980; Burg et al., 1991, 1999).

The spatial distribution of Mottled Zone rocks in Israel and Jordan is shown in Figure 1. The Mottled Zone complex has been studied best in the Hatrurim Basin west of the Dead Sea fault zone. Turonian dolomites, Santonian chalks (Menuha Formation), and Campanian cherts, marls, and phosphorites, sometimes bituminous (Mishash Formation), are the main country rocks. The Hatrurim Formation is represented by high-grade metamorphic rocks of local spotty and chaotic distribution and low-temperature hydrothermal rocks that compose the most part of Mottled Zone strata (Vapnik et al., this volume). The rocks of the Hatrurim Formation rest disconformably on the rocks of the Mishash Formation. It is assumed that metamorphic alteration of the Hatrurim Formation is related almost completely to rocks of Maastrichtian age (Ghareb Formation).

Most studies assume that combustion metamorphism was caused by the burning of bituminous rocks ("oil shales") of the Ghareb Formation (Burg et al., 1999, and references therein). Alternative mechanisms of metamorphism have been proposed (e.g., activity of magmatic or hydrocarbon fluids, underground explosions, and gas burning [Gilat, 1998; Khesin and Itkis, 2002; Khesin et al., 2005; Vapnik and Sokol, 2006]). Several different mechanisms could explain the magnetization of Mottled Zone rocks and the origin of magnetic anomalies in the Mottled Zone areas.

REVIEW OF MAGNETIZATION GENESIS

Combustion Metamorphism

The conventional concept is that Mottled Zone metamorphic rocks are a product of near-surface spontaneous combustion of protoliths that are rich in organic matter (e.g., Bentor et al., 1981; Burg et al., 1991, 1999). The extensive internal combustion of the bituminous rocks causes a high-temperature, low-pressure metamorphism in many local foci at different times. The rocks gain natural remanent magnetization (NRM) from the formation of magnetic minerals—maghemite and hematite—as a consequence of the oxidation of pyrite (Ron and Kolodny, 1992).

Hydrocarbon Flow

Upward movement of hydrocarbon fluids often leads to the formation of magnetic minerals within sedimentary rocks and the appearance of related aeromagnetic anomalies of low magnitude

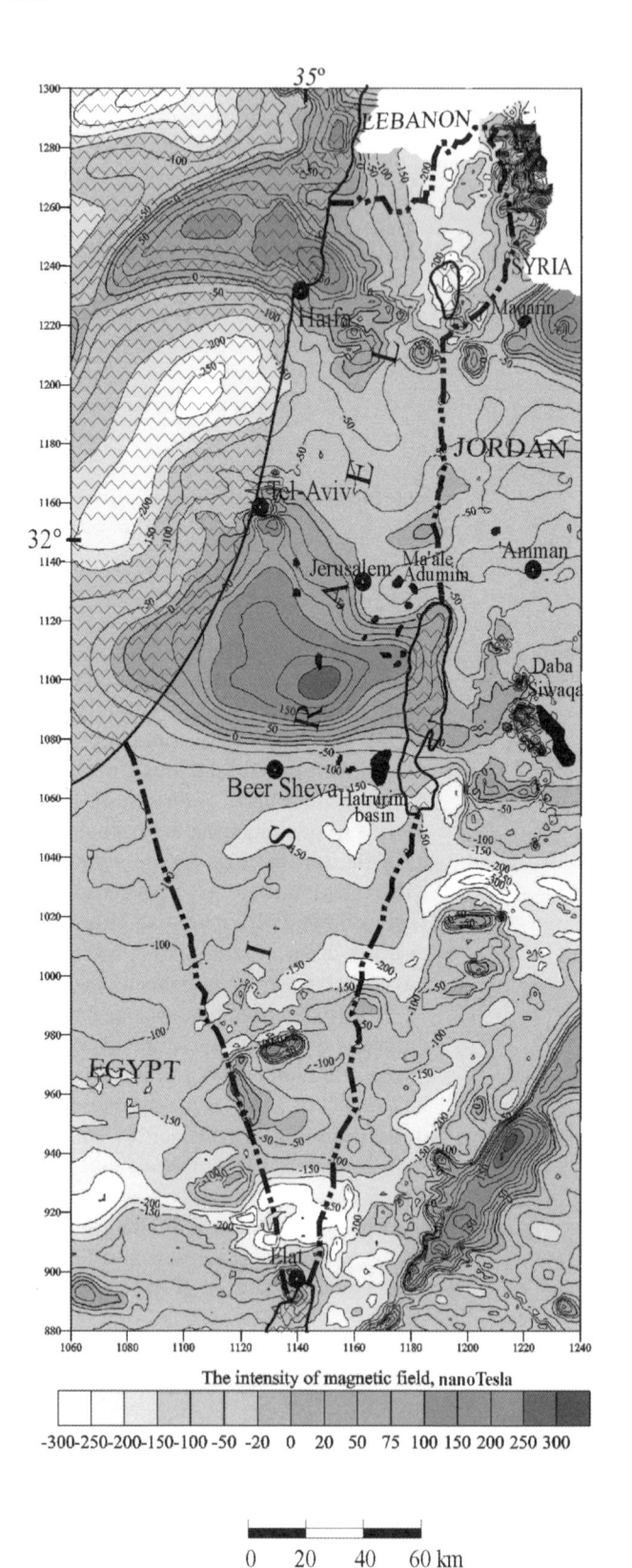

Figure 1. Mottled Zone rock outcrops (black spots, modified from Burg et al., 1999) superimposed on the magnetic anomaly map (modified from Rybakov et al., 1994) in Israel and Jordan (Israeli grid, gridline numbers represent kilometers).

(commonly 5–10 nT, sometimes up to 20 nT) and high frequency (Donovan et al., 1979). Saunders and Terry (1985) revealed the presence of diagenetic magnetite and related aeromagnetic anomalies over more than 90% of more than 100 known oil and gas fields. Usually, diagenetic magnetite occurs down to 300 m, but its generation is possible down to 1.5 km.

Foote (1996) identified maghemite and greigite as the major iron-bearing minerals in the magnetic sedimentary intervals at the anomalies that were detected by high-resolution, low-clearance aeromagnetic surveys over petroleum-bearing areas in Oklahoma, Colorado, Utah, and Alabama. According to Machel (1996), seepage of hydrocarbons from traps and migration from source rocks result in hydrocarbon-contaminated plumes in groundwater and formation waters. Such plumes are characterized mainly by a marked reduction in the redox potential; this causes the generation of magnetic ferrous iron oxides and sulfides, and the destruction of ferric iron oxides.

Peirce et al. (1998) showed the presence of magnetization related to vertical fluid flow along some fault planes in Alberta, Canada. They believed that the magnetization was caused by geochemical reactions that involved iron, sulfur compounds, and hydrocarbons. LeSchack and Van Alstine (2002) detected dipolar anomalies in high-resolution ground magnetic surveys over microseeping hydrocarbon reservoirs in western Canada. About half of the intensity of these anomalies is from NRM, which often is reversed. Wollenben and Greenlee (2002) noted that short-term ambient conditions do not affect magnetic enrichment significantly over a hydrocarbon trap; therefore, repeated positive aeromagnetic anomalies (~5 nT) were detected in different years over prospective areas in Texas.

Aldana and Costanzo-Alvarez (2003) examined the oil fields in southwestern Venezuela. Producing wells are characterized by an interval of increased magnetic susceptibility (κ) with Fe-rich spherical aggregates of authigenic origin. This magnetic contrast could be genetically associated with a reducing environment induced by the underlying hydrocarbons. Rock magnetic studies here indicate a low-coercive magnetic mineral (probably magnetite).

According to some models of oil accumulation and migration (e.g., Eventov, 2000), a relatively large secondary magnetic body may form directly over the petroleum pool within its seal and as a series of small bodies within the shallow subsurface. This model can be extended (Stone et al., 2004) by some lateral displacement of magnetic anomalies relative to hydrocarbon deposits and the annular shape of the anomalies on magnetic maps. Stone et al. (2004) noted that a chemical remanent magnetization seemed to be associated with micromagnetic deposits. This was implied in the study in southern Sudan by the north-south–striking limbs of the annular/aureole anomalies. Because the study area is located close to the geomagnetic equator, such north-south–trending features usually are not observable if they are generated by induction alone.

The partial magnetization of some combustion metamorphism areas also may be due to the influence of hydrocarbon fluids. According to Cisowski and Fuller (1987), combustion metamorphic rocks have been identified within 17 oil fields. These occurrences can be related to surface outcrops of source rocks or leakage from reservoirs. Cisowski and Fuller (1987) also noted that aeromagnetic anomalies are associated with combustion metamorphism zones only in the Marcelina Formation of northwestern Venezuela (Zulia province) and the Ghareb and overlapped Taqiye Formations of the Hatrurim Basin in Israel. In the first case (Moticska, 1978), aeromagnetic anomalies have been reported over burnt lignite beds, in Eocene brackish water deposits, which extend over an area that is 40 km long. The situation in the Hatrurim Basin is more complicated (see later discussion on aeromagnetic data).

Rising hydrocarbon gases can cause a fire at the Earth's surface. Some examples are well known in Azerbaijan. Gas is seeping to the surface at the site of the Zoroastrim Fire Temple near Baku, where it has been feeding a sacred fire since ancient times. Such a phenomenon also is observed at Yanardag, the Burning Mountain of the eastern foothills of the Greater Caucasus. Similar phenomena could be the cause of local thermal metamorphism.

Other Possible Sources

The most well-known source of magnetic anomalies is highly magnetic magmatic bodies of basic composition. Thus, the magmatic origin of magnetization and observed aeromagnetic anomalies is one of several important hypotheses.

It is impossible to exclude exotic sources of magnetization, such as lightning strikes. Nagata (1953) noted that magnetic force lines, as a consequence of lightning strikes, are concentric around the spark's point of discharge. The observed constant remanent magnetization is an isothermal magnetization at room temperature; its intensity and direction are heterogeneous and variable within intervals of a few meters, and the intensity decreases rapidly with distance from the strike point. Rusinov (1973) performed a micromagnetic survey and study of oriented samples to analyze the magnetization that was caused by atmospheric electrical discharges. He noted that average NRM directions trended to the horizon and that flow of electric current mostly coincided with fractures. Opposite magnetic polarity always was observed on both sides of the fracture, and the NRM decreased with distance from the fracture. Later, Pechersky (1985) noted that lightning strikes cause intensive normal remanent magnetization of high value, low stability, and specific spatial distribution.

METHODOLOGY OF MAGNETIC STUDY

Ground Measurements

Magnetic fields and susceptibility measurements were performed on three different scales: large, detailed, and micro, with a measurement step of 0.1–0.5 km, 5 or 10 m, and 1 or 2 m, respectively. We used a proton magnetometer M203-M for the measurements of the modulus of geomagnetic field vector (T). During the micromagnetic survey, the vertical component of the geomagnetic field (Z) was measured using a Z magnetometer. Vertical gradients of the T field were calculated from T measurements at two levels above the surface with a 1 m difference. K-2 and KT-5 magnetic susceptibility meters were used for κ measurements of samples or in outcrops. Usually, the measured κ values are lower than the true (volume) magnetic susceptibility. Therefore, the maximal value from a number of κ measurements at each station was selected. For several profiles, κ values for of bedrock (κ_{bed}) and diluvium or diluvium/alluvium (κ_{dil}) were plotted separately. A total-count radiometric survey was used for γ-ray registration in some magnetic profiles using a scintillation meter.

Qualitative Interpretation and Rapid Inversion

Several processing steps were applied to the magnetic data (Khesin et al., 1983, 1996). These included using a running average (smoothing) of the observed magnetic values with calculation of "regional" and "local" (residual) components, and upward continuation (calculation at higher level) of the observed T field.

Rapid methods of inverse problem solution were applied as well. These methods were developed for the typical environments of many regions, including Israel, which are characterized by inclined magnetization and uneven topography (Khesin et al., 1996). Modified methods of characteristic points and tangents with the method of characteristic areas permit estimation of the depth of anomaly sources and their effective magnetization using an approximation by simple geometrical bodies. The observed magnetic field T is preconditioned by the effective magnetization ($\mathbf{J}_e$), which is a vector sum of induced magnetization ($\mathbf{J}_i$) and remanent magnetization ($\mathbf{J}_n$):

$$\mathbf{J}_e = \mathbf{J}_i + \mathbf{J}_n, \tag{1}$$

where

$$\mathbf{J}_i = \kappa \mathbf{T}_0, \tag{2}$$

and $\mathbf{T}_0$ is the normal geomagnetic field (magnetized field).

Correlations between T field and altitudes of measurements (Khesin et al., 1996; Khesin, 1998) were used for a more accurate estimation of the effective magnetization of outcropping rocks.

Forward Modeling

The results of rapid inversion were used for the composition of the initial physical-geological model (Khesin et al., 1996). For subsequent iterations, we used three-dimensional forward modeling realized in the DIPOLI-3 software (Ermokhin, 1998). The program, based on the Fredholm integral equation of the second kind, allows computation of the magnetic field from an assemblage of elementary cubes. Each elementary cube is characterized by apparent κ (taking into account additional $\mathbf{J}_n$ implications) and dipole field. The program takes into account important characteristics of the regional magnetic field, including inclination of the vector and average level of the field. For each dipole, its field and the fields of all other dipoles are calculated. Thus, the demagnetization effect is taken into account.

PREVIOUS MAGNETIC DATA AND THEIR ADDITIONAL INTERPRETATION

Aeromagnetic Data

Aeromagnetic anomalies of the Hatrurim Basin are obscured in the 1:500,000 magnetic map of Israel (Rybakov et al., 1994) because of the strong smoothing out of data by automated plotting techniques (Fig. 1). It is possible to see low-amplitude local maxima in the 1:250,000 aeromagnetic map (Folkman and Yuval, 1976), which was compiled without smoothing out of the initial aeromagnetic data measured at a flight altitude of 1000 m above sea level (Domzalsky, 1967). Two local ΔT anomalies, with amplitudes of up to 20 nT, were revealed within the Hatrurim Basin (Fig. 2A). Another weak anomaly was indicated west of the basin in the Nevatim-Malhata area (east of Beer Sheva; Fig. 1), where some outcrops of Mottled Zone rocks are known. From the data of Domzalsky (1967), we derived a residual aeromagnetic map ΔT_{res} that emphasizes anomalies within the Hatrurim Basin (Fig. 2B). One anomaly forms a 7-km-long northeast-oriented zone, whereas the second is a less distinctive subisometric anomaly. The ΔT (Fig. 2A) and ΔT_{res} (Fig. 2B) anomalies of the aeromagnetic survey coincide with local structural highs (e.g., top of the Mishash Formation—Burg et al., 1991; top of the Middle Jurassic Zohar Formation—Fig. 2C). The northeast-trending fault in the structural map on top of the Zohar Formation (Fig. 2C) coincides with a maximum gradient zone of the northeast-trending aeromagnetic anomaly (Fig. 2B).

Ground Magnetic Data

We performed several thousand κ measurements on Mottled Zone rocks in situ and on more than 700 samples of high-grade metamorphic rocks and low-temperature hydrothermal rocks from sample collections. We applied T field inversion and correlation methods (Khesin, 1998) for evaluation of the rocks' $\mathbf{J}_e$ and approximate Z magnetometer measurements for the NRM ($\mathbf{J}_n$) determination in oriented samples (Khesin et al., 2005). Ground-profile measurements of magnetic field and susceptibility were performed at the Hatrurim Basin, Nevatim-Malhata, and Ma'ale Adumim areas.

Modal κ histograms of the Mottled Zone rocks (Khesin et al., 2005) suggest that rock magnetization (or demagnetization) is a superposition of several processes. It is possible to distinguish at least two modes of magnetization (in addition to demagnetization, which is related to secondary alteration and weathering). One type of magnetization is heterogeneous in strength (e.g., magnetic susceptibility reaches up to 5700×10^{-5} SI) and spatial distribution. Such magnetization is observed mainly in

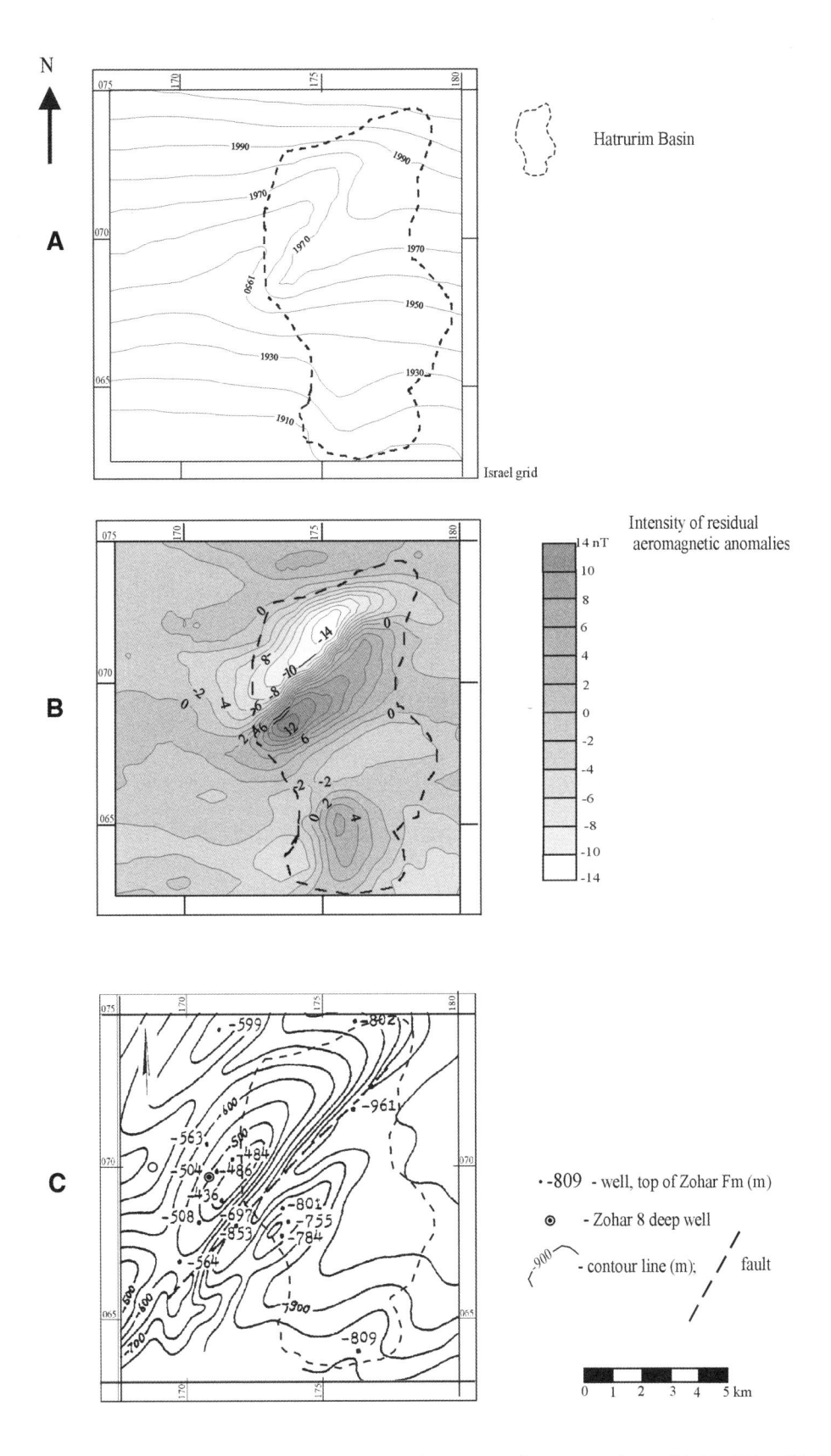

Figure 2. Comparison of magnetic and structural maps. (A) Aeromagnetic ΔT anomalies; contour interval is 10 nT (modified from Folkman and Yuval, 1976). (B) Residual anomalies of aeromagnetic survey derived from Domzalsky (1967) data; contour interval is 2 nT. (C) Structural map on top of the Jurassic Zohar Formation; contour interval is 50 m (modified from Frolov et al., 2002).

association with the foci of Mottled Zone rocks that were subjected to high-grade metamorphism. The other type of magnetization has median values that are similar to the heterogeneous type (~200 × 10^{-5} SI), but its magnitude is shorter and its distribution is wider. Such magnetization is observed mainly in association with low-temperature hydrothermal rocks, and sometimes with rocks of the underlying Mishash Formation. We have not yet successfully revealed heterogeneous-type magnetism carriers reliably, whereas stable-type magnetism carriers mainly are iron hydroxides.

The approximate field method and conventional paleomagnetic studies of oriented samples (Khesin et al., 2005) showed similar results: the NRM of high-grade metamorphic Mottled Zone rocks was heterogeneous, $\mathbf{J}_n$ values of rocks usually exceeded the $\mathbf{J}_i$ values, and both components of effective magnetization had similar direction within the Hatrurim Basin. These results are concordant with the previously published paleomagnetic data of Ron and Kolodny (1992), who noted that the $\mathbf{J}_n$ inclination of Mottled Zone rocks is close to 47°, and its magnitude varies from low values up to 6 A/m. Analogous to the magnetic susceptibility, it is possible to distinguish several rock groups according to their NRM characteristics. The magnetic characteristics of sites in the Hatrurim Basin and Ma'ale Adumim area are different. Reversed $\mathbf{J}_n$ values revealed in the Ma'ale Adumim area likely indicate another phase of metamorphism relative to the Hatrurim Basin.

There are important differences between the heterogeneous and often highly magnetized high-grade metamorphic rocks and the homogeneous magnetization of other sedimentary rocks. Magnetic bodies of low-temperature hydrothermal rocks and unmetamorphosed underlying bituminous carbonates and phosphorites are distributed more widely and show mostly weaker, but more stable, magnetization. They probably can produce marked magnetic anomalies that may be similar to recorded local aeromagnetic maxima. Locally distributed high-grade metamorphic rocks can cause only local surface magnetic anomalies. The $\mathbf{J}_e$ vector of magnetic rocks sometimes approaches horizontal. This may reveal the subhorizontal character of magnetic bodies, because the magnetization vector tends to be oriented in the direction of the long axis of source bodies.

Inversion of magnetic anomalies and application of the correlation method also show that it is impossible to explain the obtained $\mathbf{J}_e$ by $\mathbf{J}_i$ only. For example, in detailed profile 1 (Fig. 3), the effective magnetization of the source of the magnetic anomaly is 360 mA/m. It is the minimal possible value according to interpretation by the characteristic areas method, whereas inductive magnetization derived from κ data is only as high as 200 mA/m (Khesin and Itkis, 2002). The correlation method in profile 7 (Fig. 3) showed a $\mathbf{J}_e$ value of 800 mA/m, whereas the $\mathbf{J}_i$ value calculated using κ data is only 220 mA/m (Khesin and Feinstein, 2005). In both cases, to obtain the calculated $\mathbf{J}_e$ value, it is necessary to add the $\mathbf{J}_i$ value to the $\mathbf{J}_n$ value. Thus, the analyses of oriented samples and magnetic field interpretations show the important contribution of remanent magnetization to effective magnetization. On the one hand, significant NRM facilitates a study of its magnetization history. On the other, this fact requires accounting for remanent magnetization in physical-geological modeling.

The general magnetic pattern was studied on large-scale magnetic profiles (A–D and F–E profiles, Fig. 3).

These profiles cross local aeromagnetic maxima (~10, maximally 20 nT) within the Hatrurim Basin. The long-wavelength pattern for the ground measurements is consistent with aeromagnetic anomalies. The short-wave variations are attributed to the shallow patchy distribution of Mottled Zone rocks with variable levels of metamorphism and alteration. Secondary alteration and weathering lead to a decrease in magnetization. High κ values (up to 2900 × 10^{-5} SI) were observed for the rocks of the Hatrurim Formation. Increased κ values (up to 400 × 10^{-5} SI) also were obtained for the bituminous carbonates and phosphorites of the Mishash Formation, whereas Upper Cretaceous sedimentary rocks in other basins are virtually nonmagnetic (Khesin and Feinstein, 2005). Apparently, aeromagnetic anomalies are caused by more extended magnetic sources, rather than surface or near-surface local high-grade metamorphic bodies of Mottled Zone rocks.

In the Nevatim-Malhata area, where Mottled Zone rocks probably lie under the overburden (soils), the ground T maximum value is the same at different levels. The zero vertical gradient of the T field here implies a subsurface magnetic body.

Detailed profiles reveal significant T and κ variations for the Hatrurim Formation rocks; relative long- and short-wave components of the magnetic field were detected (Khesin and Itkis, 2002; Khesin and Feinstein, 2005; Khesin et al., 2005). Relatively high magnetic field and susceptibility values were recorded for high-grade metamorphic rocks and low-temperature hydrothermal rocks. Interpretation of the data suggests that magnetic anomalies are related mainly to near-surface, subhorizontal bodies. A part of the magnetic field extrema is determined by the edges of these bodies, whereas another part, in some cases, must be contributed by magnetic sources at a depth of several meters and greater (e.g., profile 1 [Khesin and Itkis, 2002] and profile 7 within the Hatrurim Basin [Khesin and Feinstein, 2005]).

NEW RESULTS OF GROUND MAGNETIC AND COMPLEMENTARY STUDIES

Large-Scale Survey

A comparison of magnetic patterns inside and outside of the Hatrurim Basin was performed for two large-scale profiles. The west-east profile B–C crosses a northeast-trending elongated anomaly inside the Hatrurim Basin (Fig. 3). The profile segment with ground magnetic anomalies broadly coincides with the aeromagnetic anomaly (Fig. 4).

Upward projection of ground measurement data to the level of the aeromagnetic survey, however, shows some shift of the aeromagnetic anomaly relative to the ground anomalies, and a great excess of recalculated field (~100 nT in Fig. 4) relative to

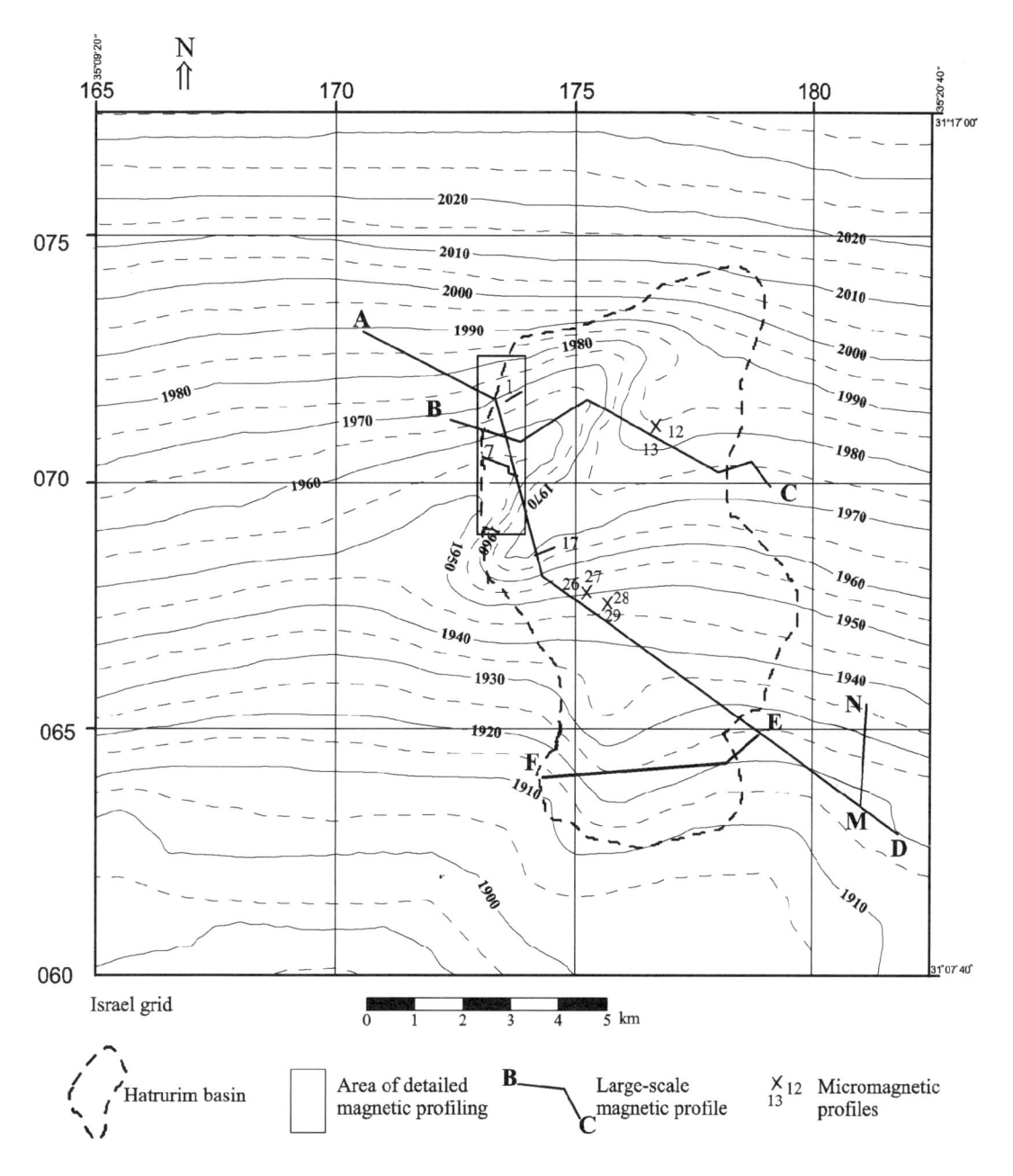

Figure 3. Location of some ground magnetic profiles and the Hatrurim Basin on the map of aeromagnetic ΔT anomalies; contour interval is 10 nT (modified from Folkman and Yuval, 1976). Micromagnetic profiles are shown bigger than they really are.

the observed aeromagnetic maximum (~15 nT). It is difficult to explain these discrepancies solely by extensive smoothing and displacement of aeromagnetic readings owing to the high speed of the aeromagnetic survey. It seems that the flight-lines (2 km spacing) with high clearance do not provide the necessary density of measurements that is necessary to represent the surface magnetic field. The ground measurements are complicated by the fact that the magnetic field can vary up to a few hundred nanoTesla (Fig. 4) within a few tens of centimeters. Such heterogeneity of surface or near-surface magnetic rocks probably would result in a near-zero magnetic effect registered at the level of an aeromagnetic survey. Thus, it is possible to suggest different origins of ground and aeromagnetic anomalies.

The south-north profile M–N is located outside of the Hatrurim Basin (Fig. 3); it crosses the Zuk-Tamrur 1 borehole, where oil-bearing deposits have been revealed in the Triassic Saharonim and Gevanim Formations at 1723–1730 m and 1860–1864 m, respectively (Goldfarb et al., 1982). The magnetic field of this profile is complicated by the regional Hebron anomaly (Rybakov et al., 1995). After removing this background, a clear local magnetic

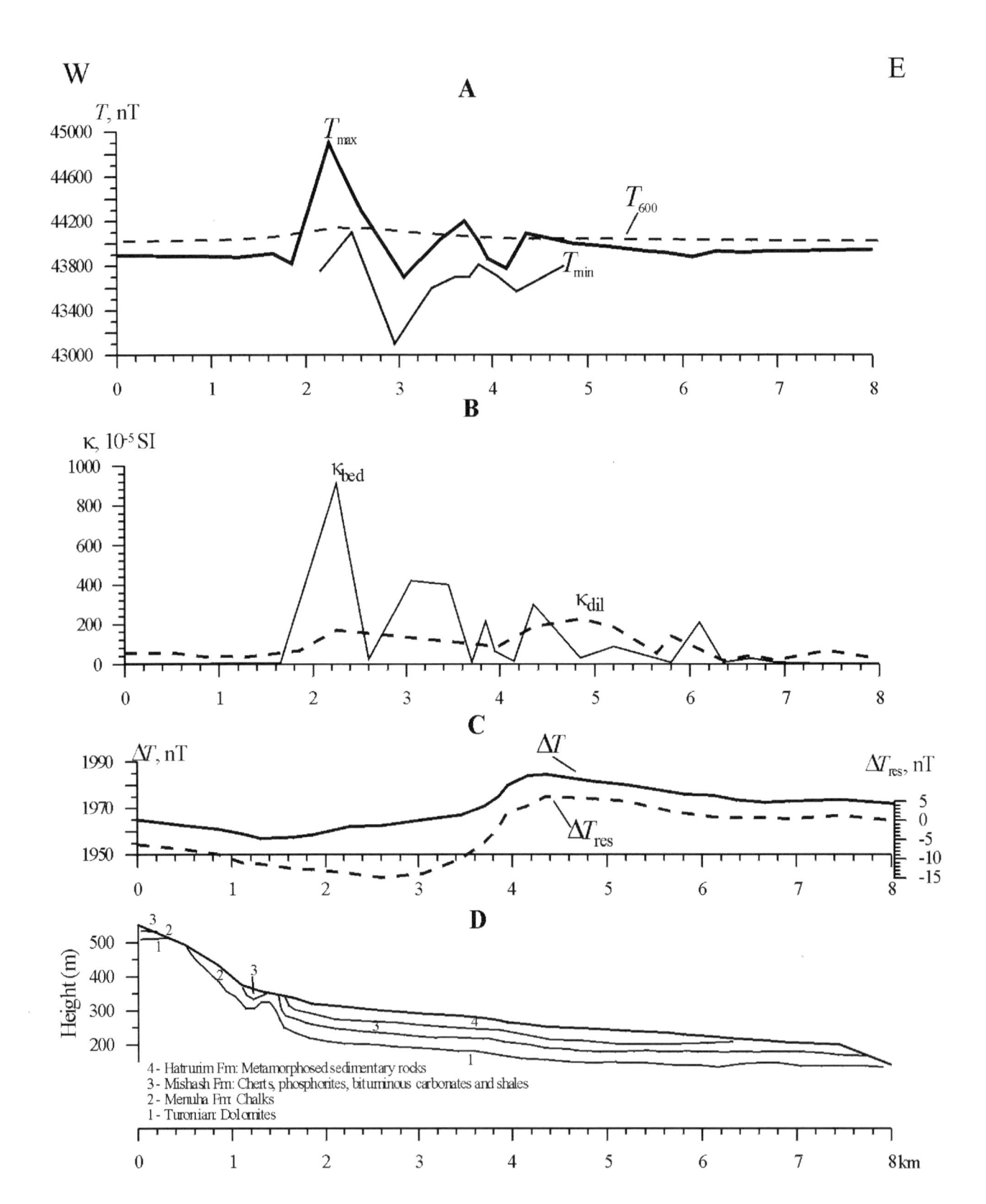

Figure 4. Large-scale profile B–C. (A) Ground T measurements. (B) Ground κ measurements. (C) Aeromagnetic data (ΔT, ΔT_{res}). (D) Geological section. T_{max} and T_{min} are the maximum and minimum of T fields measured in the vicinity of the same point of measurements of the profile, respectively. T_{600} is calculated T anomaly at 600 m above Earth's surface. κ_{bed} and κ_{dil} are magnetic susceptibilities of the bedrock and overburden (diluvium), respectively. ΔT_{res} is the difference between averaged and observed fields derived from data of Folkman and Yuval (1976). Vertical exaggeration of the geological section is 3.6.

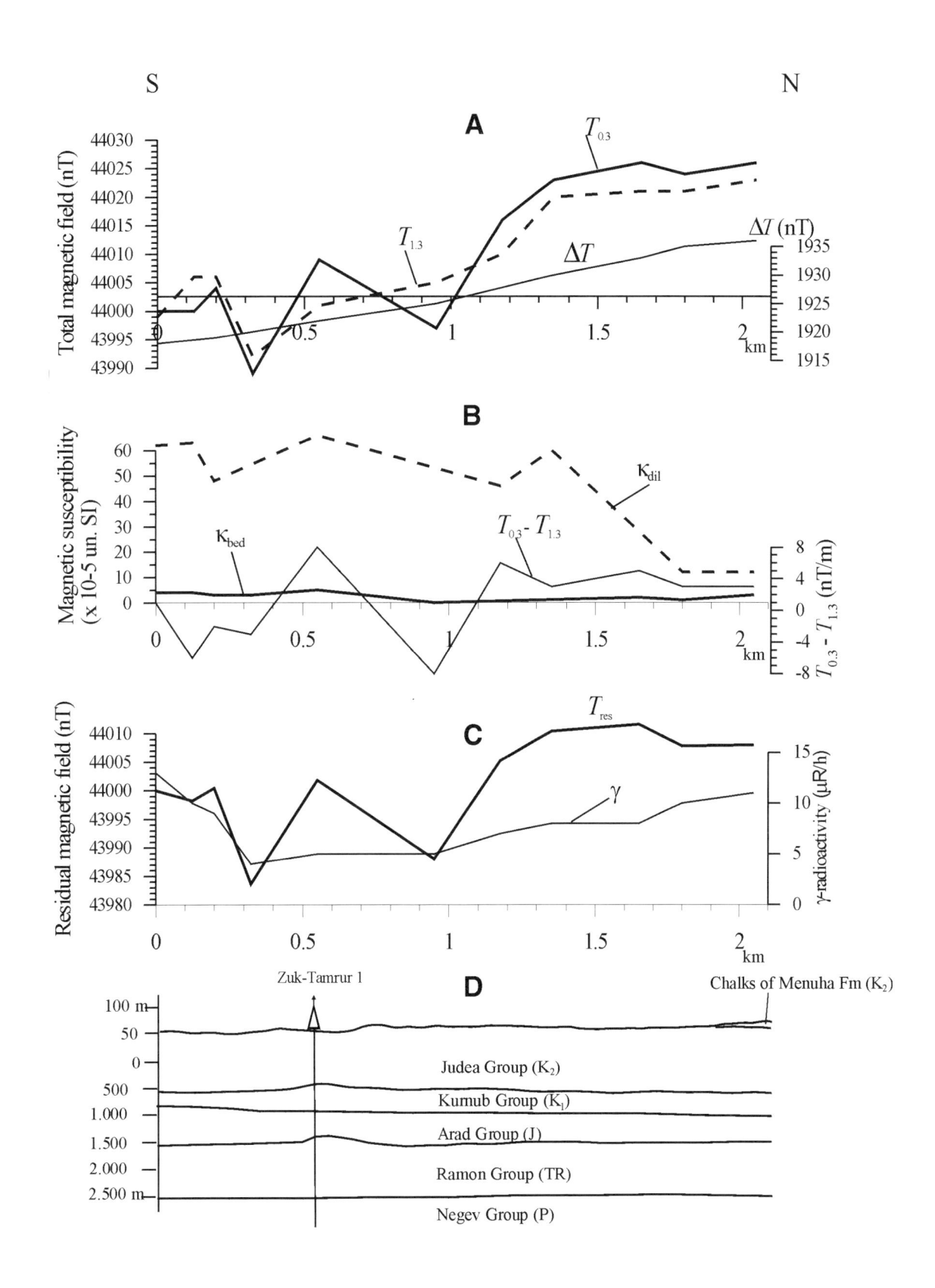

Figure 5. Large-scale profile M–N across Zuk-Tamrur 1 borehole. (A) Ground measurements at the level 0.3 m ($T_{0.3}$) and 1.3 m ($T_{1.3}$), and aeromagnetic (ΔT) field derived from the data of Folkman and Yuval (1976). (B) Magnetic susceptibilities of the bedrock (κ_{bed}) and diluvium (κ_{dil}), respectively, and vertical gradient of T field ($T_{0.3} - T_{1.3}$). (C) $\Delta T_{0.3}$ field corrected for effect of Hebron magnetic anomaly (ΔT_{res}), and γ-ray intensity (γ). (D) Geological section. Vertical exaggeration of the geological section is 5.45.

anomaly coincides with the revealed location of hydrocarbons at depth (Fig. 5D). The possible relation of this ground magnetic anomaly to hydrocarbon seepage could be supported indirectly by the association with a radioactivity minimum (Fig. 5C). Such minima are typical features of many petroleum pools (e.g., LeSchack, 1997).

Detailed Survey

The local magnetic pattern was characterized by detailed ground profiles. These profiles in the Hatrurim Basin usually detected magnetic anomalies over surface and/or near-surface Mottled Zone rocks. In profile 17, located at the epicenter of the northeast-oriented aeromagnetic anomaly (Fig. 3), we did not see deep sources (Fig. 6). The range of magnetic anomalies and susceptibility were commonly higher for high-grade metamorphic rocks and lower for low-temperature hydrothermal (LT) rocks (Fig. 6). Relatively deep sources were revealed at the Ma'ale Adumim area along profile 25 located at the "Tlat Adam" site (north of the Ma'ale Adumim–Mitzpe Jericho highway). Quantitative interpretation in this profile was performed for both ΔT maxima observed (Fig. 7). Analysis by the method of characteristic areas indicates that the centers of the magnetic sources (approximated by the horizontal circular cylinder) are at depths of 12 m and 14 m with effective magnetizations of 1100 mA/m and ~900 mA/m for the southwestern and northeastern bodies, respectively. Alternatively, the location of the well-known "Reford's point" (e.g., Khesin et al., 1996) suggests the possibility that the southwestern body also can be approximated by a thin vertical bed. The method of characteristic points shows that the upper edge of the vertical bed is at a depth of 7 m (Fig. 7), and its effective magnetization is 1600 mA/m.

Micromagnetic Survey

Further magnetic details were revealed by micromagnetic profiles. For example, profile 25 (Fig. 7) is composed of three parts (from west to east): WSW–ENE (azimuth 60°), WSW–ENE (azimuth 65°), and WNW–ESE (azimuth 110°). The central part is micromagnetic profile 25a (Fig. 8).

Profile 25a (the Ma'ale Adumim area) is located over an outcropping Mottled Zone section along a road cliff with a vertical plane height of 3.6 m. Magnetic measurements along the profile (Figs. 8A and 8B) were combined with κ measurements on a vertical plane. Magnetic susceptibility of the section was measured in situ along nine vertical lines on the road cliff (Fig. 8C). Magnetic susceptibility at surface profile 25a was in the range of 50 × 10^{-5} SI (Fig. 8B), whereas κ values increased to 1800 × 10^{-5} SI at depth (Fig. 8C). The vertical map obtained for the κ distribution and the geological section (Fig. 8D) were used for further quantitative interpretation of T anomalies. The initial model of the medium (a first approximation of the medium for subsequent iterations in the forward modeling) was composed using inversion results described already. The magnetic susceptibility of causative bodies was calculated by Equation 2, using a $\mathbf{T}_0$ value of ~44,000 nT, and $\mathbf{J}_e$ was given by the definition in the preceding subsection.

T and Z measurements in profile 25a delineated two magnetic anomalies with a magnitude of up to 900 nT (Fig. 8A). These anomalies were complicated by a local anomaly with a magnitude of up to 100 nT over the zone of highly magnetic rocks. Small T vertical gradients (Fig. 8B) reflect a deep magnetic source. The related magnetic sources on the profile appear to be buried at depth. The inversion of the data (Fig. 7) shows two bodies with effective magnetization of ~1 A/m underlying the section of profile 25a at a slightly greater depth. Such a model is defined more precisely by forward modeling that suggests ellipsoid shaped sources with vertically situated long axis (Fig. 9).

At the Hatrurim Basin, theoretical analysis of ΔZ and T fields along micromagnetic profiles of different orientation shows that the model of a horizontal circular cylinder (i.e., linear dipole) is in agreement with the measured data. The magnetic sources must be near-surface bodies; their near-surface location also is supported by the observation that the T field decreases rapidly with an increase in sensor altitude. For example, profile 22 is located along the Arad–Dead Sea highway at the Hatrurim crossroad, in the eastern end of profile 7 of the detailed magnetic prospecting (Fig. 3). Magnetic anomalies here (Fig. 10A) do not have significant correlation with local magnetic altered high-grade (AHG) metamorphic rocks (Fig. 10C). Measurements of magnetic susceptibility κ_{bed} in the section along the highway wall show that some of these AHG metamorphic rocks have κ_{bed} values up to 1800 × 10^{-5} SI, whereas κ_{dil} values are only up to 200 × 10^{-5} SI (Fig. 10B).

It is evident that local AHG metamorphic bodies are not identical to the sources of the magnetic anomalies. Therefore, we performed forward modeling of the southern part of profile 22. The magnetic sources were located within a slab with a base of 40 m × 40 m and a height of 6 m. The southern part of profile 22 is located in the center of the top of the slab. Figure 11 illustrates the magnetic field ΔT created by two source bodies that are partly exposed along this part of profile 22 (Fig. 10C). The southern body was characterized by apparent κ of 3000 × 10^{-5} SI, and the northern body was characterized by apparent κ of 800 × 10^{-5} SI. Figure 11 demonstrates the small offset between observed and calculated data that corroborates the feasibility of the model.

Thus, magnetic fields in profile 25a within the Ma'ale Adumim area suggest deep subvertical magnetic sources, whereas the magnetic anomalies in profile 22 within the Hatrurim Basin were caused by near-surface subhorizontal bodies. Such bodies, however, could be fragments of deeper sources. Vapnik and Sokol (2006) revealed explosion breccias and diatremes (pipe-like bodies) in the Hatrurim Basin. Four micromagnetic and gamma profiles (26–29; Fig. 3) across two diatremes of ~5 m in diameter show that the borders of these structures are reflected by T maxima of several hundred nanoTesla (can reach up to 1500–2000 nT), whereas minima of T and gamma-radioactivity were observed over their centers (Fig. 12). The patterns of magnetic

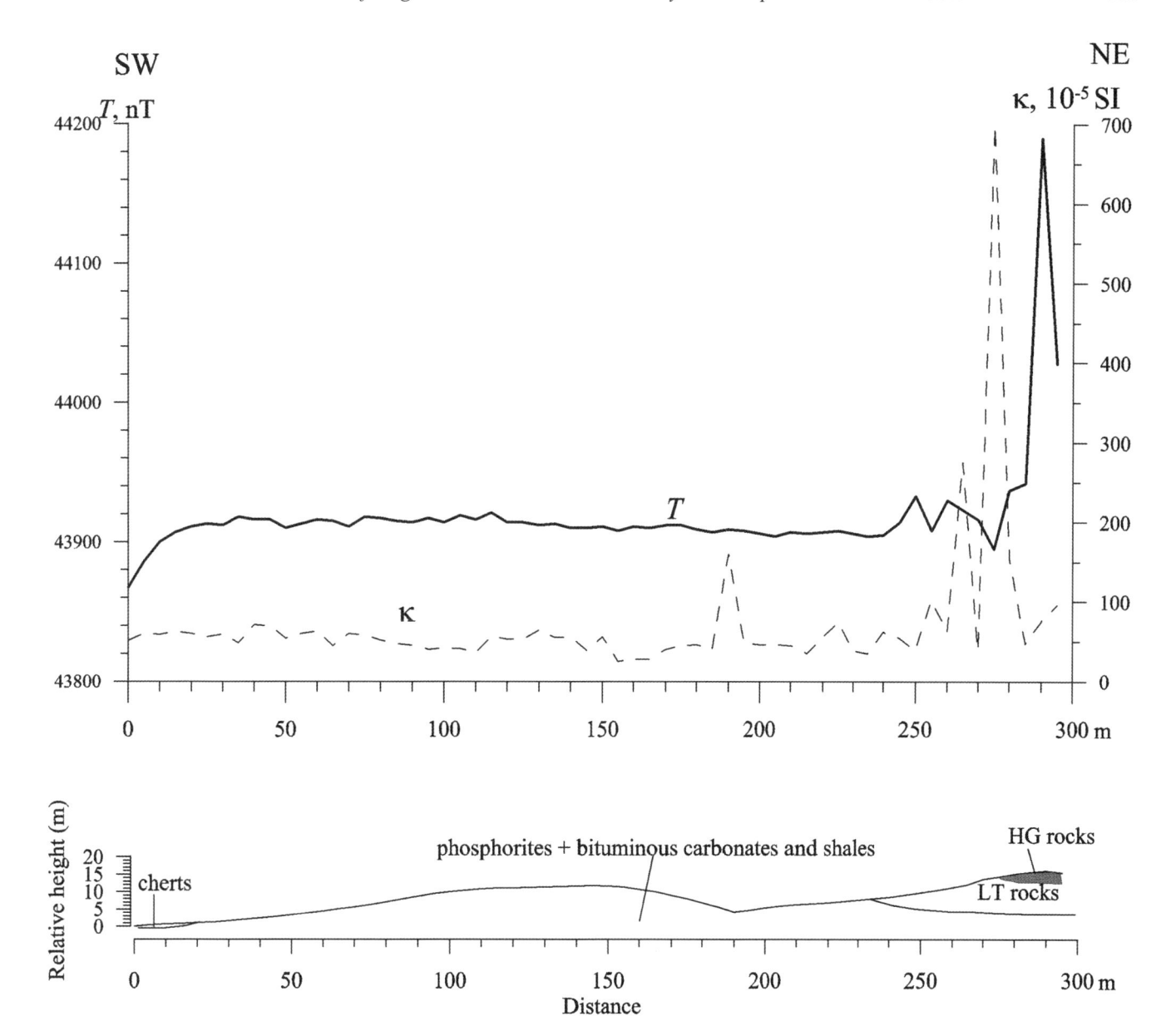

Figure 6. Comparison of magnetic and geological data along detailed profile 17. Note: Discordance of κ anomalies relative to T anomalies probably is caused by the local high-grade (HG) metamorphism relicts within low-temperature hydrothermal (LT) rocks and displacement of magnetic diluvium along the slope according to local morphology of the profile.

anomalies that were obtained over the diatremes are similar to those of pipe-like bodies in other regions. For example, the magnetic anomaly over the diatreme in profile 26 (Fig. 13A) is similar to the anomaly over a kimberlitic pipe (Fig. 13B) in the Yakut region (East Siberia), although the scales and the origins of these bodies are different.

DISCUSSION

In many combustion metamorphism zones, magnetic investigations of the strata adjacent to the burned rocks show that the clinker—the by-product of the thermal metamorphism—is magnetic (Cisowski and Fuller, 1987; Sternberg and Lippincott, 2004). Related surface anomalies are characterized by high wave numbers and magnitudes of up to 1700 nT. Such characteristics are similar to the magnetic features of Mottled Zone rocks, which were subjected to thermal metamorphism.

However, scattered, local surface and near-surface bodies of Mottled Zone rocks with high magnetization cannot be the source of the aeromagnetic anomalies that are observed at 600–700 m above Earth's surface. This source must have a more regional origin. It is possible to calculate T anomaly at the level of the aeromagnetic survey by forward modeling. Using a median κ value of 200×10^{-5} SI, $\mathbf{J}_n$ close to $\mathbf{J}_i$ of the magnetic rocks of the

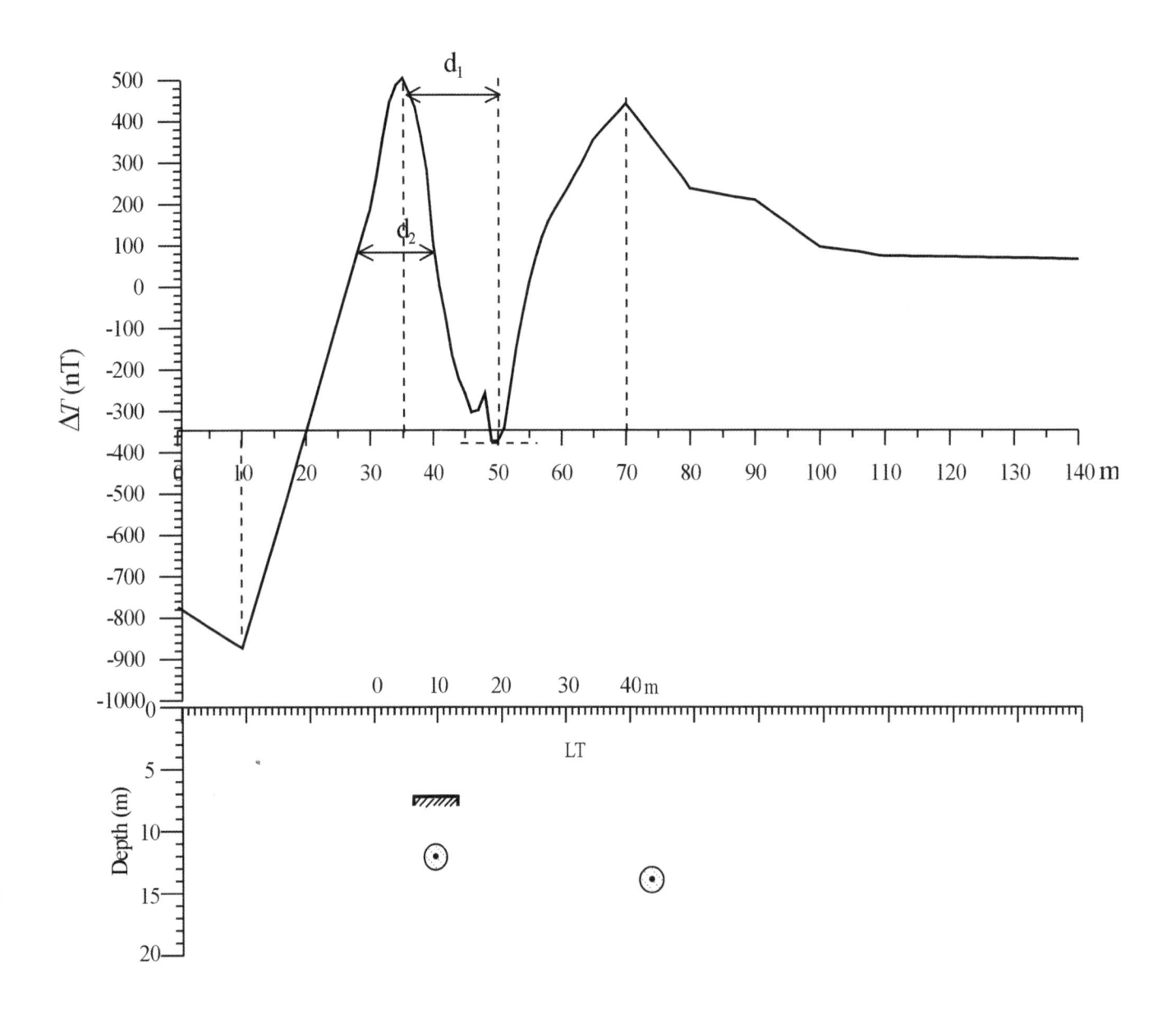

Figure 7. Increment of total magnetic field along detailed profile 25 and results of inverse problem solution (Ma'ale Adumim area). Note: Maxima of ΔT are displaced relative to the centers of magnetic sources because of their inclined magnetization. LT—low-temperature hydrothermal rocks.

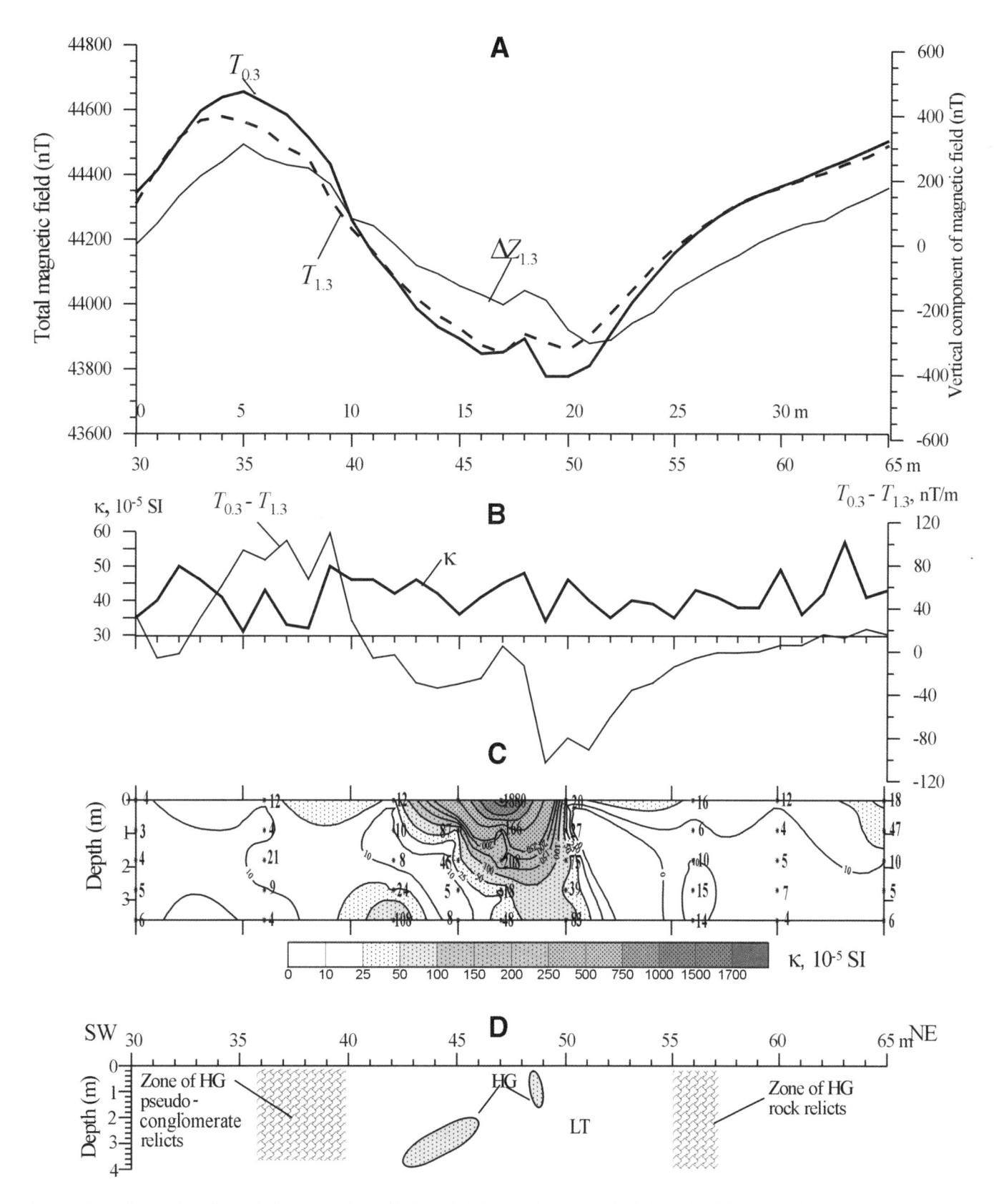

Figure 8. Central part of profile 25 (micromagnetic profile 25a). (A) Observed magnetic fields T and ΔZ. (B) T vertical gradient ($T_{0.3}-T_{1.3}$) and κ along the profile. (C) Vertical plane κ map. (D) Geological section. HG—high-grade rocks. LT—low-temperature hydrothermal rocks. Vertical and horizontal scales are equal.

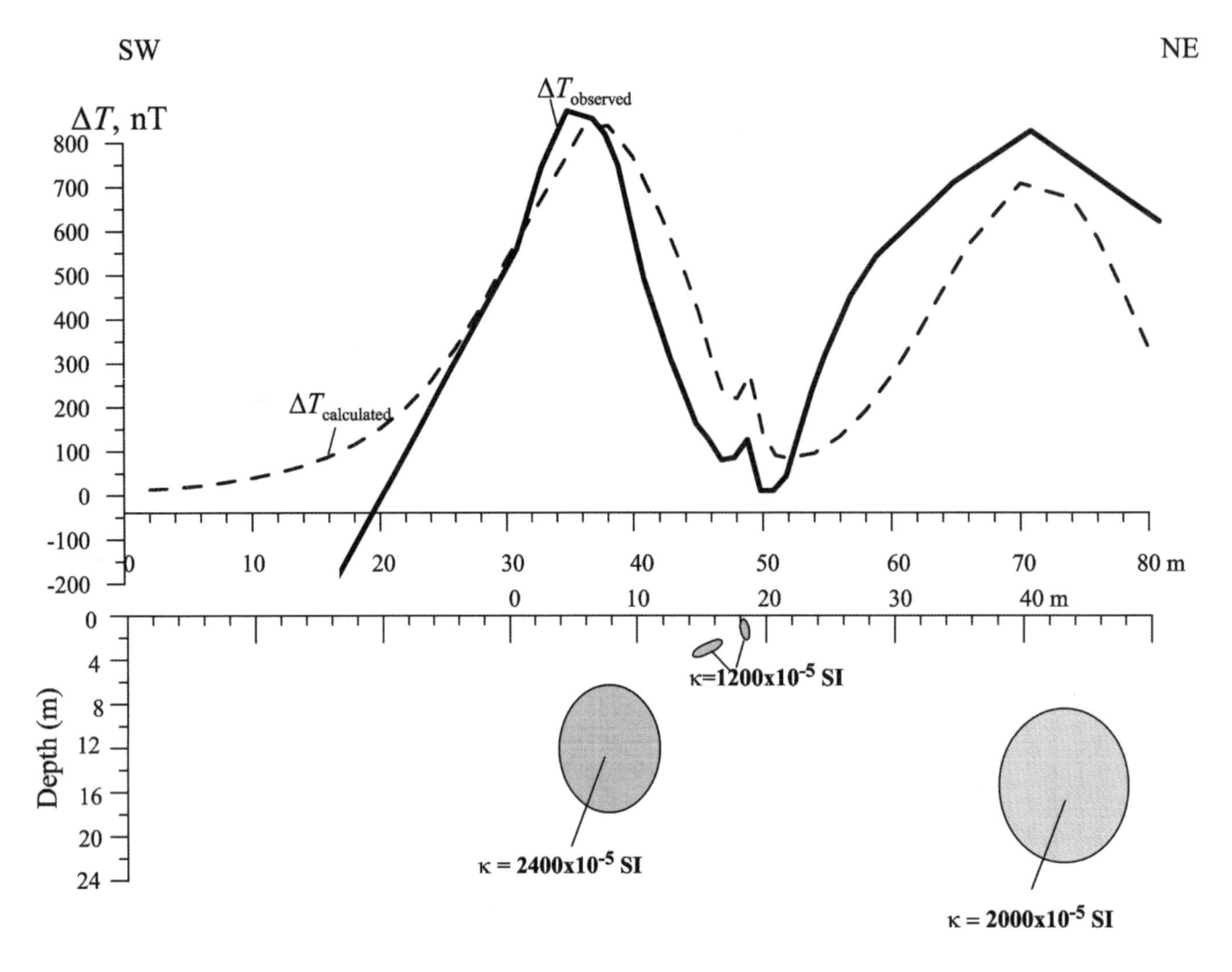

Figure 9. Subsurface magnetic model fit the observed data along profile 25a. Note: Sections of three-dimensional causative bodies are shown.

Mottled Zone (Khesin et al., 2005), and Equations 1 and 2, we calculated the J_e value as 0.2 A/m. This means that apparent κ for modeling is 400×10^{-5} SI. The source of the northeast-trending aeromagnetic anomaly can be approximated by a magnetic sheet with a length of 1800 m, a width of 800 m, and a thickness of 100 m. A magnetic sheet with such parameters can cause a T anomaly of up to 40 nT at 700 m above Earth's surface. Thus, the observed aeromagnetic anomaly can be caused by the cumulative effect of a large number of magnetic bodies that form a significant volume. The extension of possible magnetic masses to a greater depth and the substantial volume reflect large-scale magnetization processes rather than surface combustion.

Other contradictions to a hypothesis of surface combustion of bituminous rocks include scarcity of organic matter in the Hatrurim Formation. The spatial distribution of oil shales does not completely coincide with outcrops of Mottled Zone rocks. The combustion of oil shales does not provide high-temperature metamorphic process. The furnace temperature within an oil shale boiler is only 800°C (Givoni, 1991). The combustion temperature under natural conditions is lower, whereas some metamorphic minerals and associations of the Hatrurim Formation were formed at temperatures as high as 1100 °C to 1200 °C (Gross, 1977; Matthews and Kolodny, 1978; Vapnik et al., 2004).

In addition, the observed amplitudes of ground magnetic anomalies over Mottled Zone rocks are as high as a few thousand nanoTesla (Khesin and Itkis, 2002; Khesin et al., 2005). These values exceed any known anomalies caused worldwide by surface burning. Linford and Canti (2001) performed field experiments with the burning of fuel hosted by sand and clay substrates. During the burning, the clinker caused magnetic anomalies of only a few hundred nanoTesla.

Further, only one magnetization component may be explained by the conventional concept. The magnetization origin evidently is complex and requires examination of other hypotheses. The pattern of aeromagnetic anomalies suggests a possible relationship of the anomaly sources with structural highs and the

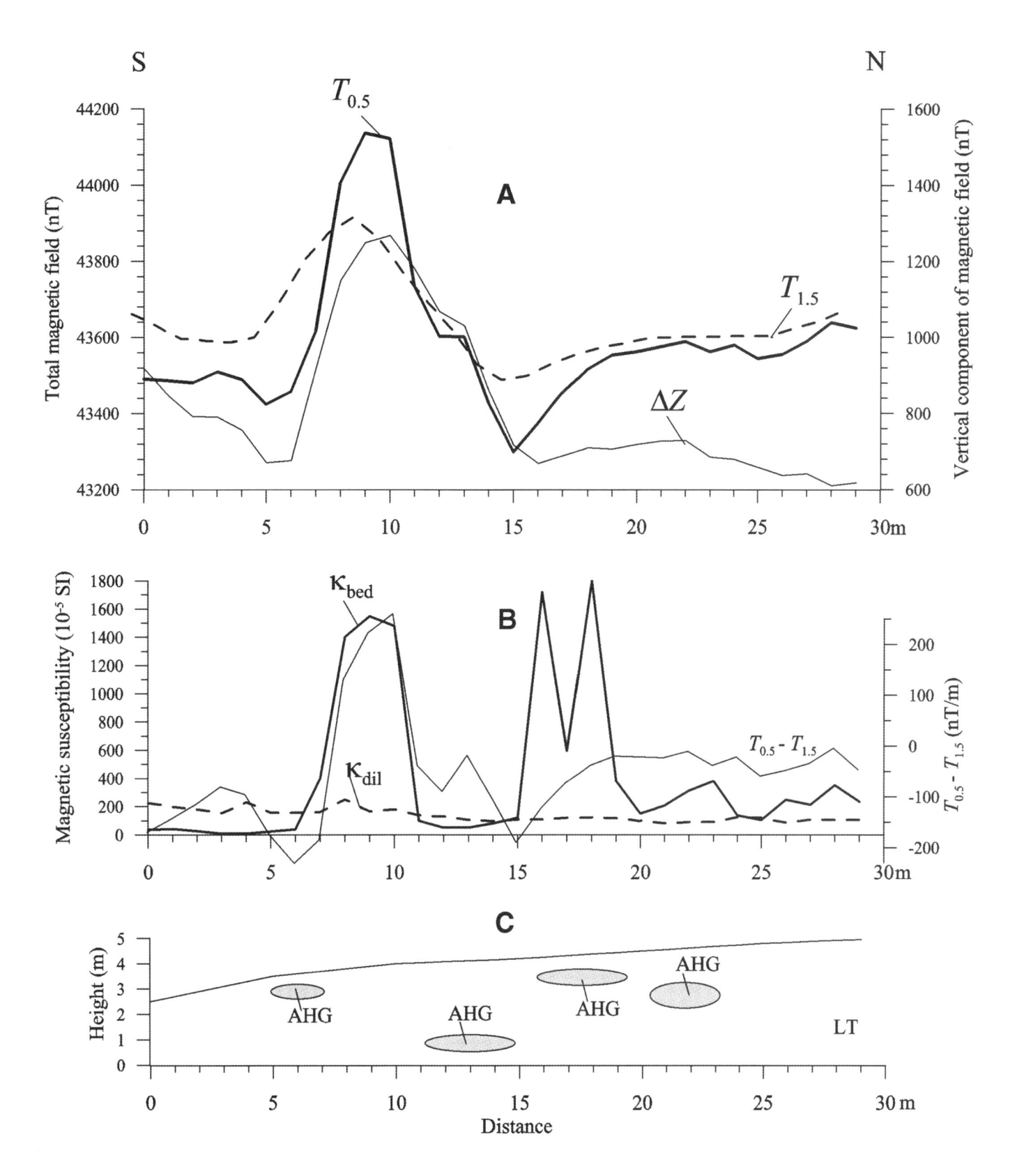

Figure 10. Micromagnetic and geological data along profile 22 (Hatrurim Basin). (A) Observed magnetic fields ($T_{0.5}$ and $T_{1.5}$ are the total fields measured at 0.5 m and 1.5 m, respectively; ΔZ is the vertical component of the magnetic field). (B) Magnetic susceptibilities of the bedrocks (κ_{bed}) and diluvium (κ_{dil}), and vertical gradient of T field ($T_{0.5}-T_{1.5}$). (C) Geological section. LT—low-temperature hydrothermal rocks, AHG—altered high-grade rocks. Vertical and horizontal scales are equal.

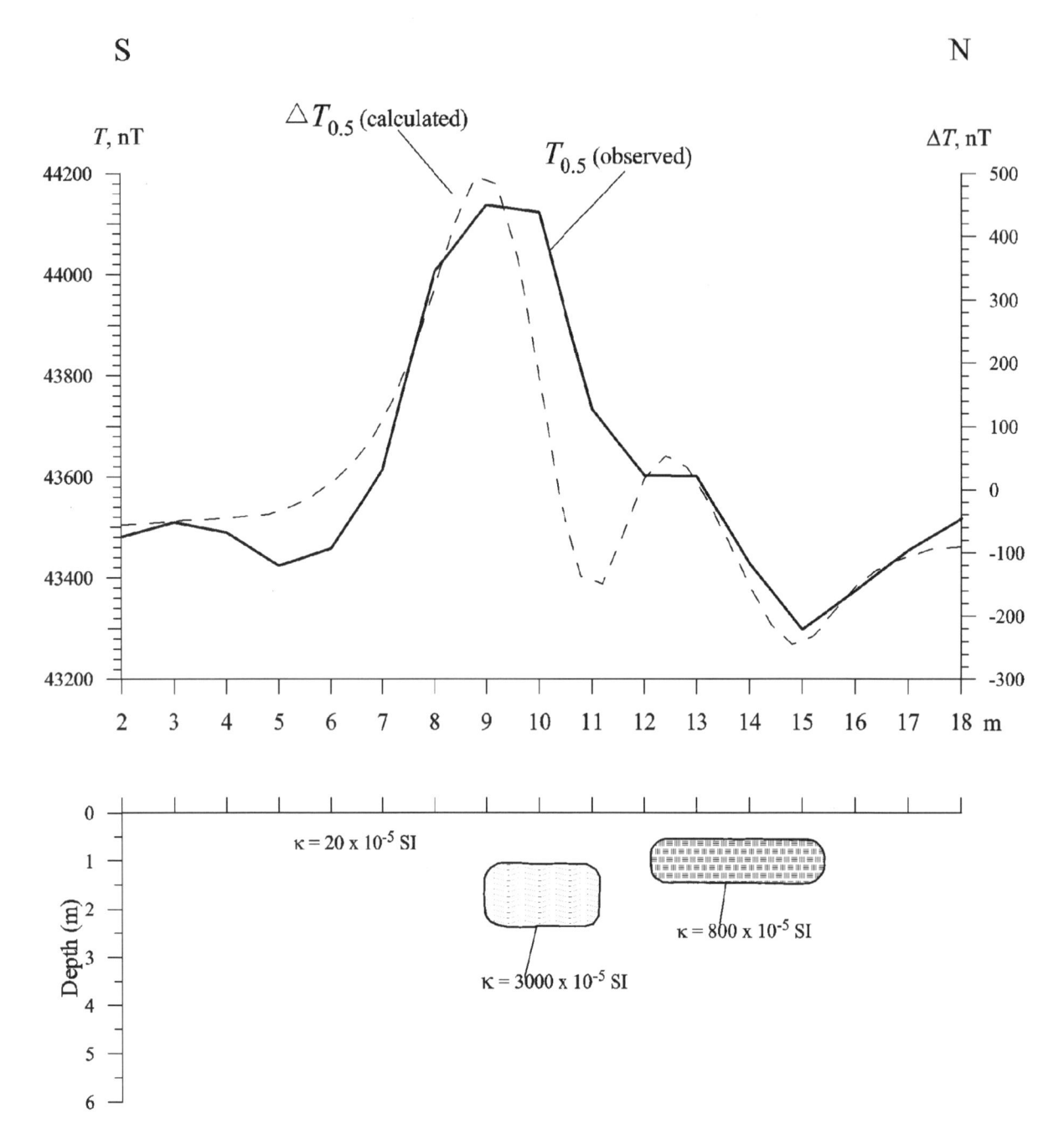

Figure 11. Subsurface magnetic model fit the observed data along profile 22. Note: Sections of three-dimensional causative bodies are shown.

fault zone; however, deep magnetic contrasts in the geological section here are unknown, so the observed anomalies cannot be explained by structural features. At the same time, these features may be hydrocarbon traps and a fluid channel, respectively. Rising flow could have caused some magnetization of the sedimentary rocks. Additional lines of evidence support this suggestion. Thermocatalytic methane has been found in the Dead Sea segment of the rift valley (Nissenbaum et al., 1999). The Zohar gas pool (Gilboa et al., 1993) within the carbonates of the Jurassic Zohar Formation coincides, in part, with the northeast-trending anomaly zone; oil and gas pools also are known in the Gurim Dome, where the aeromagnetic maximum is located.

Other natural or anthropogenic processes can produce magnetic anomalies over oil fields (Schumacher, 1996). Ellwood and Burkart (1996) documented that alternation of redox potential is typical of Mediterranean soils, where pore waters change from reducing (during cool, wet winters) to oxidizing conditions (during hot, dry summers). The conversion of iron oxides from

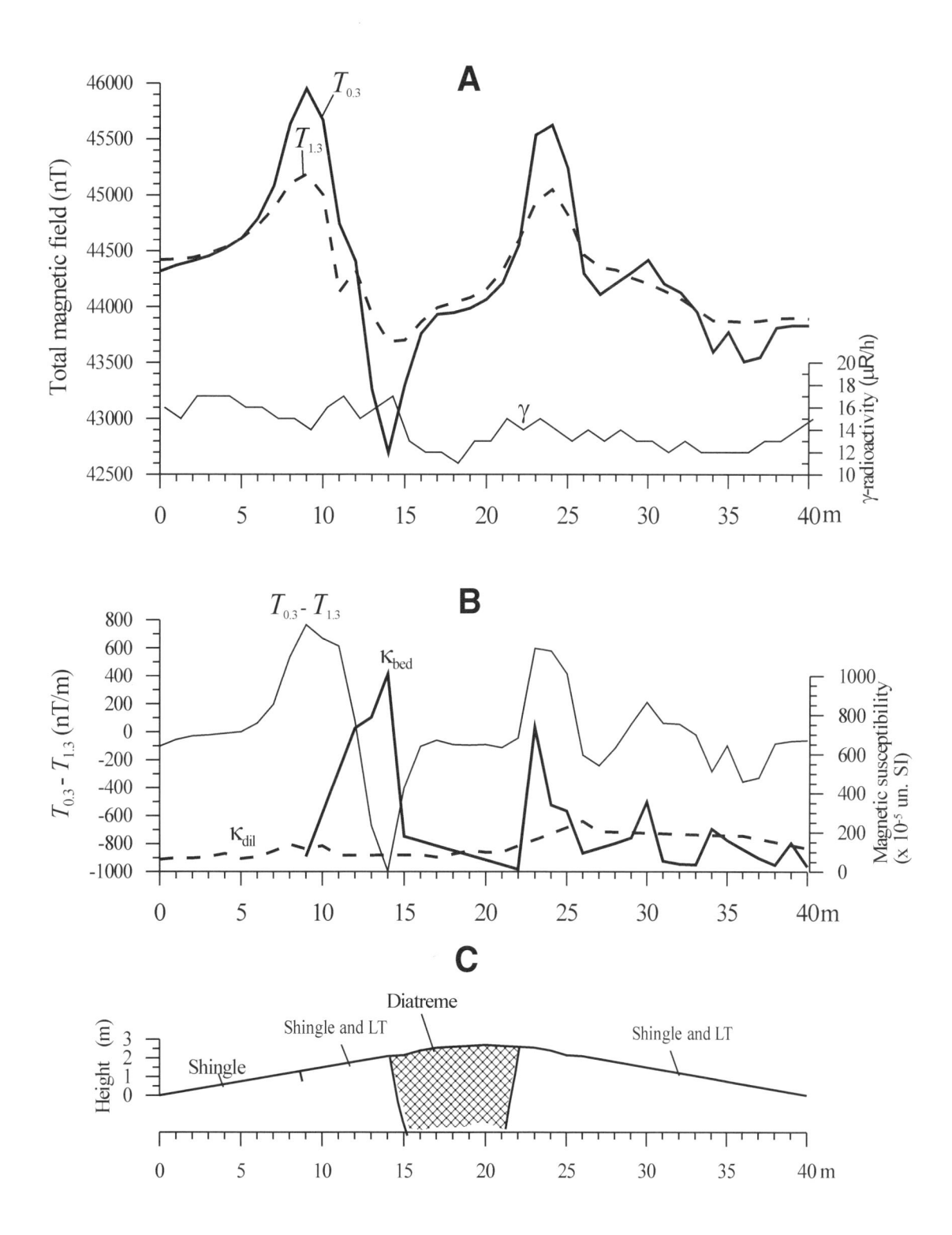

Figure 12. Micromagnetic, radioactivity, and geological data along profile 29 (Hatrurim Basin). (A) Observed magnetic fields ($T_{0.3}$ and $T_{1.3}$ are the total fields measured at 0.3 m and 1.3 m, respectively; γ is intensity of gamma [γ] rays). (B) Magnetic susceptibilities of the bedrocks (κ_{bed}) and diluvium/alluvium (κ_{dil}), and vertical gradient of T field ($T_{0.3}-T_{1.3}$). (C) Geological section. LT—low-temperature hydrothermal rocks.

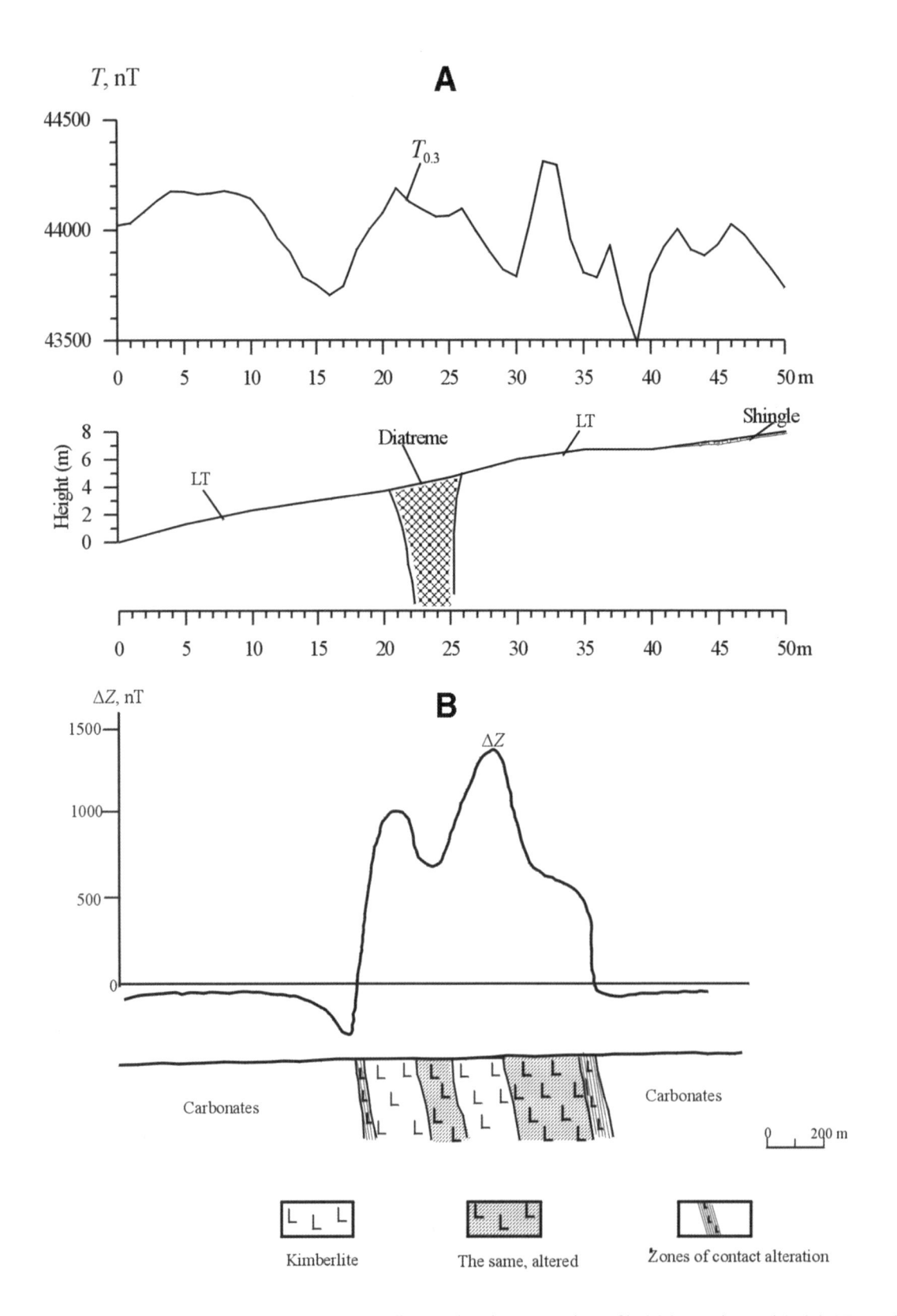

Figure 13. Comparison of observed magnetic fields. (A) T over diatreme in micromagnetic profile 26 (Hatrurim Basin). (B) ΔZ over kimberlitic pipe "Zarnitsa" ("summer lightning") in East Siberia (modified from Sokolov, 1966). LT—low-temperature hydrothermal rocks.

weakly magnetic phases, such as hematite to maghemite, causes higher magnetic susceptibility in soils compared with sedimentary bedrock. We observed such a relationship between the susceptibility of soils and bedrock of Mottled Zone areas in Israel many times during in situ κ measurements.

The direct indications of gas flows were found in the Hatrurim Basin. Vapnik et al. (2004) revealed particular rocks in the Hatrurim Basin with evident channels of high-temperature gas flow. Vapnik and Sokol (2006) identified several tens of diatremes (up to 10 m in diameter) in the Hatrurim Basin that were formed during deep-rooted gas accumulation, gas explosion, and rock brecciation. Surface or subsurface high-temperature gas combustion could provide the heat for metamorphic reactions in the Mottled Zone. Magnetic data suggest the possibility of several diatremes. Micromagnetic profile 26 (Fig. 13) shows a few magnetic anomalies (e.g., intervals 0–15 m and 39–50 m) that form a pattern similar to the anomaly over outcropping diatreme (interval 17–30 m).

Some facts support a relationship between magmatic activity and magnetization origin. The mineral assemblage of the Mottled Zone is typical for contact metamorphism zones between igneous intrusions and siliceous carbonate country rocks (Kolodny, 1979; Ron and Kolodny, 1992). The proximity of some Mottled Zone areas to outcropping (Bartov, 1990) and subsurface (Rybakov et al., 1995) magmatic bodies led to the suggestion that magnetization of the Hatrurim Formation also could be connected with magmatic processes alongside hydrocarbon seepage (Khesin and Itkis, 2002). Outcropping young basalts are widespread near Maqarin and Daba Siwaqa Mottled Zone outcrops in Jordan (Fig. 1). Nissenbaum et al. (1999) noted thermally metamorphosed carbonate rocks in the Maqarin area, where the methane and the heavier hydrocarbons could have been formed by a rapid "heat shock" of the bituminous marls by igneous intrusions. However, the Zohar 8 deep well (Fig. 2C) penetrated acid volcanics of the crystalline basement at a depth of 3.2 km (Frolov et al., 2002); such basement rocks cannot provide a source for the observed T anomaly. For a more profound evaluation of magmatic factors in the metamorphism studied, more detailed information on the spatial distribution of magnetic sources (in plane and at depth) must be obtained.

Lightning strikes are observed often in Israel. Sternberg et al. (1999) described the NRM analysis for distinguishing related magnetization at the archaeological Tel Ashkelon site in Israel. During our micromagnetic radial survey at a site in the Hatrurim Basin (Khesin et al., 2005), two oriented samples were selected at the northeastern and southwestern margins of a small outcrop of high-grade metamorphic rocks along micromagnetic profile 12 (Fig. 3). The difference in the declination of $\mathbf{J}_n$ vectors for these samples was 91°. The magnetic pattern resembles one that results from a lightning strike. However, fulgurites have yet to be found. Thus, a search for fulgurites in the Hatrurim Formation and evaluation of a possible concentric $\mathbf{J}_n$ spatial distribution in situ are required. It is also possible to determine the type of magnetization of oriented samples using coercivity spectra. Such studies can verify the hypothesis of the influence of lightning strikes on the magnetic properties of rock outcrops.

The possibility of revealing technogenic and anthropogenic magnetic sources also may be indicated. Revealed rock-forming minerals and their assemblages may be exotic and are uncommon in nature, but they are known as industrial products, whereas the magnetic patterns of some Mottled Zone and archaeological sites are similar. However, the Mottled Zone areas are too big, and no archaeological remains (e.g., furnaces) have been found.

CONCLUSIONS

Quantitative interpretation indicates diverse sources for the observed magnetic anomalies in Mottled Zone areas in Israel. The combustion metamorphism model explains some of the magnetic characteristics of the Hatrurim Formation, but it does not account completely for the complexity of the magnetic data. It is clear that additional source(s) at greater depth and of more regional extent are required to account for the complex magnetic characteristics of the Mottled Zone. A comprehensive analysis of the variable magnetic sources that are involved and their origins (e.g., hydrocarbons flow from deep traps, magmatic activity, lightning strikes) requires (1) further integrated studies of the regional distribution of magnetic sources using an aeromagnetic helicopter survey with constant small clearance on a 1:25,000–1:10,000 scale west and east of the Dead Sea fault zone (Israel and Jordan); and (2) detailed studies of the aeromagnetic anomalies using ground survey, determination of $\mathbf{J}_n$ spatial distribution, particularly at the outcrops of the high-grade metamorphic rocks, and magnetic mineralogy.

ACKNOWLEDGMENTS

The authors are grateful to the Earth Sciences Research Administration of the Ministry of National Infrastructures of Israel for supporting this study. We also thank our colleagues Y. Vapnik for active participation in major parts of this study and R. Shagam for preliminary discussion. Further, we thank M. Fuller, M. Rybakov, and an anonymous reviewer for valuable suggestions, and G.B. Stracher, editor of this book, for useful remarks and immutable support.

REFERENCES CITED

Aldana, M., and Costanzo-Alvarez, V., 2003, Magnetic and mineralogical studies to characterize oil reservoirs in Venezuela: The Leading Edge, v. 22, no. 6, p. 526–529, doi: 10.1190/1.1587674.

Bartov, Y., 1990, Israel Geological Map: Jerusalem, Survey of Israel, scale 1:500,000.

Bentor, Y.K., and Vroman, A., 1960, The Geological Map of Israel, Series A—The Negev, Sheet 16: Mount Sdom (with Explanatory Text) (second edition): Jerusalem, Geological Survey, scale 1:100,000.

Bentor, Y.K., Kastner, M., Perlman, I., and Yellin, Y., 1981, Combustion metamorphism of bituminous sediments and the formation of melts of granitic and sedimentary composition: Geochimica et Cosmochimica Acta, v. 45, p. 2229–2255, doi: 10.1016/0016-7037(81)90074-0.

Burg, A., Starinsky, A., Bartov, Y., and Kolodny, Y., 1991, Geology of the Hatrurim Formation ("Mottled Zone") in the Hatrurim Basin: Israel Journal of Earth Sciences, v. 40, p. 107–124.

Burg, A., Kolodny, Y., and Lyakhovsky, V., 1999, Hatrurim—2000, The "Mottled Zone" revisited, forty years later: Israel Journal of Earth Sciences, v. 48, p. 209–223.

Cisowski, S.M., and Fuller, M., 1987, The generation of magnetic anomalies by combustion metamorphism of sedimentary rock, and its significance to hydrocarbon exploration: Geological Society of America Bulletin, v. 99, p. 21–29, doi: 10.1130/0016-7606(1987)99<21:TGOMAB>2.0.CO;2.

Domzalsky, W., 1967, Aeromagnetic Survey of Israel: Interpretation: Lod, Israel, Institute of Petroleum Research and Geophysics Report SMA/482/67, 63 p.

Donovan, T.J., Forgey, R.L., and Roberts, A.A., 1979, Aeromagnetic detection of diagenetic magnetite over oil fields: American Association of Petroleum Geologists (AAPG) Bulletin, v. 63, no. 2, p. 245–248.

Ellwood, B.B., and Burkart, B., 1996, Test of hydrocarbon-induced magnetic patterns in soils: The sanitary landfill as laboratory, in Schumacher, D., and Abrams, M., eds., Hydrocarbon Migration and its Near-Surface Expression: American Association of Petroleum Geologists (AAPG) Memoir 66, p. 90–98.

Ermokhin, K.M., 1998, Dipole sources' method for the decision of inverse problem in electrical and magnetic prospecting: Transactions of 60th EAGE (European Association of Geoscientists and Engineers) Conference and Technical Exhibition, Leipzig, p. 27.

Eventov, L., 2000, The nature and interpretation of geophysical and geochemical anomalies over oil and gas fields: The Leading Edge, v. 19, no. 5, p. 488–490, doi: 10.1190/1.1487234.

Folkman, Y., and Yuval, Z., 1976, Aeromagnetic Map: Jerusalem, Survey of Israel, scale 1:250,000.

Foote, R.S., 1996, Relationship of near-surface magnetic anomalies to oil- and gas-producing areas, in Schumacher, D., and Abrams, M., eds., Hydrocarbon Migration and Its Near-Surface Expression: American Association of Petroleum Geologists (AAPG) Memoir 66, p. 111–126.

Frolov, B.M., Ainemer, A.I., and Shenkman, A.L., 2002, Oil habitats in the northwestern Negev: On the evaluation of Triassic hydrocarbon potential in Israel: Israel Journal of Earth Sciences, v. 51, no. 1, p. 35–53, doi: 10.1560/HR51-6C5P-2V77-84QG.

Gilat, A., 1998, Hydrothermal activity and hydro-explosions as a cause of natural combustion and pyrolysis of bituminous rocks: The case of Pliocene metamorphism in Israel (Hatrurim Formation): Geological Survey of Israel Current Research, v. 11, p. 96–102.

Gilboa, Y., Fligelman, H., and Derin, B., 1993, Zohar-Kidod-Haqanaim fields, Israel, eastern Mediterranean basin, in Foster, N.H., compiler, Treatise of Petroleum Geology, Atlas of Oil and Gas Fields, Structural Traps VIII: Tulsa, American Association of Petroleum Geologists (AAPG), p. 129–152.

Givoni, D., 1991, Oil shale utilization in Israel: The first year of combustion demonstration plant operation, in Zamkov, S., ed., Proceedings, First Energy Conference Israel–Former USSR, Ben-Gurion University of the Negev, Beer-Sheva, Israel, 13–15 May 1991: Beer-Sheva, Israel, Ben-Gurion University of the Negev, p. 16–21.

Goldfarb, A., Fligelman, H., and Spivak, D., 1982, Zuk Tamrur 1 Testing and Completion: Tel Aviv, Israel, Oil Exploration (Investments) Ltd. Report 82/57, 80 p.

Gross, S., 1977, The Mineralogy of the Hatrurim Formation, Israel: Geological Survey of Israel Bulletin no. 70, p. 1–80.

Khesin, B., 1998, Effective magnetization of the Precambrian in Sinai and southern Israel: Implication of new methods for ΔT field analysis: Israel Journal of Earth Sciences, v. 47, no. 1, p. 47–60.

Khesin, B., and Feinstein, S., 2005, Phanerozoic rock magnetization in southern and central Israel: Israel Journal of Earth Sciences, v. 54, no. 2, p. 97–109, doi: 10.1560/XW56-P42U-YPHF-XYRB.

Khesin, B.E., and Itkis, S.E., 2002, Revision of aeromagnetic data: Ground magnetic investigations of altered sedimentary rocks (Hatrurim Basin, Israel): Proceedings of the Estonian Academy of Sciences–Geology, v. 51, no. 1, p. 16–32.

Khesin, B.E., Alexeyev, V.V., and Metaxa, K.P., 1983, The interpretation of magnetic anomalies under oblique magnetization and rugged topography: Moscow, Nedra, 289 p.

Khesin, B., Alexeyev, V., and Eppelbaum, L., 1996, Interpretation of Geophysical Fields in Complicated Environments: Modern Approaches in Geophysics, Volume 14: Dordrecht, Kluwer Academic Publishers, 368 p.

Khesin, B., Feinstein, S., Vapnik, Y., Itkis, S., and Leonhardt, R., 2005, Magnetic study of metamorphosed sedimentary rocks of the Hatrurim Formation, Israel: Geophysical Journal International, v. 162, p. 49–63, doi: 10.1111/j.1365-246X.2005.02630.x.

Kolodny, Y., 1979, Natural cement factory: A geological story, in Skalny, J., ed., Cement Production and Use: Rindge, New Hampshire, Engineering Foundation, U.S. Army Research Office, Conference Proceeding, p. 203–215.

LeSchack, L.A., 1997, Results of magnetic HGI and radiometric surveys in Western Canada: Oil & Gas Journal, May 19, p. 84–89, and May 26, p. 82–87.

LeSchack, L.A., and Van Alstine, D.R., 2002, High-resolution ground-magnetic (HRGM) and radiometric surveys for hydrocarbon exploration: Six case histories in Western Canada, in Schumacher, D., and LeSchack, L. A., eds., Surface Exploration Case Histories: Applications of Geochemistry, Magnetics, and Remote Sensing: American Association of Petroleum Geologists (AAPG) Studies in Geology, no. 48, and SEG (Society of Exploration Geophysicists) Geophysical References series, no. 11, p. 67–156.

Linford, N.T., and Canti, M.G., 2001, Geophysical evidence for fires in antiquity: Preliminary results from an experimental study: Archaeological Prospection, v. 8, p. 211–225, doi: 10.1002/arp.170.

Machel, H.G., 1996, Magnetic contrasts as a result of hydrocarbon seepage and migration, in Schumacher, D., and Abrams, M., eds., Hydrocarbon Migration and Its Near-Surface Expression: American Association of Petroleum Geologists (AAPG) Memoir 66, p. 99–109.

Matthews, A., and Gross, S., 1980, Petrologic evolution of the "Mottled Zone" (Hatrurim) metamorphic complex of Israel: Israel Journal of Earth Sciences, v. 29, p. 93–106.

Matthews, A., and Kolodny, Y., 1978, Oxygen isotope fractionation in decarbonization metamorphism. The Mottled Zone event: Earth and Planetary Science Letters, v. 39, p. 179–192, doi: 10.1016/0012-821X(78)90154-1.

Moticska, P., 1978, Generacion de magmas y autometamorfismo por combustion subterranea de carbon y de limolitas carbonasas en la formacion Marcelina, Perija: Venezuela Boletin de Geologia, v. 24, p. 184–217.

Nagata, T., 1953, Rock-Magnetism: Tokyo, Maruzen Co. Ltd, 225 p.

Nissenbaum, A., Faber, E., and Gerling, P., 1999, Hydrocarbon gases in the Dead Sea–Jordan River segment of the East African Rift Valley [abs.]: Dead Sea, Israel Geological Society, Annual Meeting, p. 62–63.

Pechersky, D.M., 1985, Petromagnetism and Paleomagnetism: Moscow, Nauka, 128 p.

Peirce, J.W., Goussev, S.A., Charters, R.A., Abercrombie, H.A., and DePaoli, G.R., 1998, Intrasedimentary magnetization by vertical fluid flow and exotic geochemistry: The Leading Edge, v. 17, no. 1, p. 89–92, doi: 10.1190/1.1437840.

Ron, H., and Kolodny, Y., 1992, Paleomagnetic and rock magnetic study of combustion metamorphic rocks in Israel: Journal of Geophysical Research, v. 97, no. B5, p. 6927–6939.

Rusinov, B.S., 1973, Magnetic anomalies and rock remanent magnetization, caused by magnetic field of atmospheric origin, in Ismail-Zadeh, T.A., ed., Matter of IX Conference on problems of constant magnetic field, rock magnetism and paleomagnetism, Part I: Baku, Academy of Sciences of USSR and Academy of Sciences of Azerbaijan SSR, p. 120–121.

Rybakov, M., Goldshmidt, V., Folkman, Y., Rotstein, Y., Ben-Avraham, Z., and Hall, J., 1994, Magnetic Anomaly Map: Survey of Israel, Jerusalem, scale 1:500,000.

Rybakov, M., Fleisher, L., and Goldshmidt, V., 1995, A new look at the Hebron magnetic anomaly: Israel Journal of Earth Sciences, v. 44, p. 41–49.

Saunders, D.F., and Terry, S.A., 1985, Onshore exploration using the new geochemistry and geomorphology: Oil & Gas Journal, September 16, p. 126–130.

Schumacher, D., 1996, Hydrocarbon-induced alteration of soils and sediments, in Schumacher, D., and Abrams, M., eds., Hydrocarbon Migration and Its Near-Surface Expression: American Association of Petroleum Geologists Memoir 66, p. 71–89.

Sneh, A., Bartov, Y., Weissbrod, T., and Rosensaft, M., 1998, Geological Map of Israel: Jerusalem, Geological Survey of Israel, scale 1:200,000, 4 sheets.

Sokolov, K.P., 1966, Geophysical Prospecting Methods: Leningrad, Nedra, 464 p.

Sternberg, R., and Lippincott, C., 2004, Magnetic surveys over clinkers and coal seam fires in western North Dakota: Geological Society of America Abstracts with Programs, v. 36, no. 5, abstract 15-9, p. 43.

Sternberg, R., Lass, E., Marion, E., Katari, K., and Holbrook, M., 1999, Anomalous archaeomagnetic directions and site formation processes at archaeological sites in Israel: Geoarchaeology International Journal (Toronto, Ontario), v. 14, no. 5, p. 415–439.

Stone, V.C.A., Fairhead, J.D., and Oterdoom, W.H., 2004, Micromagnetic seep detection in the Sudan: The Leading Edge, v. 23, no. 8, p. 734–737, doi: 10.1190/1.1786893.

Vapnik, Y.A., and Sokol, E.V., 2006, Explosion breccias and diatremes as key structures in the formation of the Hatrurim Formation [abs.]: Bet-Shean, Israel Geological Society, Annual Meeting Abstracts, p. 131.

Vapnik, Y.A., Zhivkovich-Rapoport, A.Kh., Sharygin, V.V., and Sokol, E.V., 2004, Paralavas in the Hatrurim Basin of Israel: Geological Society of America Abstracts with Programs, v. 36, no. 5, abstract 90-6, p. 227.

Wollenben, J.A., and Greenlee, D.W., 2002, Successful application of micromagnetic data to focus hydrocarbon exploration, *in* Schumacher, D., and LeSchack, L.A., eds., Surface Exploration Case Histories: Applications of Geochemistry, Magnetics, and Remote Sensing: American Association of Petroleum Geologists (AAPG) Studies in Geology, no. 48, and SEG (Society of Exploration Geophysicists) Geophysical References series, no. 11, p. 175–191.

MANUSCRIPT ACCEPTED BY THE SOCIETY 7 MARCH 2007

Printed in the USA

Detecting concealed coal fires

Hartwig Gielisch*
Deutsche Montan Technologie GmbH, Am Technologiepark 1, 45307 Essen, Germany

ABSTRACT

Throughout the world, many coal fires are currently burning out of control. In the People's Republic of China, ~750 coal fires are burning and depleting a significant amount of the country's energy supply. Emissions from the smoldering fires are polluting the soil, the groundwater, and the atmosphere. To protect the environment and the natural resources, the Chinese government has taken steps to control or extinguish these fires. In fact, the People's Republic of China has been fighting these coal fires since the foundation of the country in its present form, following the Chinese fire-fighting manual from 1953.

To extinguish a fire—or hot spot, which the fire location is often called—its location must be known with a high degree of accuracy. Hot spots have been successfully located in Xinjiang and Inner Mongolia, People's Republic of China, by combining conventional and modern exploration methods. After the identification of a hot spot, phase terrain and thermal anomalies at the surface are surveyed by using the global positioning system and by thermal mapping with an infrared camera. Subsequently, detailed geological sampling and mapping provide the data to create two- and three-dimensional models of the fire. Our survey results of this initial phase revealed the location of several hot spots. The second phase concentrated on the geophysical survey of selected areas. For instance, magnetic investigations detect thermally demagnetized rocks, geoelectrical surveys measure the resistivity, which tends to increase in burned rocks, and seismo-acoustic surveys "listen to" the fires. As burning coal seams fracture along with the surrounding rock, microtremors are produced. Appropriately placed geophones can detect the source of such tremors. Investigations into coal fires include gas flux measurements and gas analyses to acquire data on air flow, air-flow velocities, and air pollution. By correlating all of the geophysical measurements and integrating them into a combined model, it is possible to determine the location and depth of a hot spot. In addition, the direction and rate of fire propagation can be calculated through interpolation and interpretation of several geophysical measurements. In this study, the results were confirmed by increasing downhole temperature measurements in holes that were drilled into the subsurface.

Keywords: coal fires, China, hot spots, extinguishing techniques, infrared, geological and geophysical exploration.

*Hartwig.Gielisch@dmt.de

INTRODUCTION

Every year, millions of tons of valuable hard coal are burnt throughout the world (Prakash and Gupta, 1999; Stracher, 2002; Prakash, 2002; Stracher and Taylor, 2004). Some ten times this amount is being contaminated chemically and destroyed physically by seam or pit fires—known as coal fires—to such an extent that it is no longer worth mining. This waste of raw materials cannot be accepted, especially when energy resources are becoming scarcer. Indeed, hard-coal–mining nations worldwide are combating these fires. Besides the United States, India, and South Africa, coal fires are threatening the energy supply of the People's Republic of China. China possesses just 2% of the world's oil and gas deposits and satisfies 67% of its energy needs with hard coal (Ellmies and Häußler, 2003). It is estimated that 750 hard-coal deposits are burning in China (Van Genderen and Haiyan, 1997), and most are located in the arid and semiarid northern part of the country. These fires result in an immense loss of easily mined near-surface coal and pollute the atmosphere through their smoldering and incomplete combustion, which creates dust and greenhouse gases (e.g., CO_2, CO, CH_4). The fires also contaminate valuable groundwater resources with hydrocarbon condensates.

The People's Republic of China has been fighting these coal fires since the nation was founded in its present form. As early as 1954 under Zhou Enlai, state organizations were tasked with stopping coal fires. Since the political and economic opening up of China, various international institutes and companies have started to participate in these activities on the Chinese market, including representatives of the German hard-coal mining industry. Deutsche Montan Technologie GmbH is one such company that has been supporting its Chinese partners since 2000 in two projects that are aimed at extinguishing coal fires.

WHAT IS A COAL FIRE?

A coal fire, as investigated in this study, is a fire started by spontaneous combustion in an unmined and in situ coal seam or in residual coal in an abandoned or operating underground mine. When tectonic movements uplift the coal so that it is exposed at the surface, oxidation of the coal and subsequent spontaneous combustion occur (Figs. 1 and 2).

Such fires probably start after the surface area of exposed coal has increased. This occurs when the seam breaks up, for example, through mining or natural rock movements (e.g., slumping or tectonic motion along fissures or faults) (Rosema et al., 1999; Gielisch and Goerlich, 2002; Zhang, 2004). The breaking up and loosening of the coal seam enables oxygen to interact with a larger coal surface, which leads to increased oxidation of the coal, and, consequently, to the buildup of heat in the seam. If this heat cannot escape, the coal seam continues to heat up until it ignites. The circumstances that lead to coal fires by spontaneous ignition are being investigated by a German-Chinese joint venture project.

CHARACTERISTICS OF COAL FIRES

A coal fire develops in those parts of a coal seam that have been broken up and loosened and where there is a consequent increase in surface area, and therefore a greater capability for the coal to oxidize; this favors the ignition of a fire. Several isolated parts of a seam can burn without necessarily forming a closed fire front, which means that the whole seam does not burn, just separate spots (Walker, 1999). The fire expands from these hot spots. In the case of a seam that has not been touched by mining, the primary hot spot lies at the surface, and the fire eats into

Figure 1. Coal fires in a Jurassic seam in ShuShiGou, Xinjiang, People's Republic of China.

Figure 2. Coal fires in a Permian seam in Wuda, Inner Mongolia, People's Republic of China. Diameter of crack is 1 m.

the seam. According to the experience of Chinese scientists, a natural coal fire can propagate to depths of up to 100 m. At this depth, the oxygen supply becomes depleted, and the fire suffocates. In the case of an abandoned mine structure, such as a ventilation shaft or a mine gallery (Figs. 3 and 4), the coal fire can burn from within the seam to the surface of the seam. There is no natural limit in terms of depth; the mine structure and a limited influx of oxygen restrict the fire. Therefore, a coal seam fire can be observed at Earth's surface only in its initial phase, when it is just starting, or in its final phase, when the fire has burnt through to the surface from underlying mine workings. Naturally, a great number of coal fires are in between these two stages and are not visible at the surface. In such cases, solely circumstantial evidence indicates the approximate location of a fire: temperature anomalies, condensed hydrocarbons (Fig. 5), sulfuric efflorescence (Fig. 6), Glauber's salt, or, in the simplest case, steam

Figure 3. Abandoned shaft of a minor mine in Liu Huang Gou, Xinjiang, People's Republic of China.

Figure 5. Condensed hydrocarbons in Wuda, Inner Mongolia, People's Republic of China. Camera case for scale.

Figure 4. Abandoned gallery of a minor mine in Wuda, Inner Mongolia, People's Republic of China.

Figure 6. Sulfuric efflorescence in Wuda, Inner Mongolia, People's Republic of China.

(Fig. 7). These indicators do not suggest how far away a coal fire is, however, because the pathways through the bedrock—by way of fissures or mine shafts—affect its appearance. The center or the hot spot of the fire is concealed from the observer, and its underground location cannot be determined by any simple means from the surface.

WHY FIND THE HOT SPOT?

The center of a fire or the hot spot is the hottest point (Bandelow and Gielisch, 2004); generally, the temperature decreases with distance from the hot spot. Once the spot has been located, it can be extinguished by cooling it down or by cutting off the supply of oxygen. If it is not possible to extinguish the fire, it can be isolated from the rest of the coal seam by injecting insulating material. Knowledge of the location of the hot spot helps in the determination of the extent of the fire and its approximate velocity of propagation (Bandelow and Gielisch, 2004).

CURRENT EXTINGUISHING TECHNIQUES AND PROBLEMS

When extinguishing coal fires, the Chinese authorities adhere strictly to the legal regulations that apply in the various parts of the country. The policy of the autonomous Uigur Republic, Xinjiang, stipulates that the following actions are necessary to extinguish a coal fire:

- Areas in which fires exist are to be surveyed, mapped, and divided into segments of approximately equal size. One after the other, the infrastructure of these segments is to be developed, and camps for engineers and workers are to be erected close by.
- Bulldozers must flatten the fire zone to create a level surface—all cavities, fissures, and morphologic barriers are to be leveled. The level surface must extend across the entire hard-coal formation, regardless of whether it is burning (Fig. 8).
- Holes are to be drilled in a dense grid of 15 m × 15 m (45 ft × 45 ft) across the entire leveled area. A stand pipe is to be installed at each grid point. Large volumes of water are to be injected through these stand pipes directly into the burning deposit. This water should cool the seam and eventually extinguish the fire.
- Each fire zone is to be sealed with loess to cut off the influx of oxygen. The loess layer can, as required, be up to 2 m (6 ft) thick. After sealing, water should continue to be injected through the stand pipes (Fig. 9).

Figure 7. Burning seam series in Wuda, Inner Mongolia, People's Republic of China.

Figure 8. Completely leveled fire zone in Liu Huang Gou, Xinjiang, People's Republic of China. Oil derrick in the lower right for scale.

Figure 9. Development of steam after the injection of water in Liu Huang Gou, Xinjiang, People's Republic of China.

- The fire is deemed to be extinguished as soon as the temperature in the deposit drops below 80 °C (176 °F). The zone is to be flattened again and the area recultivated in those parts where the climatic conditions are favorable.
- The stand pipes are to remain in a number of drill holes to allow the extinguished deposit to be controlled twice a year and to facilitate early detection in the event of re-ignition of the coal.

These fire-extinguishing methods used in China are safe and probably are the most sensible way of extinguishing coal fires. It does show deficiencies, however, which can be eliminated with the use of modern exploration techniques. For example, given that the hot spots are not surveyed, the entire area of the hard-coal formation has to be drilled and water must be injected. Fires that were burning at the surface before leveling was carried out can continue to burn underground after leveling, and they can propagate away from the area that is being controlled. The 15 m × 15 m drilling grid is costly in terms of time and money, although it does guarantee that all fires are extinguished. To improve the current Chinese method, modern techniques for locating hot spots should be used. By locating these before and after leveling, as well as during active fire fighting, areas that are burning can be differentiated from areas that are not. Thus, the number of drill holes that is required can be reduced, and time and money can be saved (Goerlich, 2004).

HOW TO FIND THE HOT SPOT: INNOVATIVE STRATEGIES TO IMPROVE COAL-FIRE MANAGEMENT

Infrared Exploration

Before starting any fire-extinguishing work, the area should be explored using infrared photography to detect any fires that are burning at the surface (Fig. 10). Recognition of even the smallest of fires provides information for preparing the area for a

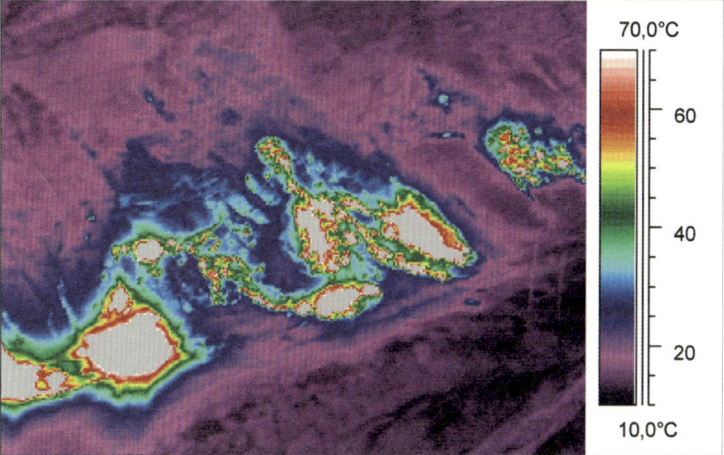

Figure 10. Thermographic evaluation of a burning seam in Tielieke, Xinjiang, People's Republic of China.

204 Gielisch

geological survey and subsequent geophysical exploration. Indeed, infrared exploration is a precondition for the proper location and definition of fire zones. By knowing the areal extent of the thermal anomalies and the temperature distribution on the surface, it is possible to draw conclusions about the amount of coal that is burning.

Geodetic Surveying

An accurate geodetic survey, independent of the topographic conditions (e.g., fixed topographical points, which can shift during earth movements), is a basic requirement for precise drilling, as well as for relocating fires that were identified before leveling. The application of the global positioning system is useful in this respect because it does not measure geodetic points with reference to a possibly changing surface topography. Instead, it uses satellite technology to determine real coordinates that are independent of the topography.

Geological Exploration

Because fires may progress during and after leveling, it is necessary to apply methods that are capable of locating hot spots even through flattened surface layers. Geophysical surveys

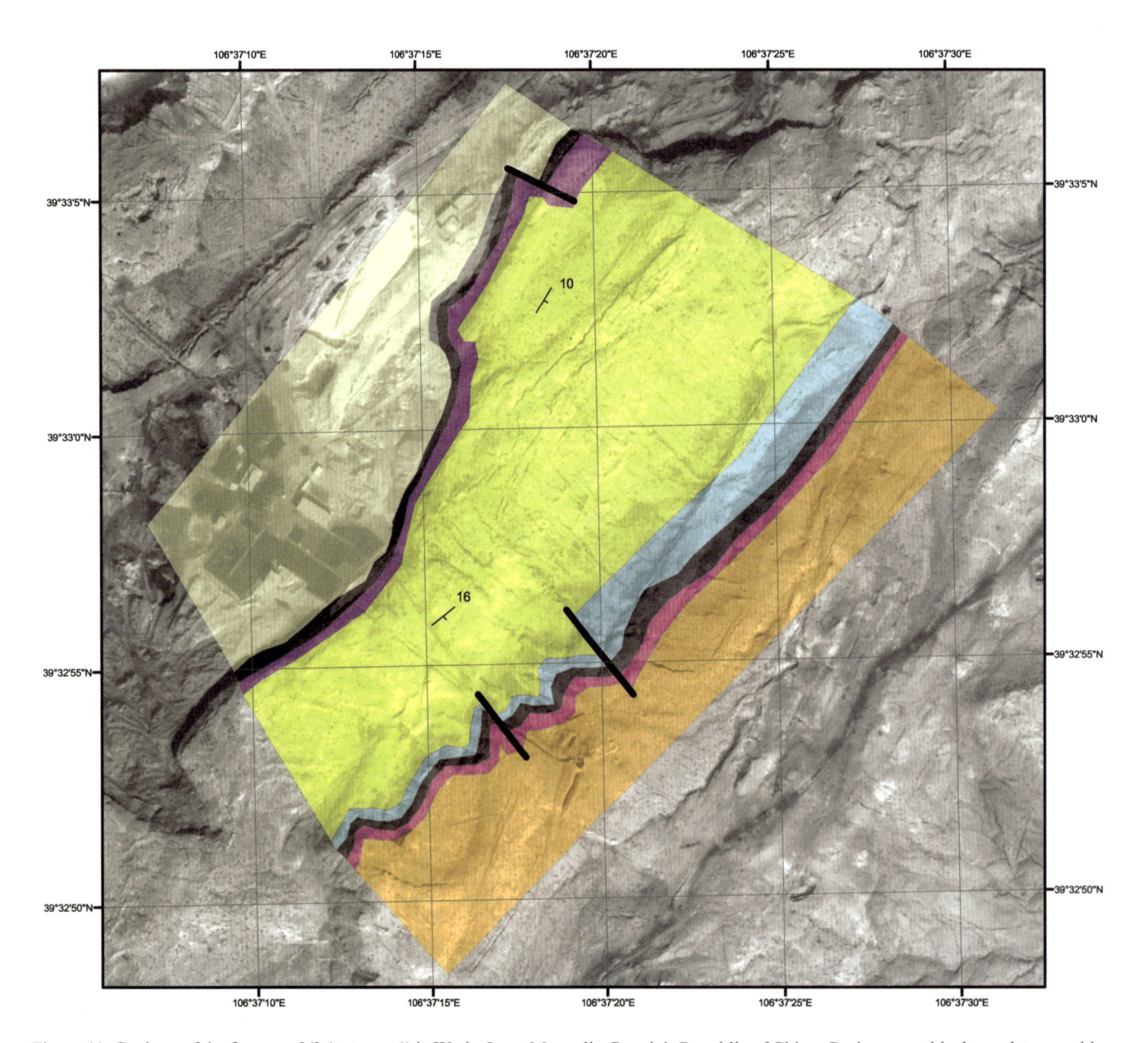

Figure 11. Geology of the fire zone 3/2 (test area 1) in Wuda, Inner Mongolia, People's Republic of China. Coal seams—black; mudstones—blue and pink; sandstones—yellow.

provide the fastest and most meaningful methods for doing so (Gielisch and Goerlich, 2002). However, the use of geophysical methods requires advanced, precise geological knowledge of the areas in question. To arrive at a meaningful statement about the geophysical characteristics of rock layers and how they change with increasing and decreasing temperature, it is essential to know the petrology, sedimentology, tectonics, and the very fabric of the rock layers. In addition, the area should be interpreted in geological terms at least in two dimensions, but preferably in three dimensions (Fig. 11). A three-dimensional geological analysis of the burning deposit forms the basis for subsequent geophysical surveying and for the planning and performance of drilling during the actual fire-extinguishing phase.

GEOPHYSICAL EXPLORATION

For geophysical surveys of fire areas, it is essential to choose the method or methods that offer the best results. Consideration must be given to the rock parameters that change as a result of fire activity and the geophysical methods that are sensitive to these parameters.

The rock parameters that are influenced by fire are electrical resistivity, magnetic susceptibility, and density (Serra, 1984). The change of electrical resistivity is based on the change of the water content in the bedrock. As the rock heats up, it dries out, and the moisture content of the rock decreases, which affects the resistivity significantly. Many rocks contain temperature-dependent magnetic minerals that lose their magnetic characteristics (Fig. 12) (Pullaiah et al., 1975) above a certain temperature (Curie temperature). When such rocks cool down, all of the minerals that had an arbitrary magnetic orientation take on the orientation of Earth's present magnetic field. These rocks have a magnetic field that can be detected geophysically. As far as rock density is concerned, the heat from a fire causes the rock to expand and break up to produce cavities, which decreases rock density locally.

Thus, it is evident that most rock parameters change under the influence of fire, and several geophysical methods can be used to observe these parameters. Electrical resistivity is measured by DC (direct current) geoelectrical methods and magneto-electrical techniques, as well as by radar. Magnetic susceptibility is observed by measurements of magnetism. Gravimetric and seismic measurements provide information about density. Geophysical methods applied by Deutsche Montan Technologie GmbH (DMT) in China include magnetic and geoelectric methods, borehole measurements, and seismo-acoustic procedures, which enable hot spots to be located precisely (Bandelow and Goerlich, 2004). Temperature-induced changes in the geophysical parameters of the rock layers can be measured and recorded. After comparing the maximum and minimum effects, such as the change in specific resistivity, conductivity, and magnetic susceptibility, the center of the fire and its hot spots can be identified. It is then possible to find fire nests and go straight to them.

Magnetic Investigations

Clastic rocks (e.g., sandstone, siltstone) that surround hard-coal deposits naturally contain higher concentrations of magnetite (Fe_3O_4) and hematite (Fe_2O_3). These iron oxides have a natural magnetism, which can be measured. The orientation of the magnetism is a document of the direction of Earth's magnetic field at the time that the rock's minerals were formed. Basalt and other rocks also have high magnetic susceptibility and can provide measurements that reflect the paleomagnetic field. Because sediments contain various erosion products, such as quartz, feldspar, and other minerals from diverse parent rocks, a uniform magnetic orientation of the iron oxides is not expected. This means that it is not possible to measure an external magnetic field of a rock complex. If the rock minerals are heated up, the natural magnetism decreases; magnetite reaches the Curie point at 578 °C, and hematite reaches it at 675 °C (see Fig. 12) (Pullaiah et al., 1975). Compared with the unheated parts of the strata, magnetic anomalies can be recognized easily by geophysical mapping (Fig. 13). It is common to use fluxgate or cesium magnetometers with advanced measuring capabilities to carry out such magnetic mapping. By making repeated measurements, it is possible to localize the hot spots and track the paths of the fires underground.

Geoelectrical Investigations

Useful procedures for determining the exact location of a fire include those that measure electrical properties or the density of the rock. In geoelectrical methods, a weak direct current is passed by way of electrodes into the ground. Probes at the surface measure the stationary potential field, which is dependent on the subsurface conductivity distribution. By making appropriate calculations, the distribution of the true specific resistivity can be

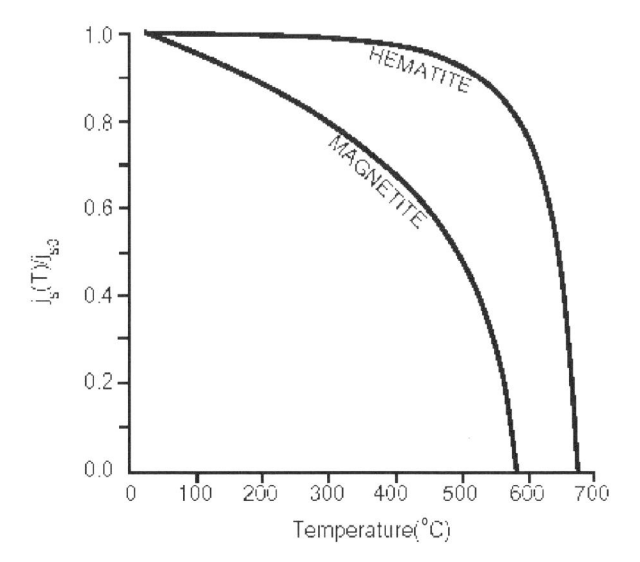

Figure 12. Curie temperature of hematite and magnetite (Pullaiah et al., 1975).

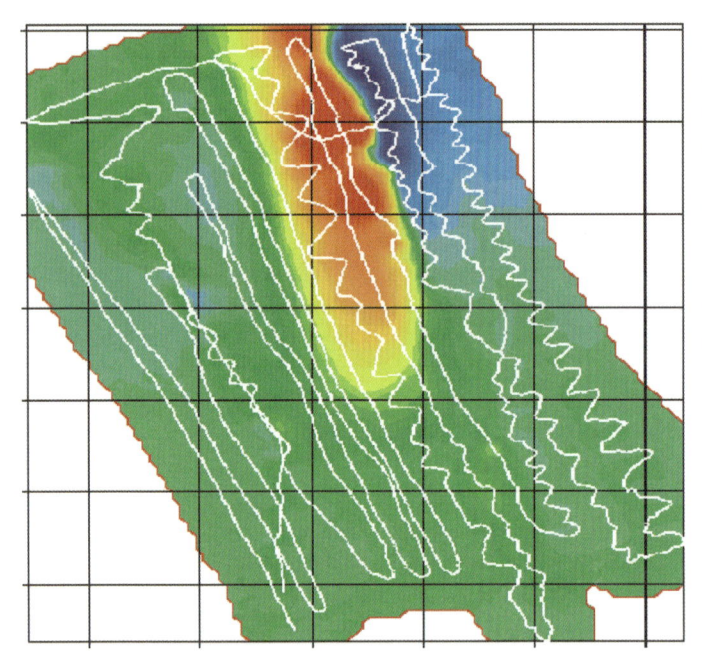

Figure 13. Magnetic anomaly of a burning hard-coal seam in Keer Jian, Xinjiang, People's Republic of China.

Figure 14. "Magnetometer" during a field survey in Liu Huang Gou, Xinjiang, People's Republic of China.

determined from the measuring points. In addition to resistivity, the IP (induced polarization) effect can be measured. In this method, the material's ability to "store" electrical energy can be determined (i.e., the ability to accept electrical charge). In this way, it is possible to differentiate between soils (e.g., clay and silt) that have similar electrical resistivity (Koefoed, 1979; Bender, 1984).

DMT also undertakes geoelectrical surveying, such as deep-depth exploration, profiling, pseudosection work, and tomography (Figs. 14 and 15). Such measurements provide information about the bedding under the survey point and about changes in resistivity along a profile or cross section.

Seismo-Acoustic Investigations

Coal and rocks are made up of a variety of organic and mineral components, each of which has a specific thermal coefficient of expansion and heat conductivity. If a coal seam and the surrounding rock are subject to a strong local increase in heat or a high temperature gradient, differential expansion causes high stresses to build up in the zone of contact. If the stress becomes too great, it causes the rock to fracture, which produces seismic and/or acoustic energy. The former can be measured as a seismo-acoustic event, and the latter can be detected as creaking and cracking (Fig. 16) (Räkers, 1989).

If geophones are laid out in a three-dimensional array at the surface and in boreholes, fracturing can be recorded at different distances from the hot spot (Fig. 17). The geometry of the system enables events to be recorded that have passed along different paths, whereas the velocity of propagation of the seismic waves can be measured (e.g., in a borehole). Based on the velocity of propagation, the different recording paths, and the geometry, the position of the hot spots can be located three dimensionally (Gibowicz and Kijko, 1994; Mendecki, 1997). From this, the underground "seat" of the fire can be located; conclusions about the type and intensity of the fire can be drawn from the character, quality, and quantity of seismo-acoustic events. In addition, by having a fixed array of geophones, it is possible to follow the progress of the fire, determine how it varies with time, and ascertain the success of fire fighting.

Borehole Surveying

Gamma-Ray Logging

The gamma-ray log measures the natural radioactivity of rock (Serra, 1984). Naturally occurring gamma rays originate from the radioactive decay of the potassium-40 isotope and the decay of uranium and thorium. Generally, potassium occurs in higher concentrations in clays, and because of their grain size, clay and shale contain certain concentrations of uranium and thorium (Hoffman and Jordan, 1982). Usually, the natural radioactivity of sedimentary rocks correlates with the amount of clay present. Thus, claystones and siltstones, which have comparatively high radioactivity, can be differentiated from sandstones

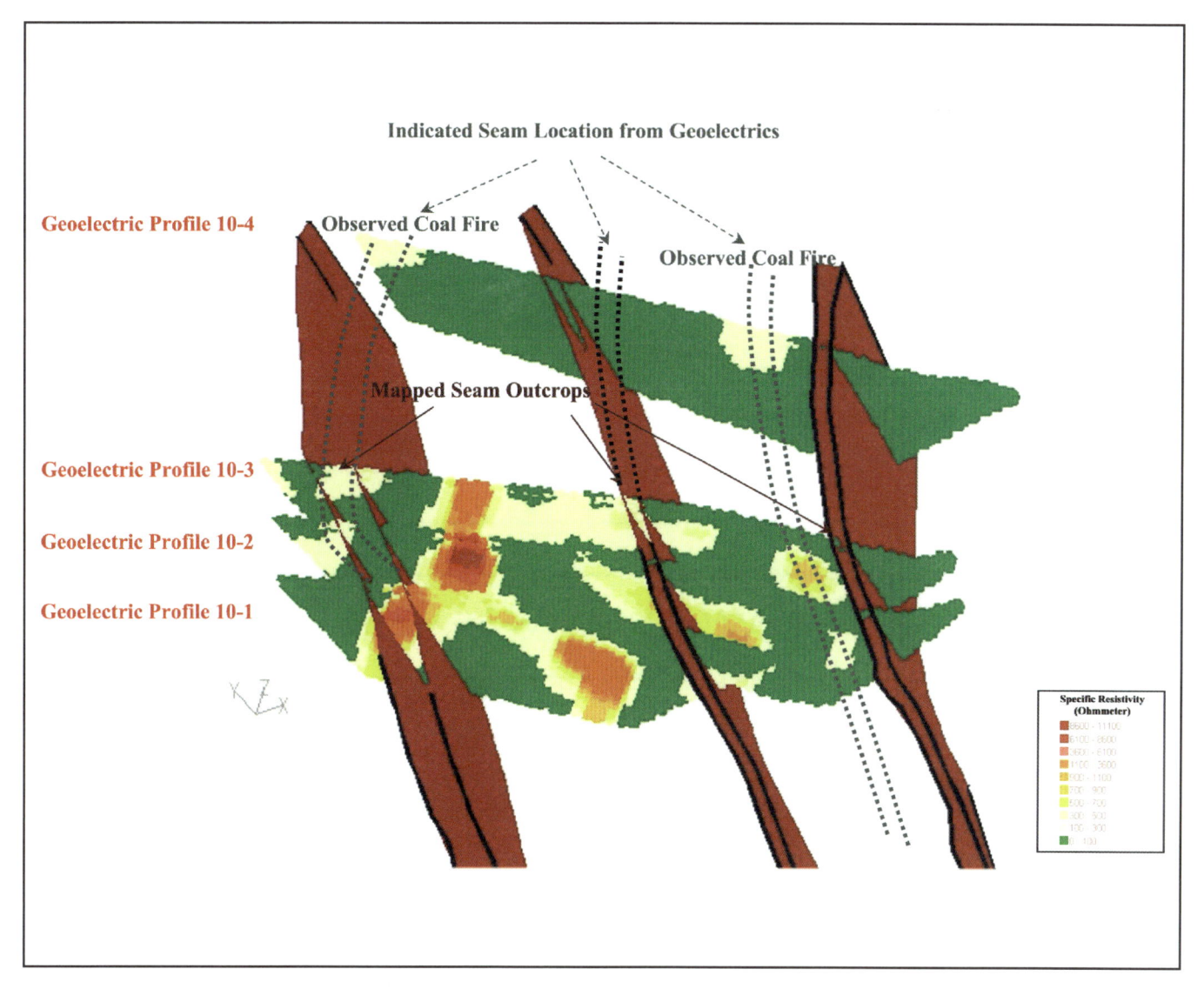

Figure 15. Anomalies of electrical resistivity in Liu Huang Gou, Xinjiang, People's Republic of China.

and coal, which contain only a small amount of clay, and, consequently, have low radioactivity (Serra, 1984), by using a gamma-ray log. If the density is measured in addition to the gamma-ray log, sandstone and coal also can be differentiated clearly (Fig. 18). Consequently, borehole surveys are ideal for surveying hard-coal deposits because they enable the surrounding clastic rock, such as claystone and siltstone, to be distinguished from the coal. These methods, which are used worldwide for surveying hard-coal deposits, also are used in boreholes for monitoring coal fires and locating seams.

Induction Logging

Induction logging is carried out in boreholes to find the electrical conductivity (in mS/m or mS/cm) or the electrical resistivity (in ohms) (Hoffman and Jordan, 1982; Serra, 1984; Bateman, 1985). The system uses a series of electrical coils to generate electromagnetic waves with frequencies of ~20 kHz, which are emitted into the surrounding rock. Depending on the conductivity in the rock layers (determined by the pore filling, permeability, porosity, and the amount of clay), secondary fields are set up as a result of eddies (Foucauld currents), and their magnetic fields induce stresses in the receiver coils.

To improve the transmitting and receiving capabilities, the signal for transmitting is focused through secondary coils. Usually, coils with two systems of receivers (ILMedium and ILDeep) are used at different distances from the emitter (Serra, 1984). Different depths of penetration can be achieved with this arrangement. This method of measurement also can be used in boreholes with oil-based mud or air.

The resolution of the method decreases rapidly with decreasing conductivity; if the conductivity decreases to less

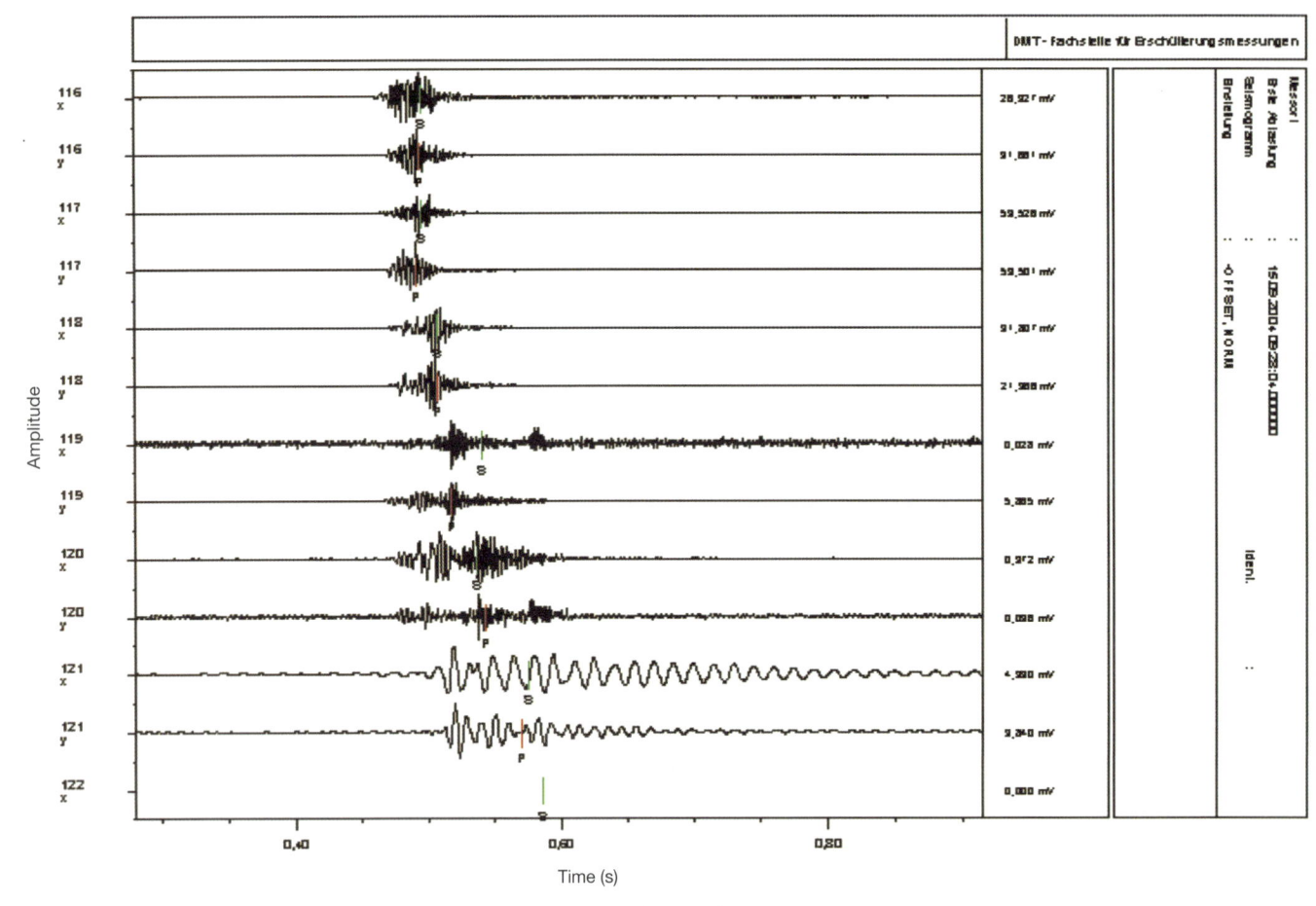

Figure 16. Microseismic events in Rujigou, Ningxia, People's Republic of China.

Figure 17. Installation of a geophone in test area 2, Wuda, Inner Mongolia, People's Republic of China.

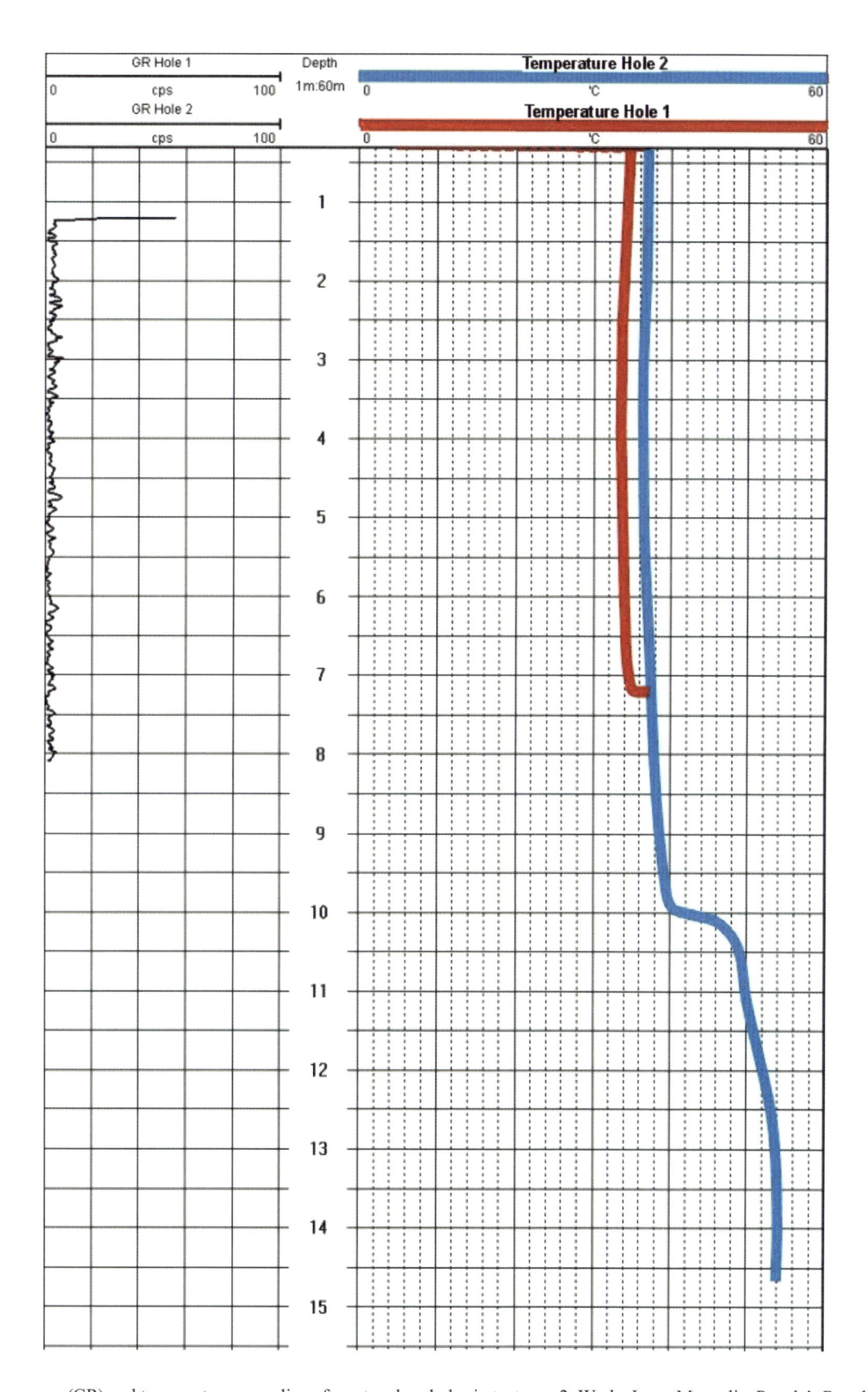

Figure 18. Gamma-ray (GR) and temperature recordings from two boreholes in test area 2, Wuda, Inner Mongolia, People's Republic of China.

than 10 mS/m (100 Ω), only approximate results can be obtained (Serra, 1984).

In China, induction logging is used to detect variations in clay layers. In such layers, there is a range from extremely low conductivity to medium conductivity in natural and "burned" layers. However, because the holes are not filled with mud, this is the only method available to obtain resistivity data for correlation with, and, if necessary, correction of, the magnetic field data. In addition, the path of the extinguishing material can be tracked after it has been injected into the surrounding rock.

Magnetic Logging

Magnetic logging is fundamentally the same as surface magnetic surveying as described earlier, with the main difference being that borehole logging is carried out vertically.

CONCLUSIONS

The methods described here make it possible to locate the hot spot of an underground coal fire and to track it while extinguishing work is being carried out. With the exception of geophysical logging, which can be performed only in existing boreholes, these methods are mobile and can be used almost anywhere without major logistical effort. Therefore, time-consuming and costly preparatory work is unnecessary. It is essential to have knowledge of the topography and of the local geology before applying geophysical methods; however, such information usually is available in regions where underground mining is being or has been carried out. In areas with natural coal fires that are not prompted by mining, such information has to be acquired in advance.

ACKNOWLEDGMENTS

The authors thank the German Ministry of Education and Research (BMBF) for funding and all of the colleagues of the BMBF project for discussions and support.

REFERENCES CITED

Bandelow, F.-K., and Gielisch, H., 2004, Modern exploration methods as key to fighting of uncontrolled coal fires in China: Geological Society of America Abstracts with Programs, v. 36, no. 5, p. 43.

Bandelow, F.-K., and Goerlich, B., 2004, Moderne Explorationsverfahren als Schlüssel zur Löschung von Kohlebränden in China: Abstract, 44. Wissenschaftliche Fachtagung des Deutschen Markscheider-Vereins in Bochum, "... aus Tradition modern" Bochum, 15–18 September 2004.

Bateman, R.M., 1985, Log quality control: International Human Resource Development Corporation: Boston, Reidel Publishing Co., 398 p.

Bender, F., 1984, Angewandte Geiwissenschaften, Band II: Stuttgart, Enke, 676 p.

Ellmies, R., and Häußler, I., 2003, China: Rohstoffwirtschaftlicher Länderstudien, XXVI Bundesanstalt für Geologie und Rohstoffe, 80 p.

Gibowicz, S.J., and Kijko, A., 1994, An Introduction to Mining Seismology: New York, Academic Press, 396 p.

Gielisch, H., and Goerlich, B., 2002, Zur Problematik brennender Steinkohleflöze am Südrand des Tian Shan im Westen der VR China (Tielieke, Autonome Provinz Xinjiang, VR China): Abstract zum 17. Sedimentologentreffen 2002, 29–31 May, Darmstadt: Schriftenreihe der Deutschen Geologischen Gesellschaft, v. 17, p. 24.

Goerlich, B., 2004, Strategien zur Löschung von Kohlenbränden in China. GTZ-Report, Project "Entwicklung von Strategien zur Löschung von Kohlenbränden in China": German Society of Technical Cooperation, unpublished.

Hoffman, G.L., and Jordan, G.R., 1982, Geophysical Borehole Logging Handbook for Coal Exploration: Edmonton, Alberta, Canada, The Coal Mining Research Centre, 270 p.

Koefoed, O., 1979, Geosounding Principles, Volume 1: Amsterdam, Netherlands, Elsevier, 276 p.

Mendecki, A.J., 1997, Seismic Monitoring in Mines: London, England, Chapman & Hall, 262 p.

Prakash, A., 2002, Coal fires—A natural or man made hazard?: http://www.gi.alaska.edu/-prakash/coalfires/introduction.html (accessed June 2007).

Prakash, A., and Gupta, R.P., 1999, Surface fires in Jharia Coalfield, India—Their distribution and estimation of area and temperature from TM data: International Journal of Remote Sensing, v. 20, p. 1935–1946.

Pullaiah, G., Irving, E., Buchan, K.L., and Dunlop, D.J., 1975, Magnetization changes caused by burial and uplift: Earth and Planetary Science Letters, v. 28, p. 133–142, doi: 10.1016/0012-821X(75)90221-6.

Räkers, E., 1989, Seismoakustische Ereignisse in Steinkohlenflözen als Hilfe zur Erkennung von Abbaubereichen mit erhöhten Gebirgsdrücken [Ph.D. dissertation]: Bochum, Ruhr University, 246 p.

Rosema, A., Guan, H., Veld, H., Vekerdy, Z., Ten Katen, A.M., and Prakash, A., 1999, Manual of Coal Fire Detection and Monitoring: Report of the project "Development and Implementation of a Coal Fire Monitoring and Fighting System in China": Netherlands Institute of Applied Geoscience Report NITG 99–221, 245 p.

Serra, O., 1984, Fundamentals of well-log interpretations, *in* Developments in Petroleum Science, Volume 15-A, I and II: Amsterdam, the Netherlands, Elsevier, 423 p.

Stracher, G.B., 2002, Coal fires: A burning global recipe for catastrophe: Geotimes, v. 47, no. 10, p. 36–37 and p. 66.

Stracher, G.B., and Taylor, T.P., 2004, Coal fires burning out of control around the world: Thermodynamic recipe for environmental catastrophe, *in* Stracher, G.B., ed., Coal Fires Burning Around the World: A Global Catastrophe: International Journal of Coal Geology, v. 59, no. 1–2, p. 7–17.

Van Genderen, J.L., and Haiyan, G., 1997, Environmental monitoring of spontaneous combustion in the North China coalfields: Final Report to European Commission: Enschede, the Netherlands, International Institute for Geo-Information Science and Earth Observation, 244 p.

Walker, S., 1999, Uncontrolled fires in coal and coal wastes: London, International Energy Agency (IEA) on Coal Research, 73 p.

Zhang, J., 2004, Spatial and Statistical Analyses of Thermal Satellite Imagery for Extraction of Coal Fire Related Anomalies [Ph.D. thesis]: Austria, Vienna University of Technology, 166 p.

Manuscript Accepted by the Society 7 March 2007

Subsurface coal-mine fires: Laboratory simulation, numerical modeling, and depth estimation

Anupma Prakash*
Antony R. Berthelote
Geophysical Institute, University of Alaska Fairbanks, 903 Koyukuk Drive, Fairbanks, Alaska 99775-7320, USA

ABSTRACT

Subsurface coal-mine fires occur in many mining regions, especially where coal has been previously excavated by "room-and-pillar" mining methods. The surface above these fires heats up to produce a thermal anomaly. The shape of the temperature profile over the fire zone holds clues to the depth of the underground fire. We simulated an underground coal-mine fire in the laboratory by burying a hot glass tube in a sandbox. The thermal anomaly over the tube was recorded using a forward looking infrared radiometer (FLIR™) camera. Numerical modeling using finite-element techniques for various combinations of tube depth and tube temperature helped to empirically derive a depth-estimation function, called the linear anomaly surface transect (LAST) function. Comparisons of the results from the LAST function with the half-anomaly-width function for depth estimation developed by Panigrahi et al. (1995) showed that the LAST function gave more accurate results for shallow subsurface coal fires ranging in depth from a few centimeters to ~10 m. For moderate-depth coal fires, ranging in depth from 10 m to 40 m, the depths estimated by the two functions were comparable.

Keywords: coal fires, simulation, modeling, depth, thermal infrared.

INTRODUCTION

Coal and Coal Fires

Coal is a stratified deposit that occurs in layers in sedimentary rocks. When these coal layers, also known as coal seams, have a continuous spatial extent and have thicknesses over tens of centimeters, they constitute mineable reserves. The coal seam may be exposed to the surface or may occur at depths ranging from a few centimeters to hundreds of meters below the surface. Both surface and subsurface (underground) coal seams are prone to catch fire, which, once started, can last for tens of years (Fig. 1A). The nature, causes, and hazards from these fires have been well documented by several researchers (Mansor et al., 1994; Saraf et al., 1995; van Genderen and Haiyan, 1997; Zhang et al., 2004; Prakash, 2005).

Surface fires are marked by very high temperatures, flames, red-hot burning materials, smoldering and smoking zones, baked rocks, rubble and ash, and they can be easily detected in field (Prakash and Gupta, 1999; Stracher, 2004). Subsurface or underground fires are, however, more difficult to detect. Surface manifestation of subsurface fires (Fig. 1B) can be seen in the

---

*prakash@gi.alaska.edu

Prakash, A., and Berthelote, A.R., 2007, Subsurface coal-mine fires: Laboratory simulation, numerical modeling, and depth estimation, *in* Stracher, G.B., ed., Geology of Coal Fires: Case Studies from Around the World: Geological Society of America Reviews in Engineering Geology, v. XVIII, p. 211–218, doi: 10.1130/2007.4118(13). For permission to copy, contact editing@geosociety.org. ©2007 The Geological Society of America. All rights reserved.

Figure 1. Field photographs of areas affected by underground coal fires in the Jharia Coalfield, India. (A) A shallow subsurface fire where mining activities have exposed the burning coal seam. Flames and smoke rising from the fires are visible. The width or the area shown on the photograph is ~35 m. (B) Surface view of an area affected by a coal fire at a depth of ~20 m. The area is locally devoid of vegetation. Note the elongated cracks and baked rocks lining these long cracks associated with the underground fire.

form of elongated cracks, smoke-emitting vents, baked rocks and ash-lined vents, and dry barren patches of land forming reflection aureoles (Gupta and Prakash, 1998). Temperatures of surface coal fire regions range from as low as 70 °C to as high as 1100 °C in areas of open flame. The surface temperature anomalies over underground fires are, however, much more subtle, with temperature differences between the hot spots and the background sometimes as low as 2 °C.

Coal Mining and Subsurface Fires

This work focuses on the subsurface fires only. Subsurface fires tend to be predominant is areas that have been previously mined out using a primitive underground mining technique known as the "room-and-pillar" mining technique. In this technique, chunks of coal are excavated leaving columns of coal behind as support pillars to prevent the roof of the mine from collapsing. Using this technique, especially in developing countries, sometimes only ~10–15% of the coal is extracted, and then the mining area is abandoned (UMWA, 2005). The abandoned underground mine is then left with a series of combustible coal pillars in contact with circulating air. The carbon from coal reacts with the oxygen from the surrounding air in an exothermic reaction, producing heat that cannot be efficiently dissipated, which generates conditions that are favorable for spontaneous combustion of coal (Rosema et al., 2001; Banerjee, 1985). Once the coal underground catches fire, it spreads to other coal pillars. Depending on air access through fractures and faults, this underground fire may spread through the entire coal seam or to a neighboring coal seam. The fire may first start at one location as a point source and then spread over large lengths of a coal seam.

In many mining areas, especially in developing countries, the room-and-pillar mining method is still commonly used for underground mining. Private and illegal mining activities complicate the situation further, as there may be no records or mining-area maps available to show which areas have been mined out in the past and are now more vulnerable to underground coal-mine fires. Abandoned underground coal mines have been used by local residents as temporary shelters in the winter months. Small fires started in these shelters for providing warmth have been known to trigger an underground fire. Illegal activities, such as distillation of alcohol in abandoned coal mines, have also been reported to start underground coal fires.

Fire Depth and Its Significance

Most illegal and private underground mining is limited to a depth of ~10 m, which makes these shallow subsurface fires a more common occurrence. Deeper coal fires, extending to depths of 40 m and more, have been reported in areas where deeper coal seams have been mined out in the past. Whether the fire starts by poor mining practice, spontaneous combustion, or other anthropogenic causes, once started and allowed to spread, it may become uncontrollable and a challenge to put out. These fires may also aggravate the problem of land subsidence commonly associated with coal-mining areas (Bell et al., 2000; Prakash et al., 2001). Accurate estimations of the fire depth are critical to the success of any fire-fighting operation.

DEPTH-ESTIMATION METHODS

Underground fires heat up the surface above them, causing a surface thermal anomaly. Shallow-depth fires produce an anomaly that has a smaller spatial extent but a higher contrast with the background temperature. Deeper fires produce an anomaly that is wider but has a more subtle contrast with the background (Prakash et al., 1997). The nature of this thermal anomaly, therefore, holds clues to the depth of the underground fire and has been the basis of most fire depth-estimation studies (Fig. 2).

Coal-fire depth-estimation techniques include borehole temperature measurement in the field (Greene et al., 1969), using geometric models based on geology and geomorphology (Saraf et al., 1995; Peng et al., 1997), and surface thermal anomalies measurements (in the field or using remote-sensing techniques) to numerically model fire depths. A brief review of fire depth estimation using numerical techniques is presented here.

Fire depth-estimation studies rely on the basic principles of heat transfer, as laid down in the first law of thermodynamics (conservation of energy) and Fourier's law of heat conduction (Incropera and DeWitt, 2002a, 2002b). A large volume of literature is also available on Earth's heat-flow and lava-cooling models, which indirectly and directly allow for depth solutions (e.g., Shaw et al., 1977; Peck et al., 1977; Dragoni, 1989; Ishihara et al., 1990; Manley, 1992; Wooster et al., 1993; Keszthelyi and Denlinger, 1996; Neri, 1998; Harris and Rowland, 2001; Patrick et al., 2002; Witter and Harris, 2007).

Quantitative estimates on the depth of subsurface coal fires were reported by Mukherjee et al. (1991). By incorporating both a theoretical temperature versus time plot for differing material thermal diffusivities and a prior knowledge of a subsurface fire, they estimated depths using the equation for linear heat flow in a semi-infinite medium. Prakash et al. (1995) took this work a step further and used a double ratioing technique to estimate fire depths. They showed that by using three sets of thermal infrared remote-sensing data, acquired at different times over the same area, fire depth could be computed without the knowledge of the fire initiation time. Rosema et al. (1999) introduced the components of surface radiation and convection along with conduction to determine a relation between the coal losses and the total heat flux. They reported a weak but useful relationship between the standard deviation of the anomaly width and the depth of the fire.

These methods either relied on a prior knowledge of the fire or on several thermal infrared measurements over the same fire area taken at different times. Alternatively, there are depth-estimation methods that draw only on the anomaly width-to-depth relationship. The half-anomaly-width model developed by Panigrahi et al. (1995) used temperature anomaly shape to estimate the depths of coal-mine fires at South Tisra Colliery of the Jharia Coalfield in India. The model was based on the belief that the temperature (T) varied inversely as the cube of the distance from the hot source, r (Fig. 3). It followed that the point on the surface with the highest temperature (T_{Max}) fell directly above the fire. Panigrahi et al. (1995) showed that if x_c is the distance from the origin to the point where the temperature equaled $T_{Max}/2$, then the fire depth, z, could be given by

$$z = \frac{x_c}{\left(2^{2/3} - 1\right)^{1/2}} \approx \frac{4x_c}{3}, \qquad (1)$$

where a parabola was fit to each anomaly to determine the actual distance x_c.

Cassells and van Genderen (1995) and Cassells (1998) attempted to quantitatively refine the half-anomaly-width method by using laboratory simulations and numerical techniques to inversely model a point source depth from its surface thermal anomaly. Our work extends the earlier work by carrying out laboratory simulations and numerical modeling work for a linear hot source, which is more representative of a mature and active underground fire.

Laboratory Simulation of a Linear Coal Fire

For the laboratory simulation, a temperature-adjustable linear source (hot fluid flowing through a thin-walled glass tube) was buried in a plastic box filled with unsorted fill material with particle size less than 4 mm. The setup allowed us to change the depth of this glass tube for different experimental runs. A forward looking infrared radiometer (FLIR™) Systems S40 video camera continuously monitored the surface above the linear hot source. The experimental setup is shown in Figure 4A, and the thermal anomaly produced by this linear hot source is shown in Figure 4B.

Finite-Element Modeling

The point source model presented by Cassells (1998) accurately determined the fire heat output as long as sufficient data about model parameters were available. However, incompatibilities in

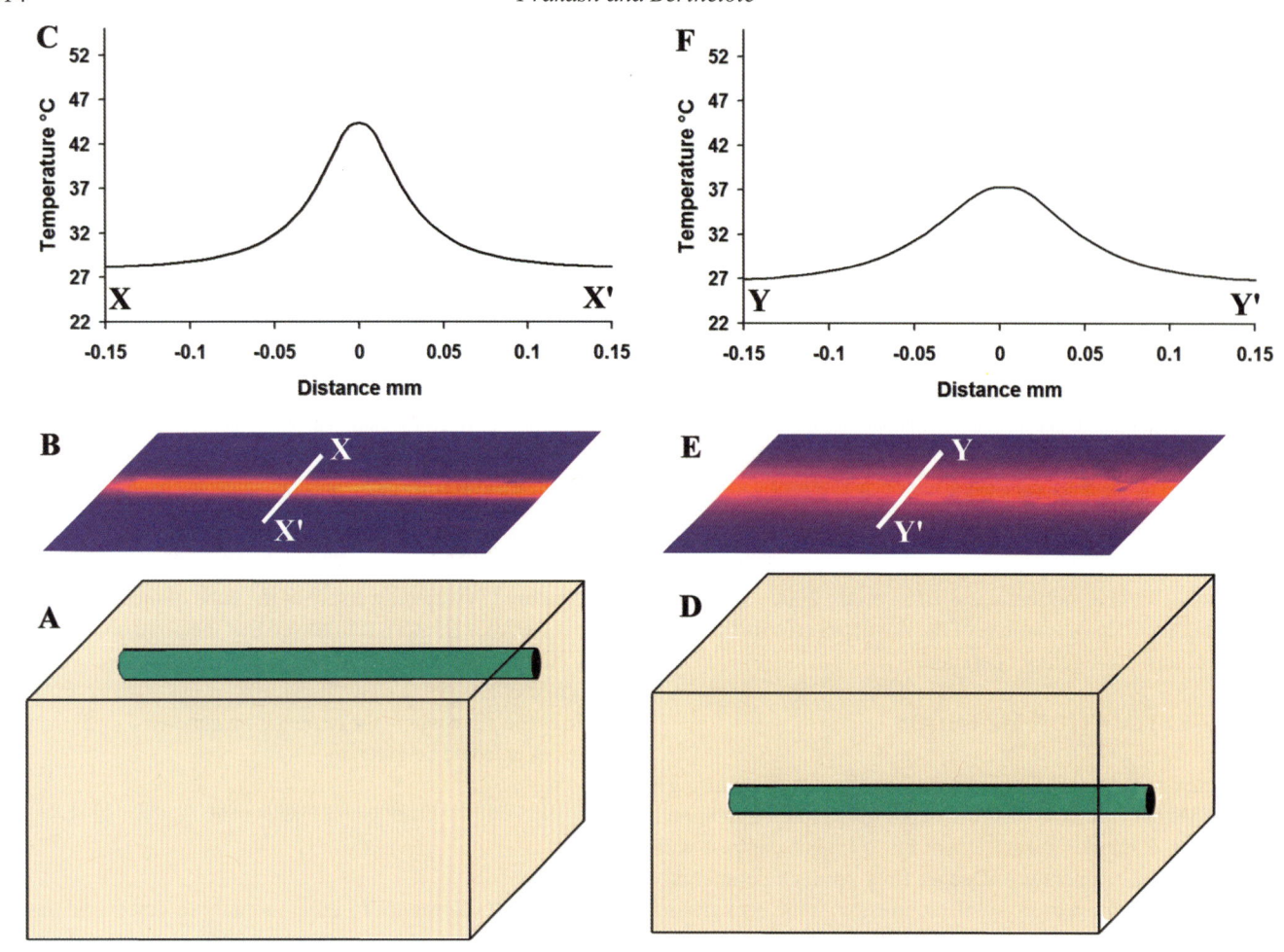

Figure 2. A conceptual diagram showing the relationship of depth of the linear hot source (subsurface coal seam fire) to its surface thermal anomaly. (A) The linear hot source is shown as a long tube buried close to the surface of the box. (B) The surface thermal anomaly produced by the shallow hot source. (C) The temperature profile along the transect line X–X′. The high temperature portion of this profile is narrow, and the profile shows a high temperature peak. (D) The linear hot source is now shown buried deeper than in A. (E) The surface thermal anomaly produced by the deeper hot source. (F) The temperature profile along the transect line Y–Y′. Compared to the temperature profile shown in C, the Y–Y′ profile is broader and less peaky, which is more characteristic of deeper hot sources.

the ANSYS software and the applied inverse modeling technique resulted in large errors in depth estimates. In the present work, the Geo-Slope Temp/W© finite-element modeling software package was used to model the laboratory experimental setup. Models were run with variable material volumetric heat capacities, source temperatures, and source width-versus-depth ratios. Figure 5 shows a typical two-dimensional modeled cross section of a 3-m-diameter tube at 6 m depth with an 1100 °C hot source and its surface temperature profile.

The Linear Anomaly Surface Transect (LAST) Depth-Estimation Function

The surface temperature profiles from these multiple models were used to develop an empirical depth-estimation function (Berthelote et al., 2007). These temperature profiles were measured in the laboratory experiment and modeled with finite-element techniques for a large range of conditions. A Lorentzian or Cauchy distribution curve was found to provide a better fit to these temperature profiles than the parabola fit used by Panigrahi et al. (1995). Data from the Lorentzian distributions were used to derive an empirical depth-estimation function from these linear surface temperature anomalies with regression techniques. The resultant linear anomaly surface transect (LAST) depth-estimation function requires a few surface temperature inputs only. The LAST depth-estimation function is defined as

$$\lambda = A\Gamma^2 + B\Gamma + C\Gamma T_{Max} + D\Gamma T_{Max} + E\Gamma T_{Max-Min} + FT\sqrt{T_{Max}^2 - T_{Min}^2}, \quad (2)$$

where λ is the estimated depth, T_{Max} is the maximum surface temperature over the hot source, T_{Min} is the minimum surface temperature representing the background temperature, and Γ is the width of the thermal profile at $T_{Max}/2$. This Γ value is referred to

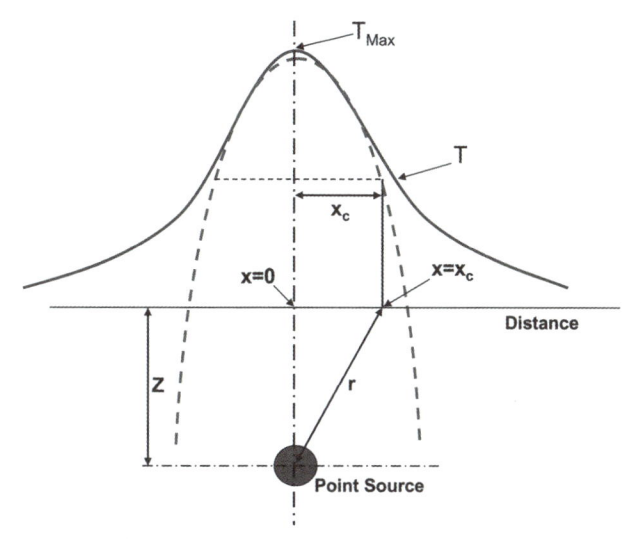

Figure 3. Basis of the half-anomaly-width method where the mean half-width of a parabola fit to a temperature anomaly represents three-quarters of the depth to its source (after Panigrahi et al., 1995).

as full width–half max (FWHM). The values of coefficients A through F are given in Table 1. This LAST function is empirical in nature and does not directly characterize a specific physical process or a mathematical characteristic.

Sensitivity and Error Analyses

Sensitivity analyses were completed for the LAST function by consecutively varying each of the three input parameters, namely T_{Max}, T_{Min}, and Γ, in the function to determine the effect on the depth

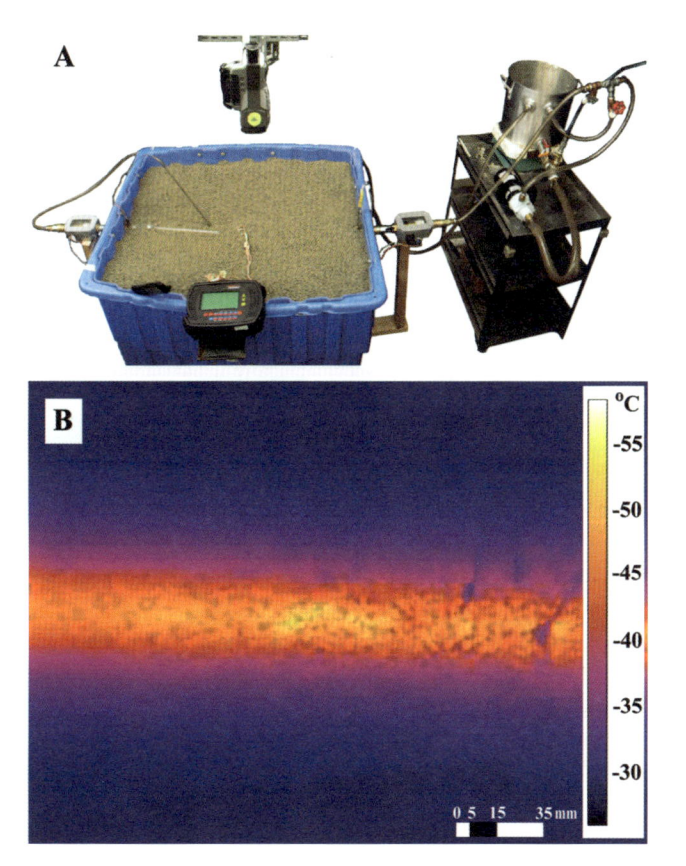

Figure 4. (A) Photograph of the laboratory experimental setup. Linear hot source was simulated by a buried glass tube (left side exposed in this photo on purpose) with hot fluid flowing through it. The fluid was heated in the pot shown on the right, and its temperature and circulation through the glass tube was carefully controlled. The forward looking infrared radiometer (FLIR™) camera used to record thermal images can be seen on top. (B) Typical FLIR™ image of a laboratory experiment with pixel dimensions of 1.235 mm × 1.235 mm. This image clearly shows the linear surface anomaly from an ~80 °C hot tube buried 3 cm deep.

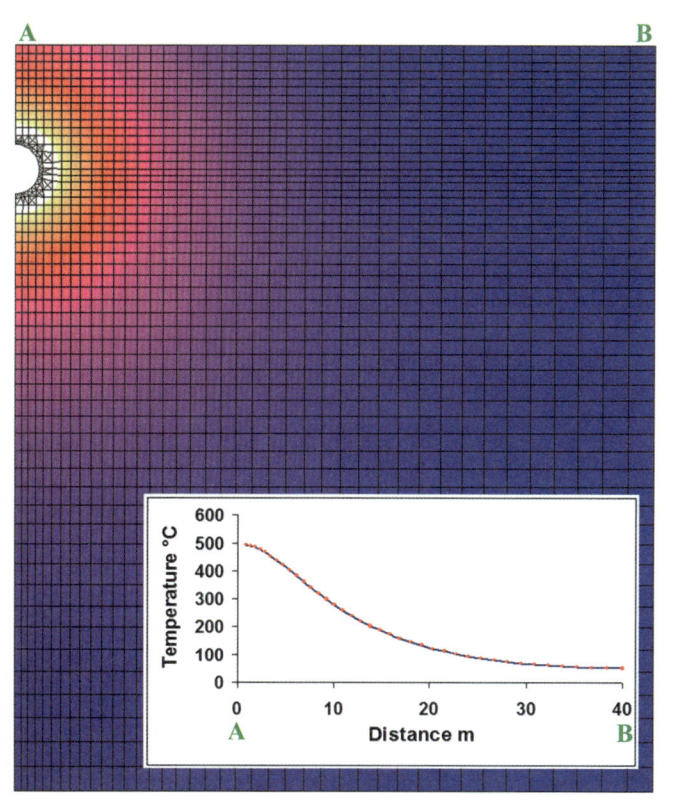

Figure 5. Geo-Slope Temp/W© temperature data derived for a two-dimensional finite-element model run. This model represents a hot source with a diameter of 3 m, buried 6 m below the surface. The model records temperature at each node within the element mesh. The subset temperature profile plots the temperature of each node along the surface A–B. The temperatures calculated for each finite-element run assisted in deriving the depth-estimation function.

TABLE 1. THE VALUES AND UNITS OF THE LINEAR ANOMALY SURFACE TRANSECT (LAST) EQUATION (2) COEFFICIENTS

Coefficient	Value	Units
A	0.008413824	m^{-1}
B	0.174742473	none
C	0.003188651	K^{-1}
D	−0.002411108	K^{-1}
E	−0.001012215	K^{-1}
F	−0.001983005	K^{-1}

estimate. Figure 6 shows that each percentage change in T_{Max} or T_{Min} results in an equivalent percent change in the error in depth estimate. The percentage error in the depth estimate has a two to one relationship with the percentage change in Γ. In general, the product of depth and change in temperature, $T_{Max} - T_{Min}$, has an inverse relationship with this error. Error analyses were completed with a Taylor series expansion of the LAST function to quantify the inevitable associated errors of measurements (Berthelote et al., 2007).

Application of the LAST Function to Known Subsurface Coal Fires

The LAST function was used to evaluate shallow hot source depths (0.01 m to 6 m) based on finite-element–modeled surface temperature transects (Fig. 7). The plot clearly shows that the LAST function greatly improves upon the half-anomaly-width method for depth estimations at scales ranging from centimeters to a few meters.

Published surface temperature data above underground coalmine fires from three sites within the Jharia Coalfield, India, were used to estimate the depth of coal fires that were known to have moderate to deep depths (Mukherjee et al., 1991; Panigrahi et al., 1995) (Fig. 8).

To properly compare the LAST method to the half-anomaly-width method, the published thermal transect used in Panigrahi's

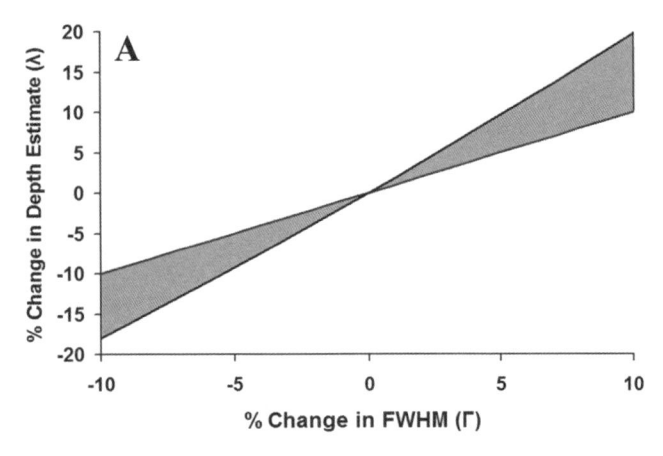

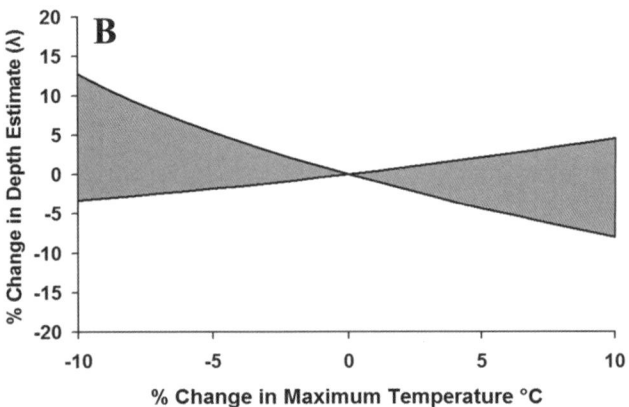

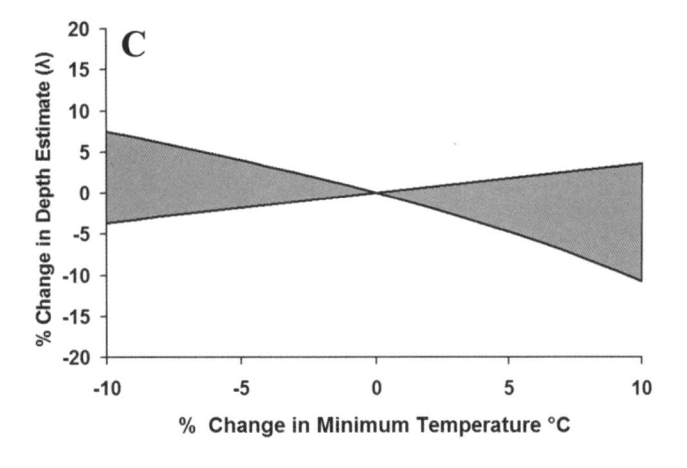

Figure 6. (A) A plot of the percent change in linear anomaly surface transect (LAST) depth estimate due to variations in the percent change in the full width–half max (FWHM) for depths ranging from 0.01 m to 45 m. Each 1% change in full width–half max produces a 1% to 2% change in the LAST depth estimate. (B) A plot of the percent change in LAST depth estimate due to variations in the percent change in the maximum temperature for depths ranging from 0.01 m to 45 m. Each 1% change in maximum temperature produces an equivalent percent change in the LAST depth estimate. (C) A plot of the percent change in LAST depth estimate due to variations in the percent change in the minimum temperature for depths ranging from 0.01 m to 45 m. Each 1% change in minimum temperature produces an equivalent percent change in the LAST depth estimate. The LAST function is twice as sensitive to changes in the full width–half max compared to changes in temperature.

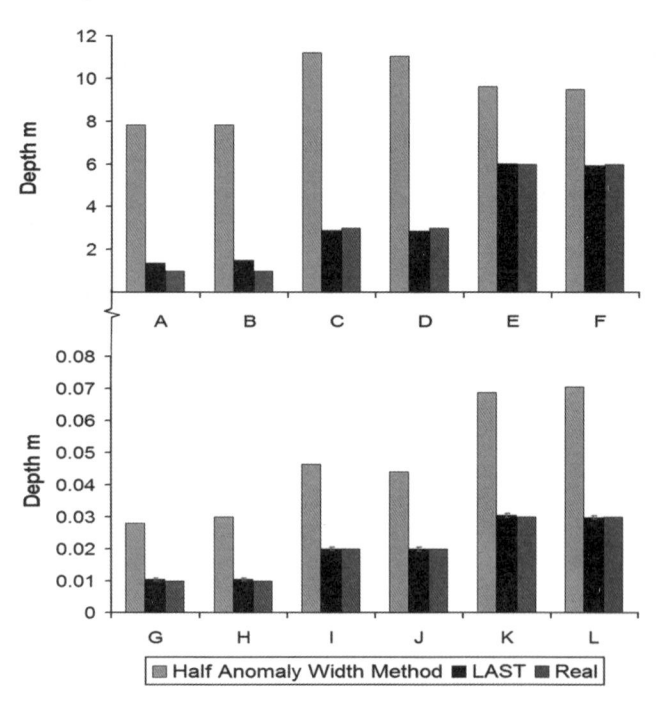

Figure 7. A comparison of depth estimates for finite-element–modeled temperature data. The linear anomaly surface transect (LAST) function produces a more accurate depth estimate than the half-anomaly-width method for depths between 1 cm and 6 m. The maximum error associated with measurement for these data was less than 0.002 m.

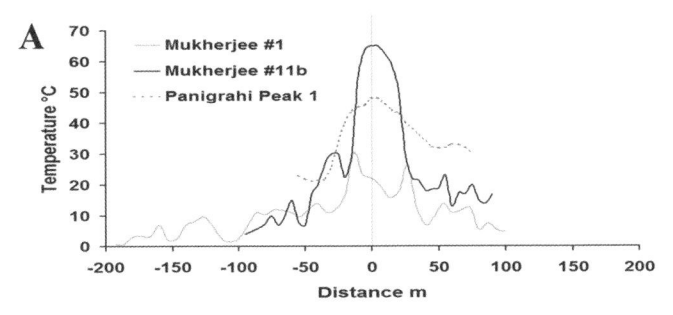

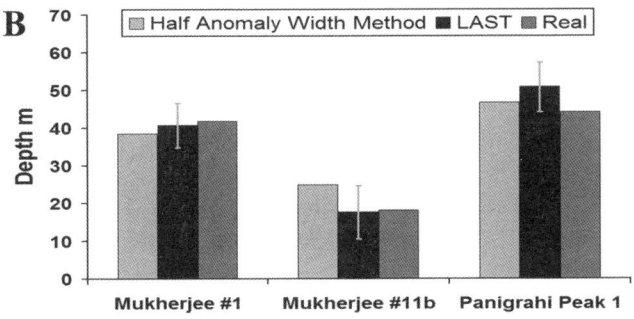

Figure 8. (A) A plot of three published temperature transects above subsurface coal fires in the Jharia Coalfield, India (after Mukherjee et al., 1991; Panigrahi et al., 1995). (B) Comparison of depth estimates for the three known subsurface coal fires. The full width–half max was determined from Lorentzian curves fit to each temperature transect. The maximum and minimum temperatures were determined from each of the published temperature transects. The error bars show the possible errors in measurements. The linear anomaly surface transect (LAST) function produces depth estimates that are similar to the half-anomaly-width method for these moderate depths.

original experiment was examined (Panigrahi et al., 1995). This pre-dawn ground transect was acquired with a handheld infrared gun (Infratrace 800) at 2.5 m intervals. The reported depth of the coal fire below Panigrahi's peak 1 was 42.2 m. The half-anomaly-width method produced a depth estimate of 46.7 m. Using these input data, the LAST method estimated a depth of 50.8 m with an error of ±6.6 m.

The transects from Mukherjee et al. (1991) were acquired during pre-dawn hours with an airborne AADS 1268 Daedalus multispectral scanner system. The two transects, namely profiles 1 and 11b (Mukherjee et al., 1991) with reported depths of 41.7 m and 18.1 m, respectively, were used for comparison. For profile 1, the half-anomaly-width method produced a 38.4 m depth estimate, while the LAST method estimated a depth of 40.6 m with an error of ±5.9 m. For profile 11b, the half-anomaly-width method produced a 25.0 m depth estimate, while the LAST method estimated a depth of 17.7 m with an error of ±7.0 m.

These three examples show that the LAST function estimates depths approaching 40 m with accuracies similar to the half-anomaly-width method of Panigrahi et al. (1995). The shallower, 18.1 m depth begins to illustrate the difference in the two methods. Depths decreasing from ~40 m are more accurately estimated by the LAST function.

LIMITATIONS

The LAST function has some limitations. The function is empirical and does not directly represent a physical process or a mathematical distribution. The LAST function has been tested on models where the hot source width to hot source depth ratio ranged from 0.6 to 1.6. Though the function works well within this range of source width to source depth, it is likely that it may not give similarly reliable results when this dimensionless ratio is exceeded. The LAST function has been tested for a wide range of temperature and depth conditions that were well outside the range of the laboratory simulations. However, the testing on real coal fires has been limited to data from the three transects that were available to the authors.

The current spatial resolution of satellite-based thermal infrared sensors may also be a limiting factor for the function. Applications of this function rely on surface temperature transects with a spatial resolution ~0.1–0.3 times the depth to the source. A sensor with a spatial resolution of 0.1–0.3 m is required for an accurate LAST depth estimation of a 1-m-deep hot source. A spatial resolution of 10–30 m is required to accurately estimate a hot source at 100 m depth. Determining the background temperature in real-world scenarios may pose an additional challenge because the width of the Lorentzian distribution, which is fit to the measured surface data, is dependent on the background temperature.

The LAST function assumes that the anomaly produced on the surface is due to one linear fire at depth. However, if the thermal anomaly is produced by fires in different seams at different depths, the function may not be able to resolve this.

CONCLUSIONS AND RECOMMENDATIONS

Estimations of the depth of underground coal fires are critical for targeting fire-fighting operations. Numerical modeling techniques for fire depth estimation are few, and all have their limitations. A laboratory simulation of subsurface coal-mine fires helps to understand the relation of surface thermal anomalies and the depth of the associated hot source. The simulation also provides a wealth of data that can be used as input for numerical modeling and for deriving a depth-estimation function. The LAST depth-estimation function empirically derived in this study is simple, easy to use, requires only a few input parameters, which can easily be derived from remote-sensing observations, and has been shown to give superior depth estimates for shallow subsurface fires compared to other known techniques.

The coal-fire modeling work could be taken a step further by designing laboratory models simulating an inclined linear hot source and testing the effect of various angles of inclination (simulating geologic dip of the strata) on the observed surface thermal anomaly. More complex models could be designed and tested to study the nature and dynamics of multiseam fires.

ACKNOWLEDGMENTS

We wish to thank Dr. Jonathan Dehn for help with the laboratory simulations, and Dr. Paul Layer for help with data analysis to derive and analyze the LAST function. We gratefully acknowledge NASA's Earth System Science Education for the 21st Century program and the Alaska Space Grant Program for funding parts of this research.

REFERENCES CITED

Banerjee, S.C., 1985, Spontaneous Combustion of Coal and Mine Fires: Rotterdam, A.A. Balekema, 168 p.

Bell, F.G., Stacey, T.R., and Genske, D.D., 2000, Mining subsidence and its effect on the environment: Some differing examples: Environmental Geology, v. 40, no. 1–2, p. 135–152, doi: 10.1007/s002540000140.

Berthelote, A.B., Prakash, A., and Dehn, J., 2007, An empirical function to estimate the depths of linear hot objects: Applied to the Kuhio Lava tube, Hawaii: Bulletin of Volcanology (press).

Cassells, C.J.S., 1998, Thermal Modeling of Underground Coal Fires in Northern China: The Netherlands, International Institute for Geo-information Science and Earth Observation, Dissertation number 51, 183 p.

Cassells, C.J.S., and van Genderen, J.L., 1995, Thermal modelling of underground coal fires in Northern China, in Curran, P.J., and Robertson, Y.C., comp., RSS95—Remote Sensing in Action, Proceedings of the 21st Annual Conference of The Remote Sensing Society, 11–14 September 1995: London, Butler & Tanner Ltd., p. 544–551.

Dragoni, M., 1989, A dynamical model of lava flows cooling by radiation: Bulletin of Volcanology, v. 51, p. 88–95, doi: 10.1007/BF01081978.

Greene, G.W., Moxham, R.M., and Harvey, A.H., 1969, Aerial infrared surveys and borehole temperature measurements of coal-mine fires in Pennsylvania, in Cook, J.J., ed., Proceedings of the Sixth International Symposium on Remote Sensing of the Environment, 13–16 October 1969: Ann Arbor, Michigan, University of Michigan, p. 517–525.

Gupta, R.P., and Prakash, A., 1998, Reflection aureoles associated with thermal anomalies due to subsurface mine fires in the Jharia Coalfield, India: International Journal of Remote Sensing, v. 19, no. 14, p. 2619–2622, doi: 10.1080/014311698214415.

Harris, A.J.L., and Rowland, S.K., 2001, FLOWGO: A kinematic thermo-rheological model for lava flowing in a channel: Bulletin of Volcanology, v. 63, p. 20–40, doi: 10.1007/s004450000120.

Incropera, F.P., and DeWitt, D.P., 2002a, Internal flow, in Incropera, F.P., and DeWitt, D.P., eds., Fundamentals of Heat and Mass Transfer (5th edition): New York: John Wiley and Sons, p. 465–555.

Incropera, F.P., and DeWitt, D.P., 2002b, Free convection, in Incropera, F.P., and DeWitt, D.P., eds., Fundamentals of Heat and Mass Transfer (5th edition): New York: John Wiley and Sons, p. 556–570.

Ishihara, K., Iguchi, M., and Kamo, K., 1990, Numerical simulations of lava flows on some volcanoes in Japan, in Fink, J.J., ed., Lava Flows and Domes: Emplacement Mechanisms and Hazard Implications: Berlin, Springer-Verlag, p. 174–207.

Keszthelyi, L., and Denlinger, R., 1996, The initial cooling of pahoehoe flow lobes: Bulletin of Volcanology, v. 58, p. 5–18, doi: 10.1007/s004450050121.

Manley, C.R., 1992, Extended cooling and viscous flow of large hot rhyolite lavas: Implications of numerical modeling results: Journal of Volcanology and Geothermal Research, v. 53, p. 27–46, doi: 10.1016/0377-0273(92)90072-L.

Mansor, S.B., Cracknell, A.P., Shilin, B.V., and Gornyi, V.I., 1994, Monitoring of underground coal fires using thermal infrared data: International Journal of Remote Sensing, v. 15, p. 1675–1685.

Mukherjee, T.K., Bandhopadhyay, T.K., and Pande, S.K., 1991, Detection and delineation of depth of subsurface coalmine fires based on an airborne multispectral scanner survey in part of Jharia Coalfield, India: Photogrammetric Engineering and Remote Sensing, v. 57, no. 9, p. 1203–1207.

Neri, A., 1998, A local heat transfer analysis of lava cooling in the atmosphere: Application to thermal diffusion-dominated lava flows: Journal of Volcanology and Geothermal Research, v. 81, p. 215–243, doi: 10.1016/S0377-0273(98)00010-9.

Panigrahi, D.C., Singh, M.K., and Singh, C., 1995, Predictions of depth of mine fire from the surface by using thermal infrared measurement, in Proceedings of the National Seminar on Mine Fires, Varanasi, India, 24–25 February 1995: Varanasi, Banaras Hindu University, p. 122–134.

Patrick, M.R., Dehn, J., and Dean, K., 2002, Numerical modeling of lava flow cooling applied to the 1997 Okmok eruption: Approach and analysis: Journal of Geophysical Research, v. 109, B03202, doi: 10.1029/2003JB002537.

Peck, D.L., Hamilton, M.S., and Shaw, H.R., 1977, Numerical analysis of lava lake cooling models: Part II. Application to Alae Lava Lake, Hawaii: American Journal of Science, v. 277, p. 415–437.

Peng, W.X., van Genderen, J.L., Kang, G.F., Guan, H.Y., and Tan, Y.J., 1997, Integrated modeling for estimating the depth of underground coal fires: Terra Nova, v. 9, p. 180–183, doi: 10.1046/j.1365-3121.1997.d01-31.x.

Prakash, A., 2005, Coal Fires: Causes and Hazards: www.gi.alaska.edu/~prakash/coal fires/causes_hazards.html (June 2007).

Prakash, A., and Gupta, R.P., 1999, Surface fires in Jharia Coalfield, India—Their distribution and estimation of area and temperature from TM data: International Journal of Remote Sensing, v. 20, no. 10, p. 1935–1946, doi: 10.1080/014311699212281.

Prakash, A., Sastry, R.G.S., Gupta, R.P., and Saraf, A.K., 1995, Estimating the depth of buried hot feature from thermal IR remote sensing data, a conceptual approach: International Journal of Remote Sensing, v. 16, no. 13, p. 2503–2510.

Prakash, A., Gupta, R.P., and Saraf, A.K., 1997, A Landsat TM based comparative study of surface and subsurface fires in the Jharia Coalfield, India: International Journal of Remote Sensing, v. 18, no. 11, p. 2463–2469, doi: 10.1080/014311697217738.

Prakash, A., Fielding, E.J., Gens, R., van Genderen, J.L., and Evans, D.L., 2001, Data fusion for investigating land subsidence and coal fire hazards in a coal mining area: International Journal of Remote Sensing, v. 22, no. 6, p. 921–932, doi: 10.1080/014311601300074441.

Rosema, A., Guan, H., Veld, H., Vekerdy, Z., ten Katen, A.M., and Prakash, A., 1999, Manual of coal fire detection and monitoring: Report of the project "Development and Implementation of a Coal Fire Monitoring and Fighting System in China": Utrecht, The Netherlands, Institute of Applied Geoscience Report NITG 99–221-C, ISBN 90-6743-640-2, 245 p.

Rosema, A., Guan, H., and Veld, H., 2001, Simulation of spontaneous combustion to study the causes of coal fires in the Rujigou Basin: Fuel, v. 80, no. 1, p. 7–16, doi: 10.1016/S0016-2361(00)00065-X.

Saraf, A.K., Prakash, A., Sengupta, S., and Gupta, R.P., 1995, Landsat TM data for estimating ground temperature and depth of subsurface coal fire in Jharia Coal Field, India: International Journal of Remote Sensing, v. 16, no. 12, p. 2111–2124.

Shaw, H.R., Hamilton, M.S., and Peck, D.L., 1977, Numerical analysis of lava lake cooling models: Part I. Description of the method: American Journal of Science, v. 277, p. 384–414.

Stracher, G.B., 2004, Coal Fires Burning Around the World: A Global Catastrophe: International Journal of Coal Geology, v. 59, no. 1–2, p. 1–6.

UMWA, 2005, United Mine Workers of America, Underground Mining: Room and Pillar Mining: http://www.umwa.org/mining/ugmine.shtml (June 2007).

van Genderen, J.L., and Haiyan, G., 1997, Environmental Monitoring of Spontaneous Combustion in the North China Coalfields: Final Report to European Commission: Enschede, International Institute for Geo-information Science and Earth Observation (ITC), 244 p.

Witter, J.B., and Harris, A.J.L., 2007, Field measurements of heat loss from skylights and lava tube systems: Journal of Geophysical Research, v. 112, B01203, doi: 10.1029/2005JB003800.

Wooster, M.G., Huppert, H.E., and Sparks, R.S.J., 1993, The crystallization of lava lakes: Journal of Geophysical Research, v. 98, no. B9, p. 15,891–15,901.

Zhang, J., Wagner, W., Prakash, A., Mehl, H., and Voigt, S., 2004, Detecting coal fires using remote sensing techniques: International Journal of Remote Sensing, v. 25, no. 16, p. 3193–3220, doi: 10.1080/01431160310001620812.

MANUSCRIPT ACCEPTED BY THE SOCIETY 7 MARCH 2007

Remotely sensed land-cover changes in the Wuda and Ruqigou-Gulaben coal-mining areas of China

Claudia Kuenzer*
Jianzhong Zhang
Institute of Photogrammetry and Remote Sensing, Vienna University of Technology, Gusshausstrasse 27-29, A-1090 Vienna, Austria

Stefan Voigt
German Remote Sensing Data Center (DFD), German Aerospace Center (DLR), Postfach 1116, D-82230 Wessling, Germany

Wolfgang Wagner
Institute of Photogrammetry and Remote Sensing, Vienna University of Technology, Gusshausstrasse 27-29, A-1090 Vienna, Austria

ABSTRACT

Medium-resolution Landsat5 TM (Thematic Mapper) and Landsat7 ETM+ (Enhanced Thematic Mapper) satellite data were used to investigate multitemporal land-cover changes in the Wuda and Ruqigou-Gulaben coal basins of China between 1987 and 2002. The surficial distribution of coal as detected by remote sensing may be an important indicator of mining activity. Coal that is exposed on the surface may appear as seams in openpit mines; the coal also may occur in storage and waste piles, as refuse at the entrance to a mine, or as coal dust covering areas adjacent to the exposed coal. Because both coal basins are affected by coal fires, monitoring surficial features, including the expansion of coal, is crucial for evaluating the risk for a coal fire. This risk increases where new, often private, mines develop.

Remote-sensing data reveal that the increasing number of exposures of coal in the Wuda and Ruqigou-Gulaben coal basins between 1987 and 2002 is correlative with an increase in mining activity there and the associated surficial coal dust adjacent to mining-transport networks. The data also reveal an increase in small-scale private coal mines, especially in the eastern part of the Wuda Coalfield. Such mines often suffer from inadequate mining and environmental regulations and are at high risk for coal fires. Satellite-based detection and in situ verification of new coal fires during field campaigns in 2003 and 2005 confirmed that land-cover analysis is useful for the identification of potential locations for coal fires. The changes that were observed in all other land cover in the Wuda and Ruqigou-Gulaben coal basins represent the more general dominant spatial dynamics in the two regions.

Keywords: coal fires, remote sensing, change detection, land-cover changes, mining.

*ck@ipf.tuwien.ac.at

INTRODUCTION

Different land-cover types have different spectral reflectance characteristics over the visible and infrared wavelength range of the electromagnetic spectrum. Hence, reflectance satellite data can be used to classify different land-cover types. This paper focuses on multitemporal land-cover analyses in the mining areas of Wuda and Ruqigou-Gulaben in north-central China. The two study areas are located near the border between northern Ningxia Province and Inner Mongolia Province (Fig. 1). Changes in this area over 15 yr (1987–2002) reflect an expansion in coal-mining activities. Analyses can be used to concentrate thermal coal-fire monitoring in the appropriate locations.

A comparison of two single classified satellite data sets was used to detect multiclass changes. A maximum likelihood classification (MLC) approach was used for the classification of remote-sensing data. This classification approach was supplemented with information that was gathered during several field campaigns and by additional geospatial data. Hence, each land-cover class was extracted from the data with the most possible detail. The Landsat satellite data sets were processed thoroughly to account for the geometric-, atmospheric-, and illumination-related disturbances that exist in the raw data. We are aware that MLC methods are state-of-the-art classification approaches, but they have yielded land-cover classification of great detail and have attained the highest accuracies in numerous studies. The purpose of this article is not to introduce the latest—and, perhaps, not yet validated—remote-sensing methods to the general geoscience audience, but to underscore and strengthen the possibility of using remote sensing as an analytical tool for the early recognition of developing coal-fire risk areas.

STUDY AREAS

The first study area, Wuda, is located in the Inner Mongolia Autonomous Region. The city borders a structural syncline that hosts a large coal-fire area that has been investigated over the past several years. It is located on the western side of the Yellow River, north of the Helan Shan mountain range. The first dunes of the Badain Jaran Desert are located 10 km to the west of Wuda. Wuda city center is located at ~39.51°N, 106.60°E. The nearest big cities are Wuhai, located 20 km to the northeast, Shizuishan-Dawukou, located 75 km to the southwest, and Yinchuan, the capital of Ningxia Hui Autonomous Region, which is located 150 km to the south. The elevation in this region varies from 1010 m to 1980 m above sea level.

The second study area, Ruqigou-Gulaben, is located in the northern part of the Helan Shan mountain range in Ningxia Hui Autonomous Region. The Helan Mountains strike from southwest to northeast for 200 km along the northwestern border of Ningxia and Inner Mongolia. Their average elevation is 2000 m; the highest peak, Helan Mountain, reaches 3557 m. The mining area is centered at ~39.07°N, 106.12°E. The nearest large cities are Shizuishan-Dawukou, located 25 km to the east, and Yinchuan, located 70 km southeast of Ruqigou. Wuda is located 75 km to the northeast. Elevations in this region vary from 1400 m to 2640 m above sea level.

Both areas have a semiarid to arid climate, with an annual precipitation of less than 220 mm. Temperatures vary between –31 °C and 41 °C, and average 8.6 °C in Wuda and 6.4 °C in Ruqigou-Gulaben. Soil genesis progresses very slow, and vegetation cover is sparse. Coal has been mined in both areas since the late 1950s. The first coal fires in the Wuda area were discovered in 1961. Today 20 major surface and subsurface fires are consuming coal in the structural syncline located west of the city. The fires cover more than 4 km^2, and the annual coal loss is estimated to exceed 200,000 tons. The mines in the Ruqigou-Gulaben area are affected by more than 20 coal fires, most of which are being extinguished. The total area affected by coal fires in the Ruqigou-Gulaben region is assumed to cover more than 2 km^2 (Chen, 1997; Kuenzer, 2005; Kuenzer et al., 2007a). The terrain and geographic features of both areas are depicted in Figure 1.

SATELLITE DATA PROCESSING

Selection of Imagery and Preprocessing

Two cloud-free Landsat data sets (Landsat5 TM [Thematic Mapper] and LANDAT7 ETM+ [Enhanced Thematic Mapper]) were analyzed (i.e., path 129, row 33 of 20.09.1987 and of 21.09.2002). Landsat TM and ETM+ have a spatial resolution of 30 m/pixel in the multispectral (visible, near infrared, and mid-infrared) bands. The data set of 2002 was georeferenced on the basis of global positioning system (GPS) data that were acquired during a field campaign in September of 2002. The GPS data consisted of tracked roads as well as distinct points that are stable over time (e.g., intersections). We referenced the scene of 1987 to the master scene of 2002. A 25 m digital elevation model (DEM) that was derived from ERS-2 SAR (European Remote Sensing Satellite-2 Synthetic Aperture Radar) data was resampled from 25 m to 30 m resolution. This terrain data set was used to orthorectify both Landsat scenes. Orthorectification accounts for spatial shifts of individual pixels that occur as a result of terrain-induced object displacement. The overall root mean squared error, which is a quality measure for the orthorectification process, was less than 30 m (subpixel accuracy) for each scene.

Satellite data are most accurate when reflection changes that are detected by the sensor are caused only by surface characteristics. However, the reflected signal that arrives at the satellite sensor is influenced by relief-induced illumination differences and atmospheric gasses and aerosols. To account for such effects, the data were corrected radiometrically using the ATCOR-3 program (Schott, 1997; Richter, 1998; Song et al., 2001). The code calculates at-sensor radiances for each pixel in each channel (sensor calibration). Then, atmospheric transmittance, direct and diffuse solar flux, and path radiance—as needed for the atmospheric correction—are extracted from an atmospheric database compiled with MODTRAN-4 radiative transfer

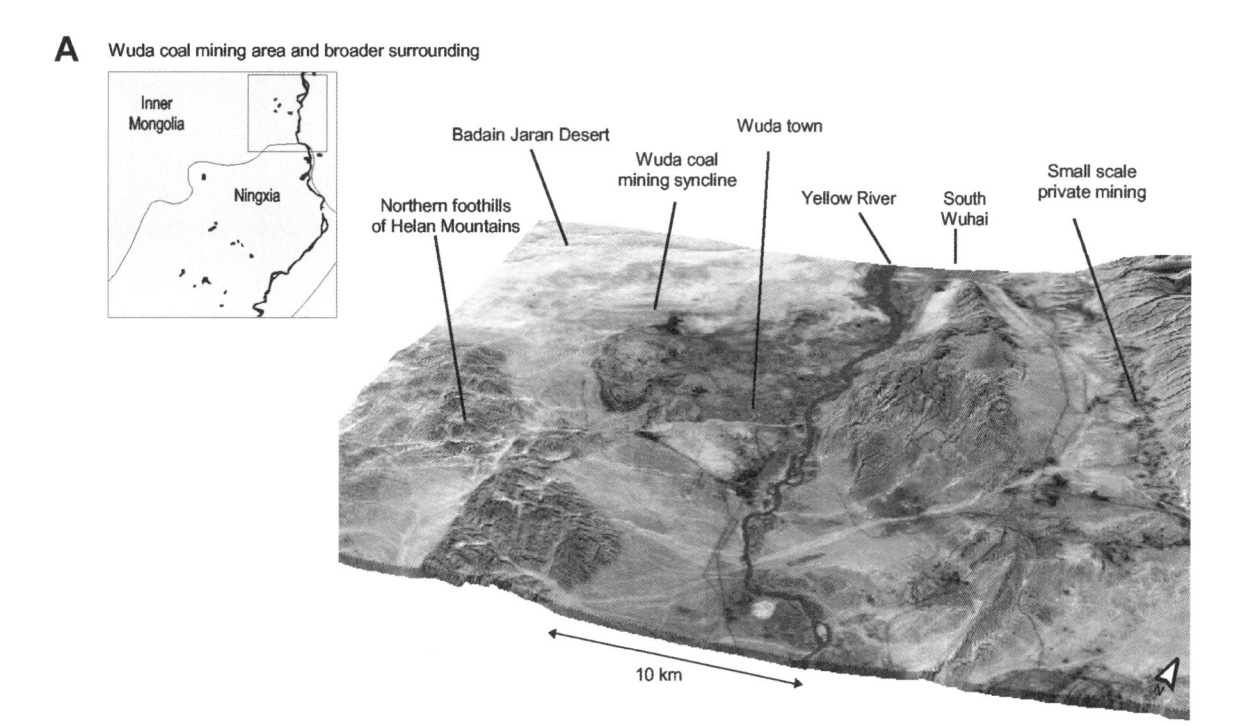

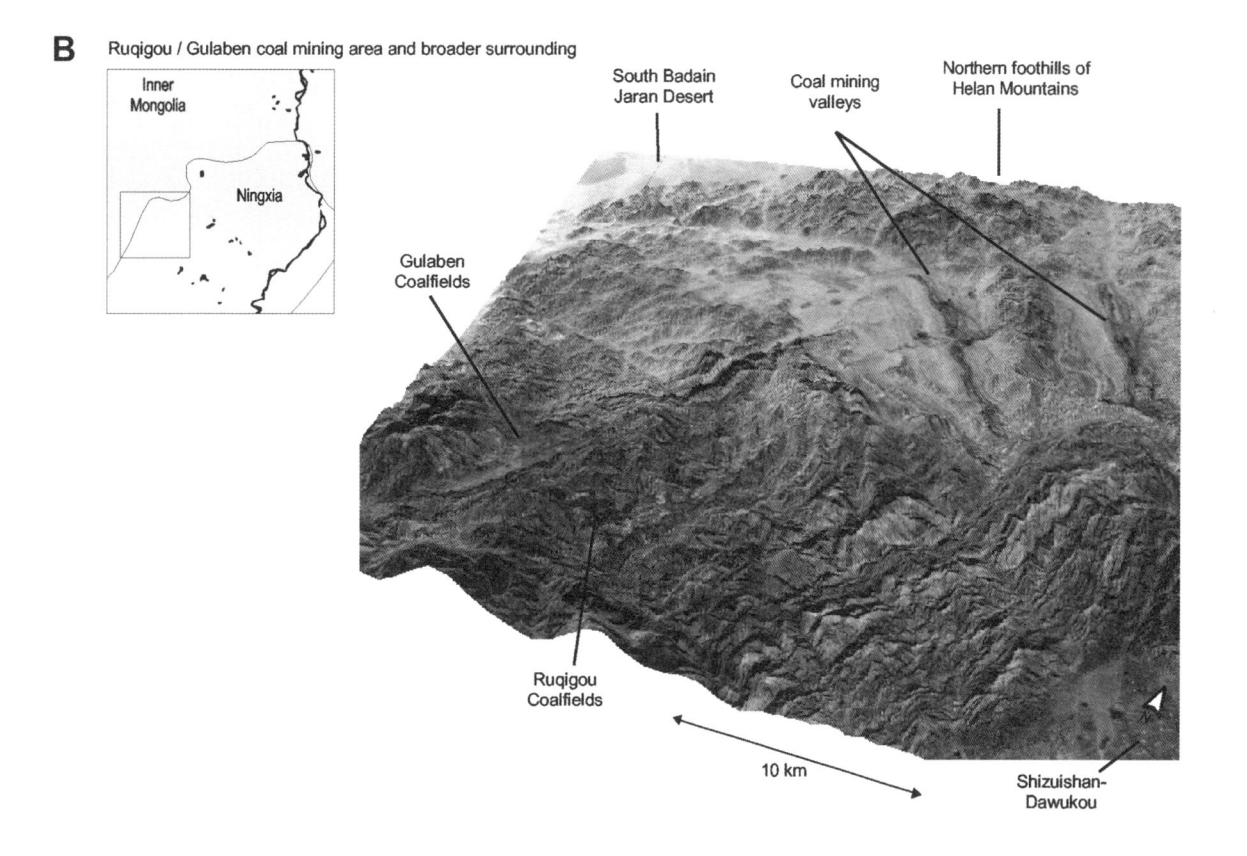

Figure 1. General characteristics and topography of the study areas Wuda (A) and Ruqigou-Gulaben (B) as seen in Landsat 7 ETM+ satellite data from September 2002 combined with a digital elevation model. Both coal-mining areas are located in the border region of northern Ningxia Province and Inner Mongolia Province in north-central China.

code. This database represents a wide range of atmospheric conditions that are based on manifold cross-combinations of water-vapor content (8 classes, 0.4–4.1 cm water-vapor column), aerosol types (5 classes), ground elevations (6 classes, 0–2.5 km), solar zenith angles (8 classes, 0–70°), and visibility. It allows for an approximated definition of atmospheric conditions for the satellite data without the need to recalculate the whole radiative transfer for each case. For details on the calculations within ATCOR-3, see Richter (1998). This preprocessing is also shown in the flowchart of Figure 2.

Supervised Satellite Image Classification

To classify a satellite data set, so-called "training areas," which represent a certain land-cover class, need to be provided to the classification algorithm. Special care was given to the selection of such training data, which were mapped in situ. Areas of a specific land-cover class were mapped so as to be highly representative for a spectral class, to be representative for the variance within an object class, to contain an appropriate number of pixels, and finally not to contain pixels of other adjacent classes. The principle adopted was that the minimum number of pixels in a training area should exceed $10n$, where n is the number of bands of the satellite image (Congalton and Green, 1999).

The data were classified into 25 major classes for the Wuda area and 15 classes for the Ruqigou area. All available bands were used as input channels. The maximum likelihood algorithm was the main algorithm applied for classification in both areas. However, the whole data set was not classified simultaneously. First, vegetation classes were separated using thresholds in the soil-adjusted vegetation index and normalized-differential vegetation index derivatives. DEM information was included to distinguish desert from mountainous vegetation. We sorted the geologic classes in a last step by masking all classified vegetated areas and applying the maximum likelihood algorithm based on training samples (Fig. 2).

The most frequently used MLC belongs to the class of parametric classifiers, which assumes that the data are distributed according to a previously defined probability model that has parameters derived automatically through the statistics of the input training data. Each pixel, $x_j R^d$, is classified independently by assigning it to the class, $y_j\{\omega_1,\ldots,\omega_m\}$, that has the highest a posteriori probability $P(y_j = \omega_k|x_j)$. The Bayes' rule is used to calculate these probabilities:

$$P(y_j = \omega_k|x_j) = [P(x_j|y_j = \omega_k)p(\omega_k)] / [\Sigma_i P(x_j|y_j = \omega_i)p(\omega_i)], \quad (1)$$

where $P(x_j|y_j = \omega_k)$ is the probability that x_j is observed for the given class ω_k, and $p(\omega_k)$ denotes the a priori probability of the class ω_k, which is known approximately from ground checks. To calculate $P(x_j|y_j = \omega_k)$, a multivariate Gaussian distribution with mean vector μk and covariance matrix Σ_k is assumed for each class ω_k:

$$P(x_j|y_j = \omega_k) \propto \left[(2\pi)^d |\Sigma k|\right]^{\frac{1}{2}} \cdot \exp\left[-\frac{1}{2}(x_j - \mu_k)^2\right]. \quad (2)$$

The resulting a posteriori probability (Eq. 1) is maximized for a given data point x_j when it is assigned the label ω_{k^*} according to the classification rule:

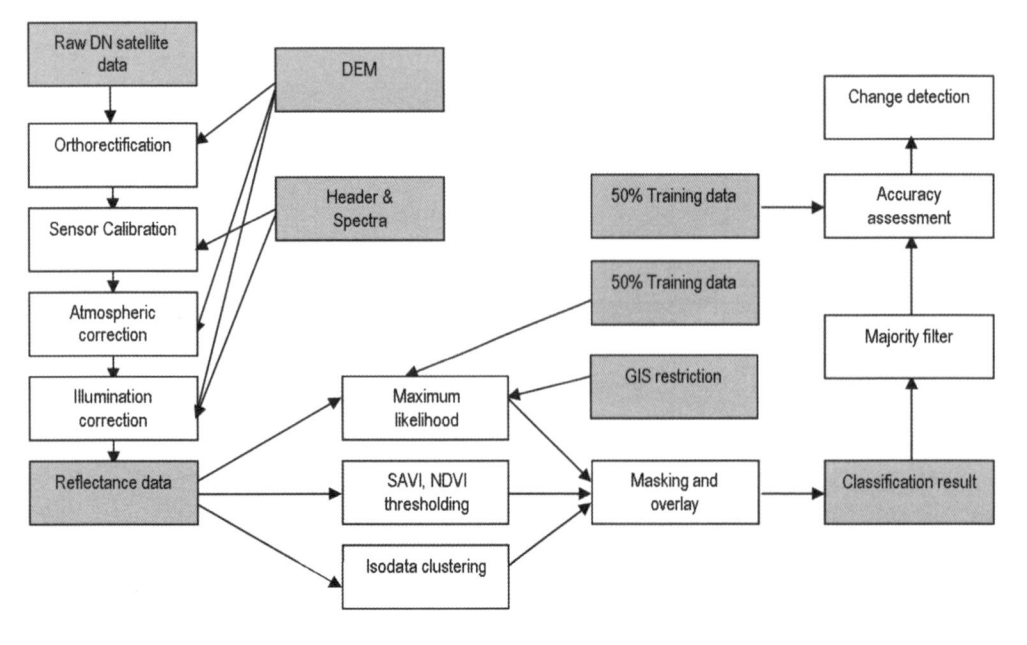

Figure 2. Flowchart of the land-cover classification scheme, including detailed pre- and postprocessing. Vegetation indices, elevation data, and in situ knowledge supported the automated process. DEM—Digital Elevation Model, DN—Digital Number, GIS—Geographic Information System, NDVI—Normalized Difference Vegetation Index, SAVI—Soil Adjusted Vegetation Index.

$$k^* = \arg\max_{k=1,\ldots,m} -\frac{1}{2}(x_j - \mu_k)^T \sum_k^{-1}(x_j - \mu_k) - \frac{1}{2}$$
$$\ln(|\sum_k|) + \ln[p(\omega_k)]. \tag{3}$$

This classification rule leads to quadratic decision boundaries between the single classes. The corresponding class-dependent parameters μ_k and Σ_k are estimated from the training data vectors x_i available for class k:

$$\mu k = \frac{1}{n_k}\sum_{i=1}^{nk} x_i \text{ and } \sum k = \frac{1}{n_k}\sum_{i=1}^{nk}(x_i - \mu_k)^T. \tag{4}$$

Here, n_k is the number of training samples for class k (Schroeder-Brzosniowsky, 2000; Pal et al., 2001; Keuchel et al., 2003). T—transpose. Therefore, MLC quantitatively evaluates the variance and the covariance of the individual classes when assigning an unknown pixel. A cluster of a defined class is described by its mean vector and its covariance matrix. Given these parameters, the statistical probability of a pixel belonging to a particular predefined class can be computed. After evaluating the probability of membership for each of the predefined training classes, the pixel is assigned to the class with the highest probability (Richards, 1986). MLC yields better results when the class means are separated farther from each other and when there is a smaller variance within the predefined classes (Kraus, 1990). However, Patrono (1996) states that it often is impossible to classify an entire image without segmenting or masking the original data.

Settlement extraction was supported through additional Geographic Information System (GIS) information that restricted the MLC for this land-cover class (Watson and Wilcock, 2001). Thus, the classification process is a hierarchical one, where certain parts of the image are masked. The final images were majority filtered with a 3 × 3 matrix to exclude single scattered misclassified pixels and to smooth the overall appearance of the classification results.

RESULTS

Quantitative Changes in the Wuda Area

Figure 3 shows the results of the land-cover classification for the broader Wuda study area for 1987 and 2002. Distinct changes can be noticed in settlement development, coal surfaces, water surfaces, and agricultural areas. Changes from 1987 to 2002 were quantified as the percentage of change with respect to 1987. Although both scenes are from the same month and nearly the same day, variations in natural vegetation cannot be compared on a quantitative basis. Annual variations in precipitation, temperature, and runoff patterns are responsible for different phenological developments. In 2002, which was an unexpectedly wet year (Y. Jia, 2003, personal commun.), a much larger area was covered with the typical desert shrub vegetation (mainly *Artemisia*). This led to changes in the geologic classes, which remain stable and only are covered with vegetation. The main agricultural area north of Wuda has increased substantially (~130%). The city of Wuda (in the center of the classified subset), as well as the southern margins of the city of Wuhai (at the northern rim of the subset), has expanded significantly. Furthermore, industrial areas east of the Yellow River and directly south of Wuda have grown. In total, the settled or surface-sealed area has increased by more than 150%. Information from statistical yearbooks confirms that the population of greater Wuda increased from 40,000 in 1987 to 125,000 in 2002.

Other classes that were extracted are deep and shallow water of the Yellow River. The river is extremely rich in sediment and algae. Hence, the spectra show a distinct peak in the near infrared range, which allows for good separation from clearer lake water. Water depth can be differentiated through a higher signal in the visible part of the spectrum for shallow-water areas. The water level of the Yellow River has declined. The river arm north of Wuda (1987) was dried out in 2002. Deep river areas have declined by 25%, whereas 43% of shallow river areas have been lost compared with 1987. According to local authorities and farmers, the withdrawal of water for agricultural irrigation has increased dramatically in recent years. This is of great concern to farmers. In dry years, the water demand can hardly be met, and the once-majestic river appears like a small, shallow stream (Y. Jia, 2003, personal commun.).

Coal mining in the Wuda syncline has increased rapidly, as has the development of new and often private small mines (Kuenzer et al., 2007d). This can be seen, especially in the valley east of the central limestone complex on the eastern side of the Yellow River. Here, a lot of private mines, which occur as "black little dots" along the valley, indicate new small-scale mining activity. New coal fires were detected during ground checks in this area in 2004. Additionally, industrial areas with coal waste and storage piles have spread out in the southeastern portion of the subset as well as directly southeast of Wuda.

Because the accuracy of satellite-based classification is not 100%, the percentage changes give only estimates of the magnitude of change. The accuracy of the classification was assessed using ground truth polygons mapped with GPS during the 2002 field campaign. The overall classification accuracy, error of omission, and error of commission also represent reliability measures for the analysis. The overall classification accuracy for the Wuda region was 89.41%.

Quantitative Changes in the Ruqigou-Gulaben Area

Similar changes over the 15 yr period were investigated in the greater Ruqigou study area. According to classification results, coal-mining areas increased by more than 300%. This

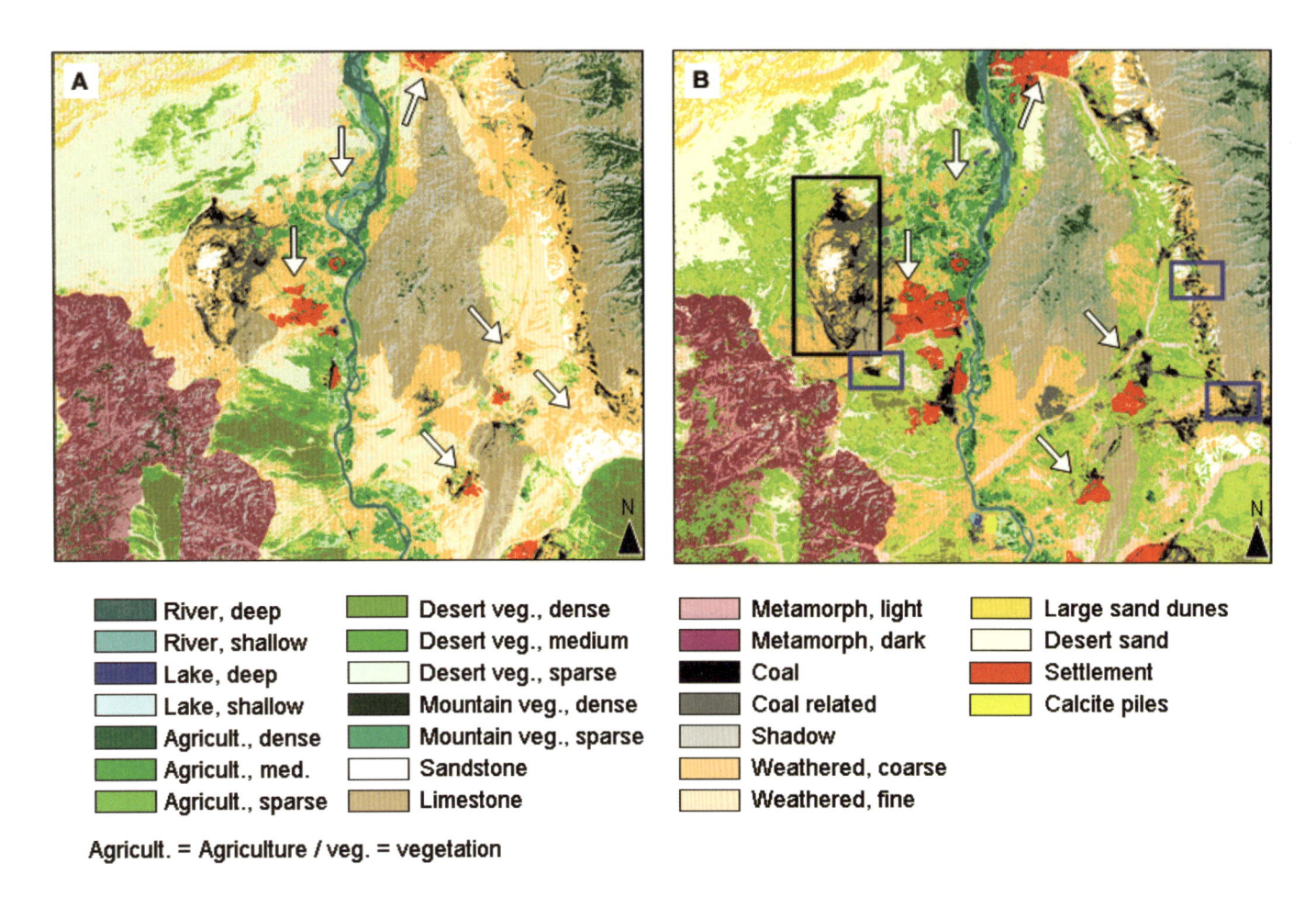

Figure 3. Land-cover classification for Wuda in 1987 (A) and in 2002 (B) from radiometrically corrected Landsat5 TM and Landsat7 ETM+ data. The arrows indicate areas of dominant change. Black box—known coal-fire area. Blue boxes—newly developed coal fires that were validated in 2003 and 2004. Blue box west of the Yellow River—coal storage pile fire. Blue boxes east of the Yellow River—subsurface coal fires. The big red settlement area in the center of the image is the city of Wuda. The subset equals the area of the local topographic map 1:100,000 and covers 35 km (W–E) × 42 km (N–S).

increase resulted from the expansion of the Ruqigou and Baijigou coal mines (already known coal-fire areas), as well as from the mining that started in the two parallel valleys of Hulusitai and Shitanjing in the early 1990s. Expansion of coal surfaces in this mountainous area has been greater than in Wuda. There are fewer indicators of the rapid sprawl of private mining, instead there has been significant industrialized expansion (Fig. 4).

Less expansion of settlement has occurred in Ruqigou compared with Wuda. This is because Ruqigou is an isolated mountain village where people make their living exclusively from coal mining and some tertiary sector activities. In Wuda, employment is available in other primary, secondary, and tertiary activities. Therefore, the Wuda region experienced more migration from outside areas than did Ruqigou, which had 20,000 inhabitants in 1987 and 35,000 in 2002. It is estimated that coal-mining activities in Ruqigou will last for only 50 more years. Therefore, the government has started to move people slowly from Ruqigou to Shizuishan-Dawukou.

Land-Cover Information as a Support to Coal-Fire Detection

Coal fires in both study areas occur to a depth of ~150 m. Below this depth, groundwater exists or the oxygen supply is not sufficient to sustain a fire. Furthermore, in China, mining technology is less advanced, and mines are not deep. Depending on the dip of a coal seam, the surface expression of a coal fire usually occurs in close vicinity to coal on the surface. In the case of a burning waste or coal storage pile, the thermal anomaly will occur "on top" of the surface class "coal." This concept was used by Zhang (2004) and Kuenzer (2005), who combined the results of remote-sensing–based thermal anomaly extraction from thermal data and the demarcation of coal-fire risk areas (Kuenzer et al., 2007b).

Zhang (2004) developed a moving-window–based algorithm for the automated extraction of thermal anomalies from thermal satellite images. Taking into account that the surface "coal" only

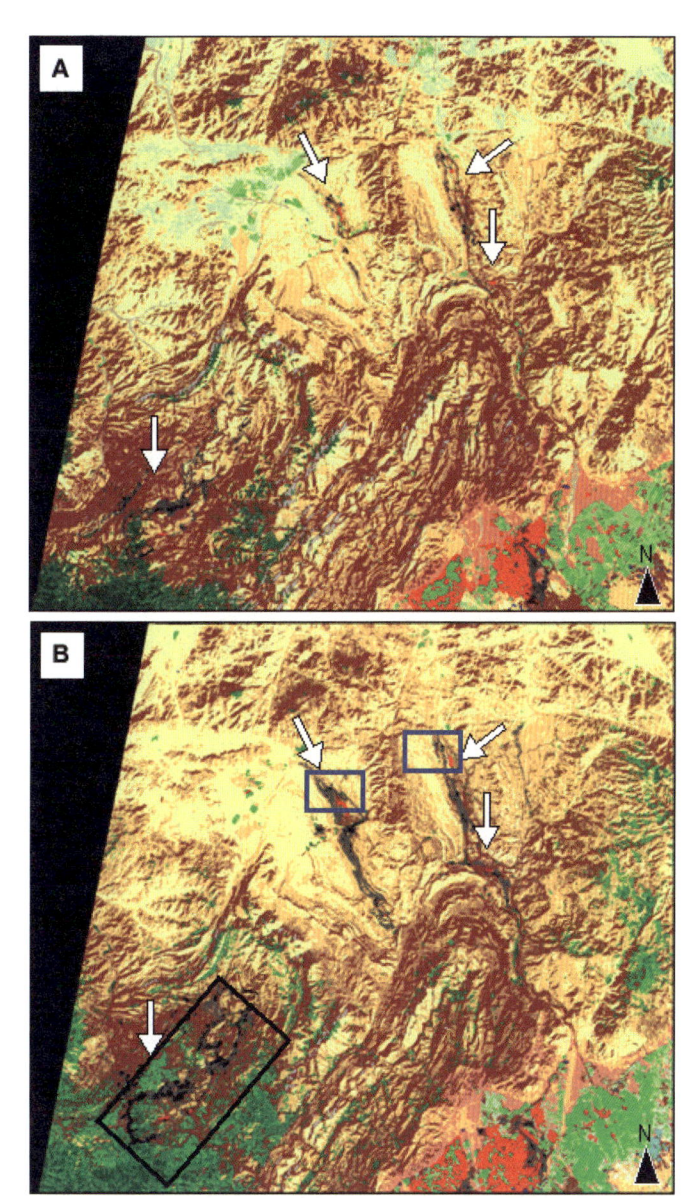

Figure 4. Land-cover classification for Ruqigou in 1987 (A) and in 2002 (B) from radiometrically corrected multispectral Landsat5 TM and Landsat7 ETM+ data. The arrows indicate areas of dominant change. Black box—known coal-fire area near Ruqigou-Gulaben. Blue boxes—newly developed coal fires that were validated in 2003. Blue box on the right side—coal storage pile fires near Hulusitai. Blue box on the left side—subsurface coal fires near Shitanjing. The subset equals the area of the local topographic map 1:100,000 and covers 35 km (W–E) × 42 km (N–S).

occupies a small percentage of a whole Landsat image, the land-cover information can help to exclude false alarms in thermal anomaly extraction and can focus thermal coal-fire investigations to the appropriate locations. By applying an automated surface extraction algorithm based on partial unmixing, Kuenzer (2005) showed that, on average, more than 80% of thermal anomalies that are extracted from thermal Landsat daytime data can be rejected because they are not related to coal fires but result from industry, biomass fires, or solar effects. In thermal Landsat nighttime data, with a smaller influence of uneven solar heating, still more than 50% of automatically extracted thermal anomalies are not related to coal fires. An automated coal-fire risk area demarcation scheme, which is based on the automated extraction of the surface class coal, can focus thermal monitoring to the right places. Automated thermal anomaly extraction and automated coal-fire risk area demarcation have a strong synergetic effect for the detection of new coal fires.

Detection and Verification of Unknown Coal Fires in 2003 and 2004

In September of 2003, a field trip was organized to the remote area shown in Figure 5. The fire-fighting team of Wuda, who joined the survey, did not know about the area visited beforehand, nor did they have any information on possible coal fires in that region. The area is settled very sparsely and is only accessible via dirt roads. Information from the local inhabitants revealed that the region once was a prospering coal-mining area;

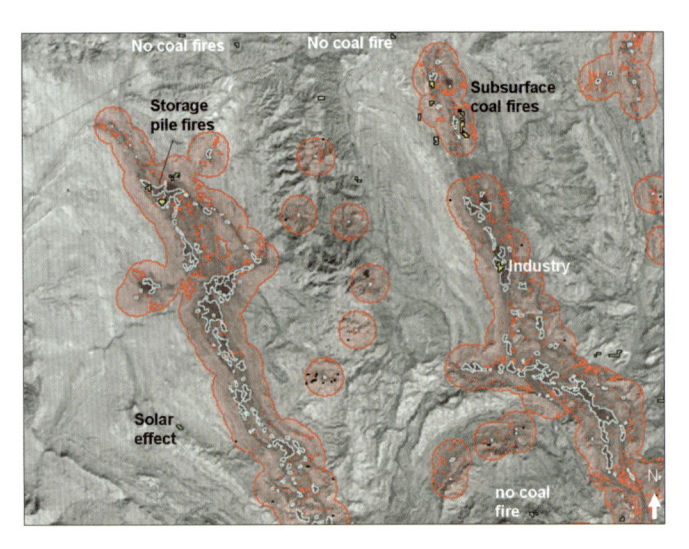

Figure 5. Newly detected coal fires based on land-cover information and thermal anomaly detection in the two parallel valleys of Hulusitai and Shitanjing, 30–50 km northeast of Ruqigou-Gulaben and 50 km southwest of Wuda. The coal fires were not known outside of the local community. Extracted coal is displayed as a superimposed vector layer in light blue; thermal anomalies are displayed in yellow (within the demarcated risk area) and green (outside of the demarcated risk area); and demarcated coal-fire (risk) areas are displayed in red. Center coordinate: 611355°E, 4342156°N, UTM, Z48N.

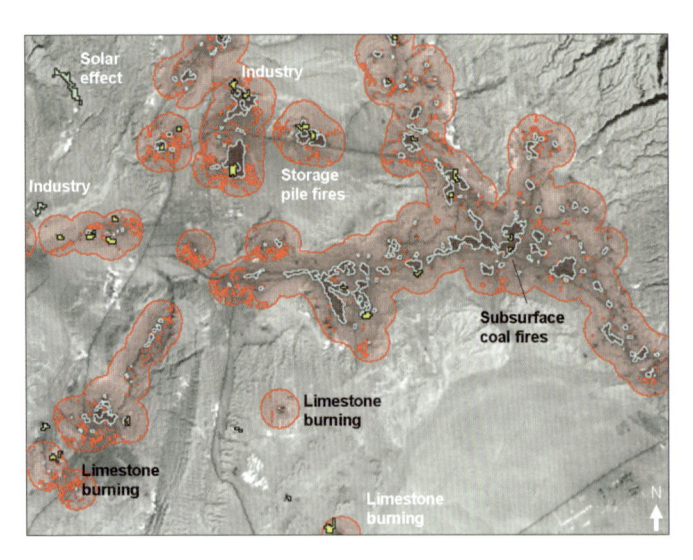

Figure 6. Newly detected coal fires based on coal-fire (risk) area demarcation and thermal anomaly detection 25 km southeast of Wuda. The coal fires were known only to a few local inhabitants. Extracted coal is displayed as a superimposed polygon vector layer in light blue; thermal anomalies are displayed in yellow (within the demarcated risk area) and green (outside of the demarcated risk area); and demarcated coal-fire (risk) areas are displayed in red. Center coordinate: 669169°E, 4365414°N, UTM, Z48N.

however, production had decreased drastically because of financial problems. With the help of GPS, the satellite data thermal anomalies were located to an area of very rugged terrain of former coal waste piles, abandoned mines, and heavily disturbed coal outcrops. Five of the six suspicious thermal anomalies (as extracted from thermal satellite data) were fires in a coal seam, whereas the other anomalies resulted from fires in a coal waste pile. Therefore, all six anomalies could be verified as coal-fire anomalies. In this area, the fire temperatures ranged from 170 °C to 340 °C.

In June of 2004, another field trip was taken to the eastern side of the Yellow River, ~25 km southeast of Wuda (Fig. 6). Based exclusively on remote-sensing data, this area was shown to host at least one coal fire. The thermal anomalies that were extracted and located within the demarcated coal-fire areas proved to be fires in underground coal seams (Kuenzer et al., 2007b). The local inhabitants knew about the fires; however, they were unknown to the authorities of Wuda and Wuhai city (Y. Sun, 2004, personal commun.). According to a local worker, the fire had developed in private coal mines that were abandoned and sealed improperly. The local people had just recently started to seal these mines with sand and loess. Coal-fire–fighting was initiated, and a trench was used in an attempt to separate the burning portion of the coal seam from the unaffected part (Fig. 7). To date the fire is not under control, and it spreads along the dip and strike directions.

DISCUSSION

The automated, and, thus, standardized demarcation of coal-fire risk areas and the detection of coal-fire–related thermal anomalies combine to be challenging task in remote-sensing science. Coal-fire–related thermal anomalies always will occur in the vicinity of coal surfaces. Fire-induced vegetation degradation and the remote-sensing–based detection of pyrometamorphic rocks (rocks that have changed in color and texture as a result of the adjacent heat) help to demarcate coal-fire risk areas (Kuenzer, 2005). However, many coal-fire risk areas do not show thermal anomalies yet. Furthermore, thermal anomalous clusters—even when located within risk areas—may result from "false alarms," such as industry, settlement, or biomass burning. Hence, a thorough field check is mandatory before new observations derived from remote-sensing data are announced. This study was the first time that unknown coal fires were detected first using remote-sensing data and then were validated subsequently in the field.

CONCLUSION

Satellite-based multitemporal land-cover analysis helped to visualize the development of coal mining in the two study areas that were investigated. A strong increase in mining-related surfaces could be observed in both areas. Especially for Wuda, an increase in mining activity was accompanied by substantial population growth and a resultant expansion of

Figure 7. Newly detected coal fires based on coal-fire (risk) area demarcation and thermal anomalies 25 km southeast of Wuda. Note the coal-fire–related smoke (A) and the improperly sealed private mines (B). The trench was dug to prevent the spreading of the fire. These are the same fires as in Figure 5. Photos are by (A) C. Kuenzer and (B) J. Zhang, June 2004.

settlement. This was framed by expanding agricultural areas and increased water withdrawal from the Yellow River. Multitemporal land-cover analysis emphasized areas where new coal-mining operations had developed, especially east of the Yellow River. In the Ruqigou-Gulaben area, coal mining had expanded northeast of the existing mines as well as in the two valleys of Hulusitai and Shitanjing.

Combinations of thermal anomalies retrieved from thermal satellite data with the occurrence of coal deposits support the separation of coal-fire–related thermal anomalies from false alarms.

With the failure of the Landsat7 satellite in May 2003, the demand for new sensors for multispectral and thermal coal-fire analysis is increasing. Low-resolution MODIS (Moderate Resolution Imaging Spectroradiometer) data (1 km pixel) are suitable only for the detection of very hot coal-fire anomalies on the surface; they are not suitable for the spatially detailed extraction of coal outcrops or coal-covered surfaces (see the article by Hecker and Kuenzer, this volume; Kuenzer et al., 2007c). Therefore, the Advanced Spaceborne Thermal Emission and Reflection Radiometer (ASTER) satellite sensor is considered to be a major source for future multispectral and thermal data. With a 30 m pixel resolution in the multispectral bands, and 90 m spatial resolution in the thermal band, its accuracy is sufficient for the detection of smaller coal accumulations and thermal anomalies that result from subtle subsurface coal fires.

ACKNOWLEDGMENTS

The authors thank Gao Yan for help with in situ mapping of land-cover classes. Further thanks go to Yaorong Jia and Yulin Sun from the Wuda mine for their support during field campaigns. The comments of the reviewers as well as a grammatical reworking by Stacia Spaulding have greatly improved the quality of the manuscript.

REFERENCES CITED

Chen, L., 1997, Subsidence Assessment in the Ruqigou Coalfield, Ningxia, China, Using a Geomorphological Approach (master's thesis): Enschede, the Netherlands, International Institute for Geo-Information Science and Earth Observation, 107 p.

Congalton, G., and Green, K., 1999, Assessing the Accuracy of Remotely Sensed Data: Principles and Practices: Boca Raton, Florida, CRC/Lewis Press, 137 p.

Keuchel, J., Naumann, S., Heiler, M., and Siegmund, A., 2003, Automatic land cover analysis for Tenerife by supervised classification using remotely sensed data: Remote Sensing of Environment, v. 86, no. 4, p. 530–541, doi: 10.1016/S0034-4257(03)00130-5.

Kraus, K., 1990, Fernerkundung Band 2: Auswertung Photographischer und Digitaler Bilder: Bonn, Germany, Dümmler Press, 314 p.

Kuenzer, C., 2005, Demarcating Coal Fire Risk Areas Based on Spectral Test Sequences and Partial Unmixing Using Multi-Sensor Remote Sensing Data [Ph.D. thesis]: Vienna, Austria, Technical University Vienna, 199 p.

Kuenzer, C., Zhang, J., Tetzlaff, A., Voigt, S., Van Dijk, P., Wagner, W., and Mehl, H., 2007a, Uncontrolled coal fires and their environmental impacts: Investigating two arid mining environments in north-central China: Applied Geography, v. 27, pp. 42–62.

Kuenzer, C., Zhang, J., Li, J., Voigt, S., Mehl, H. and Wagner, W., 2007b, Detecting unknown coal fires: Synergy of coal fire risk area delineation and improved thermal anomaly extraction: International Journal of Remote Sensing, v. 28, no. 20, p. 4561–4585, doi: 10.1080/01431160701250432.

Kuenzer, C., Hecker, C., Zhang, J., Wessling, S., and Wagner, W., 2007c, The potential of multi-diurnal MODIS thermal bands data for coal fire detection: International Journal of Remote Sensing, doi: 10.1080/01431160 1352147 (in press).

Kuenzer, C., Bachmann, M., Mueller, A., Lieckfeld, L. and Wagner, W., 2007d, Partial unmixing as a tool for single surface class detection and time series analysis: International Journal of Remote Sensing, doi: 10.1080/0143116 0701469107 (in press).

Pal, S.K., Bandyopadhyay, S., and Murthy, C.A., 2001, Genetic classifiers for remotely sensed images: Comparison with standard methods: International Journal of Remote Sensing, v. 22, no. 13, p. 2545–2569, doi: 10.1080/01431160152497727.

Patrono, A., 1996, Synergism of remotely sensed data for land cover mapping in heterogeneous alpine areas: An example combining accuracy and resolution: Enschede, The Netherlands, International Institute for Geo-Information Science and Earth Observation (ITC) Journal, no. 2, p. 101–109.

Richards, J.A., 1986, Remote Sensing Digital Image Analysis—An Introduction: Berlin, Germany, Springer-Verlag, 281 p.

Richter, R., 1998, Correction of satellite imagery over mountainous terrain: Applied Optics, v. 37, no. 18, p. 4004–4015.

Schott, J.R., 1997, Remote Sensing: The Image Chain Approach: New York, Oxford University Press, 394 p.

Schroeder-Brzosniowsky, M., 2000, Stochastic Modeling of Image Content in Remote Sensing Image Archives [Ph.D. thesis]: Zürich, Switzerland, Swiss Federal Institute of Technology–Zürich, 177 p.

Song, C., Woodcock, C., Seto, K., Lenney, M., and Macomber, S., 2001, Classification and change detection using Landsat TM data: When and how to correct atmospheric effects?: Remote Sensing of Environment, v. 75, p. 230–244, doi: 10.1016/S0034-4257(00)00169-3.

Watson, N., and Wilcock, D., 2001, Preclassification as an aid to the improvement of thematic and spatial accuracy in land cover maps derived from satellite imagery: Remote Sensing of Environment, v. 75, p. 267–278, doi: 10.1016/S0034-4257(00)00172-3.

Zhang, J., 2004, Spatial and Statistical Analysis of Thermal Satellite Imagery for Extraction of Coal Fire Related Anomalies [Ph.D. thesis]: Vienna, Austria, Vienna University of Technology, 166 p.

MANUSCRIPT ACCEPTED BY THE SOCIETY 7 MARCH 2007

Remote-sensing–based coal-fire detection with low-resolution MODIS data

Christoph Hecker*

International Institute for Geo-Information Science and Earth Observation (ITC), Box 6, 7500 AA, Enschede, The Netherlands

Claudia Kuenzer
Jianzhong Zhang

Institute of Photogrammetry and Remote Sensing, IPF, Vienna University of Technology, Gusshausstr. 27-29, A-1040 Wien, Austria

ABSTRACT

Remote-sensing imagery is often used for detecting and monitoring coal fires. The Landsat7 Enhanced Thematic Mapper Plus (ETM+) sensor and its predecessors of the Landsat family were frequently utilized for that purpose. With Landsat5 quickly approaching the end of its lifetime and the partial malfunction of Landsat7 in 2003, other potential sensors, including Moderate Resolution Imaging Spectroradiometer (MODIS), merit investigation. One kilometer MODIS data were successfully acquired and analyzed to detect coal fires in China during one summer and two winter night scenes. Band ratios of MODIS bands 20/32 enhanced subpixel-sized hot spots over background values, and an automated thermal anomaly algorithm was an asset in extracting potential coal-fire locations. For areas with known subsurface fires, between 0% and 17% were correctly detected in the three images. Areas with surface fires had success rates of 42% to 49%. These results indicate that MODIS is potentially useful for monitoring large areas for newly developing surface coal fires. Most subsurface coal fires, however, remain undetected.

Keywords: coal fires, remote sensing, thermal infrared, MODIS, Landsat, automated detection algorithm.

INTRODUCTION

Many studies have shown that remote sensing is a suitable tool for detecting and monitoring coal fires (Prakash and Gupta, 1999; Rosema et al., 1999; Tetzlaff, 2004; van der Meer et al., 2004; Zhang, 2004; Zhang et al., 2004; Kuenzer, 2005). Before 1990, most coal fire research was based on aerial scanner surveys (Moxham and Greene, 1967; Fisher and Knuth, 1968; Knuth et al., 1968; Greene and Moxham, 1969; Rabchevsky, 1972; Ellyett and Fleming, 1974). Only when Landsat Thematic Mapper (TM) data became more widely available was spaceborne coal-fire research initiated. For example, Saraf et al. (1995) as well

---

*hecker@itc.nl

Hecker, C., Kuenzer, C., and Zhang, J., 2007, Remote-sensing–based coal-fire detection with low-resolution MODIS data, *in* Stracher, G.B., ed., Geology of Coal Fires: Case Studies from Around the World: Geological Society of America Reviews in Engineering Geology, v. XVIII, p. 229–238, doi: 10.1130/2007.4118(15). For permission to copy, contact editing@geosociety.org. ©2007 The Geological Society of America. All rights reserved.

as Prakash et al. (1997), Prakash and Gupta (1998), and Gupta and Prakash (1998), studied the coal fires in the Jharia Coalfield, India, using Landsat5 TM thermal data with 120 m spatial resolution. Because it has been over 20 years since its launch in 1984, Landsat5 is swiftly approaching the end of its fuel reserves, and it may stop operating within the next 1–2 yr. In 1999, Landsat7 was launched. Compared to its predecessors, its thermal band had an increased spatial resolution of 60 m. In 2003, Landsat7 encountered technical problems that could not be entirely fixed, and this left the sensor acquiring images at lower quality than before. In addition, a continuation mission for the Landsat program is not expected to be launched before the year 2010 (LDCM, 2005). Hence, alternatives to the inadequately functioning Landsat7 sensor have to be found for operational coal-fire detection and monitoring.

Sensors with mid-wave infrared (MWIR) and thermal infrared (TIR) bands are used to detect coal fires based on their heat-emission signature. Most satellites that record MWIR and TIR have a medium to coarse spatial resolution. Data from those with a coarser spatial resolution than that of Landsat7 or Landsat5 (spatial resolution in the thermal band 60 m and 120 m, respectively) have rarely been used for coal fire research and have not yielded promising results (Mansor et al., 1994; Zhang, 1998). Although Mansor et al. (1994) could detect very hot surface coal fires in Jharia, Zhang showed that the spatial resolution of the National Oceanic & Atmospheric Administration's Advanced Very High Resolution Radiometer (NOAA-AVHRR) data (1 km spatial resolution, one thermal band) in general was too low to detect subsurface coal fires in northwest China. In a related application, Gumbricht et al. (2002) attempted to use Moderate Resolution Imaging Spectroradiometer (MODIS) and NOAA-AVHRR data to detect subsurface peat fires. They concluded that only intense surface fires could be detected with these data.

A second major focus in recent coal-fire research is the development and application of automated methods for coal-fire detection and monitoring. Such methods should be suitable for large areas, such as the entire northern Chinese coal belt. The main focus is the long-term monitoring of known coal-fire areas in order to assess the development of fires over time and to detect newly ignited fires at an early stage. Furthermore, remote sensing of coal fires can monitor the success of extinction activities and may help to detect new unknown coal fires in adjacent regions of already known fires. To implement an early warning and monitoring system on a national to global scale, an automated detection algorithm must be used that does not rely on a priori knowledge and uses only minimal operator input.

In the thermal bands with pixels of a spatial resolution of 1 km or coarser, the subsurface coal-fire anomalies are usually not large or hot enough to have a measurable impact on the overall thermal signal measured. However, through use of band ratios, small emittance differences between coal-fire hot spots and their surrounding background temperature can be enhanced. In this work, we will demonstrate that ratioing results show enough contrast for coal fires to be visually picked up. We applied a detection algorithm to test the suitability of the MODIS ratios for automated extraction of potential coal fires in a regional detection and monitoring program.

CHARACTERISTICS OF MODIS

With a 1 km × 1 km pixel size at nadir, MODIS is part of the group of low-spatial-resolution satellites (LSRS) that are unable to resolve fine spatial details that would be visible in a typical Landsat7 scene (thermal band 6 of Landsat7 has a 60 m × 60 m resolution). Swath widths and scan angles of LSRS are usually very large. At their maximum scan angles, near the periphery of the swath, spatial resolution is even more diminished; at full scan angle of 55° to either side of the flight path, a single pixel for MODIS represents a ground surface area of almost 2 km × 5 km (Masuoka et al., 1998).

However, the design of LSRS also produces advantages over the higher-resolution imagers. The large swath width multiplies the revisiting frequency over an area of interest. This improves the chances of having cloud-free acquisitions over a study area. LSRS also acquire data continuously, even during the night cycle. This increases their effectiveness in detecting hot spots during advantageous night acquisitions and leads to large, complete data archives that can be queried years later for comparison. As signal strength greatly decreases toward the thermal infrared, the increased instantaneous field of view (IFOV) of LSRS also boosts the potential signal-to-noise ratio and the radiometric resolution of the detector.

While several of the LSRS could be tested for their suitability in coal-fire detection, we focused in this study on the MODIS sensor. Its increased radiometric and spectral resolution as well as its low noise levels compared to other LSRS satellites (for example, AVHRR) made a successful detection of coal fires more likely.

STUDY AREAS AND DATA SETS

The main study areas, Wuda (39.51°N, 106.60°E) and Ruqigou (39.07°N, 106.12°E), are located in the border region of Ningxia and Inner Mongolia Autonomous Regions in north-central China (Fig. 1). While the elevation in the broader Wuda surroundings, located on the floodplains of the Yellow River, averages 1500 m, Ruqigou, within the northern outreach of the Helan Shan mountains, is above 2000 m elevation and is characterized by alpine terrain. Both areas have a semiarid to arid climate, and vegetation cover is sparse (Kuenzer et al., 2005).

Coal has been mined there since 1958 (Wuda) and 1960 (Ruqigou). The first coal fires in the area of Wuda were discovered in 1961. Today 20 major fires are destroying the coal in the structural syncline located west of Wuda city. Most fires occur in the subsurface, and temperature anomalies are dominated by cracks and vents emitting hot gases. To the north of the syncline, fires are known from coal stockpiles. The overall coal-fire area has an extent of over 4 km², and recent annual coal loss is estimated to exceed 200,000 tons (Kuenzer et al., 2005).

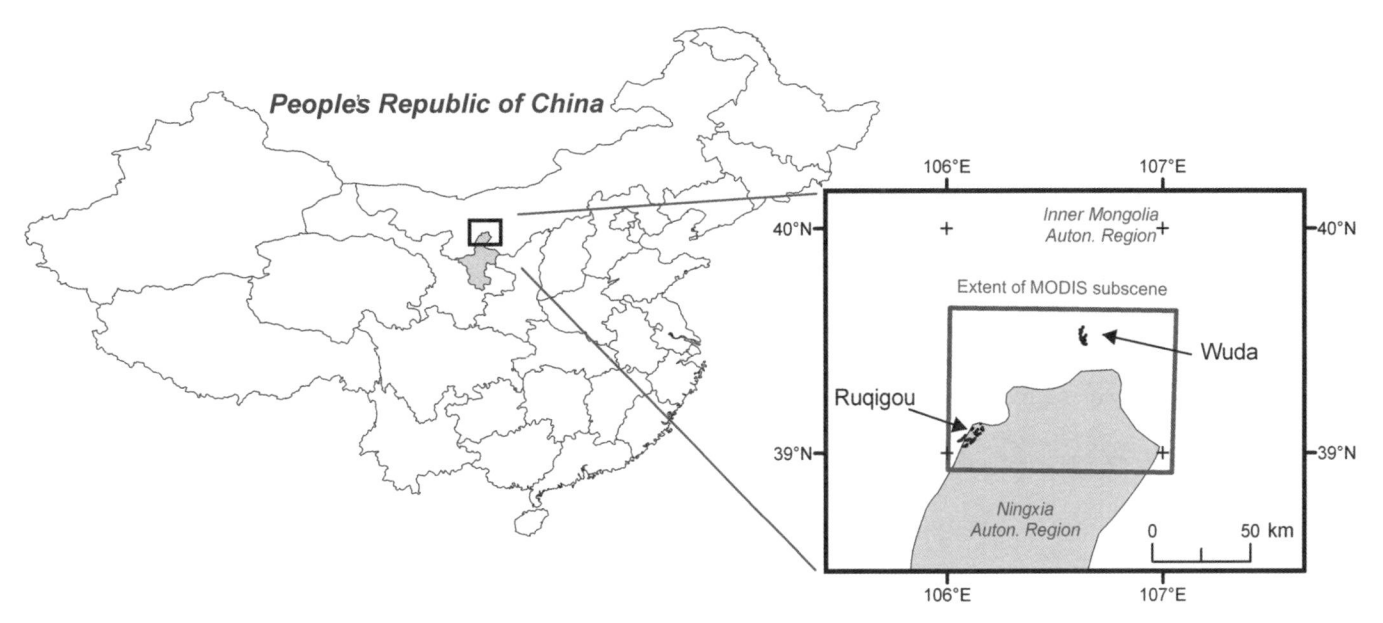

Figure 1. Location of Ruqigou and Wuda study areas. Gray box indicates extent of Figures 3 and 4.

In the Ruqigou area, the mines are at present affected by over 20 coal fires, most of which are currently undergoing the process of extinction. Coal fires occur on the surface as well as underground and create large temperature anomalies at the surface. Most fires are related to former or present small-scale commercial mining. The total area affected by coal fires in this location is assumed to cover more than 2 km^2.

Selection of Imagery

Success rates for thermal hot spot detection greatly vary with seasonal and diurnal ambient temperature fluctuations. The larger the temperature contrast between hot spot and background temperature, the higher the success rate will be. MODIS data are acquired daily from the platform TERRA around 10 a.m. and 11 p.m. local time over the study area as well as from the platform AQUA, with acquisition times of 2 p.m. and 3 a.m. over the study area. For this study, we selected scenes acquired during the 3 a.m. predawn cycle. Predawn data are known to be best for thermal anomaly detection, since Earth's surface has had time to cool during the night and solar heating effects are minimized.

We selected one winter and one summer MODIS scene close to the 2002 ground measurement campaign for seasonal comparison (Table 1). A recent winter scene (February 2005) was included to check for the most recent fire developments.

DATA PROCESSING AND ENHANCEMENTS

Multiband Thermal Information

Virtually all coal-fire–related temperature anomalies observed on the ground are considerably smaller than a MODIS pixel. Therefore, temperatures represented in a single pixel will always be an integrated average over the entire ground area of the IFOV during acquisition. The temperature difference between a homogeneous background pixel and one containing a background as well as a coal fire component is very small compared to the spatial variability of the background temperature itself (Table 2; Fig. 2). Hence, detection of coal fires based on a single band cannot give satisfying results.

One of the big advantages of MODIS, however, is the fact that it registers thermal radiation in several wavelengths of the electromagnetic spectrum. This characteristic of MODIS data allows temperature distributions to be compared not only spatially (i.e., change of radiant temperature with location) but also spectrally (i.e., change of radiant temperature for same location with change of wavelength). In order to achieve this, the idea of the bispectral technique was adapted, which was introduced by Dozier (1981). While a homogeneously warm background pixel ideally shows identical radiant temperatures in all thermal bands, the mixed pixel of a colder background

TABLE 1. SPECIFICATIONS OF MODIS IMAGES USED IN THIS STUDY

Name	Local acquisition date (mm/dd/yyyy)	Local acquisition time	GMT acquisition time	Data level	Product code	Sensor	Platform
Summer	09/22/2002	03:00	19:00	L1b	021KM	MODIS	AQUA
Winter1	02/13/2003	03:00	19:00	L1b	021KM	MODIS	AQUA
Winter2	02/02/2005	03:00	19:00	L1b	021KM	MODIS	AQUA

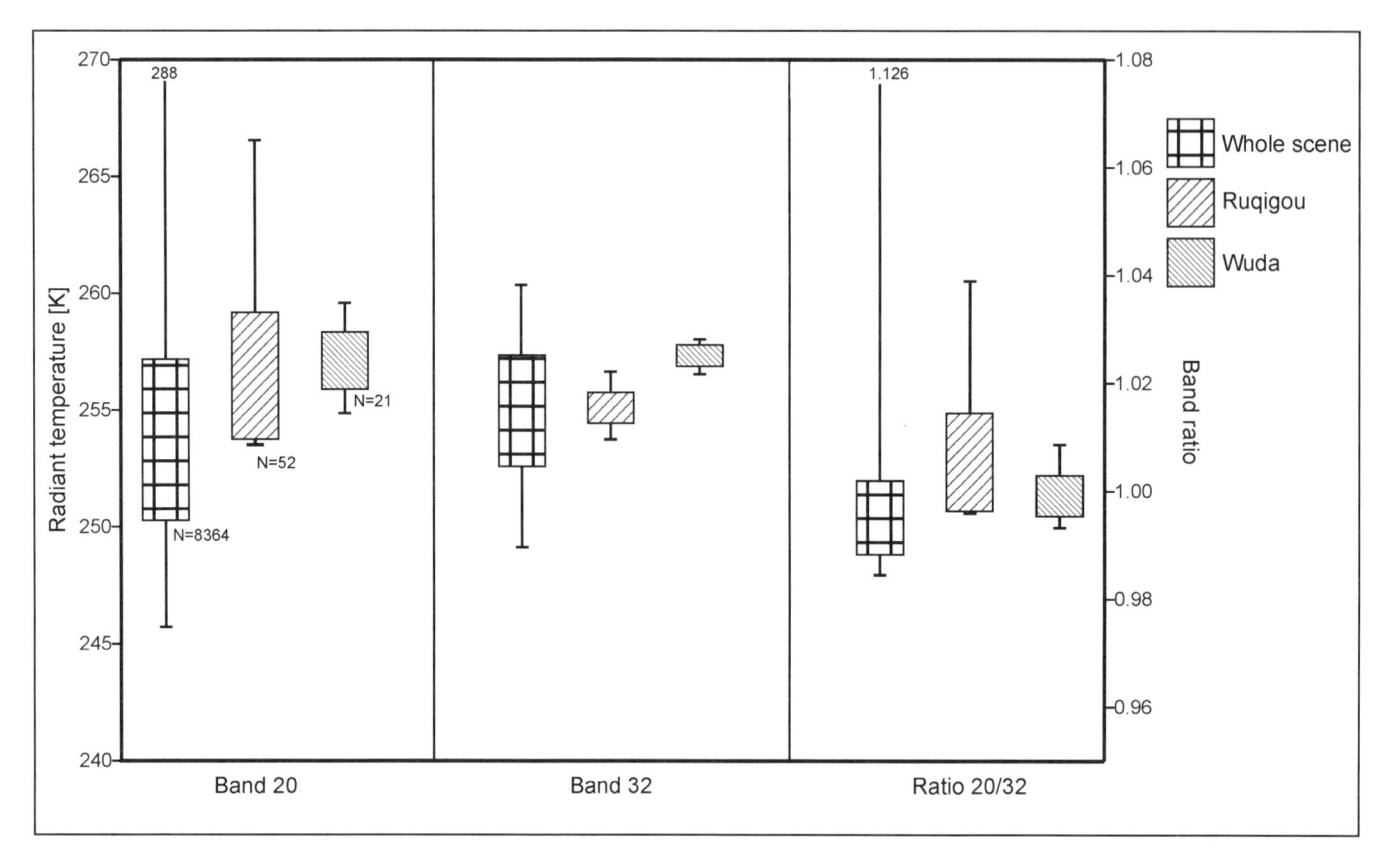

Figure 2. Plot showing recorded radiant temperatures and band ratio values for the whole February 2005 scene (including fires), for the fires in Ruqigou, and for the fires in Wuda. The increase in measured radiant temperatures for the fires is small compared to the overall range of background temperatures. Boxes show the range of mean ± 1σ. The whiskers indicate minima and maxima.

TABLE 2. THEORETICAL LINEAR MIXTURES OF BACKGROUND AND COAL-FIRE TEMPERATURES

MODIS band	Type	Radiant temperature (K)	Planck radiance contribution (Wm^{-2}sr^{-1}μm^{-1})	Fraction of pixel	Linear radiance combination (Wm^{-2}sr^{-1}μm^{-1})	Mixed radiant temperature (K)
Homogeneous background cold						
Band 20*	Background	273	0.0754	1.00	0.0754	273.00
Band 32†	Background	273	5.9349	1.00	5.9349	273.00
Homogeneous background warm						
Band 20*	Background	300	0.2869	1.00	0.2869	300.00
Band 32†	Background	300	8.8062	1.00	8.8062	300.00
Background cold with 5% coal fires						
Band 20	Background	273	0.0754	0.95	1.3670	339.20
	Fire	450	25.9072	0.05		
Band 32	Background	273	5.9349	0.95	7.3608	287.12
	Fire	450	34.4529	0.05		
Background cold with 1% coal fires						
Band 20	Background	273	0.0754	0.99	0.3337	303.40
	Fire	450	25.9072	0.01		
Band 32	Background	273	5.9349	0.99	6.2201	299.83
	Fire	450	34.4529	0.01		

Note: Colder background pixels with a small percentage covered by coal fires show almost identical mixed radiant temperatures compared to a slightly warmer background pixel. Mixed radiant temperatures of heterogeneous pixels vary with wavelength and can be enhanced in a ratio image.
*Band 20 corresponds to a central wavelength of 3.55 μm.
†Band 32 corresponds to a central wavelength of 12.22 μm.

with a small hot spot shows variable integrated radiant temperatures for different wavelengths. Table 2 illustrates the effect of a subpixel-sized hot coal fire on the total radiant temperature recorded at the sensor. Radiant temperatures are calculated for theoretical mixed pixels containing background temperatures as well as 1% and 5% pixel area covered by coal fires of 450 K. With only a small fraction of a pixel covered by coal fires, the overall integrated temperature for the whole pixel only increases by a few degrees. Compared to the overall variability in the radiant temperature images (e.g., Winter 2005, band 32 in Fig. 2), this increase in temperature may not be significant enough to be detected. However, Table 2 also shows how radiant temperatures become wavelength-dependent for mixed pixels. Homogeneously warm background pixels theoretically do not vary their measured radiant temperature with wavelength (emissivity and atmospheric effects neglected). Hence, ratios of radiant temperature images in different wavelengths will suppress variation in background temperature and enhance the contribution of small patches of hotter temperature. This effect can be exploited for detection of pixels containing subpixel-sized coal fires.

Image Processing

MODIS measures radiance in six mid-wave infrared (MWIR) and 10 thermal infrared (TIR) bands. In order to select the appropriate MODIS bands in the emittance maxima of background and coal fires, their expected temperatures have to be taken into account using Planck's radiation law and Wien's displacement law (hotter surfaces emit radiation at shorter wavelength).

Nighttime background temperatures were derived from the images themselves and ranged around 250 K in winter to 280 K in summer images. Band 32 (centered at 12.22 μm) was selected to represent the background temperatures because the emission maximum for 250 K lies at 11.6 μm using Wien's displacement law. As an alternative, band 31 (11.03 μm) could be used instead, but with band 32 being at a longer wavelength, it more strictly represents pure background and contains less contribution from the fires.

Mean recorded ground temperatures for typical coal fires in the study area ranged from 50 °C and 850 °C, where the very hot temperatures were often restricted to surface fires or small cracks and vents over a shallow subsurface fire. Maximum emittance for these temperatures is in the MWIR. Bands 24 and 25 were strongly influenced by atmospheric absorption and were not used. Bands 20–23 were all suitable. In order to enhance the hottest anomalies, band 20 (centered at 3.55 μm) was selected to represent the coal fires.

The 1 km MODIS imagery (MODIS data product MYD021KM; DAAC, 2005) was first georeferenced using MODIS geolocation files (MODIS data product MYD03; DAAC, 2005). Radiance was transformed into radiant temperatures using Liam Gumley's IDL™ routines (Gumley, 2000).

Ratios were calculated between band 20 and band 32. In this ratio, all homogeneously warm pixels will show values around 1.0 (see Table 2). Mixed pixels containing hot spots in addition to the background temperature contribute relatively greater radiance in the shorter wavelength (band 20) than in the longer wavelength (band 32). While this may not be conspicuous in bands 20 or 32 themselves, the effect is enhanced in the ratio image, showing hot spots with ratio values above 1.0. Atmospheric corrections and temperature/emissivity separation were not performed in order to demonstrate the speed and simplicity of the approach for successful hot spot detection.

Automated Anomaly Extraction

To automatically extract thermal anomalies from the MODIS data, the coal-fire detection algorithm developed by Zhang (2004) was applied. This algorithm for automated thermal anomaly extraction from thermal satellite bands can run on raw satellite data (i.e., byte values) as input for subimage statistical analysis. As we ran the algorithm on ratio images, the ratio numbers were rescaled into byte range for further processing. Within a moving window of varying size (starting with 11 × 11 and ending with 25 × 25), the histograms for these subsets of the scene were investigated to detect the occurrence of thermally anomalous pixels. Image histogram statistics of coal fires from different known coal-fire areas were studied in detail. The average statistical parameters of coal-fire clusters within thermal images (such as minimum, maximum, mean values; standard deviation) were determined. We used the position of the first local minimum following the global maximum within the subset histograms as the cutoff value. We found this threshold to be most suitable for separating thermally anomalous pixels from background pixels. Due to the concept of a repeatedly applied moving window with increasing kernel size, each pixel within the scene was sampled several times. If a pixel was regarded as thermally anomalous in 70% or more of all checks executed by the moving windows algorithm, it was flagged as a thermal anomaly. The advantage of this approach is that extracted pixels are anomalous compared to other pixels in their surroundings. This is contrary to the approach where an overall temperature threshold is used on an entire image.

The thermally anomalous pixels were then clustered based on an eight-pixel neighborhood (i.e., a connectivity analysis that considers two anomalous pixels to be neighbors, even if they only touch at one corner). These thermally anomalous clusters were statistically investigated for their minimum, maximum, and mean value, their standard deviations, and their spatial coverage. The statistics for each cluster were then compared to typical values that were determined from known coal-fire areas. As an example, warm water surfaces can be rejected based on the cluster's temperature variance. While water surfaces show a very low temperature variance, coal-fire clusters are characterized by high variances (Zhang, 2004; Kuenzer, 2005). Coal-fire areas also do not extend for many kilometers. If a river is detected as thermally anomalous, it can be rejected based on the cluster size.

In this way, the final output image only contains thermal anomalies of small size, which have a reasonable chance of being coal fires. Nevertheless, automated statistical analysis of thermal anomalies can still lead to the extraction of thermally anomalous pixels not related to coal fires.

The algorithm for thermal anomaly detection was applied to the band ratios 20/32 of the three MODIS images. The processing for each MODIS subset, containing just over 8000 pixels, took only several minutes each. The results were binary output images, indicating background area (0) and thermally anomalous pixels (1).

RESULTS

Visual Inspection

Three MODIS scenes were processed and analyzed in this study. The results are demonstrated on the winter scene of 2005 (Figs. 3 and 4). Visual inspection of the radiant temperature images revealed no evidence of coal fires in any of the images for the longer wavelength (band 32, 12.22 µm; Fig. 3B). This was to be expected since virtually all of the radiance reaching the sensor in this band came from the background itself. The contribution of coal fires to the measured radiance in this band could be neglected because the fires were considerably smaller than the

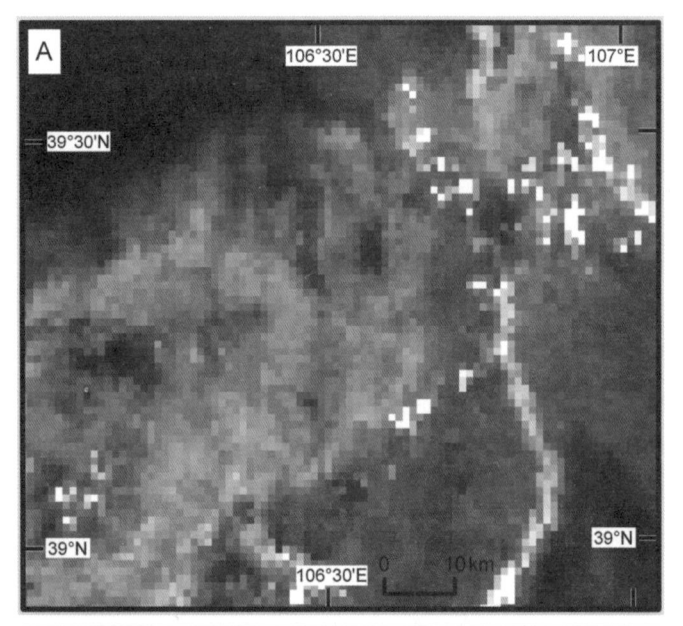

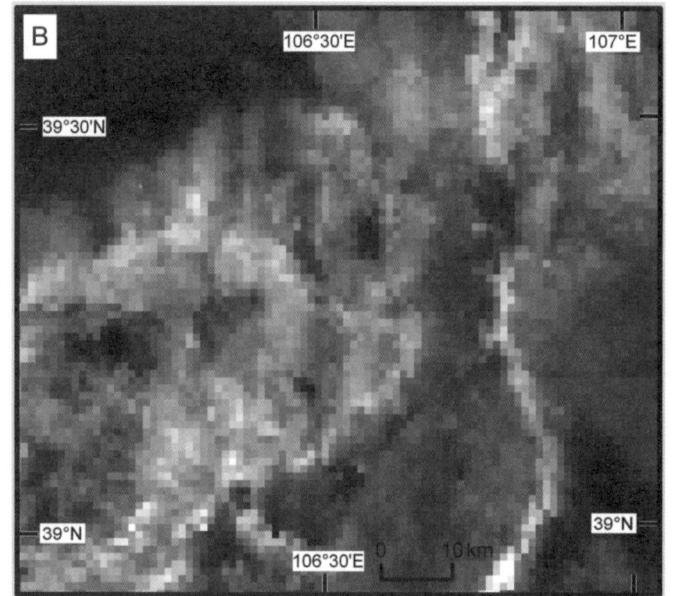

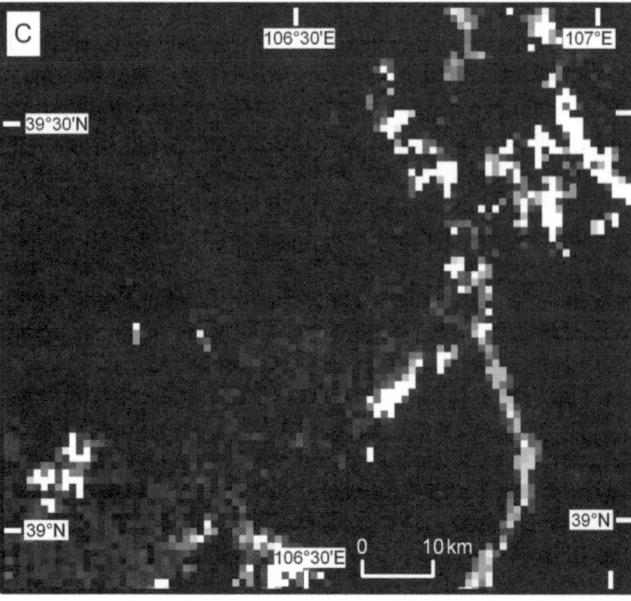

Figure 3. Comparison of MODIS radiant temperature in band 20 (A), band 32 (B), and ratio image (C) for predawn 02 February 2005. Both radiant temperature images show strong variations in background temperature. Brighter pixel values indicate higher temperatures. The ratio image enhances thermal anomalies as bright pixel values. Extent of image: 80 km × 90 km.

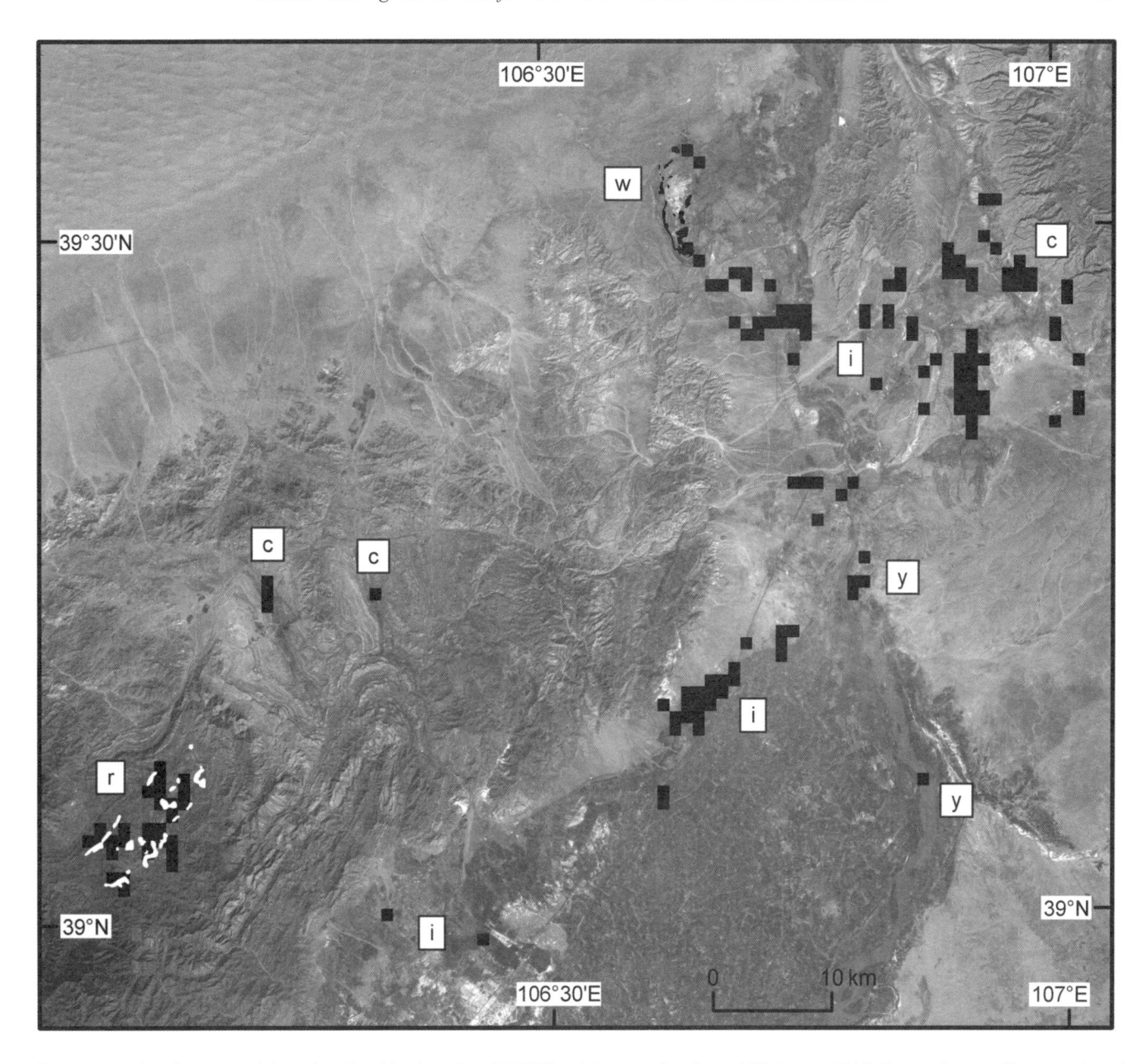

Figure 4. Results of automated detection algorithm based on MODIS ratio image of predawn 02 February 2005. Detected anomalies are assigned to their sources: r—Ruqigou coal fires, w—Wuda coal fires, c—other confirmed coal fires, i—industrial installations, y—Yellow River. Ground extent of Ruqigou and Wuda coal fires are indicated as fine lines in white and dark gray, respectively. Background image: Landsat7, band 1. Extent of image: 80 km × 90 km.

pixel size and their maximum emittance occurred at much shorter wavelengths than band 32. In the radiant temperature images of the shorter wavelength (band 20; 3.55 μm), we did, however, observe patterns of pixels with elevated temperatures (Fig. 3A). Several of these anomalously warm pixels could be attributed to known locations of coal fires, but not all known affected areas were actually visible. Only large or very hot fires contributed enough energy to the total radiation budget of a pixel to display sufficient contrast for it to stand out over the natural variation in the background temperature itself. If we compare the radiant temperature images (Figs. 3A and 3B) with the ratio image (Fig. 3C), it becomes evident that the effect of varying background temperature has been strongly reduced and the thermal hot spots stand out clearly above the surrounding pixels (Fig. 3C). The comparatively hot coal fires in Ruqigou show up very distinctly. As for the study area in Wuda, the results are not as clear, although the hottest fires at the northern and southern end of the syncline are quite distinctly visible. These thermal anomalies result from hot surface coal fires and burning coal waste piles of the major mines located within Wuda syncline. The more subtle subsurface fires in between show up faintly and could be overlooked in a visual inspection without knowledge of the location of the fires.

Apart from the known coal-fire areas, the ratio image also enhanced other heat sources, such as lime factories and coal coking plants. The valley of the Yellow River shows a high concentration of hot spots on the ratio image that can be attributed to industrial- and settlement-related heat sources. In night images, the Yellow River itself also shows a slightly elevated band ratio compared to the surrounding landmass. Nevertheless, the conspicuous shape of a river makes confusion with coal fires very unlikely.

Automated Detection Results

Figure 4 shows the results of the automated extraction algorithm. Pixels that were determined to be thermally anomalous are represented in black. For the coal-mining areas of Wuda and Ruqigou, we can compare the results to ground data that were acquired during September 2002 and 2003 field campaigns. The ground information consists of maps showing the extent of thermal anomalies above burning coal seams in the Wuda and Ruqigou coal-mining areas (Fig. 4). It also contains information on hot spots in other areas that were detected on thermal imagery prior to the field work and were then systematically checked on the ground to confirm the type of heat source.

For Ruqigou and Wuda coal-fire areas, we assessed the success of MODIS in identifying coal-fire pixels by comparing the outcome of the detection algorithm to the true extent of the fire anomalies on the ground. Typically, coal-fire–related temperature anomalies are considerably smaller than the 1 km^2 spatial resolution of the MODIS sensor. Depending on the exact position of the individual image pixels, some will end up having only a minimal fraction of the sensed ground surface actually covered by a mapped coal-fire anomaly. As it was our intention to demonstrate the usefulness of MODIS for detection of coal fires without any a priori knowledge, we decided not to use a threshold for the pixel fraction covered by a coal-fire anomaly on the ground.

Table 3 shows a significant difference between detection results in Wuda as compared to Ruqigou. The influence of summer and winter acquisitions is somewhat less pronounced but contrary to our expectations (i.e., highest success in Ruqigou in summer image).

In Wuda, the subsurface coal fires along the coal seams were poorly detected. Success rates ranged from 0% to 17%, with the lowest success rate in the summer image. In the 2005 winter image, some of the hotter subsurface fires in the southern tip of the Wuda syncline show up in the detection image. However, most subsurface coal fires did not get picked up by the algorithm, which led to the very low success rate in two of the three images. Several hot spots were repeatedly detected just north of the Wuda mining area. Ground checks revealed that they were caused by large fires in coal waste piles in a coal-processing facility. Another cluster of hot spots extended eastward from the southern end of the Wuda Coalfield (Fig. 4). These hot spots

TABLE 3. SUCCESS RATE OF DETECTION ALGORITHM FOR THREE MODIS SCENES AND TWO STUDY AREAS

	22 September 2002	13 February 2003	02 February 2005
Wuda area*			
Detected	0	4	1
Not detected	24	20	20
Success rate (%)	0.0	16.7	4.8
Ruqigou area†			
Detected	25	18	22
Not detected	26	25	31
Success rate (%)	49.0	41.9	41.5

Note: Except for percentages, all numbers express amount of true coal-fire pixels that were correctly detected/not detected as such by the detection algorithm. All fire-hosting pixels were considered true coal-fire pixels, even if the pixel fraction covered by fires was minimal. Variations in total pixel counts are due to changes in viewing geometry for different MODIS overpasses. Inside the two study areas, virtually all detected pixels could be attributed to a hot source (natural or manmade).
*Wuda area contains mainly subsurface coal fires with small temperature anomalies at surface.
†Ruqigou area contains hot surface and subsurface coal fires.

were caused by industrial installations along an access road to the Wuda coal mines.

In Ruqigou, the automated detection algorithm achieved detection success rates between 42% and 49%, where the highest value was observed in the summer image. Beyond the burning coal seams, no additional hot spots were detected in Ruqigou because the coal-mining area is surrounded by steep terrain with few industrial installations.

Apart from the two main study areas in Ruqigou and Wuda, coal fires were also detected in two valleys 50 km southwest of Wuda and along a hillside 25 km southeast of Wuda (Fig. 4). These fires were confirmed in the field in 2004 (Kuenzer, 2005).

DISCUSSION

Detection success rates were determined by comparing how many true coal-fire pixels (confirmed by mapping) were detected/not detected as such by the detection algorithm.

The relatively low success rate was partially attributed to the way the detection results were validated. All image pixels that were covered by coal fires, even on a marginal fraction of their surface area, were also included in the calculation and added up as not-detected fire pixels. In this way, the results were kept rather conservative, and we expect that unsuccessful detection would drop upon the use of a minimum fraction threshold.

How do these results compare to studies done with Landsat? Kuenzer (2005) showed that for Ruqigou, more than 90% of all known fires did show up in the thermal band of Landsat7. For Wuda, this count was 55% of all fires. These success numbers, however, only express that of several LANDSAT pixels covering an individual fire; at least one was successfully flagged as a fire pixel. Several others may have gone undetected without showing in the statistics as a false negative. In MODIS imagery, we do not have several pixels covering one fire, but rather one pixel covering several fires, as well as a large pixel fraction with background temperatures. Instead of counting the fires that were (partially) detected, this study evaluated the amount of all correctly identified fire pixels. A comparison of the LANDSAT with the MODIS results based on these numbers alone does not seem to be appropriate because the methods they were achieved with are not comparable. It is evident, however, that coal-fire detection based on a properly operating Landsat7 ETM+ sensor was more reliable and gave more spatial details that could be used to study the temporal development of the fires.

MODIS results for Ruqigou and for Wuda are distinctly different from each other. This is interpreted to be due to the existence of mainly subsurface coal fires in Wuda as compared to a mix of surface and subsurface in Ruqigou. In the Wuda syncline, only stockpile fires and, sporadically, the hottest of the subsurface fires were detected in winter images. In the summer image, the stockpile fires were still extracted, but none of the in situ coal fires could be detected. These results show that surface coal fires and some of the hotter subsurface coal fires can be detected in winter predawn images.

Surprisingly, for Ruqigou, the highest detection result came from the 2002 summer scene. This is opposite to our expectations and also opposite to the results for Wuda. There are two arguments to explain this result: (1) the 2003 winter scene had an ~1.5 pixel georeference shift that came with the MODIS georeference file, and (2) the 2005 winter scene was acquired 2.5 yr after the ground truth was recorded in the field. Although coal fires often burn stably for long periods of time on a 1 km^2 scale, many fires in Ruqigou are known to have changed or have been smothered in that period. For these reasons, both winter scenes performed worse than expected as compared to the summer scene.

The error of commission (i.e., pixels that are wrongly flagged) is difficult to quantify. In the study area of Wuda and Ruqigou, virtually all flagged pixels were either true coal fires or hot spots caused by other known heat sources, such as heavy industry (coking plants, lime burning, power plants, etc.) or settlements. These all have large heat emissions in a small area and are enhanced by the ratioing and picked up by the detection algorithm. Few genuinely wrong pixels (flagged as thermally anomalous but could not be linked to known heat source) were recorded inside and along the bank of the Yellow River (Fig. 4). They were individual pixels or small groups of pixels that did not get rejected by the algorithm's variance and size checks.

Thermally anomalous pixels caused by heat sources other than coal fires cannot necessarily be distinguished from true coal-fire anomalies in thermal data alone. Through the combination of thermal anomaly detection (Zhang, 2004) and coal-fire risk area demarcation, as presented by Kuenzer (2005), many "false" anomalies can be eliminated. If industrial installations are located within areas that are considered to have coal-fire risk, high-resolution imagery (e.g., IKONOS, QUICKBIRD) or field checks may sometimes be the only viable solution to distinguish between new coal fires and other new heat sources.

CONCLUSIONS

Since many of the coal-fire–related hot spots are one or two orders of magnitude smaller than the MODIS pixel size, most researchers would question the usefulness of MODIS in coal-fire detection. However, in this work, we demonstrated that by using band ratioing on predawn MODIS images, we could considerably enhance the contrast between pixels that were of homogeneous temperature and those that contained an additional heat source (e.g., coal fires). We employed the automated extraction algorithm from Zhang (2004) to detect likely coal-fire pixels in the ratio images. The results from the Wuda site showed that areas with subsurface fires could only be detected if the thermal anomalies they caused on the surface were strong enough and if they had a large enough extent. For most subsurface coal fires in Wuda, this was not the case, and MODIS did not produce reliable results. The results for the Ruqigou site showed that MODIS can in fact be used to reliably and repeatedly detect coal fires as long as they burn at the surface or create a thermal surface anomaly.

OUTLOOK

Especially with respect to the failure of Landsat7 in 2003, the results presented here are of great relevance to future detection of coal fires. However, surface and hot subsurface coal fires, such as those in the Ruqigou coal-mining area and the few surface coal fires in Wuda, are not the standard case (Zhang, 2004). Most coal fires are deeper, subsurface fires. These subsurface fires obviously pose a greater challenge for detection and quantification by means of remote sensing. The very subtle thermal anomalies from subsurface coal fires are rarely detected with MODIS data. In this case, alternatives such as the thermal channels of the Advanced Spaceborne Thermal Emission and Reflection Radiometer (ASTER) satellite, which have a spatial resolution of 90 m, are promising. Furthermore, the China-Brazil Earth Resources Satellite (CBERS), which has a spatial resolution of 160 m for its thermal band of the Infrared Multispectral Scanner (IR-MSS) instrument, is currently being investigated for its potential in supporting coal-fire detection.

In the future, the use of automated extraction algorithms, such as that introduced by Zhang (2004), will become more important when expanding techniques from small study areas to operational coal-fire detection and monitoring systems on a regional to national scale. While MODIS imagery may not have the spatial resolution to monitor coal-fire movements or aid successful fire-fighting efforts on the ground, it does have a strength in the large spatial coverage and high temporal resolution. Both of these attributes could play a vital role in a future automated early warning system on a nationwide scale that could focus further attention to areas showing new development of potential surface coal fires.

ACKNOWLEDGMENTS

The authors would like to thank Yaorong Jia and Yulin Sun from the Wuda mine for their support with ground truth activities during field campaigns. Comments by M. van der Meijde, J. Hyde Hecker, and J. Horn have greatly improved the quality of this manuscript. The authors would also like to thank reviewers R.D. Wheate and R.K. Vincent for their comments and corrections, which greatly helped clarify this paper.

REFERENCES CITED

DAAC (Distributed Active Archive Center), 2005, NASA Goddard Earth Sciences Data Information Service Center: http://daac.gsfc.nasa.gov/MODIS/products.shtml (accessed 28 September 2005).

Dozier, J., 1981, A method for satellite identification of surface temperature fields of subpixel resolution: Remote Sensing of Environment, v. 11, p. 221–229, doi: 10.1016/0034-4257(81)90021-3.

Ellyett, C.D., and Fleming, A.W., 1974, Thermal infrared imagery of The Burning Mountain coal fire: Remote Sensing of Environment, v. 3, no. 1, p. 79–86, doi: 10.1016/0034-4257(74)90040-6.

Fisher, W.J., and Knuth, W.M., 1968, Detection and delineation of subsurface coal fires by aerial infrared scanning: Abstracts for 1967: Geological Society of America Special Paper 115, p. 67–68.

Greene, G.W., and Moxham, R.M., 1969, Aerial infrared surveys and borehole temperature measurements of coal mine fires in Pennsylvania, in Proceedings of the 6th Symposium on Remote Sensing of Environment: Ann Arbor, University of Michigan, p. 517–525.

Gumbricht, T., McCarthy, T.S., McCarthy, J., Roy, D., Frost, P.E., and Wessels, K., 2002, Remote sensing to detect sub-surface peat fires and peat fire scars in the Okavango Delta, Botswana: South African Journal of Science, v. 98, no. 7–8, p. 351–358.

Gumley, L.E., 2000, IDL Routines for MODIS: Madison, University of Wisconsin–Madison: ftp://origin.ssec.wisc.edu/pub/MODIS/IDL (accessed 28 October 2002).

Gupta, R.P., and Prakash, A., 1998, Reflectance aureoles associated with thermal anomalies due to subsurface mine fires in the Jharia Coalfield, India: International Journal of Remote Sensing, v. 19, no. 14, p. 2619–2622, doi: 10.1080/014311698214415.

Knuth, W.M., Fisher, W.J., and Stingelin, R.W., 1968, Detection, Delineation and Monitoring of Subsurface Coal Fires by Aerial Infrared Scanning: State College, Pennsylvania, Geographer HRB-Singer, Inc., a subsidiary of the Singer Company, p. 877–881.

Kuenzer, C., 2005, Demarcating Coal Fire Risk Areas Based on Spectral Test Sequences and Partial Unmixing Using Multisensor Remote Sensing Data [Ph.D. thesis]: Vienna, Austria, Technical University Vienna, 199 p.

Kuenzer, C., Voigt, S., and Moerth, D., 2005, Investigating land cover changes in Chinese coal mining environments using partial unmixing, in Ersami, S., Cyffka, B., and Kappas, M., eds., Remote Sensing & GIS for Environmental Studies. Göttinger GIS and Remote Sensing Days (GGRS 2004), 7–8 October 2004, Göttinger Geographische Abhandlungen 113: Goettingen, Germany, p. 31–37.

LDCM (Landsat Data Continuity Mission), 2005, U.S. Geological Survey Landsat Data Continuity Mission Website: http://ldcm.usgs.gov/ (accessed 26 July 2005).

Mansor, S.B., Cracknell, A.P., Shilin, B.V., and Gornyi, V.I., 1994, Monitoring of underground coal fires using thermal infrared data: International Journal of Remote Sensing, v. 15, no. 8, p. 1675–1685.

Masuoka, E., Fleig, A., Wolfe, R.E., and Patt, F., 1998, Key characteristics of MODIS data products: IEEE Transactions on Geoscience and Remote Sensing, v. 36, no. 4, p. 1313–1323, doi: 10.1109/36.701081.

Moxham, R.M., and Greene, G.W., 1967, Infrared Surveys of Coal Mine Fires in the Anthracite and Bituminous Fields, Pennsylvania: U.S. Geological Survey–U.S. Bureau of Mines Inter-Agency Report BM-2, 67 p.

Prakash, A., and Gupta, R.P., 1998, Land-use mapping and change detection in a coal mining area—A case study in the Jharia Coalfield, India: International Journal of Remote Sensing, v. 19, no. 3, p. 391–410, doi: 10.1080/014311698216053.

Prakash, A., and Gupta, R.P., 1999, Surface fires in Jharia Coalfield, India—Their distribution and estimation of area and temperature from TM data: International Journal of Remote Sensing, v. 20, no. 10, p. 1935–1946, doi: 10.1080/014311699212281.

Prakash, A., Gupta, R.P., and Saraf, A.K., 1997, A Landsat TM based comparative study of surface and subsurface fires in the Jharia Coalfield, India: International Journal of Remote Sensing, v. 18, no. 11, p. 2463–2469, doi: 10.1080/014311697217738.

Rabchevsky, G.A., 1972, Determination from Available Satellite and Aircraft Imagery of the Applicability of Remote Sensing Techniques to the Detection of Fires Burning in Abandoned Coal Mines and Unmined Coal Deposits Located in North-Central Wyoming and Southern Montana: Allied Research Associates Inc., Contract No. SO 211087, Report No. 8G86-F, 21 p.

Rosema, A., Guan, H., Veld, H., Vekerdy, Z., Ten Katen, A.M., and Prakash, A., 1999, Manual of Coal Fire Detection and Monitoring: Netherlands Institute of Applied Geosciences (NITG) Report 99–221-C, 245 p.

Saraf, A.K., Prakash, A., Sengupta, S., and Gupta, R.P., 1995, Landsat-TM data for estimating ground temperature and depth of subsurface coal fire in the Jharia Coalfield, India: International Journal of Remote Sensing, v. 16, no. 12, p. 2111–2124.

Tetzlaff, A., 2004, Coal Fire Quantification Using Aster, ETM and Bird Satellite Instrument Data [Ph.D. thesis]: Munich, Germany, Ludwig Maximilians University, 155 p.

van der Meer, F.D., van Dijk, P.M., Gangopadhyay, P.K., and Hecker, C.A., 2004, Remote-sensing GIS based investigations of coal fires in northern China: Global monitoring to support the estimation of CO_2 emissions from spontaneous combustion of coal, in Vibulresth and Soo, D.N., eds., Proceedings of the 1st Asian Space Conference (ASC), Chiang Mai, Thailand: Bangkok, Geo-Informatics and Space Technology Development Agency, p. 6.

Zhang, J., 2004, Spatial and Statistical Analysis of Thermal Satellite Imagery for Extraction of Coal Fire Related Anomalies [Ph.D. thesis]: Vienna, Austria, Technical University Vienna, 166 p.

Zhang, J., Wagner, W., Prakash, A., Mehl, H., and Voigt, S., 2004, Detecting coal fires using remote sensing techniques: International Journal of Remote Sensing, v. 25, no. 16, p. 3193–3220, doi: 10.1080/01431160310001620812.

Zhang, X., 1998, Coal Fires in Northwest China—Detection, Monitoring, and Prediction Using Remote Sensing Data [Ph.D. thesis]: Delft, the Netherlands, Technical University Delft, 235 p.

MANUSCRIPT ACCEPTED BY THE SOCIETY 7 MARCH 2007

Application of remote sensing in coal-fire studies and coal-fire–related emissions

Prasun K. Gangopadhyay*
International Institute for Geo-Information Science and Earth Observation (ITC),
P.O. Box 6, 7500 AA, Enschede, The Netherlands

ABSTRACT

Coal fires are one of the most common geohazards in most coal-producing countries, such as India and China. Combustion can occur spontaneously or due to anthropogenic causes, either within underground coal seams or in exposed layers of coal on Earth's surface. Once started, coal fires are difficult to extinguish and sometimes cannot be controlled. In addition to burning millions of tons of coal, the fires have an enormous negative impact on the local and global environments. Coal fires produce large quantities of greenhouse gases, such as CO, CO_2, CH_4, SO_x, and NO_x, which have a direct impact on the local and global atmospheric composition. Since the preindustrial era, the concentration of CO_2, a major greenhouse gas that contributes to global warming, has increased from 280 ppm to 375 ppm. Land subsidence is an associated problem in areas that are affected by coal fires. Coal fires also create operational difficulties in existing mines and endanger human safety. After the first use of remote sensing to study a coal fire in the early 1960s, this technology became a useful and convenient tool for the detection and monitoring of additional coal fires. Several air- and spaceborne thermal remote sensors are available for studying coal fires. Coal-fire–related emissions have not been studied extensively; ground-based methods mainly use CO_2 detection instruments or other indirect calculations (e.g., amount of coal burnt). Few attempts are being made to estimate coal-fire emission using remote sensing.

Keywords: coal fire, remote sensing, greenhouse gas, emission.

INTRODUCTION

Coal Fires: Problems and Occurrences

Minerals constitute the backbone of the economic growth of any nation, and coal is one of the easily accessible minerals that is used as a primary energy resource in most parts of the world. Total recoverable reserves of coal around the world are estimated at 1083 billion tons, which is enough to last ~210 yr at current consumption levels. Although coal deposits are distributed widely, 60% of the world's recoverable reserves are located in three countries: the United States (25%), the former Soviet Union (23%), and China (12%). Australia, India, Germany, and South Africa account for an additional 29% (Goddard Space Flight Center, 2006). One of the most primitive uses of coal by humans (~75,000 yr ago) was found in France (Théry et al., 1996). One of the earliest coal fires was revealed by Goldammer and Seibert (1989), who used the thermoluminescence of baked rocks close to a coal seam in East Kalimantan to obtain a date of ca. 13 ka.

---

*prasun@itc.nl

Gangopadhyay, P.K., 2007, Application of remote sensing in coal-fire studies and coal-fire–related emissions, *in* Stracher, G.B., ed., Geology of Coal Fires: Case Studies from Around the World: Geological Society of America Reviews in Engineering Geology, v. XVIII, p. 239–248, doi: 10.1130/2007.4118(16). For permission to copy, contact editing@geosociety.org. ©2007 The Geological Society of America. All rights reserved.

Subsurface and surface coal fires are a serious problem in many coal-producing countries. Combustion can occur within the coal seams (underground or surface), in piles of stored coal, or in spoil dumps at the surface. The fires in coal seams can be initiated spontaneously under certain conditions, where air, heat from the sun, and water vapor are the main constituents. Lightning and forest/bush fire also can initiate a coal fire. In some cases, human negligence, a mine accident, or a human-induced heat source could be the reason for a coal fire (Sinha, 1986). Coal fires are common in many coal-producing countries, such as China, India, Indonesia, the United States, Australia, South Africa, and Russia (Fig. 1; Ellyett and Fleming, 1974; Saraf et al., 1995; van Genderen et al., 1996; Walker, 1999; Tetuko et al., 2003; Stracher and Taylor, 2004).

Significant environmental and economic problems are associated directly and indirectly with coal fires. Some of the impacts of coal fires include:

- Coal fire pollutes the immediate environment by emitting toxic gases, such as CO, CO_2, NO_x, SO_x, and CH_4. Among these noxious gases, CO_2 and CH_4 contribute to global warming.
- Land subsidence is a common event in areas that are affected by subsurface coal fires, which leads to a change in the local drainage pattern.
- Heat radiation from a coal fire increases the local temperature.
- A coal fire increases the cost of production by making existing mining operations difficult.
- A coal fire consumes the most used and potentially valuable energy resource.

Several researchers have suggested that the concentration of CO_2 in the atmosphere has increased unambiguously since the Industrial Revolution (Etheridge et al., 1996; Keeling and Whorf, 1999). Derived from historical data, based on energy source, the total CO_2 emission in the global atmosphere has increased at an alarming rate (Andres et al., 1999). Figure 2 shows a reconstruction of past emissions of CO_2 from fossil-fuel combustion since ca. 1750 (Energy Information Administration [EIA] 1, 2006; Marland et al., 2006). Other researchers have confirmed that the current CO_2 concentration has increased to 375 ppm, which has increased the radiative forcing to 1.46 W/m^2 (Intergovernmental Panel on Climate Change [IPCC], 2001).

In addition to CO_2, CH_4 and NO_2 are significant greenhouse gases that are emitted from coal fires. Both of these greenhouse gases have increased at a significant rate. Atmospheric methane (CH_4) concentration has increased by ~150% (1060 ppb) since 1750, whereas the NO_2 concentration has increased by 16% (46 ppb) (IPCC, 2001).

The focus here is on the global distribution of coal fires with emphasis on remote sensing used for the detection, monitoring, and quantification of coal-fire–related emissions.

Properties and Burning Process of Coal

Coal, the most-used fossil fuel, is a readily combustible rock that contains more than 50%, by weight, of carbonaceous material, and it is formed by the compaction and induration of variously altered plant remains. Initially, these remains were deposited in a swampy basin in the form of peat. Unpredictable amounts of other chemicals (e.g., sulfur, chlorine, sodium) and other minerals can be found in coal. The physical properties of coal, such as color, specific gravity, and hardness, vary considerably. This variance depends on the composition and the nature of preservation of the original plant material that formed the coal; the amount of impurities in the coal, derived from soil and silt that were deposited with the parent material; and the amount of time, heat, and pressure that has affected the coal since it was formed. Time, heat, and pressure also determine the degree of maturation of the sequence, which, according to the increasing amount of carbon, is classified as lignite, subbituminous coal, bituminous coal, or anthracite. Rank is another index of coal quality. This is a measure of the brightness (reflectivity) of the coal as measured microscopically, and it is a function of the vitrinite content of one of the microlithotypes in coal (D.J. Williams, 2005, personal commun.).

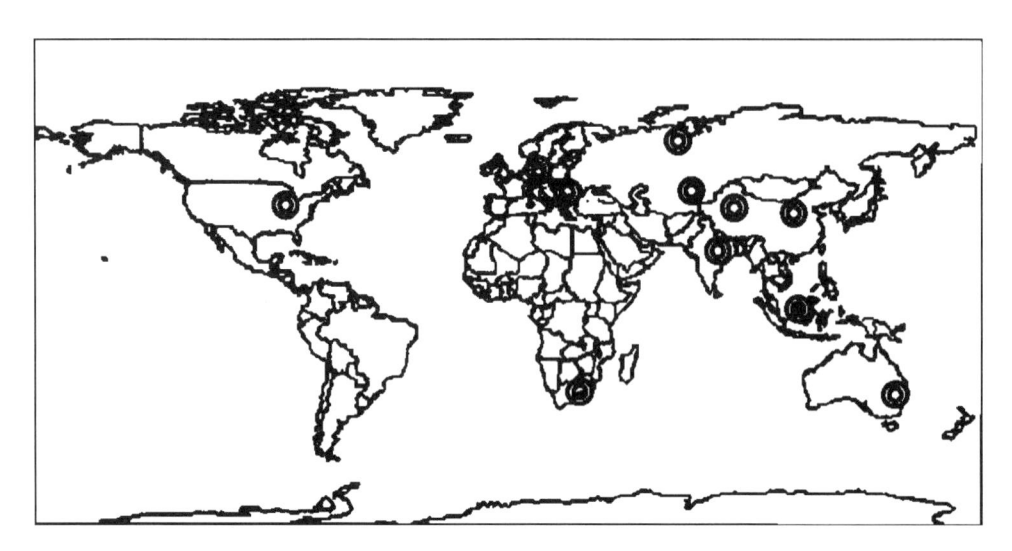

Figure 1. Worldwide occurrences of coal fires.

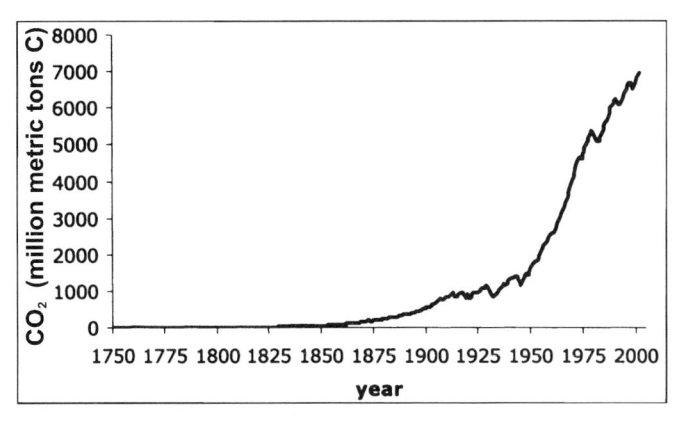

Figure 2. Historical CO_2 emissions from fossil-fuel consumption. (Source: Energy Information Administration [EIA] 1, 2006; EIA 2, 2006; Marland et al., 2006.)

Combustion of coal is a chemical process that can be defined in a simplified form as:

$$Coal + O_2 \rightarrow CO_2 + energy. \quad (1)$$

Practically, it is more complicated and may consist of different stages, which also depend on the presence of other substances (e.g., water [vapor], pyrite). For dry coal, the overall reaction was described by Schmal (1987), referencing Kok (1981) as:

$$C_{100}H_{74}O_{11} + 113O_2 \rightarrow 100CO_2 + 37H_2O + 4.2 \times 10^8 \text{ J/kmol } O_2. \quad (2)$$

The first part of the reaction, which consists of the chemical absorption of oxygen by the coal surface, can be presented in the following equation:

$$C_{100}H_{74}O_{11} + 17.5O_2 \rightarrow C_{100}H_{74}O_{46} + 2.5 \times 10^8 \text{ J/kmol } O_2. \quad (3)$$

Several factors can accelerate or decrease the speed of the reactions (e.g., oxygen content of air, exposed surface area of the coal, temperature, composition of coal).

Spontaneous Combustion of Coal

The potential for coal to spontaneously combust depends on its aptitude of oxidization at ambient temperature in certain conditions (Fig. 3). This occurs through the absorption of oxygen at the surface of the coal, and it is an exothermic reaction. The temperature of the coal may start to increase. If the temperature reaches what often is called the "threshold" temperature, somewhere between 80 °C and 120 °C, a steady reaction results in the production of gaseous products, such as CO_2. The temperature of the coal almost certainly will continue to increase until, somewhere between 230 °C and 280 °C, the reaction becomes rapid and strongly exothermic (i.e., the coal reaches "ignition" or "flash" point and starts to burn).

Not all types of coal are equally susceptible to spontaneous combustion. High-ranking coals (high carbon content) are more fire prone than lower-ranking coals; however, the exact reasons are unclear. Another important factor is the size of the particles—the larger the effective area of the coal (fine particles), the more rapidly the reaction can proceed.

External factors also play a role in the oxidation reaction. Oxidation requires an adequate supply of air; cracks, fissures, and the porosity of rock and soil over the coal seams may encourage underground coal fires by allowing oxygen to reach the coal. A very rapid air flow may remove the heat so rapidly that the coal never reaches the combustion temperature. Conduction through the coal itself and the surrounding rocks also can remove heat. Nevertheless, in general, the thermal conductivity of coal is quite poor. The removal of heat is the only likely way to stop the process in the initial stages of the oxidation reaction. Once the threshold temperature is reached and the second stage of the reaction begins, heat usually is generated at too great a rate, and the temperature of the coal will continue to increase until combustion occurs.

The presence of water (vapor) also has some important effects. Coal with very high or very low moisture content tends to exhibit a low oxidation rate. However, a minimum amount of water is considered necessary for the reaction to continue (after Banerjee, 1985; Schmal, 1987).

Depending on the influencing factors, a coal fire can initiate in an outcrop of coal seams at the surface. Rosema et al. (1999) defined an open fire as a coal fire that burns in direct contact with the atmosphere, usually with visible flames. Other than in exposed seams of coal, fire can occur in stockpiles

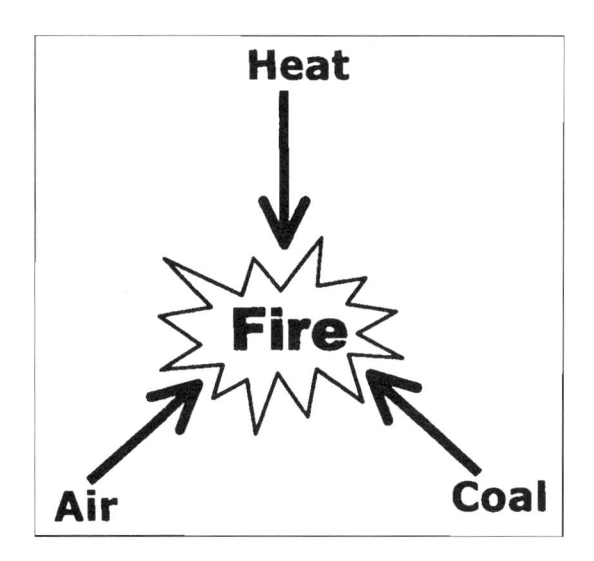

Figure 3. Factors that contribute to spontaneous combustion.

or spoil dumps. In the case of a subsurface coal fire, the required oxygen enters through cracks/fissures at the surface or mine shafts. However, the coal fire can cause subsidence as it voids the support (coal seam) beneath the overburden rock, which will make sufficient passage for breathing to continue combustion.

COAL-FIRE DETECTION AND MONITORING

It is obvious that a surface coal fire can be detected easily; however, a subsurface coal fire can burn over time without any surface activity. Before the introduction of remote sensing for the study of coal fires, borehole measurement was one of the popular methods used to identify coal fires. Although this method can be used to measure the temperature of a coal seam that is suspected to be affected by fire very accurately, it is not feasible for use over a large area.

Despite the fact that remote sensing already was being used for Earth observation, it was first used in the early 1960s for a coal-fire study with an airborne thermal scanner. With time, the technology and quality of thermal remote sensing has improved and so has its use. Some established geophysical methods also are used for coal-fire study.

Remote Sensing and Coal Fires

Aerial photography based on remote sensing was introduced by the end of the nineteenth century and was used mostly for reconnaissance purposes. Satellite-based remote sensing began in 1960, but it was not accessible to civilians until 1972. With time, remote sensing has evolved dramatically. Remote sensing involves the following components:
- Energy source—illuminates or provides electromagnetic energy to the target of interest.
- Radiation—energy travels from its source to the target.
- Interaction with the target—depends on the properties of the target and the radiation.
- Interaction with atmosphere—reflected and radiated energy from the target is transmitted through, and interacts with, the atmosphere.
- Recording of energy at the sensor—a sensor collects and records the transmitted electromagnetic radiation.
- Reception and processing—the data recorded by the sensor are received at a ground station where they are processed into an image and distributed.
- Interpretation and analysis—the processed image is interpreted, visually and/or digitally, to extract information about the target that was illuminated.
- Application—the final element of the remote-sensing process is achieved when it reveals new information or assists in solving a particular problem.

The 3 μm to 14 μm region of the electromagnetic spectrum (EMS) is referred to as the thermal infrared region (Lillesand and Kiefer, 2000). Thermal remote sensing uses atmospheric windows in the 3 μm to 5 μm and 8 μm to 14 μm regions only, because these parts of the EMS are not affected as much by atmospheric interaction.

Thermal infrared sensing exploits the fact that everything above absolute zero (−273 °C) emits radiation in the thermal infrared range of the EMS. The thermal infrared radiation of an object is controlled mainly by its emissivity, geometry, and temperature. Thermal infrared sensors record differences in the received infrared radiation from various objects. Because these differences often are considerable, an infrared image can exhibit a wide range of contrasts. The sensors that are carried by aircraft or spacecraft, which are sensitive to this (infrared) region, provide a possible means of making synoptic measurements of land-surface temperatures. The relation between the radiated energy that is recorded by the sensor and the temperature of the surface can be drawn by Planck's distribution function.

Many commercial and research scanners (air- and satellite-borne) are acquiring data in the thermal infrared region (3–5 μm and 8–14 μm) all over the world. Numerous thermal airborne scanners (e.g., Daedalus [8.5–15.5 μm, USA], TIMS [8.2–12.2 μm, USA], Aries [France], Digital Thermal Linescanner [8–14 μm, UK], TABI-320 [8–12 μm, Canada], OMIS1[1] [8.08–12.4 μm, China]) and several other thermal scanners are operating in different countries with many applications, such as coal, forest, or bush-fire detection and monitoring, surveillance, and reconnaissance. Airborne operations are not as cost effective as satellite-based operations for monitoring a broad area. Satellite-based Earth observation thermal scanners operate from two types of platforms: polar orbiting and geostationary. Among polar-orbiting satellites, Landsat7 ETM+ (Enhanced Thematic Mapper) band 6 thermal data has 60 m spatial resolution, which is the highest of the commercially available satellites. Presently, Landsat7 is nearly nonfunctional because of scanner failure. Another useful spaceborne-thermal scanner is ASTER, on board the TERRA satellite. With five channels in the thermal infrared region (8.125–11.65 μm), it provides an opportunity to estimate emissivity values from the satellite data directly, which is crucial for the quantitative thermal study of Earth's surface. For a large area, the AVHRR (Advanced Very High Resolution Radiometer) instrument (on the NOAA [National Oceanic and Atmospheric Administration] series of polar-orbiting satellites) is recording Earth's surface thermally with a spatial resolution of 1.1 km. Among the small satellite series, Bi-spectral Infra-Red Detection (BIRD, Deutsche Zentrum für Luft- und Raumfahr, Germany) has two scanners to detect hot spots on Earth's surface: one in the mid-infrared range (3.4–4.2 μm) and another in the TIR (thermal infrared) range (8.5–9.3 μm). Although the spatial resolution is coarse (290 m) but with

---
[1]TIMS—Thermal Infrared Multispectral Scanner, TABI—thermal airborne broadband imager, OMIS—Optical Module Imaging System.

a high quantization level (14 bits), BIRD data are quite useful for detecting temperature anomalies over Earth's surface. MODIS (National Aeronautics and Space Administration) is acquiring data in the thermal infrared region (spatial resolution 1 km), which are being used in many applications in meteorology and Earth observation. Many geostationary satellites, such as Geostationary Operational Environmental Satellites (NASA), Eumetsat and Meteosat (ESA), Indian National Satellite (India), Geostationary Operational Meteorological Satellite (Russia), Geostationary Meteorological Satellite (Japan), and Feng-Yun (China), are gathering data about cloud-top temperature for weather prediction. Some of these satellites also are being used for forest fire monitoring because of their very high temporal resolution (B. Maathuis, 2005, personal commun.).

When present as flaming combustion on the surface, coal fires emit significant thermal energy that is easy to detect by any thermal remote-sensing scanner. However, the surface heating is comparatively subdued with a subsurface coal fire, and it may be masked by daytime solar heating. In that case, it is necessary to use nighttime remote-sensing data to reveal and measure the extent of heating. There are three steps in the use of remote sensing to detect a coal fire:

1. Acquisition of a thermal image (preferably night) of the area under investigation using remote sensing and processed digitally to create a surface temperature map to reveal the temperature anomalies,
2. Acquisition of information about the local geological setting, temperatures of coal fire vents, and different land cover types through field survey,
3. Use of the geological and other field knowledge to eliminate anomalies, other than coal fires, to produce a final temperature map that is calibrated with temperatures that were collected in the field.

The atmosphere between the object (coal fire) and the receiver (remote sensor) plays a significant role in the accuracy of the surface-temperature estimation, especially in quantitative studies. The temperature of the surface over a coal fire depends on several factors, such as inherent properties (emissivity, fire temperature) and the conditions of the surrounding areas (soil type, topography, local atmosphere, crack or fissures on the surface, depth of fire).

Soil moisture and wind also affect the surface temperature. The thermal properties of rock also are significant factors in thermal remote sensing. For example, some high-thermal-inertia rocks can appear as coal fires in nighttime images because they trap the heat longer compared with other rocks. However, predawn data acquisition and a good knowledge of the local geological setting can exclude these false anomalies.

Although satellite remote sensing is established as a matured technology in the field of coal-fire detection and monitoring, few constraints (e.g., direct comparison between satellite-derived temperature and field measurements) have been overcome to much of an extent. In general, coal fires are a very local phenomenon, and, in many cases, they are not large enough to saturate a whole pixel to appear as an anomaly in comparison with the background. The aggregated temperature of a pixel depends on the location, spread, surface type, and temperature of the fire/crack and its surroundings. For example, an ASTER pixel is large enough (90 m) to accommodate a few cracks with active fire, local rocks, and (sometimes) sparse vegetation. These different types of land cover may have different temperature ranges to influence each other and finally appear as an anomaly or background with a certain pixel-integrated temperature value. Also, the viewing angle, foot print, field of view of sensor, distance from the object, and operating range (wavelengths) can determine the final pixel-integrated temperature.

Geophysical Methods

In addition to borehole measurements and remote sensing, some established geophysical methods are being used for coal-fire detection and monitoring. However, these methods are point-based measurements, and errors are to be expected during their interpolation.

The radioactive method is based on the fact that sedimentary rocks contain radioactive elements, such as uranium ($^{235}U_{92}$, $^{238}U_{92}$) and thorium ($^{232}Th_{90}$). These radioactive elements emit α particles during decay. During this process, they are transformed into radon ($^{222}Rn_{86}$, $^{220}Rn_{86}$, $^{219}Rn_{86}$), which has a half-life of 3.96 s to 3.825 d. The concentration of α particles depends on the temperature (i.e., if the temperature is higher, then the transportation of α particles is higher). Pressure, porosity, and water content also influence the transportation.

Another method that is used is the resistivity method. The resistance of rock is calculated by using a few electric poles, measuring the resistance in ohms (Ω) per meter, and comparing these with the standard value. Under normal conditions, the resistance of sedimentary rock is 600–800 Ω/m, but in burnt rock, it increases to 1200–3000 Ω/m, because of high porosity, cracks, and low water content.

WORLDWIDE OCCURRENCE OF COAL FIRES AND USE OF REMOTE SENSING

Coal fires, which are a common problem in most coal-producing countries, have a very long history. The oldest coal fire is nearly 2 m.y. old (early Quaternary) and is found near Urumqi in northwest China (Zhang et al., 2004). There also are some reports about a historic coal-seam fire in Indonesia, which probably was started by a lightning strike. In countries where coal mining is a major industry, coal fires are reported inevitably—some were responsible for mine disasters and were short-lived, others burned for considerable periods of time. Major fires, largely uncontrolled, are burning in China, India, and the Unites States. The following sections describe the worldwide occurrence of coal fires and remote-sensing–aided coal-fire detection and monitoring research.

Coal Fires in Asia and the Pacific Region

Coal Fires in India

The Raniganj coal belt, West Bengal, which is the oldest and largest coalfield in India, belongs to the Gondwana Supergroup 1. With an area of ~1260 km^2, the Raniganj coal belt is located ~200 km west of Calcutta, West Bengal. Underground fires have been known there for the past 100 yr or more (Fig. 4). Some surface fires also are evident in some opencast mines. Based on Landsat5 thermal data and local knowledge, the coal-fire areas have been identified in the Raniganj coal belt (Gangopadhyay, 2000).

The neighboring sickle-shaped Jharia Coalfield, which is the only coking coal source in India, lies in the Dhanbad district in the Jharkhand state of India. With an area of 700 km^2, it is located ~250 km northwest of Calcutta. Coal fires are a common phenomena in Jharia that increase the temperature of the surrounding areas and render the area dangerous for the inhabitants and unfit for cultivation (Prakash et al., 1997).

Aided by remote sensing, many researchers, such as Bhattacharya et al. (1991), Cracknell and Mansor (1992), Reddy et al. (1993), Saraf et al. (1995), and Prakash et al. (1997), have studied the Jharia coal fires. In 1991, Bhattacharya et al. used airborne predawn thermal infrared and daytime multispectral images to distinguish the coal fires from the background. Another attempt to detect coal fires was made by Mukherjee et al. (1991), using predawn airborne thermal data. They also attempted to estimate the depth of the fire using a linear heat-flow equation. Cracknell and Mansor (1992) first used Landsat-5 TM and NOAA-9 AVHRR data, and found that nighttime NOAA data were quite useful to isolate the warm areas from the background. Reddy et al. (1993) used the short-wave infrared (SWIR) region of the EMS, which is covered by Landsat TM bands 4, 5, and 7. Using Landsat TM bands 6 and 7, Saraf et al. (1995) established a relation between high-temperature events and surface fires. Later, Prakash et al. (1997) used the Landsat TM TIR and SWIR bands to identify surface and subsurface fires separately. Based on a dual-band approach using TM data, Prakash and Gupta (1999) attempted to calculate the area of surface fires.

Coal Fires in China

Coal fires in China originate at the outcrop of coal in the surface and underground and have natural and manmade origins (Fig. 5). It is estimated that up to 200 million tons of high-quality coal are lost every year, mostly in the northern half of China (Rosema et al., 1999). In 1991, Huang et al. (1991) studied the extent of coal fires using Daedalus data and presented an alarming picture of Chinese coal fires. Since 1986, several researchers have worked on coal fires in the Xinjinag and Ningxia Hui regions. In 1995, Yang (1995) identified several coal fires in these areas using predawn airborne thermal scanner data. Later, Wan and Zhang (1996) carried out a detailed study in the same area. They used daytime Landsat TM band 6 data to estimate the relative amount of solar illumination during the overpass time, which was used to correct for the effect of terrain. To detect smaller coal fires from Landsat 5 thermal band data (spatial resolution 120 m), Zhang et al. (1997) used a subpixel temperature estimation method. Cassells (1998) attempted to model an underground coal fire in the Kelazha area of northern China with input from a three-dimensional geological model. By analyzing the SWIR spectra of rocks, Zhang (1996) identified the burnt rocks, which

Figure 4. Smoke from an underground coal fire in Raniganj, India. Width of the crack is 5 m.

Figure 5. Burnt outcrop of coal (white layer, middle of photo) in Wuda, Inner Mongolia Autonomous Region of northern China.

also are an indication of coal fire. In 1997, Wang (2002) identified areas that were affected by coal fires with ASTER and Landsat TM data in Xinxiang province.

In the Wuda mining region, in Inner Mongolia, an extensive study was done on coal fire using satellite-derived emissivity, which returns a more reliable surface temperature (Gangopadhyay, 2003; Gangopadhyay et al., 2005a). Because emissivity values vary with land cover, satellite-derived emissivity values can increase the accuracy of a quantitative study of coal fires.

Coal Fires in Australia

Although Burning Mountain in Australia has been known for a long time, the first recorded observation was in 1828 after its discovery by a local farmer (Anonymous, 1972). In 1829, and again in 1831, these fires were mapped with a detailed description by the surveyor general, T.L. Mitchell (Mitchell, 1839). In 1918, Abbott recorded detailed information about this phenomenon concerning the movement of the main vent area (Abbott, 1918). In addition, Bunny (1967) and Rattigan (1967a, 1967b) made detailed and careful contributions on the study of the Burning Mountain coal fire. A notable study was produced by Fleming (1972), which suggested that the fires could have been burning since the Pleistocene. Later, Ellyett and Fleming (1974) did an extensive study using a Daedalus thermal airborne scanner that operates in the 8 µm to 14 µm region. Today, that fire is more than 152 m underground and is still burning the coal slowly. Fires also occur spontaneously in opencast coal mines in many locations, such as Hunter Valley (New South Wales) and the lignite mines in Victoria and South Australia (Williams, 2005).

Coal Fires in Indonesia

Slash and burn (forest clearing by fire) is a popular and easy method by which to claim cultivation land from forests in Kalimantan (Borneo), Indonesia; these fires sometimes burn out of control and ignite the coal seams that are exposed nearby. These fires can be very difficult to extinguish because they often ignite in the peat layer. In the same area, Tetuko et al. (2003) studied the burnt coal seams to estimate the thickness of the fire scar using synthetic aperture radar. In east Kalimantan, baked rocks were found close to a coal seam, which could be evidence of an ancient coal fire. The existence of coal fires in southern Sumatra also is familiar to the community that researches coal fires. It has been assumed that these ancient coal fires were ignited by lightning. A combination of forest fire, frequent lightning, and warm climate create a favorable situation for spontaneous combustion in coal. A recent study by Whitehouse and Mulyana (2004) estimated that between 760 and 3000 coal fires are burning in east Kalimantan.

Coal Fires in Russia and the Former Soviet Union States

Like other coal-producing countries, Russia also has the problem of coal fires. In addition to spontaneous combustion, some coal fires are initiated by explosion of trapped methane and human errors (Interfax News Bulletin, 2004; Reuters, 2005). In 1998, 74 coal fires were reported in Russia (Walker, 1999). Kyrgyzstan, a former Soviet Union state, also has been stricken by coal fires, which is evident in the Issyk-Kul region (Reuters, 1998). However, no open literature is available that reports on the application of remote sensing to the detection and monitoring of coal fires in Russia and the former Soviet Union states.

Coal Fires in the Americas

Coal Fires in the United States

Coal mining started in Pennsylvania mainly to make coke for iron smelting. The first coal fire was reported in 1772; in 1869, it turned into a major disaster and claimed the lives of many miners (Glover, 1998; Stracher and Taylor, 2004). Finally, it extinguished itself about a year after an attempt to pour water into the mine failed. The Pennsylvanian fire adversely affected the local flora and fauna and was a major acid rain producer in the United States (Glover, 1998). Many underground coal fires continue to remain poorly documented because they are unprofitable, intangible, and unpredictable. The Centralia mine fire in Pennsylvania has been burning since 1962 and is one of the worst mine fires in the United States (Geissinger, 1990; Memmi, 2000).

In 1962, the U.S. Bureau of Mines reported 223 coal fires all over the United States (Slavecki, 1964). The United States was the first country to apply remote sensing to coal-fire detection. Using the "Reconofax" thermal scanner on an airborne platform, Slavecki (1964), Fisher and Knuth (1968), and Greene et al. (1969) studied fires on waste coal and subsurface coal fires in Pennsylvania, a state where coal fire remains a serious problem. Greene et al. also studied the depth of fire, and they classified fires into three types according to their depth: shallow fires (≤10 m deep), intermediate fires (10–30 m deep), and deep fires (>30 m deep).

Coal Fires in Canada and South America

Bustin and Mathews (1982, 1985) reported a burning coal seam in southeastern British Columbia, Canada. They found that the upper 3 m of the coal seam is being consumed and the fire is advancing along the strike of the coal seam. They also found that some vents in the burnt-out zones are being used as air intakes.

Some coal fires in Venezuela and Argentina, which were initiated mostly by human errors, have been reported by news agencies (Reuters, 2004). Other than these news agency reports, no open scientific documents exist that confirm coal-fire study using remote sensing in these countries.

Coal Fires in Europe

The occurrence of coal fires in mines and stockpiles in many coal-producing countries, such as Germany, Poland, Bulgaria (Walker, 1999), and Romania (R□dan and R□dan, 1998), has been reported by different researchers. In addition, some coal-seam fires were reported by Walker (1999) near the Poland–Czech Republic border. One of the biggest coal-fire–related accidents in Serbian (then Yugoslavia) mining history happened in 1989 and claimed 92 lives. This coal fire was initiated by a trapped methane explosion that took a long time to extinguish (AAS, 1989). Coal fire is a known phenomenon in Ukrainian coal mines. One of the biggest coal-fire–related accidents claimed 33 lives in recent years (Interfax News Bulletin, 2002). A very recent coal fire was reported in the Svea Nord mine on the Arctic island of Spitsbergen, off Norway, which was initiated by human negligence. The fire had a width of 1 km and took several weeks to extinguish (APN, 2005). An unconfirmed web-based source reports about the use of a thermal (airborne) scanner in Germany for monitoring (stockpile) coal fires, but there is no reviewed report available.

Coal Fires in Africa

The occurrence of fires in South African underground coal mines has been observed for a long time. In addition to South Africa, there are some web-based reports about stockpile or spoil dump fires in Zimbabwe, Botswana, Mozambique, and Zambia. An underground coal fire in one of the Anglo coal mines had been burning for the last few years, but was extinguished recently (M. Kooij, 2005, personal commun.).

GREENHOUSE GAS EMISSIONS FROM COAL FIRES

Over the past two centuries, the anthropogenic emissions of greenhouse gases have increased, which has led to an alarming situation. This steady increase in greenhouse gases in the atmosphere acts as a blanket that retains solar radiation in the atmosphere and has led to global warming. Among all of the greenhouse gases, CO_2 plays a significant role in this phenomenon. Since the preindustrial era, the concentration of CO_2 has increased from 280 ppm to 375 ppm (IPCC, 2001). In this linear increase in the concentration of CO_2, coal fires have contributed a significant amount. The greenhouse gases, emitted from all sources, have increased the global mean surface air temperature by ~0.3 °C and 0.6 °C since the late nineteenth century and have caused serious consequences for low-lying coastal areas (EIA 2, 2006).

The impact of coal fires on climate change and their contribution to global warming are getting increasing amounts of expert attention. Recent coal-fire studies on China, one of the major producers of coal, estimate that the country contributes 0.3% (Voigt et al., 2004) to the total world annual output of CO_2 that is caused by fossil fuels. Some previous studies have suggested this amount to be 3% of the world's total (Cassells, 1998), neither of which is a negligible amount. However, the aforementioned estimates are based on indirect methods, such as the total coal burnt in a certain area. The Energy Technology department of the Commonwealth Scientific and Industrial Research Organisation, Australia, established an empirical relation between airborne infrared thermography and greenhouse gases emission from Australian coal mines (Williams, 2005). Some hyperspectral remote-sensing–based methods are being developed that exploit the absorption features of CO_2, in a particular part of the EMS, to quantify the CO_2 emission from coal fires (van der Meer et al., 2004; Gangopadhyay et al., 2005b).

CONCLUSIONS

Coal fires are a widespread problem in most coal-producing countries. Remote sensing can play a significant role in the detection and monitoring coal fires, which may prevent huge economic losses and environmental disasters. In developing countries, such as China and India, coal is the most available and economic energy source. To ensure the proper use of this prime and nonrenewable energy resource, better planning and management are required, with a special emphasis on environmental rehabilitation for sustainable development. Although methods for coal-fire detection and monitoring that are based on remote sensing are well established, some constraints, such as pixel-integrated temperature (with respect to spatial resolution) and acquisition time need to be considered much more cautiously. Most of the research on coal fires has been concerned with detection and monitoring; the greenhouse gases that are emitted from coal fires need to be considered more seriously because they have a significant, adverse impact on global climate. To reduce the steadily increasing greenhouse gases in the atmosphere, emissions that are related to coal fires should be studied more effectively. The countries that are affected most by coal fires (e.g., the United States, China, and India) are concerned about their impact on the environment. However, in most cases, the efforts to restrain these fires are limited if they do not have an immediate economic impact.

ACKNOWLEDGMENTS

The author duly acknowledges the critical comments from Ben Maathuis, International Institute for Geo-Information Science and Earth Observation, the Netherlands, Glenn B. Stracher, East Georgia College, USA, other anonymous reviewers, and Stacia A. Spaulding, USA, that have helped to improve this manuscript.

REFERENCES CITED

Abbott, W.E., 1918, Mount Wingen and the Wingen Coal Measures: Sydney, N.S.W., Australia, Angus and Robertson, 26 p.
Andres, R.J., Fielding, D.J., Marland, G., Boden, T.A., Kumar, N., and Kearney, A.T., 1999, Carbon dioxide emissions from fossil-fuel use, 1751–1950: Tellus, v. 51B, p. 759–765.
Anonymous, 1972, The Burning Mountain: Wingen, New South Wales, Australia, Scone Shire Council Archive.
Associated Press Newswires (APN), 2005, Coal mine fire in northern Norway could burn for weeks: Associated Press, 2 August 2005.
Austin American-Statesman (AAS), 1989, Coal mine fire in Serbia kills 92 workers: Austin American-Statesman, 19 November 1989.
Banerjee, S.C., 1985, Spontaneous Combustion of Coal and Mine Fires: Rotterdam, Holland, A.A. Balekema, 18 p.
Bhattacharya, A., Reddy, C.S., and Mukherjee, T., 1991, Multi-tier remote sensing data analysis for coal fire mapping in Jharia Coalfield of Bihar, India, in Proceedings, 12th Asian Conference on Remote Sensing: Singapore, National University of Singapore, p. 22-1 to 22-6.
Bunny, M.R., 1967, The Burning Mountain, Wingen: New South Wales: Australia Geological Survey Report GS1967, no. 063.
Bustin, R.M., and Mathews, W.H., 1982, In situ gasification of coal, a natural example: History, petrology, and mechanics of combustion: Canadian Journal of Earth Sciences, v. 19, p. 514–523.
Bustin, R.M., and Mathews, W.H., 1985, In situ gasification of coal, a natural example: Additional data on the Aldridge Creek coal fire, southeastern British Columbia: Canadian Journal of Earth Sciences, v. 22, p. 1858–1864.
Cassells, J.S.C., 1998, Thermal Modelling of Underground Coal Fires in Northern China [Ph.D. thesis]: Enschede, The Netherlands, International Institute for Geo-Information Science and Earth Observation, 183 p.
Cracknell, A.P., and Mansor, S.B., 1992, Detection off sub-surface coal fires using Landsat Thematic Mapper data: International Archives of Photogrammetry and Remote Sensing, v. 29, no. B7, p. 750–753.
Ellyett, C.D., and Fleming, A.W., 1974, Thermal infrared imagery of the Burning Mountain coal fire: Remote Sensing of Environment, v. 3, p. 79–86, doi: 10.1016/0034-4257(74)90040-6.
Energy Information Administration 1, 2006, Greenhouse gases, climate change, and energy: http://www.eia.doe.gov/oiaf/1605/ggccebro/chapter1.html (accessed 8 December 2006).
Energy Information Administration 2, 2006: http://www.eia.doe.gov/oiaf/ieo/coal.html (accessed 8 December 2006).
Etheridge, D.M., Steele, L.P., Langenfelds, R.L., Francey, R.J., Barnola, J.M., and Morgan, V.I., 1996, Natural and anthropogenic changes in atmospheric CO_2 over the last 1000 years from air in Antarctic ice and firn: Journal of Geophysical Research, v. 101, p. 4115–4128, doi: 10.1029/95JD03410.
Fisher, W.J., and Knuth, W.M., 1968, Detection and delineation of subsurface coal fires by aerial infrared scanning: Geological Society of America Bulletin, v. 115, p. 67–68.
Fleming, A.W., 1972, Investigations into Permian Geology and the Burning Mountain Coal fire [Bachelor's thesis]: Newcastle, Australia, University of Newcastle.
Gangopadhyay, P.K., 2000, Surface Temperature Mapping and Detection of Surface & Subsurface Coal-Mine Fires of the Raniganj Coalbelt (West Bengal) Using Remote Sensing and GIS: Dehradun, India Institute of Remote Sensing, 56 p.
Gangopadhyay, P.K., 2003, Coal Fire Detection and Monitoring in Northern China: A Multi-Spectral and Multi-Sensor TIR Approach [Master's thesis]: Enschede, The Netherlands, International Institute for Geo-Information Science and Earth Observation, 72 p.
Gangopadhyay, P.K., Maathuis, B., and van Dijk, P., 2005a, ASTER derived emissivity and coal fire related surface temperature anomaly: A case study in Wuda, North China: International Journal of Remote Sensing, v. 26, no. 24, p. 5555–5571, doi: 10.1080/01431160500291959.
Gangopadhyay, P.K., van der Meer, F.D., and van Dijk, P.M., 2005b, Atmospheric modeling using high resolution radiative transfer codes and identification of CO_2 absorption bands to estimate coal fire related emissions, in Zagajewski, B., and Sobczak, M., eds., Imaging Spectroscopy: New Quality in Environmental Studies: Warsaw, Warsaw University Press, p. 373–381.
Geissinger, J., 1990, Mapping underground mine fires: The case of Centralia, Pennsylvania: The Pennsylvania Geographer, v. 28, no. 2, p. 22–26.
Glover, L., 1998, Underground mine fires sparks residents' fear: Tribune Review: Greensburg, Pennsylvania, p. A-1, A-10.
Goddard Space Flight Center, 2006, Global, regional, and national carbon dioxide (CO_2) emission estimates (1751–2000) from CDIAC: http://gcmd.nasa.gov/records/GCMD_CDIAC_CO2_EMISS_MODERN.html (accessed 8 December 2005).
Goldammer, J.G., and Seibert, B., 1989, Natural rain forest fires in eastern Borneo during the Pleistocene and Holocene: Die Naturwissenschaften, v. 76, p. 518–520, doi: 10.1007/BF00374124.
Greene, G.W., Moxham, R.M., and Harvey, A.H., 1969, Aerial infrared surveys and borehole temperature measurements of coal mine fires in Pennsylvania, in Environmental Research Institute on Michigan, ed., Proceedings of the Sixth International ERIM Symposium on Remote Sensing of Environment: Ann Arbor, University of Michigan, p. 517–525.
Huang, Y., Huang, H., and Chen, W., 1991, Remote sensing approaches for underground coal fire detection, in Proceedings of the Beijing International Conference on Reducing Geological Hazards: Beijing, p. 634–641.
Interfax News Bulletin, 2002, Coal mine fire in Ukraine kills 33—Govt.: Interfax International Ltd., 8 July 2002.
Interfax News Bulletin, 2004, 11 injured in coal mine fire in Kuzbass: Interfax Information Services, B.V., 9 June 2004.
Intergovernmental Panel on Climate Change (IPCC), 2001, Climate Change 2001: The Scientific Basis, in Houghton, J.T., et al., eds., Climate Change 2001: Cambridge, UK, Cambridge University Press, 881 p.
Keeling, C.D., and Whorf, T.P., 1999, Atmospheric CO_2 records from sites in the SIO air sampling network: La Jolla, California, Carbon Dioxide Information Analysis Center, University of California: http://cdiac.esd.ornl.gov/trends/co2/sio-mlo.htm (accessed 07 July 2006).
Kok, A., 1981, Spontaneous Heating of Coal: KEMA Report, WSK/2780-55, 159 p. (in Dutch).
Lillesand, T.M., and Kiefer, R.W., 2000, Remote Sensing and Image Interpretation (fourth edition): New York, John Wiley and Sons Inc., 724 p.

Marland, G., Andres, B., and Boden, T., 2006, Global CO_2 emissions from fossil-fuel burning, cement manufacture, and gas flaring: 1751–2003: Oak Ridge, Tennessee, Carbon Dioxide Information Analysis Center, Oak Ridge National Laboratory, http://cdiac.ornl.gov/ftp/ndp030/global.1751_2003.ems (accessed 8 December 2006).

Memmi, J., 2000, Cooking Centralia: A recipe for disaster: Geotimes, v. 45, no. 9, p. 26–27.

Mitchell, T.L., 1839, Three Expeditions into the Interior of Eastern Australia with Descriptions of the Recently Explored Region of Australia Felix and the Present Colony of N.S.W., Eastern Australia, Volume 1: London, T. and W. Boone.

Mukherjee, T.K., Bandopadhyay, T.K., and Pande, S.K., 1991, Detection and delineation of depth of subsurface coalmine fires based on an airborne multispectral scanner survey in a part of the Jharia Coalfield, India: Photogrammetric Engineering and Remote Sensing, v. 57, p. 1203–1207.

Prakash, A., and Gupta, R.P., 1999, Surface fires in Jharia Coalfield, India—Their distribution and estimation of area and temperature from TM data: International Journal of Remote Sensing, v. 20, no. 10, p. 1935–1946, doi: 10.1080/014311699212281.

Prakash, A., Gupta, R.P., and Saraf, A.K., 1997, A Landsat TM based comparative study of surface and subsurface fires in the Jharia Coalfield, India: International Journal of Remote Sensing, v. 18, no. 11, p. 2463–2469, doi: 10.1080/014311697217738.

Rădan, S.C., and Rădan, M., 1998, Rock magnetism and paleomagnetism of porcelanites/clinkers from the western Dacic Basin (Romania): Geol Carpathia, v. 49, p. 209–211.

Rattigan, J.H., 1967a, Phenomena about Burning Mountain, Wingen, New South Wales: Australian Journal of Science, v. 30, p. 183–184.

Rattigan, J.H., 1967b, Occurrence and genesis of halloysite, Upper Hunter Valley, N.S.W., Australia: The American Mineralogist, v. 52, p. 1795–1805.

Reddy, C.S.S., Srivastav, S.K., and Bhattacharya, A., 1993, Application of Thematic Mapper short wavelength infrared data for the detection and monitoring of high temperature related geoenvironmental features: International Journal of Remote Sensing, v. 14, p. 3125–3132.

Reuters, 1998, Five die in coal mine fire in Kyrgyzstan: Reuters Limited, 16 July 1998.

Reuters, 2004, Six dead in Argentina coal mine fire, 8 missing: Reuters Limited, 17 July 2004.

Reuters, 2005, Nine missing after Russian coal mine fire: Reuters Limited, 26 January 2005.

Rosema, A., Guan, H., Veld, H., Vekerdy, Z., Ten Katen, A.M. and Prakash, A., 1999, Manual of Coal Fire Detection and Monitoring: Utrecht, Netherlands Institute of Applied Geosciences, NITG Publications, 245 p.

Saraf, A.K., Prakash, A., Sengupta, S., and Gupta, R.P., 1995, Landsat TM data for estimating ground temperature and depth of subsurface coal fire in the Jharia Coalfield, India: International Journal of Remote Sensing, v. 16, p. 2114–2124.

Schmal, D., 1987, A Model for the Spontaneous Heating of Stored Coal [Ph.D. thesis]: Delft, The Netherlands, Technology University of Delft, 167 p.

Sinha, P.R., 1986, Mine Fires in Indian Coalfields: Energy, v. 11–12, p. 1147–1154, doi: 10.1016/0360-5442(86)90051-4.

Slavecki, R.J., 1964, Detection and location of subsurface coal fires, in Environmental Research Institute on Michigan, ed., Proceedings of the 3rd Symposium on Remote Sensing of Environment, Ann Arbor: Ann Arbor, Michigan, University of Michigan, p. 537–547.

Stracher, G.B., and Taylor, P.T., 2004, Coal fires burning out of control around the world: Thermodynamic recipe for environmental catastrophe, in Stracher, G.B., ed., Coal Fires Burning Around the World: A Global Catastrophe: International Journal of Coal Geology, v. 59, no. 1–2, p. 7–17.

Tetuko, S.S.J., Tateishi, R., and Takeuchi, N., 2003, A physical method to analyse scattered waves from burnt coal seam and its application to estimate thickness of fire scars in central Borneo using L-Band SAR data: International Journal of Remote Sensing, v. 24, no. 15, p. 3119–3136, doi: 10.1080/0143116021000021215.

Théry, I., Gril, J., Vernet, J.L., Meignen, L., and Maury, J., 1996, Coal used for fuel at two prehistoric sites in southern France: Les Canalettes (Mousterian) and Les Usclades (Mesolithic): Journal of Archaeological Science, v. 23, no. 4, p. 509–512, doi: 10.1006/jasc.1996.0048.

van der Meer, F.D., van Dijk, P.M., Gangopadhyay, P.K., and Hecker, C.A., 2004, Remote-sensing GIS based investigations of coal fires in northern China: Global monitoring to support the estimation of CO_2 emissions from spontaneous combustion of coal, in Vibulresth and Soo, D.N., eds., Proceedings of the 1st Asian Space Conference, Chiang Mai, Thailand: Bangkok, Geo-Informatics and Space Technology Development Agency, p. 6.

van Genderen, J.L., Cassells, C.J.S., and Zhang, X.M., 1996, The synergistic use of remotely sensed data for the detection of underground coal fires: International Archives of Photogrammetry and Remote Sensing, v. XXXI, no. 7, p. 722–727.

Voigt, S., Tetzlaff, A., Zhang, J., Künzer, C., Zhukov, B., Strunz, G., Oertel, D., Roth, A., van Dijk, P., and Mehl, H., 2004, Integrating satellite remote sensing techniques for detection and analysis of uncontrolled coal seam fires in North China: International Journal of Coal Geology, v. 59, p. 121–136, doi: 10.1016/j.coal.2003.12.013.

Walker, S., 1999, Uncontrolled Fires in Coal and Coal Wastes: London, International Energy Agency, 73 p.

Wan, Y.Q., and Zhang, X.M., 1996, Using a DEM to reduce the effect of solar heating on Landsat TM thermal IR images and detection of coal fires: Asia-Pacific Remote Sensing and GIS Journal, v. 8, p. 65–72.

Wang, C., 2002, Detection of Coal Fires in Xinjiang (China) Using Remote Sensing Techniques [Master's thesis]: Enschede, The Netherlands, International Institute for Geo-Information Science and Earth Observation, 93 p.

Whitehouse, A.E., and Mulyana, A.A.S., 2004, Coal fires in Indonesia, in Stracher, G.B., ed., Coal Fires Burning Around the World: A Global Catastrophe: International Journal of Coal Geology, v. 59, no. 1–2, p. 91–97.

Yang, H., 1995, Detection of Areas of Spontaneous Combustion of Coal Using Airborne and TM Data in Xinxiang, China [Master's thesis]: Enschede, The Netherlands, International Institute for Geo-Information Science and Earth Observation, 73 p.

Zhang, J.Z., 1996, SWIR Spectra of Rocks in Areas Affected by Coal Fires, Xinjinag Autonomous Region P.R. of China [Master's thesis]: Enschede, The Netherlands, International Institute for Geo-Information Science and Earth Observation, 74 p.

Zhang, X.M., van Genderen, J.L., and Kroonenberg, S.B., 1997, A method to evaluate the capability of Landsat-5 TM band 6 data for sub-pixel coal fire detection: International Journal of Remote Sensing, v. 18, p. 3279–3288, doi: 10.1080/014311697217080.

Zhang, X.M., Kroonenberg, S.B., and de Boer, C.B., 2004, Dating of coal fires in Xinjiang, north-west China: Terra Nova, v. 16, no. 2, p. 68–74, doi: 10.1111/j.1365-3121.2004.00532.x.

MANUSCRIPT ACCEPTED BY THE SOCIETY 7 MARCH 2007

Three-dimensional thermal-imaging methodology for detecting underground coal fires

Zhang Jianmin*
China Shenhua Energy Company Limited, Beijing 100011, China

Huan Zhongdan
Sun Yujing
Tian Yuan
Beijing Normal University, Beijing 100875, China

Stefan Voigt
German Remote Sensing Data Center (DFD), German Aerospace Center (DLR), Postfach 1116, D-82230 Wessling, Germany

Zhao Xuejun
China University of Mining Technology, Beijing 100083, China

ABSTRACT

The study of underground coal fires in China began in the 1960s. The huge loss of coal resources and the ecological disasters caused by coal fires in northern China promoted the study of these fires. Various remote-sensing methods are used to detect ground anomalies due to underground fires. However, locating these fires using remote-sensing data is a difficult task. Ground thermal anomalies are useful for locating underground coal fires. Thermal-geological models link ground thermal anomalies to underground fires. A method of point-source inversion is applicable to a simplified model for the inverse locations of underground coal fires. When tested with data from the Wuda area in the Inner Mongolia Autonomous Region of China, this method exhibits encouraging results.

Keywords: underground coal fires, 3-D detection, steady-thermal model, point sources, dynamic-monitoring thermal model.

INTRODUCTION

Underground coal fires are worldwide disasters that can be started, for example, by manmade fires, lightning strikes, forest fires, or spontaneous combustion. The large-scale development of coal mines in northern China has been accompanied by underground coal fires, especially in the case of fully mechanized longwalls retreating along strike.

*zhangjm@shenhuachina.com

Underground coal fires cause many physical changes to various environmental elements, such as rock, coal, vegetation, and soil. The changes can be described in terms of thermal stress, remnant magnetism, thermoelectricity, and thermal radiance. Thermal stress deforms the structure of rock by developing fissures so that heat transmits conductively and convectively. The heat effect of underground coal fires creates temperature anomalies on the surface. Besides responding to spatial states of underground coal fires, thermal radiance dynamically records evolutionary processes of underground coal fires such as spreading out and vanishing. Underground coal fires may be considered as dynamic processes in coal seams constrained by environmental factors.

Thermal remote sensing can help detect and map coal fires. The technique was first adopted for detecting coal fires in coal refuse in Scranton, Pennsylvania (Slavecki, 1964). In the decades afterward, numerous studies about the detection of coal fires have been conducted in many countries. In the last decade, the Sino-Euro (Guan and van Genderen, 1997) and Sino-Dutch coal-fire projects (Rosema et al., 1999) have been carried out successfully in the Xinjiang and Ningxia Regions of China. Other investigations include coal fires in the Indian Jharia Coalfield (Mukherjee et al., 1991; Huang et al., 1991; Prakash and Gupta, 1999) and coal fires near Centralia, Pennsylvania (Chaiken et al., 1998).

All previous studies based on space- and airborne thermal-scanning data have delineated the coal-fire area using the density slicing method (Fisher and Knuth, 1968; Bhattacharya et al., 1991, 1996; Bhattacharya and Reddy, 1994; Cassells and van Genderen, 1995; Cassells et al., 1996; Jianmin, 1998; Rosema et al., 1999; Xiangmin et al., 1997). It is still unknown how to isolate a coal-fire system by remote-sensing data. The main obstacle is how to interpret underground coal fires in terms of ground thermal anomalies detected by satellite, airborne, or ground scanner.

The fire-extinguishing program in northern China, including coal-fire areas in Wuda and Rujigou, challenged the UCF (Underground Coal Fire) study to provide models that mimic dynamic processes of coal fires and to design effective methodologies for monitoring the processes that are crucial for the engineering design of fire-extinguishing methods and relevant monitoring. In the research of underground coal fires in Rujigou, Jianmin (1999a, 1999b) proposed using the thermal-imaging method and established the imaging profile of temperature anomalies in the Beisan and Dafeng coal-fire areas by ground portable measurement, in which the wavelet method was used. The results, checked by subsequent borehole measurements and rock removal, revealed relative spatial locations and influenced scopes of underground coal fires.

In the present study, the focus was to locate positions of underground coal fires by ground thermal anomaly. Thermal-geological models are proposed for relations between underground coal fires and their ground thermal anomaly. The method of point-source inversion (MPSI) is proposed for a simplified model to find inverse locations of underground coal fires from ground thermal anomalies. The validity of MPSI was tested with data from the Wuda area in the Inner Mongolia Autonomous Region of China. The results are encouraging.

The article is organized as follows: in the first section, thermal models are established with detailed explanations and analysis; the inversion method (MPSI) is the topic of the second section; and inversion results at the site by MPSI are presented in the third section.

THERMAL MODELS AND ANALYSIS

Several thermal models are demonstrated in this section. Formulae for solutions to simplified models in homogeneous media are established. A detailed discussion on relations between ground thermal fields and physical properties of media for point sources is presented.

Surface Temperature Related to Coal Fires

It is a common observation that underground coal fires create ground thermal anomalies; namely, there are differences between surface and background temperatures where underground coal fires occur. However, the detection or measurement of the anomaly is generally a very subtle task, since it not only depends on survey conditions such as observation time, location, weather, etc., but also on geological parameters such as depth, volume of thermal source, fissures, and ground cracks.

Surface temperatures were measured in the coal-fire areas by portable infrared thermometers (Cassells et al., 1996) and by ground thermal-scanning devices (Rosema et al., 1999; Jianmin and Guan, 2004) in Xinjiang, Rujigou, and Wuda. The differences ranged from 3 °C to over 800 °C above background temperature, depending on how close or far the measurement was from the cracks. Measurements at a fracture in the Burning Mountain chimney, Australia, through which hot fumes escaped, showed a horizontal gradient in damp soil of 3.8 °C/cm (Ellyett and Fleming, 1974). In the Mukunda area, Jharia Coalfield, India, the average ambient temperature was ~12 °C during measurement (Prakash et al., 1995a, 1995b), with a 4 °C difference between the highest anomaly ground temperatures.

Thermal Models Describing Underground Coal Fires

Our target of study and detection is the heat effect resulting from underground coal fires. More precisely, underground coal fires are understood as follows: (1) The heat delivered from a burning source(s) in coal seams releases to the air from the ground through media (rock and coal seams); and (2) the thermal phenomenon (anomaly) can be detected, as shown next (Fig. 1).

Based on Fourier's law—namely, the conductional direction is opposite to that of thermal gradient with coefficient λ—and assuming that specific heat c and density of medium ρ are independent of position and temperature, and heat-source intensity $\omega(x, t)$ is independent of temperature, we have the following equation from the law of thermal conservation

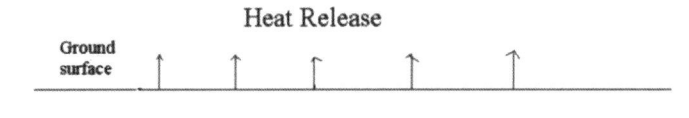

Figure 1. Thermal processes of underground coal fires.

$$\frac{\partial u}{\partial t} = \sum_{i=1}^{3} \frac{\partial}{\partial x_i}\left(a(x,u)\frac{\partial u}{\partial x_i}\right) + f(x,t), \quad (1)$$

where $\alpha(x, u) = \lambda(x, u)/(c\rho)$ and $f(x, t) = \omega(x, t)/(c)$. Please note that u, x, and t, stand for temperature, spatial position, and time, respectively.

Considering the fissures and cracks that occur in the process of coal-fire development as well as heterogeneity in medium, we should employ a generalized model

$$\frac{\partial u}{\partial t} = \frac{1}{c(x,u)\rho(x,u)} \sum_{i,j=1}^{3} \frac{\partial}{\partial x_i}\left(a_{ij}(x,u)\frac{\partial u}{\partial x_j}\right) + f(x,t,u,\nabla u). \quad (2)$$

Now, specific heat c and density might depend on both position and temperature and heat-source intensity might depend on temperature and its gradient, if considering material migration. The matrix

$$A(x,u) = \begin{pmatrix} a_{11}(x,y) & a_{12}(x,y) & a_{13}(x,y) \\ a_{21}(x,y) & a_{22}(x,y) & a_{23}(x,y) \\ a_{31}(x,y) & a_{32}(x,y) & a_{33}(x,y) \end{pmatrix} \quad (3)$$

is called a diffusion tensor. This model formally includes many simplified versions and gives more choices. The drawback is lack of understanding of geological, physical, and chemical processes during development of coal fires, which is crucial for modeling diffusion tensors and heat sources.

For the choice of boundary condition on the ground surface, it is natural to select the Robin condition (or Newton's law of cooling) due to our observation of the process; namely, if $R^3_+ = \{x = (x_1, x_2, x_3)| x_3 > 0\}$ is a working region $x_3 = 0$, the ground boundary condition is

$$\left(-\frac{\partial u}{\partial x_3} + \sigma u\right)(x',0,t) = \sigma u_1(x',t), \quad (4)$$

where $\sigma = \alpha/\lambda$, α is the heat-transfer coefficient, λ is the thermal conduction coefficient, u_1 is the environmental temperature, and $x' = (x_1, x_2)$. We also have to know the temperature underground at a specific time, that is, the initial condition $u(x, 0) = u_0(x)$.

Since the development of underground coal fires is a relatively long process and variations of temperature anomaly relative to time are slow in general, the corresponding steady model is an alternative. For example,

$$\begin{cases} \frac{-1}{c(x,u)\rho(x,u)} \sum_{i,j=1}^{3} \frac{\partial}{\partial x_i}\left(a_{ij}(x,u)\frac{\partial u}{\partial x_j}\right) = f(x,u,\nabla u), x \in R^3_+ \\ \left(-\frac{\partial u}{\partial x_3} + \sigma u\right)(x',0) = \sigma u_1(x'), \quad x' = (x_1, x_2) \in R^2 \end{cases} \quad (5)$$

or

$$\begin{cases} \frac{-1}{c(x)\rho(x)} \sum_{i,j=1}^{3} \frac{\partial}{\partial x_i}\left(a_{ij}(x)\frac{\partial u}{\partial x_j}\right) = f(x), \quad x \in R^3_+ \\ \left(-\frac{\partial u}{\partial x_3} + \sigma u\right)(x',0) = \sigma u_1(x'), \quad x' = (x_1, x_2) \in R^2 \end{cases} \quad (6)$$

In this article, we choose models in homogeneous media because environmental changes are small relative to the detection process, the stratum above the coal seam is simple, and the available mathematical theory is limited. In this case, the thermal-conduction coefficient (λ), specific heat (c), density (ρ), and heat-transfer coefficient (α), are constants, so that the dynamic model possesses the following form

$$\begin{cases} \frac{\partial u}{\partial t} - a^2 \Delta u = f(x,t), \quad x \in R^3_+, t > 0 \\ u(x,0) = u_0(x), \quad x \in R^3_+ \\ \left(-\frac{\partial u}{\partial x_3} + \sigma u\right)(x',0,t) = \sigma u_1(x',t), \quad x' = (x_1, x_2) \in R^2, t > 0 \end{cases} \quad (7)$$

The steady model is

$$\begin{cases} -a^2 \Delta u = f(x), \quad x \in R^3_+ \\ \left(-\frac{\partial u}{\partial x_3} + \sigma u\right)(x',0) = \sigma u_1(x'), \quad x' = (x_1, x_2) \end{cases} \quad (8)$$

where $a^2 = \lambda/(c\rho)$, $\sigma = \alpha/\lambda f(x) = \omega(x)/(c\rho)$.

For these simplified models, we have formulae for their solutions. Here, we only present the formula for the solution **u** (thermal field) to the model (Eq. 8). The thermal field **u** is decomposed into two parts: the contribution of the source, v, and, w, that of environmental thermal field u_1, where

$$v(x) = \int_{R_+^3} \left[\Gamma(x-y) + \Gamma(x-\bar{y})\right] f(y) dy$$
$$-2\sigma \int_{R_+^3} \Gamma(x-\bar{y}) \int_0^{y_3} f(y',s) e^{-\sigma(y_3-s)} ds dy \quad (9)$$

and

$$w(x) = u_1(x') + \int_{R_+^3} \left[\Gamma(x-y) + \Gamma(x-\bar{y})\right] a^2 \Delta u_1(y') dy$$
$$-2 \int_{R_+^3} \Gamma(x-\bar{y}) a^2 \left(1 - e^{-\sigma y_3}\right) \Delta u_1(y') dy, \quad (10)$$

where $y_1, y_2), \tilde{y} = (y', -y_3)$, and $\Gamma(x) = \dfrac{1}{4\pi a^2} \dfrac{1}{|x|}$.

The establishment of the formula follows from properties of Newton potential and method of extensions. Therefore, the ground thermal field $\varphi(x')$ determined by the steady model can be written as $\varphi(x') = v(x', 0) + \omega(x', 0)$. The part by source $v(x', 0) = \varphi(x') - \omega(x', 0)$ and that by environment $\omega(x', 0)$ are called the anomaly field and the natural field of the ground thermal field, respectively.

Thermal Modeling Analysis

Since environmental temperature u_1 can be quite precisely simulated so that the natural field $\omega(x', 0)$ can be calculated by the formula, in this study we concentrated on the behavior of the anomaly field $v(x', 0)$ of point sources with the steady model in a homogeneous medium.

Suppose that there is a point source underground (space) with intensity ω at (b, h), where $b = (b_1, b_2)$ and h are called the ground coordinates and the depth of the source, respectively. For convenience, $\omega/(\rho c)$ is called the intensity, denoted by c (forgive us for notation abuse), $\lambda/(\rho c)$ is the thermal-conduction coefficient, denoted by a^2, and α/λ is the heat-transfer coefficient, denoted by σ. Assuming that the environmental temperature is zero, that is, $\omega \equiv 0$, the ground thermal field is

$$\varphi(x') = \frac{c}{2\pi a^2 \sqrt{|x'-b|^2 + h^2}} - \frac{c\sigma}{2\pi a^2}$$
$$\int_0^\infty \frac{e^{-\sigma s} ds}{\sqrt{|x'-b|^2 + (s+h)^2}}, \quad (11)$$

where $x' = (x_1, x_2)$.

It is not hard to verify that (expression 1) is symmetric about b and achieves a maximum at b; (expression 2) is linear for intensity c; (expression 3) $\varphi(x')$ decreases as the heat-transfer coefficient h or thermal-conduction coefficient a^2 increases; (expression 4) $\varphi(x')$ decreases and the temperature graph becomes flat as the depth h increases (Figs. 2, 3, and 4). In the multi–point-source case, the linearity of the model implies that the ground thermal field is the superposition of all single sources. When the intensities of point sources are comparable and the distances in between are not small, each point source corresponds to a local maximum value, close to the projected ground coordinate of the source. On the other hand, when some of the intensities among point sources are larger, or some of the distances in between are small, the correspondence might fail.

In fact, a point source is uniquely determined by its ground thermal field—namely, two different point sources cannot produce the same ground-temperature anomaly field.

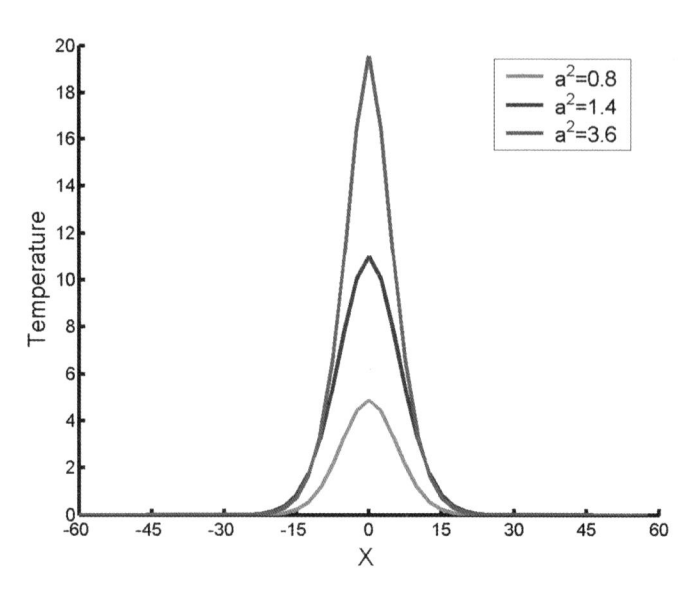

Figure 2. Graph of ground-temperature anomaly (°C/m) of a single source with different thermal-conduction coefficients, where $\sigma = 0.5$ m^{-1}, $h = 20$ m, and $c = 1000$ °C/h.

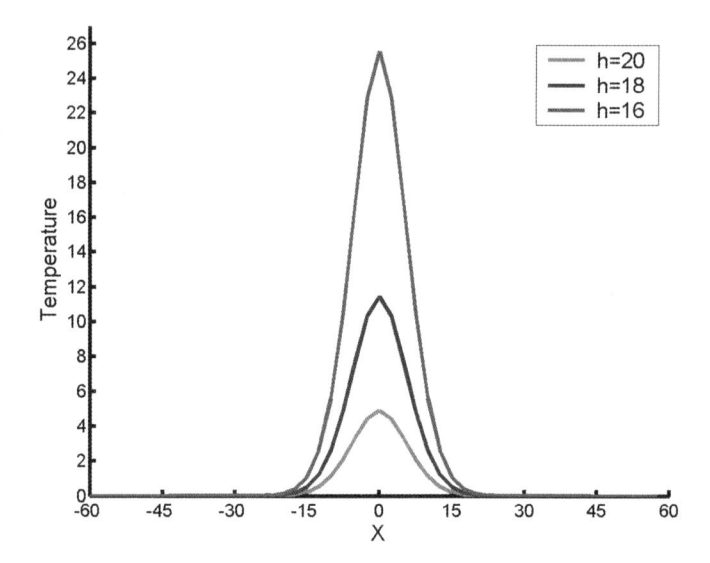

Figure 3. Graph of ground-temperature anomaly (°C/m) of a single source with different heat-transfer coefficients, where $a^2 = 0.8$ m^2/h, $h = 20$ m, and $c = 1000$ °C/h.

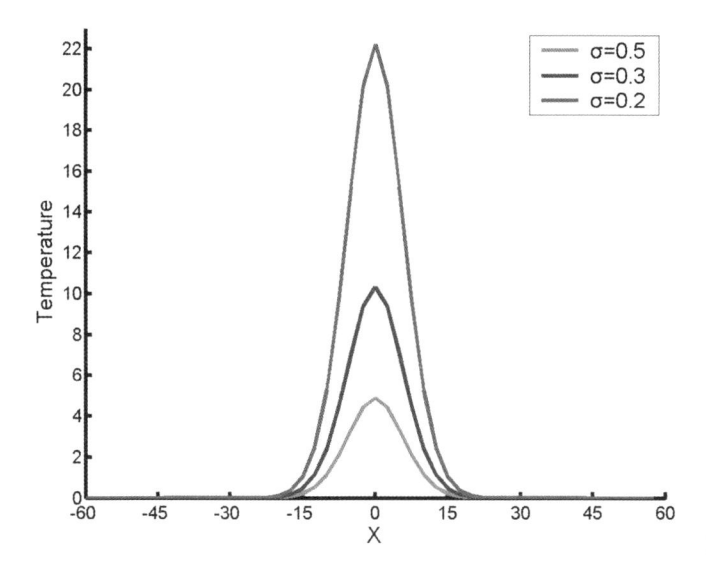

Figure 4. Graph of ground-temperature anomaly (°C/m) of a single source with different depths, where σ = 0.5 m^{-1}, a^2 = 0.8 m^2/h, and c = 1000 °C/h.

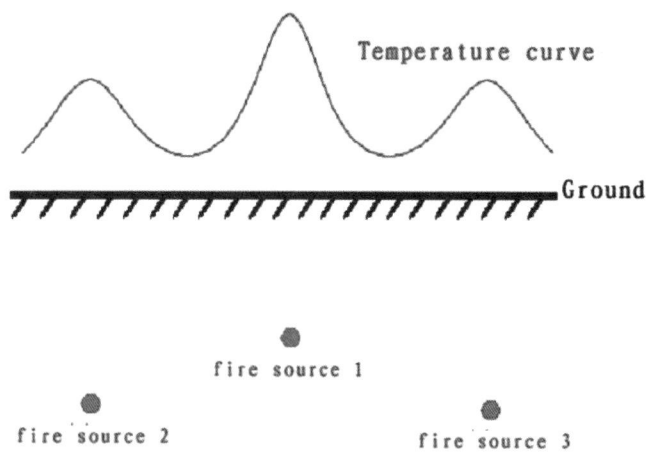

Figure 5. Ground thermal fields of multipoints.

Based on the theory established here for the anomaly field in a rock medium, the continuous measurability on site, and observations on cross sections of coal fires at the Rujigou region, Ningxia, we can think of the anomaly field generated by spontaneous combustion in a coal seam as a contribution of several point sources. These analyses of point sources in homogeneous media provide an effective method to recover the positions and intensities of underground coal fires, demonstrated in the third section by numerical results comparing data at the site.

AN INVERSION MODEL IN TERMS OF POINT SOURCES

Tracing the thermal source is a very difficult task in the detection of underground coal fires, since the complexity of evolution of underground coal fires makes it impossible to have a precise geological and physical model, which is crucial to building a reasonable mathematical model. Taking care of such a circumstance and noting that the spatial size of a fire source is rather small relative to that of the considered region, we simplify a fire source to one or several single-point sources (Fig. 5). For these types of fire sources, we can obtain loci and intensities of heat sources by an inversion method for steady models. If we have enough information, we can predict the evolution of the sources; the corresponding dynamic model allows us to present the underground and ground thermal fields.

Numerical Analysis Based on Point-Source Model

In this subsection, we present the effectiveness of MPSI for various loci and intensities of tripoint sources with similar depths.

Three tripoint sources were chosen: the first one with the same intensity and reasonable distances from each other, the second one with different intensities and reasonable distances from each other, and the third with same intensity and small distances. We chose values of σ = 0.5 and a^2 = 1.

In the figure for each group, the upper part displays the thermal-anomaly field (a surface with thermal isopleths), and the lower part shows the temperature profile. The temperature scale is represented by colors.

Group 1

The loci of three point sources are: P_1 = (25, 10, 15), P_2 = (−25, 10, 15), and P_3 = (0, 0, 14), and all three point sources have the same intensity: c_j = 1000 (where j = 1, 2, 3). This is a case of a tripoint source with same intensity and reasonable distances from each other.

It is expected that there are three clearly separated and remarkable local maxima in the ground thermal-anomaly field. This is confirmed by the three remarkable peaks in Figure 6. Moreover, the temperature profile shows that the thermal anomaly possesses three distinguishable centers in the temperature profile.

In this type of situation, for each local maximum, one point source contributes much more than the others, so that we might think the thermal anomaly around that local maximum was produced by that point source.

Group 2

The loci of three point sources are: P_1 = (25, 10, 15), P_2 = (−25, 10, 15), and P_3 = (0, 0, 14), and the intensities are $c_1 = c_2$ = 400, and c_3 = 1000, respectively. This is a case of a tripoint source with different intensities and reasonable distances from each other. The intensity of one point source is larger the others (by a ratio of 2.5:1).

In this situation, we still expect that there will be three clearly separated local maxima in the ground thermal-anomaly

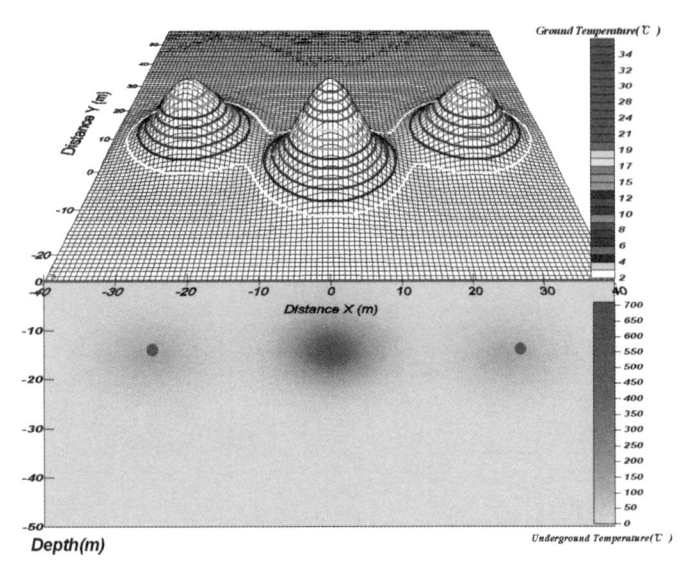

Figure 6. Composite graph with isothermal surface and section graph of temperature anomaly of group 1 (point-source positions P_1 [25, 10, 15], P_2 [–25, 10, 15], P_3 [0, 0, 14] and intensities $c_j = 1000$, where $j = 1, 2, 3$).

field and that only the point source with the highest intensity will yield a remarkable maximum. This is clearly verified in Figure 7, even though the temperature profile also displays three distinguishable centers.

Note that the consideration for Group 1 is still applicable for the largest local maximum (it is a global maximum!). The other local maxima might not be treated this way. We might not be able to determine if such maxima in a real situation are a result of noise. More precisely, small local maxima might be ignored.

Group 3

The loci of three point-sources are: $P_1 = (5,10,15)$, $P_2 = (-5,10,15)$, and $P_3 = (0,0,14)$, and all three point sources have the same intensity: $c_j = 1000$ (where $j = 1, 2, 3$). In this case, the point sources in the tripoint source possess the same intensity and are close to each other.

We cannot expect that there will be three local maxima in the ground thermal-anomaly field now. In fact, it (Fig. 8) displays only one maximum, and even the temperature profile shows that the centers are not very clearly separated.

Thus, in this case, the correspondence between point sources and local maxima may not be one-to-one any more. If we consider noise in real situations, we cannot expect to recover each point source from local maxima in the ground thermal-anomaly field. On the other hand, if we could determine a position close to each point source, we still could obtain much information about the source.

Inversion Method Based on Point-Source Model

Consider a point source at point $P = (b_1, b_2, h)$ with intensity c, i.e., the source $f(x) = c\delta_p(x)$. The anomaly field generated by the point source is

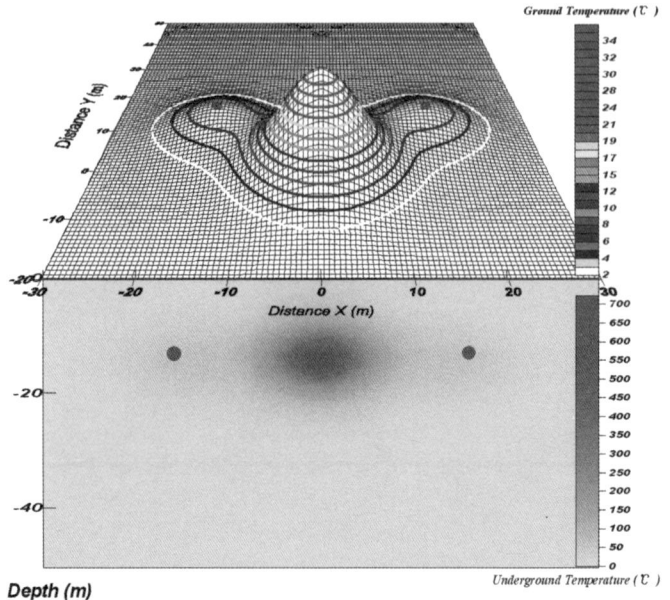

Figure 7. Composite graph with isothermal surface and section graph of temperature anomaly of group 2 (point-source positions P_1 [25, 10,15], P_2 [–25, 10, 15], P_3 [0, 0, 14] and intensities $c_1 = c_2 = 400$, $c_3 = 1000$).

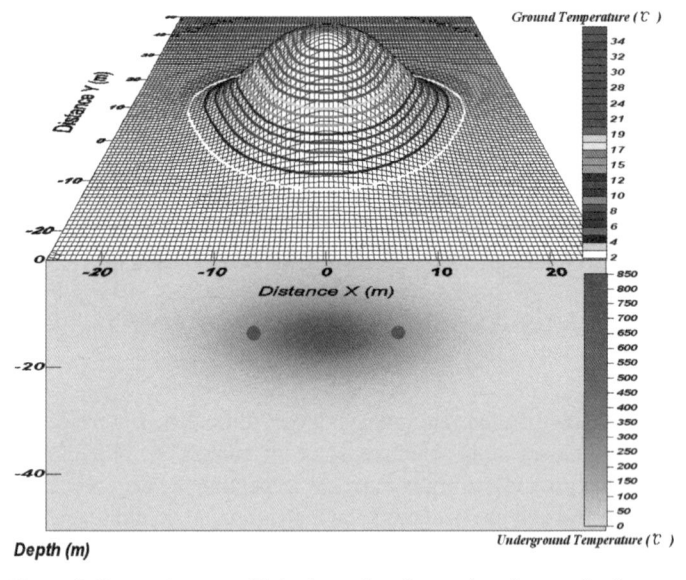

Figure 8. Composite map with isothermal surface and section graph of temperature anomaly of group 3 (point-source positions P_1 [5, 10, 15], P_2 [–5, 10, 15], P_3 [0, 0, 14] and intensities $c_j = 1000$, where $j = 1, 2, 3$).

$$\varphi(x') = \frac{c}{2\pi a^2 \sqrt{|x'-b|^2 + h^2}} - \frac{c\sigma}{2\pi a^2} \int_0^\infty \frac{e^{-\sigma s} ds}{\sqrt{|x'-b|^2 + (s+h)^2}}. \quad (12)$$

So, the ground coordinates $b = (b_1, b_2)$ of the source can be obtained by figuring out the maximum point of the anomaly field $f(x) = (x_1, x_2)$. Therefore, the goal of the inversion is to calculate the depth h and intensity c of the source. This goal can be achieved by comparing the peak with some other point nearby.

Method of Point-Source Inversion (MPSI)

Given an anomaly field with k peaks corresponding to k point sources, each of which approximately has the same ground coordinates as the related peak, we can calculate its depth and intensity for each source. There are four basic steps:

Step 1. For each peak, we assume that the surrounding anomaly field is generated by a point source with the same ground coordinates as the peak. More precisely, let the jth point source be at $P_j = (b_j, h_j)$, where $b_j = (b_{j1}, b_{j2})$, and select β_j close to b_j, $j = 1,...,k$. Solve the following equation for r_j and obtain an approximate $\rho_j = 1/r_j$ to the jth depth h_j:

$$\frac{\varphi(\beta_j)}{\varphi(b_j)} = \frac{\frac{1}{\sqrt{1+r_j^2|\beta_j-b_j|^2}} - \int_0^\infty \frac{\sigma e^{-\sigma s} ds}{\sqrt{(1+r_j s)^2 + r_j^2 |\beta_j - b_j|^2}}}{\int_0^\infty \frac{\sigma r_j s e^{-\sigma s} ds}{1 + r_j s}}. \quad (13)$$

Denote the anomaly field generated by the point source with unit intensity at (b_j, ρ_j) by

$$\varphi_j(x') = \frac{1}{2\pi a^2 \sqrt{|x'-b_j|^2 + \rho_j^2}} - \frac{\sigma}{2\pi a^2}$$
$$\int_0^a \frac{e^{-\sigma s} ds}{\sqrt{|x'-b_j|^2 + (s+\rho_j)^2}}, \quad \text{where } j = 1,...k \quad (14)$$

Step 2. Minimize the equation

$$\min_{\substack{c_j \in R \\ 1 \leq j \leq k}} \left\| \sum_{j=1}^k c_j \varphi_j(x') - \varphi(x') \right\| \quad (15)$$

to determine the intensities c_j, $j = 1...,k$, where $\| \; \|$ is the usual Euclidean norm.

Step 3. Utilize the modified anomaly value

$$\varphi(x') - \sum_{l \neq j} c_l \varphi_l(x') \quad (16)$$

to recalculate ρ_j for the jth depth h_j as in step 1, and then use the new jth depth value h_j to repeat step 2.

Step 4. Analyze the errors:

$$e(x') = \varphi(x') - \sum_{l=1}^k c_l \varphi_l(x'). \quad (17)$$

To make a modification to the ground coordinates, go back to step 1, if necessary, until a satisfactory result is obtained.

Inversion of Theoretical Model Data

In this section, we will carry out the inversions for the three groups used in the previous section (Fig. 5) using MPSI. Units of length are in meters. For Group 1, the number of peaks is equal to that of point sources. Table 1 shows the result of steps 1 and 2. For Group 2, the number of peaks is still the same as that of point sources, but the third point source is much stronger than the others, and its peak might cover the others in practice. For Group 3, there is only one peak, which is close to the center of the triangle with vertices b_j ($j = 1, 2, 3$). The results verify the predictions in the first section. The results shown in Table 1 illustrate the power of MPSI.

INVERSION RESULTS AT THE SITE

In this section, we present inversion results by MPSI for a site in the No. X coal-fire area of the Wuda Coalfield, Inner Mongolia Autonomous Region, China. The results are satisfactory.

The Landscape of the Test Area

The test area is in Wuda Coalfield, Wuhai, Inner Mongolia Autonomous Region, China (Fig. 9). The Wuda Coalfield covers an area of 35 km² to the west of Wuhai city (Fig. 9). There are 19

TABLE 1. RESULTS OF STEPS 1 AND 2

Test group	Parameter source point	Coordinate (m) Assumed	Coordinate (m) Inversed	Depth (m) Assumed	Depth (m) Inversed	Intensities Assumed	Intensities Inversed
Group 1	P1	(25, 10)	(23.6, 10)	15.0	16.35	1000	992.57
	P2	(−25, 10)	(−23.6, 10)	15.0	16.35	1000	943.96
	P3	(0, 0)	(0, 0.6)	14.0	15.97	1000	1178.97
Group 2	P1	(15, 10)	(16.5, 11.5)	15.0	15.17	400	263.28
	P2	(−15, 10)	(−16.5, 11.5)	15.0	15.17	400	321.57
	P3	(0, 0)	(0, 0.85)	14.0	15.67	1000	1336.12
Error	Group 1 + Group 2	Absolute distance error: 1.41 m		Relative depth error: 7.59%		Relative intensity error: 16.53%	

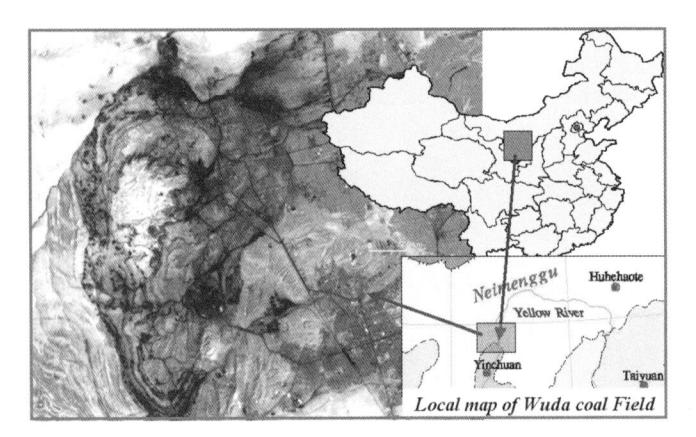

Figure 9. Sketch map of the Wuda Coalfield, Inner Mongolia Autonomous Region, China. The coalfield is located within the Wuda geosyncline and trends N-NE, with total area of 35km², 10 km from north to south and 3–5 km from west to east, which is geographically within 106°34'41"–106°38'41" in longitude and 39°27'00"–39°34'04" in north latitude.

mineable coal seams in the coalfield, of which seams 1, 2, 4, 6, 7, and 8 are mined out and seams 9, 10, and 13 are currently being mined. Sixteen coal-fire fields have been determined, the total area of which is 3,496,000 m². The combustion depth is as much as 84 m. The active coal-fire areas still covered over 422,000 m² in 2004, so the loss of coal is huge.

The coal-fire areas grew faster from 1996 to 2004. The growth continued vertically and horizontally, which threatened the utilization of surrounding coal sources and mining safety. To slow this trend, the Wuhushan and Suhaitu mines started successive fire-extinguishing projects, which have not been effective so far. One of the difficulties is in determining the depth of a fire source.

The No. X coal-fire area of the Wuda Coalfield is a typical coal-fire area (Fig. 10). In July 2004 and January 2005, we collected multiphase geology and ground-temperature data for the region. In this study, we utilize the data at nonfissure observation points as our basic data.

Inversion Results of Field Data

The base image of the No. X coal-fire area map (Fig. 10) in which measured positions are marked is the high-resolution satellite image (Quickbird, USA) with a pixel size of 0.6 m × 0.6 m. Spots with dark-green color in the image represent areas of sparse grass. Measurements at the sites, marked with the number (+), include temperature, fissures, etc., in which positions (Δ) were chosen for inversion. Position and surrounding temperature were acquired using a portable thermal-infrared thermometer, setting radiation rate as 0.92.

In order to understand general features of the coal-fire distribution, temperature was measured by a DL-700A thermal-infrared-scanning device, with a temperature resolution of 0.1 °C and spatial resolution of 1 m × 1 m. The temperature data of the selected point, the delineated area, and survey profile images could be captured using postprocessing software of thermal-scanning data. High thermal anomalies generally correspond to active coal fires with developing cracks or fissures induced by underground coal fires. Due to the irregular spatial size of the fire, photogrammetry in-step was carried out at the same time for convenient comparison, so as to select the inversion data required by the inversion model. A mosaic image of thermal-infrared scanning was made for interpretation (Fig. 11). The field photo at the top corresponds to the thermal image at the bottom which shows thermal anomalies in gray (~21–21.5).

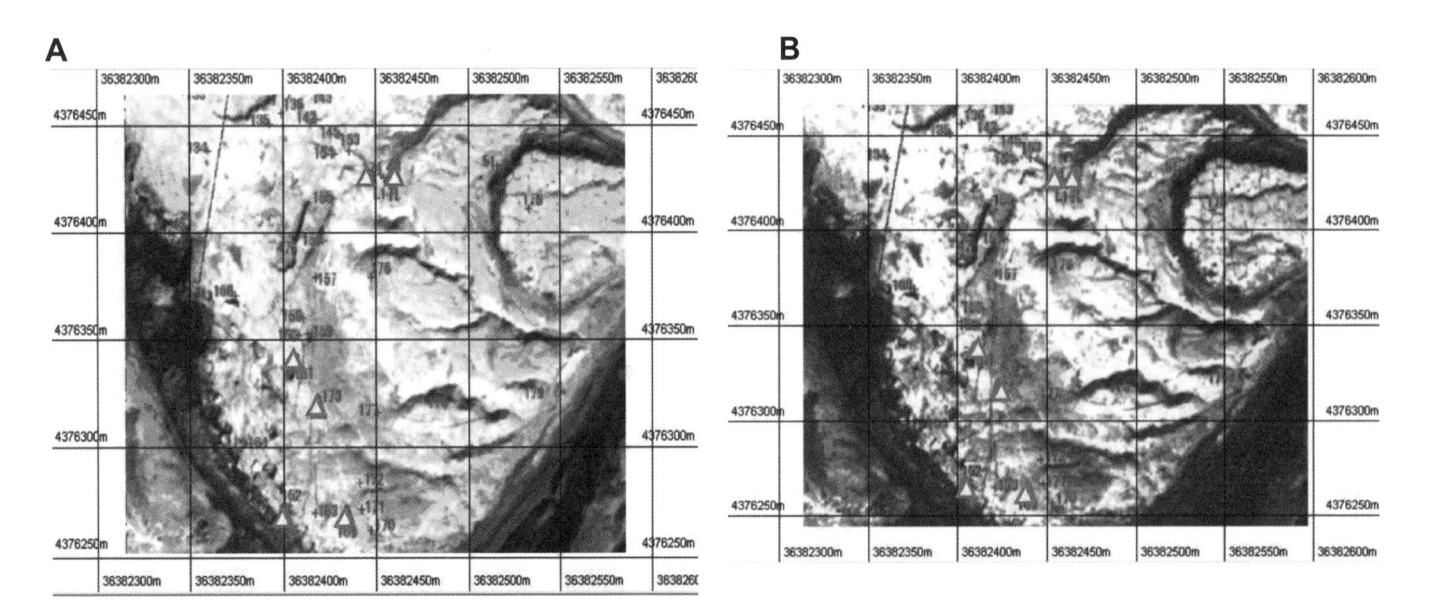

Figure 10. Quickbird image of No. X coal-fire area with observation points (+) and inversion points (Δ), June 2005. (A) Image after linear-stretch processing; and (B) image after equalization-stretch processing.

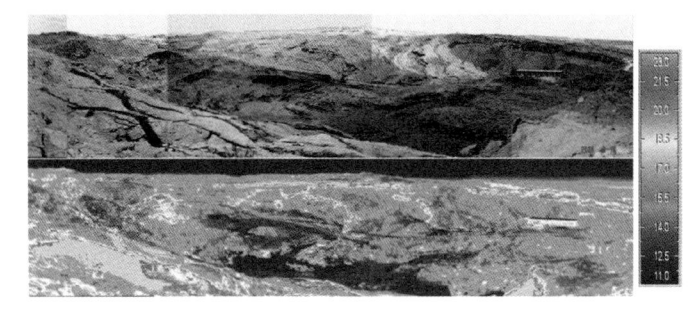

Figure 11. Ground thermal-infrared-scanning image of No. X coal-fire area (time: 3:34 a.m.–3:42 a.m., 16 June 2004).

Since the design of MSPI is based on thermal conduction in a homogeneous medium, it is important to ensure that the requirements of the model be fulfilled as much as possible. We selected inversion points where the rocks were bare in the thermal-anomaly zone. The inversion points and their relevant data are listed in Table 2.

From the data, three candidate peaks were chosen: observation points 48, 169, and 173. For these peaks, three adjacent observation points 174, 166, and 161 were chosen, respectively. A heat transfer coefficient $\sigma = 0.03$ and a thermal-conduction coefficient $a^2 = 0.5$ were selected for the area.

Table 3 summarizes the results by MPSI. For fire sources are in coal seams, the real depth was chosen to be the thickness of the roof above the coal seam. The roof thickness was acquired from the three-dimensional (3-D) visualized geologic model, which was established based on drill hole, coal seam, mined-out area, mining engineering, and other data.

The results are rather satisfactory compared with the theoretical calculations (Table 3). The depths at point 48 and 169 are surprisingly good. The error of the depth of point 173 is a little bit large, but still decent.

The predicted temperature values at the observation points computed from the theoretical formula are presented in Table 3. The root mean square error was 1.032 °C, and relative error was 5.33%. The root mean square depth error was 1.39 m, and relative depth error was 8.3%.

DISCUSSION

The models and methods proposed in this article are based on ground thermal anomalies. After simplifying geological conditions and the structure of fire sources, we formulated a workable method (MPSI) to calculate depths and intensities of sources. In applying the method, besides obtaining relevant geological parameters, topography and background corrections have to be applied to ground-temperature data, and fissure-observation data should be removed.

The theoretical-numerical results based on the steady linear model in a homogeneous medium demonstrate the following:

(1) In the single-point source case, the magnitude and scope of the thermal anomaly depend on depth and intensity of the source and on heat transfer and the thermal-conduction coefficients of the medium. For example, when the source is shallow, the peak of the anomaly is high; namely, the gradient near the ground-projective center of the source is large, so that the source is easy to recognize; on the other hand, as the source gets deeper, the peak gets flatter, and the gradient near the center is smaller, so that the anomaly may disappear in the background, and the source may be hard to recognize.

(2) In the multi–point source case, the anomaly is affected not only by the parameters of each point source and medium but also by the relative spatial locations of the point sources. The anomaly is a superposition of all the point sources. The theoretical-numerical analyses show that as the point sources get closer, the compounding affect is larger, so that it gets hard to separate the effect for each point source, while as the distances among the point sources get farther apart, the joint anomaly gets weaker, and each point source can be recognized individually.

TABLE 2. OBSERVATION DATA OF NO. X COAL-FIRE AREA*

No. of measurement position	Relative coordinate x (m)	y (m)	Measured ground temperature (°C)	Thermal anomaly (°C)	Local height (m)
48	460.946	6426.72	49	27	117.35
174	459.38	6427.70	47	25	117.62
166	407.18	6268.10	26	4	118.91
169	437.78	6266.90	46	24	114.01
161	413.18	6339.50	35	13	120.00
173	423.38	6326.30	45	23	120.00

Note: Surrounding temperature was 22 °C.
*Data acquired in June 2004.

TABLE 3. COMPARISON BETWEEN OBSERVATION AND INVERSION DATA

No. of observation inversed points	Temperature anomaly (°C) Observed	Inversed	Thermal source depth (m) Real depth	Inversed depth	Thermal source intensity inversed (°C/h)
48	27	26.045	19.21	18.64	814.99
174(I)	25	25.857			
169	24	24.002	10.76	10.636	369.87
166(I)	4	4.547			
173	23	21.824	20.36	16.22	948.35
161(I)	13	14.750			

Note: Points like 166(I) were used as inversed computation, 48 as inversed point.

The numerical-inversion results, both using theoretical data and data at the site using MSPI, showed the following:

(1) MSPI can determine the depth and intensity for a single-point source. The accuracy depends on the depth and intensity; accuracy is high for shallow sources with strong intensity but low for deep sources with weak intensity.

(2) For a multi–point source, the compound action among point sources gets stronger as the distances among point sources become smaller, so that the depths inversed by MSPI possess larger error. If the distances are very small, MSPI can only inverse the depth of certain geometric centers of the point thermal source group. As the distances increase, the compound action weakens: each point source can be dealt with independently, and MSPI still can provide accurate positions of each point source.

The experimental results from Wuda show that MSPI can inverse the combustion depth and scope by using the data from thermal-infrared measurement. Because of the complex geological structure at the Wuda Coalfield, which has multiseam coal and mined-out areas left by longwall mechanized mining technology, it is not realistic to accurately determine the depth. It is more practical to determine the depth by combining MSPI and other available information such as geologic structures and heat sources at the site.

Ground thermal anomalies on the surface obtained by satellite, airborne-thermal scanning, or ground thermal scanning measurements, are direct indicators of active coal fires. Using the thermal-imaging method, it is possible to create an image that shows the special situations of underground coal fires in three dimensions. Such an image can be used as an aid in the engineering design of fire-extinguishing methods in a 3-D geographic information system (GIS) environment. By verifying the loci of results, the MSPI method could be used as a way to interpret and outline the spatial locations and relative intensities of underground coal fires.

ACKNOWLEDGMENTS

This research was supported through the "Research on Key Techniques Using Remote Sensing Methods for Detecting Underground Coal Fires" project from the National 863 research program through "Innovative Technologies for Exploration, Extinction and Monitoring of Coal Fires in North China," supported by the Sino-German governments. We thank China Shenhua Energy Limited for its financial support and Guan Haiyan and Gu Dazhao from the Shenhua Group for their contribution and exchange of ideas. In addition, we thank John van Genderen of the International Institute for Geo-Information Science and Earth Observation, Enschede, The Netherlands, and Huang Haiyang of Beijing Normal University, as well as Edward Heffern of the U.S. Bureau of Land Management in Cheyenne, Wyoming, and Glenn B. Stracher of East Georgia College, Swainsboro, Georgia, USA, for their review of this manuscript.

REFERENCES CITED

Bhattacharya, A., and Reddy, S., 1994, Underground surface coal mine fire detection in India's Jharia Coalfield using airborne thermal infrared data: Asian-Pacific Remote Sensing Journal, v. 7, p. 59–73.

Bhattacharya, A., Reddy, S., and Mukherjee, T., 1991, Multi-tier remote sensing data analysis for coal fire mapping in Jharia Coalfield of Bihar, India: Asian Conference on Remote Sensing, Singapore, v. 1, p. P-22–1 to P-22–6.

Bhattacharya, A., Reddy, S., and Dangwal, M., 1996, Coal mine fire inventory monitoring in Jharia Coalfield, Bihar, India, using Thematic Mapper thermal IR data, in Proceedings of the Eleventh Thematic Conference and Workshops on Applied Geologic Remote Sensing: Las Vegas, Nevada, p. 27–29.

Cassells, C.J.S., and van Genderen, J.L., 1995, Thermal modeling of underground coal fires in northern China, remote sensing in action, in Proceedings of the 21st Annual Conference of the Remote Sensing Society, September 11–14, 1995: Southampton, UK, Remote Sensing Society, p. 544–551.

Cassells, C.J.S., van Genderen, J.L., and Xiangmin, Z., 1996, Detection and measuring underground coal fires by remote sensing, in Proceedings of the 8th Australian Remote Sensing Conference, March 28, 1996, Volume 2: Canberra, Australia, Remote Sensing and Photogrammetry Association of Australia, p. 90–101.

Chaiken, R.F., Brennan, R.J., Heisey, B.S., Kim, A.G., Malenka, W.T., and Schimmel, J.T., 1998, Problems in the Control of Anthracite Mine Fires: A Case Study of the Centralia Mine Fire, in Report of Investigations 8799 of the United States Department of the Interior: Pittsburgh, Pennsylvania, U.S. Department of the Interior Report of Investigations 8799.

Ellyett, C.D., and Fleming, A.W., 1974, Thermal infrared imagery of the Burning Mountain coal fire: Remote Sensing of Environment, v. 3, p. 79–86, doi: 10.1016/0034-4257(74)90040-6.

Fisher, W., and Knuth, W.K., Jr., 1968, Detection and delineation of subsurface coal fires by aerial infrared scanning: Abstracts for 1967: Geological Society of America Special Paper 115, p. 67–68.

Guan, H., and van Genderen, J.L., 1997, Report on environment monitoring of spontaneous combustion in the coal fields of North China: Xi'an, China, Aerophotogrammetry and Remote Sensing Bureau of China Coal and the International Institute for Aerospace Survey and Earth Sciences (Enschede, The Netherlands) [in Chinese].

Huang, Y., Huang, H., Chen, W., and Li, Y., 1991, Remote sensing approaches for underground coal fire detection, in Proceedings of the International Conference on Reducing of Geological Hazards, 20–25 October: Beijing, China, Ministry of Geology and Mineral Resources, p. 634–641.

Jianmin, Z., 1998, Study on 3D Dynamic Monitoring of Spontaneous Combustion of Coal Seams [Ph.D. thesis]: Beijing, China University of Mining and Technology, 102 p.

Jianmin, Z., 1999a, Study on 3D imaging method for detection of underground coal fires, application to Rujigou Coalfield of Ningxia, China, in Proceedings of 13th International Remote Sensing Conference on Geology, Volume 2: Vancouver, British Columbia, ERIM International, 1–3 March 1999, p. 142–149.

Jianmin, Z., 1999b, Study on 3-D dynamic monitoring system for spontaneous combustion of coal seams, in Mining Science and Technology: Rotterdam, Holland, A.A. Balkema, p. 97–100.

Jianmin, Z., and Guan, H., 2004, Study on the four-layers remote sensing detection system for underground coal fire monitoring: Remote Sensing for Land and Resources, v. 4, p. 50–53.

Mukherjee, T.K., Byopadhyay, T.K., and Pande, S.K., 1991, Detection and delineation of depth of subsurface coalmine fires based on an airborne multispectral scanner survey in a part of the Jharia Coalfield, India: Photogrammetric Engineering and Remote Sensing, v. 57, p. 1203–1207.

Prakash, A., and Gupta, R.P., 1999, Surface fires in Jharia Coalfield, India—Their distribution estimation of area temperature from TM data: International Journal of Remote Sensing, v. 20, p. 1935–1946.

Prakash, A., Saraf, A.K., Gupta, R.P., Dutta, M., and Sundaram, R.M., 1995a, Surface thermal anomalies with underground fires in Jharia coal mine, India: International Journal of Remote Sensing, v. 16, p. 2105–2109.

Prakash, A., Sastry, R.G.S., Gupta, R.P., and Saraf, A.K., 1995b, Estimating the depth of buried hot features from thermal IR remote sensing data:

A conceptual approach: International Journal of Remote Sensing, v. 16, p. 2503–2510.

Rosema, A., Guan, H., Vekerdy, Z., Katen, Z.A., and Prakash, A., 1999, Four level data collection, *in* Manual of Coal Fire Detection and Monitoring: Utrecht, The Netherlands, Netherlands Institute of Applied Geoscience Report NITG 99-221-C, p. 115–123.

Slavecki, R.J., 1964, Detection and location of subsurface coal fires, *in* Proceedings of the 3rd Symposium on Remote Sensing Environment, October 14–16, 1964: Ann Arbor, University of Michigan, p. 537–547.

Xiangmin, Z., Cassells, C., and van Genderen, J.L., 1997, Mapping underground coal fires using Remote Sensing and GIS techniques, *in* Proceedings of the 10th International Congress of the International Society for Mine Surveying, November 2–6, 1997: Fremantle, Western Australia.

MANUSCRIPT ACCEPTED BY THE SOCIETY 7 MARCH 2007

Comparison of Pennsylvania anthracite mine fires: Centralia and Laurel Run

Melissa A. Nolter*
Service Access & Management, Inc., 1 S. 2nd St., Pottsville, Pennsylvania 17901, USA

Daniel H. Vice
Pennsylvania State University, Hazleton Campus, 76 University Drive, Hazleton, Pennsylvania 18202, USA

Harold Aurand Jr.
Pennsylvania State University, Schuylkill Campus, 200 University Drive, Schuylkill Haven, Pennsylvania 17972, USA

ABSTRACT

Centralia is in Pennsylvania's western middle anthracite field, a large synclinorium in Columbia and Schuylkill Counties. Centralia residents set fire to a landfill at the edge of town in 1962, thereby igniting the Buck Mountain coal bed. Laurel Run is in Pennsylvania's northern anthracite field, on the northwest-dipping limb of the Wyoming Valley syncline. In 1915, a miner's abandoned carbide lamp started a fire at Laurel Run, igniting the Red Ash, Top Red Ash, and Bottom Ross coal beds.

The Centralia and Laurel Run fires are burning out of control. Subsidence and the venting of toxic gases have destroyed large sections of each community. Because the Centralia fire started in the hinge zone of an anticline separating two synclines in the Western Middle Field, it spread in four directions. The Laurel Run fire occurred on one limb of a syncline, limiting its spread to two directions. At Centralia, the steeper-dipping beds permitted the fire to reach a greater depth more rapidly than at Laurel Run. In addition, the point of origin and steeper dip at Centralia make this fire more difficult to control, even though only one coal bed is burning.

A historical and sociological comparison of both communities shows that the people of Laurel Run had greater access to political power and more experience as a community in dealing with crises. Laurel Run secured more government support in combating the fire than Centralia did and so emerged from the fire as a more socially intact community. The present state of each fire further underscores how different geologic settings and social conditions can lead to different outcomes.

Keywords: Centralia, Pennsylvania, Laurel Run, coal fires, coal-bed fires, mine fires.

---

*bumrat00@ptd.net

INTRODUCTION

Centralia and Laurel Run are small towns in Pennsylvania's anthracite coal fields (Fig. 1). In 1962, both were affected by uncontrolled underground mine fires that destroyed large sections of their respective communities. Of the two fires, Centralia has received more scholarly and popular attention, particularly during the late 1970s and 1980s (e.g., studies by GAI Consultants, Inc., 1983; Kroll-Smith and Couch, 1990; DeKok, 2000). In fact, the sociological literature often treats Centralia's experience as a model for how communities and governments respond to underground mine fires and other manmade environmental disasters (Kroll-Smith and Couch, 1990). This paper will examine the Centralia and Laurel Run mine fires, efforts at fire abatement and control, movement of the fires, response of the two communities to their disasters, and provide new data on the present status of the fires. The two fire experiences are compared to show how they are similar and different, and the extent to which Centralia is an effective model for these types of events.

CENTRALIA

Centralia, has achieved fame, or at least notoriety, for being almost completely destroyed by a coal fire. The fire started in May 1962, when members of the community decided to burn the accumulated municipal garbage in an abandoned strip-mining pit (Fig. 2) on the edge of town. The burning garbage ignited the Buck Mountain anthracite coal bed. The fire spread to the underground mine tunnels in the Buck Mountain bed beneath the town. On 15 August 1962, state officials closed the coal mines beneath Centralia because of the danger from asphyxiation by carbon monoxide gas (Kroll-Smith and Couch, 1990, p. 3). The fire attracted peak media attention in the early 1980s when the ground subsided beneath Todd Domboski on 14 February 1981. Fortunately, he was rescued before disappearing beneath the surface (Dekok, 2000, p. 152).

Early attempts to fight the fire included flushing with water-rock slurry, construction of fly ash barriers, and trenching. These efforts failed because insufficient material was used and extensive fracturing of the coal and bedrock made it very difficult to get a complete seal. In addition, the extensive fracturing and attitude of the rock layers served to direct water away from the fire. Local politics and history exacerbated problems the community had working with state and federal bureaucrats.

Consequently, the state of Pennsylvania decided to buy out the nearly 1000 residents of Centralia, providing them with relocation funds appropriated by the U.S. Congress (Dekok, 2000, p. 269–270), while permitting the fire to burn. The first residents were bought out in 1969 followed by the majority in the early 1980s, however, some residents refused to leave. Today, less than 15 people reside in Centralia (S.R Couch, 31 October 2003, oral commun.).

Although the Centralia coal fire attracted scientific and popular attention during the late 1970s and 1980s (e.g., studies by GAI Consultants, Inc., 1983; Kroll-Smith and Couch, 1990), it has received little detailed attention since then.

Geologic Setting of the Centralia Fire

Centralia, located in the central Appalachian Mountain section of the Valley and Ridge Province (Faill and Nickelsen, 1999), is part of the Western Middle Field. This field is one of four major anthracite coalfields of Pennsylvanian age in northeastern Pennsylvania (Eggleston et al., 1999). Each field is a complexly folded and faulted synclinorium that is separated from the others by anti-

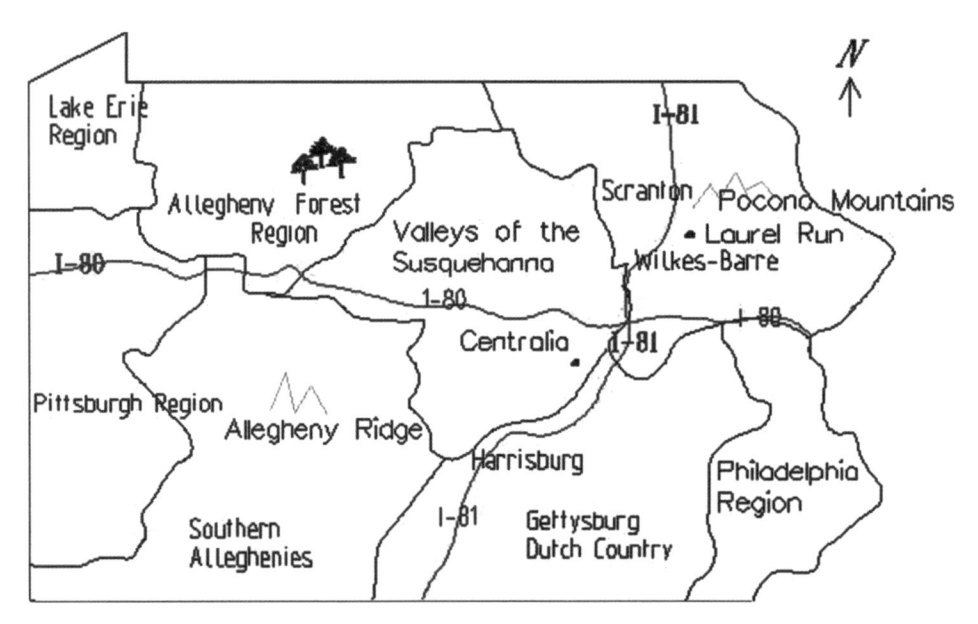

Figure 1. The location of Centralia and Laurel Run in Pennsylvania (Centralia is 40°48′ 15″ N, 76°20′ 25″ W and Laurel Run is 41°13′ 20″ N, 75°51′ 46″ W).

clinoria (Arndt, 1971; Eggleston et al., 1999). The general trend of these structures is northeast-southwest (Eggleston et al., 1999).

The Western Middle Field consists of a complex series of east-west–trending asymmetric synclines and anticlines (Arndt, 1971). The field covers 243 km^2 (94 mi^2) in Columbia, Northumberland, and Schuylkill Counties (Chaiken et al., 1980). A series of anticlines and synclines divides this field into six major basins. The central basin of the Western Middle Field consists of a syncline, the Centralia syncline, which extends through Centralia to Mount Carmel (Arndt, 1971). It is bounded on the north by the Centralia thrust fault and on the south by the Locust Mountain anticline (Fig. 2).

The Pottsville and the Llewellyn Formations (Pennsylvanian) contain the anthracite coal beds and are restricted to the synclinoria (Eggleston et al., 1999). These formations consist predominantly of conglomerate, sandstone, siltstone, claystone, and shale. The Pottsville Formation includes a distinctive conglomerate, which has a framework of large, white quartz pebbles and a dark-grayish, organic-rich sandy matrix. Much of the shale/claystone in the Llewellyn Formation is medium to dark gray in color.

Only the lowest four beds of 33 coal beds in the Llewellyn Formation outcrop in the Centralia area. From oldest to youngest, these beds are the Buck Mountain, Seven Foot, Skidmore, and Mammoth. The stratigraphically higher coal beds have been eroded (Chaiken et al., 1980, p. 4 and 5). These coal beds resemble a series of concentric coaxial folds separated from one another by ~10–100 m of shale and sandstone (Chaiken et al., 1980, p. 4), where the Buck Mountain bed is the lowest. The rim of each fold is the surface outcrop of the coal bed. These beds dip 22°N (Arndt, 1971) in the fire area (Fig. 3). The Buck Mountain bed crops out close to the southern margin of the Centralia syncline and is the only bed burning. The regional folding that formed the Western Middle Field has fractured both the coal and adjacent rock, permitting air access to the subsurface.

The Centralia fire began on the north limb of the Locust Mountain anticline in the Buck Mountain coal bed (Fig. 2). From this location, near the Odd Fellow Cemetery, the fire has advanced through underground mine workings in four directions; these are referred to herein as "fronts." The first or "cemetery" front (Figs. 2 and 3) has advanced west past St. Ignatius Cemetery and is continuing toward Mount Carmel. The second front has advanced southwest and crossed Pennsylvania Route 61, necessitating the construction of a detour around the highway. The third front has advanced south-southeast through Brynesville, while the fourth front has advanced east toward Big Mine Run.

Observation of the Fire Fronts

We observed visual evidence of an active cemetery front (Fig. 3), including gas issuing from surface vents, warm ground, the absence of vegetation, snow melt around vents, mineral

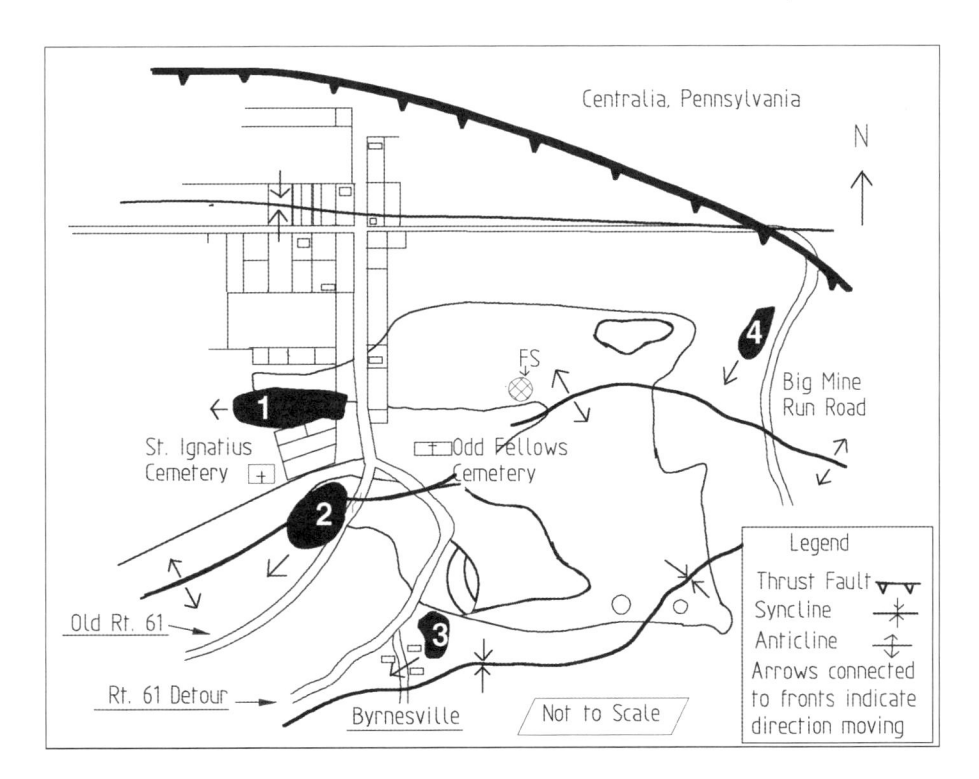

Figure 2. A map showing the syncline and anticline in Centralia; FS indicates the start of the fire (from Stracher et al., 2006, with modifications). Used with permission of the Geological Society of America.

Figure 3. Centralia fire front 1 in Figure 2. (Field of view is 40 ft [12.9 m].)

deposition around some vents, and baked rock immediately adjacent to the hottest vents. Mineral deposition has occurred at several sites on the cemetery front and includes a frothy, light-gray mineral as well as a yellowish crust-like material. Shale from the Llewellyn Formation has been baked to a reddish color in the area of the hottest vents.

Using remote-sensing data acquired by Shallenberger (1993) and current data acquired by compass and pace methods, we located and re-examined the four fire fronts (Fig. 3). The cemetery front has advanced ~400 m since 1982, for an average rate of 19 m/yr. This rate of advance for the coal fire at the cemetery front compares with the average rate of advance of 23 m/yr for that front observed by the Pennsylvania Department of Environmental Protection since 1962 (Jones, 2002).

At the second front, snowmelt, steam, and gas emissions were present during visits on 7 December 2002, 14 March 2003, and 29 December 2003, indicating that this front was still active at that time. During subsequent visits on 24 April 2004, 1 May 2004, and 31 December 2004, steam, additional subsidence, absence of vegetation, mineral deposition, and gases were present, indicating that it was still active. Subsidence has occurred on Pennsylvania Highway 61 (Figs. 2 and 4), which has necessitated the closure of a section of the road and required a detour to be built. We measured a temperature of 128 °C at one vent; however, rate of advancement was not determined due to insufficient data. Reddish and/or light-gray rock, similar to that observed at the first front, was observed at two other locations. Mineral deposition had also occurred on this front.

At the third front, we found no gas emission, steam, or visible fire effects on the vegetation, suggesting that this front is inactive (Fig. 2). At the fourth front, steam emissions are visible from an old mine tunnel on Big Mine Run Road. There is no apparent effect on vegetation in the area and only slight mineral deposition on the rock adjacent to the old tunnel. We consider this front to be weakly active.

On 21 and 22 June 2003, during the filming of the Centralia fire for part of a television documentary on wildfires, the compositions of gases being emitted from vents on both the cemetery and second fronts were measured by Stracher and Taylor. The preliminary results from three sites showed carbon dioxide concentrations of up to 2200 ppm and carbon monoxide concentrations of up to 1000 ppm at the cemetery front. Preliminary results from one site at the second front showed concentrations of 2200 ppm carbon dioxide and 1000 ppm carbon monoxide.

The reddish and/or light-gray rock observed at both the cemetery and the second front is similar to the clinker found in much of the Powder River Basin of Montana and Wyoming (Coates, 2003). Raymond (1986, p. 13) indicated that at temperatures of 450 °C, an irreversible mineralogical change occurs in clay, and it forms a hard, impermeable material. The temperatures that the authors measured on the cemetery front are in the range that Raymond (1986, p. 13) provided for this mineralogical change, suggesting that clinker is now forming at Centralia (Coates, 2003).

Geological Factors That Hindered Attempts to Control the Centralia Fire

Underground coal fires are suppressed or controlled by starving them of air, cooling the coal below its ignition point, and/or forming a barrier to prevent the fire from advancing (Bannerjee, 1985; Chaiken et al., 1980, p. 5 and 6). The Centralia fire has not been controlled due to the geology, previous mining, and cultural factors (Chaiken et al., p. 5–8). For example, the moderately dipping beds (22°N) on the north limb of the Locust Mountain anti-

Figure 4. Subsidence and cracking of asphalt on Pennsylvania Highway 61, fire front 2 in Figure 2. (Field of view is 72 ft [21.95 m].)

cline (Arndt, 1971) allowed the fire to quickly reach depths that were very costly to excavate and isolate using a barrier. The location of the starting point for the fire, on the nose of an anticline, allowed the fire to spread in multiple directions.

In addition to leading the fire deeper into the ground, the dip of the beds served to form a "self-propagating" convection cell between the fire and the incoming air within the coal bed (Chaiken et al., 1980, p. 9). The fire can draw air to itself from both the fractures in the coal and in the surrounding bedrock, while hot gases from the fire can escape updip within the coal bed. A 1983 study by McElroy showed that these convection cells provided the fire with a steady supply of air and allowed it to propagate both laterally and downdip on the coal bed (Dekok, 2000, p. 31). Both the coal and the intervening bedrock are heavily fractured because of the folding that occurred during the formation of the Appalachian Mountains (Faill and Nickelsen, 1999). Extensive fracturing of the coal and bedrock makes surface sealing impossible for controlling the fire because of the inability to locate all surface and lateral openings. In addition, late subsidence could fracture the material used as a seal and permit air to reach the fire (Chaiken et al., 1980, p. 8).

With fractured bedrock, it is very difficult to cool the coal with water because the fractures can lead some of the water away from the fire. In addition, the Buck Mountain coal bed reaches the surface near the top of the Locust Mountain anticline; water used to flood the coal bed would not reach the outcrop but would run off (Chaiken et al., 1980, p. 7). Attempts to use fly ash to form a barrier rather than cooling the coal with water were similarly affected in that the fractured coal and bedrock prevented a complete seal (Chaiken et al., 1980, p. 29–32). Dekok (2000) also suggested that the fly ash barrier failed because insufficient material was used.

Nonscientific Factors That Caused Problems with Control of the Centralia Fire

In addition to the geological factors, a number of nonscientific factors also hindered efforts at control. One was the cultural heritage in much of the anthracite region that made people inclined to mistrust authority, which made it difficult for Centralia to believe what the state and federal bureaucrats were saying.

Other problems included an uncoordinated response by local officials who had to deal with several government agencies with overlapping jurisdictions, and who lacked experience in dealing with state and federal bureaucracy. This combination of factors led to a start-stop response in fighting the fire (Kroll-Smith and Couch, 1990, p. 90–93). Similarly, the Bureau of Mines, Office of Surface Mining, and other federal and state agencies failed to develop an effective, coherent plan to deal with the Centralia fire, in part because of location, and because of cost (Kroll-Smith and Couch, 1990; Chaiken et al., 1980, p. 21–35).

This combination of factors led some people in the anthracite region to believe that the state wanted all attempts to extinguish the fire to fail, so that the people would be forced to relocate and the coal under their property could be seized for profit (Dekok, 2000, p. 83). Although it is hard to judge the relative significance of the different geological and nonscientific factors, the latter may have been a more significant factor in the inability of the government and local residents to control the Centralia fire. The geological factors added cost but did not preclude controlling the fire.

Summary and Discussion of the Centralia Mine Fire

Centralia is an example of a worst-case scenario of the effect of a coal fire on a community. This fire destroyed the physical structure of an entire town and displaced a community. Dipping coal beds and heavily fractured rock enclosing the coal made the fire expensive and difficult to control. The inexperience of local community leaders in working with state bureaucracy (Kroll-Smith and Couch, 1990, p. 10–11) when trying to extinguish the fire was responsible for periods during which the fire was permitted to burn. After the decision was made by the state to buy out homeowners, there has been no attempt to extinguish the fire. It is currently being monitored by the Pennsylvania Department of Environmental Protection. The two fronts that currently appear to be active will continue to burn for many years. The most active front, the first or cemetery front appears to continue to advance at a rate of ~19 m/yr.

The Centralia fire continues to emit carbon dioxide, carbon monoxide, and other gases into the atmosphere, while consuming the valuable coal resource and prohibiting the use of land for agricultural, industrial, or recreational use. Minerals condensed from the gas and the gases themselves may have detrimental effects on human health. Dekok (2000) stated that two people in Centralia suffered from carbon monoxide poisoning. The presence of carbon monoxide gas was confirmed by preliminary measurements of gas from vents by Stracher and Taylor (2003) in June of 2003. They recorded concentrations of up to 2200 ppm of carbon monoxide at three sites on the cemetery front, and concentrations of 1000 ppm carbon monoxide at one site on the second front. Fire-induced subsidence has locally been a problem where it has damaged Pennsylvania Highway 61 (Fig. 4).

INTRODUCTION TO THE LAUREL RUN MINE FIRE

Laurel Run, a small town in the anthracite region of eastern Pennsylvania, is an example of a community that responded successfully to a major underground mine fire (Fig. 1). The fire began in the Red Ash Coal Mine on 6 December 1915, when a careless miner left a lit carbide lamp attached to a mine timber. The owners of the Red Ash Mine had recently fired their night watchman, so the fire was allowed to burn undetected over the weekend. When work resumed, the company became aware of the fire, and they tried to cut off the air supply by plugging openings with concrete and flushing sand into the immediate area (Randolph, 2002). They claimed the fire was under control. In 1921, the fire was still burning beyond the company's containment area (Ashmead, 1922). At that point, the company began erecting a series of temporary seals, designed to contain the fire while allowing other sections of the mine to be worked. When active mining ceased in 1957, so did efforts to control the fire. It was not until September 1962 when a Laurel Run resident was forced to abandon her home on South Dickerson Street due to gases and subsidence, that the community knew it faced an uncontrolled underground mine fire (Randolph, 2002).

By April 1966, a plan was created to deal with the fire. A series of boreholes were drilled to determine the geographic extent of the burn. It was found that part of the fire was under the borough of Laurel Run. Mine operators had robbed the pillars, weakening the overhead strata, and leaving the ground susceptible to subsidence. Gases from the fire escaped through fractures into people's homes, threatening their health. An urban renewal project under the auspices of the Luzerne County Redevelopment Authority and the U.S. Department of Housing and Urban Development was initiated, relocating 850 residents. Another section of the fire was burning beneath the Georgetown section of Wilkes-Barre (Fig. 1). The pillars had not been robbed, and the fire and its effects were kept from the surface by a solid layer of rock. Dealing with the fire below Georgetown required only blocking off access tunnels that fed oxygen to the fire (Dierks et al., 1971).

Geologic Setting of Laurel Run

The Laurel Run coal fire has a regional setting similar to the Centralia fire. The Laurel Run fire occurred on the south limb of the Wilkes-Barre synclinorium, which is ~6.4 km (4 mi) wide. This structure is part of the Valley and Ridge Province of the Appalachian Mountains. There is a broad, shallow-dipping shelf that does not have any sub-basins as in Centralia. Unlike Centralia, three coal beds are burning at Laurel Run; Ross, Top Red Ash, and Bottom Red Ash (Dierks et al., 1971). The Red Ash coal bed is stratigraphically equivalent to the Buck Mountain bed in Centralia (Eggleston et al., 1999, p. 461). The Llewellyn Formation occurs above the Red Ash and the Pottsville Conglomerate occurs below it, within the northern anthracite field.

Based on the cross section in Dierks et al. (1971, p. 13), the coal beds at the Laurel Run fire dip between 10°N and 15°N. The structural setting of the Laurel Run fire is less complex than the Centralia fire. Major faults near the Laurel Run fire could allow communication between the coal beds, but no major folds would split the fire (Sevon, 1976). Thus, the Laurel Run fire has been limited to movement along the strike of the coal beds (to the northeast and/or southwest) and downdip (to the northwest).

The limited aerial extent of the Red Ash Coal Company Mine and adjacent mine boundaries provided barriers that prevented the spread of the fire in either a northeast or a southwest direction. The fire did spread from the Red Ash mine into the Stanton-Empire mine and, in doing so, spread downdip under the Georgetown section of Wilkes-Barre. It also threatened Interstate 81 as it moved toward Wilkes-Barre.

Observation of the Laurel Run Fire Front

Visual evidence of the mine fire at Laurel Run is sparse (Figs. 5–9). Some of the area impacted by the fire, where homes and businesses were torn down, has been rebuilt, often with larger homes and yards than in the older section of the community. Areas that have not been rebuilt are often visually indistinguishable from other local areas that have not been involved in underground or surface mining.

The only actively burning area is under the Laurel Run Estates Mobile Home Park driveway. We observed gas issuing from surface vents, and a few examples of fractured, baked rock, and slight mineral deposition around the vents (Figs. 5–9). There is no widespread lack of vegetation. Areas near hot spots have different plants, such as the lichen *Cladonia cristatella* and *Hypericum gentianoides*, which normally thrive in dry, rocky, sun-baked soils (Fig. 9). Unlike Centralia, where the fire has caused a subsidence on Pennsylvania Highway 61, necessitating the closure of the road and a detour, the less-intense fire in Laurel Run has required the owners of the mobile home park to make only minimal repairs to keep their driveway open. The ~9.14-m-long section of the road that is located above the fire has required extra paving but lacks visible cracks.

Geologic Factors Enhancing Fire Control

Although the Laurel Run Mine fire was a challenge due to its size, several factors made its abatement a fairly straightforward engineering problem. The Bottom Ross, Top Red Ash, and Bottom Red Ash coal beds of coal were burning. All three beds cropped out east of town, where a surface-mining operation exposed them. This action provided a barrier to the fire's advance. To the west, the mine pool under the city of Wilkes-Barre provided another barrier. With the exception of the area directly beneath the borough of Laurel Run, where pillar robbing caused subsidence and cracking, the rest of the fire burned under solid rock. There were only a few possible routes for oxygen to reach the fire (Dierks et al., 1971, p. 13).

The Laurel Run Mine fire started in 1915. For forty years, the Red Ash Coal Mine and neighboring concerns used temporary seals to protect active mine workings. The mine operators were able to pass information about the fire on to the government agencies involved in remediation. Planners came into Laurel Run with a much better understanding of what they were up against than those in Centralia.

Nonscientific Factors Aiding Abatement Efforts

In addition to the geological factors that made the Laurel Run Mine fire less difficult to suppress than the Centralia fire, there were a number of nonscientific factors that also provided the town an edge. The Wyoming Valley was unique among areas

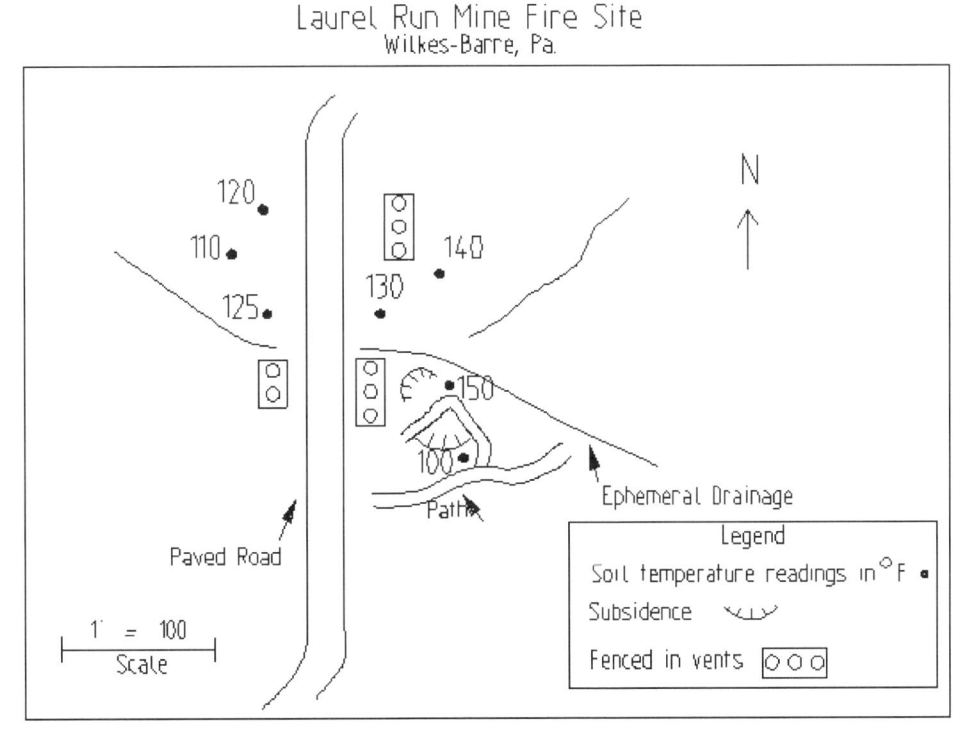

Figure 5. A map of the Laurel Run Mine fire site.

Figure 6. Surface crack caused by the Laurel Run Mine fire. (Field of view is 6 ft [1.83 m].)

in the anthracite coal region of Pennsylvania because of the communities' access to political power (Davies, 1985). In other parts of northeastern Pennsylvania, anthracite mines were initially started by local entrepreneurs. Later, when heavily capitalized railroads and other corporations entered and bought up the coal land and existing mine operations, they pushed these local leaders aside. Political power passed to an absentee ownership class. Residents felt powerless and alienated from power and had little faith in either economic or political elites.

In the Wyoming Valley, on the other hand, the original local mine owners successfully merged their companies with the new railroads and corporate interests. Members of the local elite often sat on the board of directors of these concerns and gave the community a perceived voice in what was going on. People in the Wyoming Valley felt less alienated and were more willing to approach and work with government authorities (Davies, 1985, passim).

Political culture also played a role. Most of the original settlers in the Wyoming Valley came from Connecticut. They brought with them a way of life that emphasized political involvement and encouraged people to hold office and participate in public meetings. This was very different from what prevailed in the rest of Pennsylvania, where a tradition of minimalist local government, with few local officials or opportunities for input had been created. When the mine fire at Laurel Run became uncontrolled, the borough's inhabitants were able to organize, come together, and work with their elected officials. In other parts of the coal region, people did not have the experience to do this as effectively (Aurand, 1998, p. 10–18).

Finally, the residents of Laurel Run benefited from having Daniel J. Flood as their congressman. Flood, a prominent member of the House Appropriations Committee, was able to make sure his constituents received the help they needed. Between June and August 1967, for example, Flood personally escorted Dr. Walter J. Hibbard Jr., director of the U.S. Bureau of Mines, J. Cordell Moore, Assistant Secretary of the Interior, Joe W. Fleming, Federal Co-Chair of the Appalachian Regional Commission, and Stewart L. ("Mo") Udall, Secretary of the Interior, to Laurel Run. Each pronounced the mine fire to be a major disaster and promised to do their best to help. The presence of Interstate 81 and a major railroad near the mine fire was an additional reason for the government to be very concerned. In fact, the mine fire and relocation project at Laurel Run was so well-funded it came in 7.7 million dollars under budget (Randolph, 2002).

Summary and Discussion of the Laurel Run Mine Fire

The mine fire at Laurel Run destroyed a large part of the town's physical structure. Because the fire was located directly under the borough's streets, it received abundant oxygen through the fractured rock strata. People were displaced, and the community radically changed. What was not destroyed was the people's faith in society's abilities to handle such disasters. When the fire became uncontrolled, the community of Laurel Run came together around their leaders, received the government aid they needed, and created a plan to control the fire and prevent its spread to other settled areas. In this regard, they were successful. While the fire is still burning, it is in an area where no one's life or property is threatened.

CONTRASTS BETWEEN THE CENTRALIA AND LAUREL RUN MINE FIRES

The local geologic structure at Laurel Run and Centralia is significantly different. In Centralia, the fire began on the nose of the Locust Mountain anticline, which allowed it to spread in four directions. The Laurel Run fire began on a broad shelf on the regional syncline and lacked the folds that would have directed the fire in multiple directions. The local mines at Laurel Run were separated from each other by barriers, which served to slow the movement of the fire, allowing the U.S. Bureau of Mines to reinforce the barriers. The surface manifestation of each fire is significantly different. In Laurel Run, the absence of clinker and altered vegetation indicate less heat at the surface. At Centralia,

Figure 7. A vent caused by the Laurel Run Mine fire. (Field of view is 7 ft [2.13 m].)

clinker can be found forming at the surface, as can extensive areas of subsidence and no vegetation. Three coal beds are burning in Laurel Run—the Ross, Top Red Ash, and Bottom Red Ash—while at Centralia, only the Buck Mountain coal bed is burning.

The communities of Centralia and Laurel Run had very different levels of experience in dealing with the government and access to political power. As a result, Laurel Run was able to get the help they needed, while Centralia was not. The effect of this experience left the people of Centralia feeling powerless and convinced that neither the government nor society could fight coal fires. In Laurel Run, the experience led to the opposite conclusion because fire-control efforts were mostly successful in that they moved people out of harm's way and controlled the fire so residents could move back.

CONCLUSIONS

What the authors discovered by comparing the Centralia and Laurel Run mine fires was that all mine fires are not alike. Different geologic and social factors can lead to different outcomes in fire abatement and community impact. Attempts by sociologists to use the Centralia experience as a model for how communities respond to mine fires, and, in fact, all manmade natural disasters (Kroll-Smith and Couch, 1990), are premature. Further, because of the variety of factors that can influence efforts to control mine fires, the best way to understand abatement efforts is in an interdisciplinary fashion, rather than focusing on the purely geologic and engineering aspects of fire control, or the sociological and historical issues that limit the ability of a community to deal with crises.

Figure 8. Manmade vents along the road to the Laurel Run Estates Mobile Home Park. (Field of view is 30 ft [9.14 m].)

Figure 9. The vegetation above the Laurel Run Mine fire. Notice the small area of discolored vegetation in the middle of the photograph. (Field of view is 72 ft [21.95 m].)

ACKNOWLEDGMENTS

We thank Glenn Stracher for his assistance with our field research and manuscript, Shelton Alexander for providing us with the Shallenberger study, and Stephen Couch for providing us with a copy of his book and the GAI Consultants reports. We also thank Joseph T. Nolter Jr. for the many hours he spent creating our maps. Steve Jones provided us with information about the rate of movement of fires. Amy Deuink, Marianne Sciler, and Rosanne Chesakis, librarians at Pennsylvania State University–Schuylkill, assisted us with the acquisition of information for our study. Finally, we would like to thank Nan Lindsley-Griffin, Steve Renner, and David Philbin for reviewing our manuscript.

REFERENCES CITED

Arndt, H.H., 1971, Geologic Map of the Ashland Quadrangle, Columbia and Schuylkill Counties, Pennsylvania: U.S. Geological Survey Map GQ-918, scale 1:24:000.

Ashmead, D.C., 1922, Red Ash Company's mine fire, thought to be slushed out, blazes up, threatening nearby properties: Coal Age, 11 May 1922, p. 769–770.

Aurand, H., Jr., 1998, Regional Origin and Political Culture on the Upper Susquehanna Frontier, 1750–1800 [Unpublished dissertation]: Minneapolis, University of Minnesota, 286 p.

Bannerjee, S.C., 1985, Spontaneous Combustion of Coal and Mine Fires: Rotterdam, Holland, A.A. Balkema, 168 p.

Chaiken, R.F., Brennan, R.J., Heisey, B.S., Kim, A.G., Malenka, W.T., and Schimmel, J.T., 1980, Problems in the Control of Anthracite Mine Fires: A Case Study of the Centralia Mine Fire (August 1980): U.S. Bureau of Mines Report of Investigations 8799, 93 p.

Coates, D., 2003, Geologic history of coal-bed fires, Powder River Basin, U.S.A.: American Association for the Advancement of Science Annual Meeting Abstracts with Programs, CD-ROM (13–18 February 2003), p. A-29.

Davies, E.J., 1985, The Anthracite Aristocracy: Leadership and Social Change in the Hard Coal Region of Northeast Pennsylvania, 1800–1930: DeKalb, Northern Illinois University Press, 277 p.

DeKok, D., 2000, Unseen Danger: A Tragedy of People, Government, and the Centralia Mine Fire: Philadelphia, University of Pennsylvania Press, 299 p.

Dierks, H.A., Whaite, R.H., and Harvey, A.H., 1971, Three mine fire control projects in Northeastern Pennsylvania: U.S. Department of the Interior, Bureau of Mines, Information Circular 8524, 58 p.

Eggleston, J.R., Kehn, T.M., and Wood, G.H., Jr., 1999, Anthracite, in Shultz, C.H., ed., The Geology of Pennsylvania: Pennsylvania Geological Survey Special Publication 1, p. 458–469.

Faill, R.T., and Nickelsen, R.P., 1999, Appalachian Mountain section of the Valley and Ridge Province, in Shultz, C.H., ed., The Geology of Pennsylvania: Pennsylvania Geological Survey Special Publication 1, p. 269–285.

GAI Consultants, Inc., July 1983, Engineering Analysis and Evaluation of the Centralia Mine Fire: U.S. Department of the Interior, Office of Surface Mining, p. 17–40.

Kroll-Smith, J.S., and Couch, S.R., 1990, The Real Disaster Is Above Ground: A Mine Fire and Social Conflict: Lexington, University Press of Kentucky, 200 p.

Randolph, A., 2002, Overview of the Laurel Run Mine Fire, Borough of Laurel Run, Luzerne County, Pennsylvania: Wilkes-Barre, Pennsylvania, Luzerne County Historical Society, 7 p.

Raymond, R., 1986, Out of the Fiery Furnace: State College, Pennsylvania State University Press, 274 p.

Sevon, W.D., 1976, Wilkes-Barre East Quadrangle: Atlas of Preliminary Geologic Quadrangle Maps of Pennsylvania: Pennsylvania Geological Survey Map 61 (published 1981), scale 1:62,500.

Shallenberger, P.J., 1993, The Use of Remote Sensing to Track the Progression of the Underground Coal Mine Fire at Centralia, Pennsylvania [Unpublished senior thesis]: State College, The Pennsylvania State University, 77 p.

Stracher, G.B., Nolter, M.A., Schroeder, P., McCormack, J., Blake, D.R., and Vice, D.H., 2006, The great Centralia mine fire: A natural laboratory for the study of coal fires, in Pazzaglia, F.J., ed., Excursions in Geology and History: Field Trips in the Middle Atlantic States: Geological Society of America Field Guide 8, doi: 10.1130/2006.fld008(03).

MANUSCRIPT ACCEPTED BY THE SOCIETY 7 MARCH 2007

The Geological Society of America
Reviews in Engineering Geology, Volume XVIII
2007

Congressional response to coal fires: Illustrating transitions in the policy process

Karen M. McCurdy*

Department of Political Science, Georgia Southern University, Statesboro, Georgia 30460-8101, USA

ABSTRACT

The conduct of the elected representatives of the Eleventh Congressional District of Pennsylvania in dealing with two fires at two separate times permits greater understanding of the transition zone between inaction and action in congressional policy making. In 1984, Congress passed a bill that provided funds to relocate the citizens of Centralia, Pennsylvania, while ignoring the citizens in Laurel Run, Pennsylvania. Both communities were plagued by structural damage from the coal fires burning under their towns. In contrast, citizens from both communities had received government assistance in 1966 in the form of Department of Housing and Urban Development (HUD) grants during earlier flare-ups of both fires. Both Pennsylvania senators and the Eleventh Congressional District representative all had less seniority in the 1980s than had been the case for the individuals holding those seats in the 1960s. Divided government made policy influence less certain and consensus more difficult to achieve in the 1980s, and the range of policy templates available for adaptation was more restricted in the 1980s. In 1966, a senior representative operating within a solid majority government could convince officials at HUD to adapt urban renewal legislation to assist constituents whose houses were structurally unsound because of the fires. In 1984, the policy templates had changed to toxic waste and environmental protection, affording inexperienced representatives working in a divided government no suitable argument to justify a bureaucratic solution to the coal fires. The resulting legislative-policy solution was limited to the constituents in acute distress in Centralia.

Keywords: coal fires, public policy, environment, congressional decision making.

INTRODUCTION

Public policy issues are perceived differently by different audiences. Coal fires, for example, interest geologists, chemists, engineers, mine operators and owners, government employees, elected officials, and private landowners for different reasons, i.e., various aspects of the physical phenomenon gain the attention of the various actors in the political process differentially, much like the blind man "seeing" the elephant. Scientists, engineers, and mine owners often view coal fires as naturally occurring phenomena; they will collect information to understand the origin of the fire, to verify its current extent and likely expansion, and to analyze these data to select the best means either to extinguish or to manage the fire in the long run. Private landowners respond to the impact a coal fire may have on the prior and future use of their land. A mine owner whose reserves are endangered, a farmer

*kmccurdy@georgiasouthern.edu

McCurdy, K.M., 2007, Congressional response to coal fires: Illustrating transitions in the policy process, *in* Stracher, G.B., ed., Geology of Coal Fires: Case Studies from Around the World: Geological Society of America Reviews in Engineering Geology, v. XVIII, p. 271–278, doi: 10.1130/2007.4118(19). For permission to copy, contact editing@geosociety.org. ©2007 The Geological Society of America. All rights reserved.

whose crop is burned, and a retired person faced with home damage as a result of subsidence caused by an underground coal fire are likely to have very different views of the problem. Likewise, their requests for government assistance will be quite different. While the farmer may file a claim for government-subsidized crop insurance, and the mine operator may claim a business loss from the fire on that year's corporate tax return, private landowners are more likely to request government assistance of an unspecified nature. They may have a problem for which there is no clearly responsible party. Elected representatives are frequently contacted for assistance to find public solutions to these sorts of problems. Depending on a variety of factors, government officials may respond to the requests for help broadly, narrowly, or not at all.

Federal responses to coal fires provide case studies of a frequently described public policy problem: why, with a given set of circumstances, does Congress act comprehensively at some times and in some places, incrementally in others, or not at all. Two coexisting coal fires in the same geologic setting in eastern Pennsylvania illustrate this seeming paradox in public policy. In 1984, the United States Congress acted to fund relocation for citizens in Centralia, Pennsylvania, while ignoring the nearby Laurel Run fire (Nolter et al., this volume) and most other fires burning in the nation at the time. The next sections of the paper will examine the political conditions in the country and in Pennsylvania in which the congressional response to these coal fires emerged, review agenda-setting models, and examine the policy process in action.

The Eleventh Congressional District of Pennsylvania

The Laurel Run coal fire is located near Wilkes-Barre, Luzerne County, Pennsylvania. The Centralia coal fire is 55 mi (88.5 km) southwest in Columbia County. Following reapportionment from the 1980 census, when Pennsylvania lost two House of Representatives seats, both counties remained in Pennsylvania's Eleventh Congressional District (Fig. 1). While the physical setting of both fires is similar, the political and social structure is different (Nolter et al., this volume). Table 1 summarizes the election margins and terms of office for the national and state political actors involved in the coal-fire policy making from 1976 to 1984 (Barone et al., 1979; Barone and Ujifusa, 1981, 1983, 1985).

In the early to mid-twentieth century, Pennsylvania voters traditionally chose Republican candidates in national and statewide offices, and the state legislative majority was generally held by the Republican Party, although this outcome was in doubt for national elections and statewide office in the 1980s. Democratic Party candidates were gaining ground in Pennsylvania because of the economic disruptions and hardship associated with the decline of the steel and coal-mining industries and the general economic recession of the early 1980s. Ronald Reagan (R, 1981–1989) carried the state in the 1980 presidential election only by 50% and anticipated an equally tight race for reelection in November 1984.

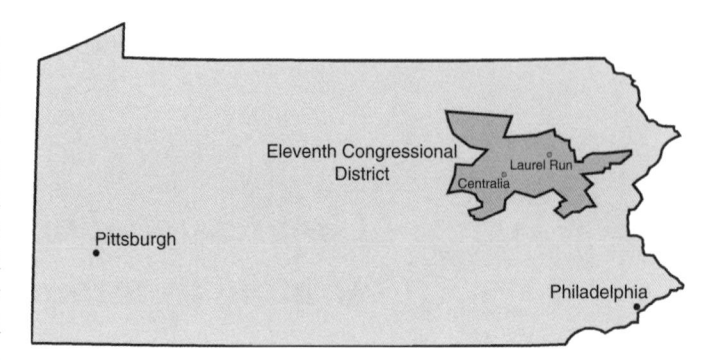

Figure 1. Pennsylvania Eleventh Congressional District. (Source: U.S. Bureau of the Census, 3 March 1982 boundaries.)

The Pennsylvania governor, Richard Thornburgh (R, 1978–1986), was elected to an unprecedented second term in 1982 with 51% of the votes. The senior senator, H. John Heinz III (R, 1976–1991), elected to his first term with an electoral margin of 52%, faced re-election in 1982. The junior senator, Arlen Specter (R, 1980–present), won his first election with 50% of the vote. The statewide politicians had little seniority and thin electoral margins from which to govern. Locally, the situation was worse.

The Eleventh Congressional District was in turmoil in the early 1980s, electing four freshman representatives in as many elections after having been represented for thirteen consecutive terms and sixteen terms overall from 1945 to 1980 by Democrat Daniel J. Flood. Flood was re-elected in 1978 with 58% of the vote after being indicted for bribery. After the election, the Democratic Caucus in the House of Representatives stripped him of his powerful and influential position as chairman of the Appropriations Subcommittee for Labor, Health, Education, and Welfare. He resigned his seat in January 1980 after being convicted of the bribery charge. Democrat Ray Musto won the special election to fill the vacancy in April 1980. He lost the November 1980 general election to Republican James Nelligan, who in turn lost the 1982 general election to Democrat Frank Harrison, who then lost the 1984 Democratic Party primary to Paul Kanjorski (who continues to represent the district following the 2006 congressional elections). These four men had all contested the special election in 1980, finishing in what would become their order of service in the House: Musto (27%), Nelligan (23%), Harrison (17%), and Kanjorski (16%). The turnstile election record in the Eleventh Congressional District over four years reflects divisions in the district brought to the surface with the unexpectedly early departure of Representative Flood from Congress. The persistence these four men showed in contesting the seat over the four elections also indicates differences of opinion about the future of the district held strongly enough by financial backers, party regulars, and campaign workers that significant amounts of time, effort, and funds were spent over the span of four elections to seek an equilibrium.

TABLE 1. UNITED STATES AND PENNSYLVANIA POLITICAL CONDITIONS, 1974–1984

Political office	Incumbent	Political party	Years of service*	Electoral margin	Next election
United States president	Ronald Reagan	Republican	1981–1989	50% of Pennsylvania in 1980	1984
Pennsylvania governor	Richard Thornburgh	Republican	1978–1986	51% in 1982	Limited to two terms
Senior United States senator from Pennsylvania	John Heinz III	Republican	1976–1991	52% in 1976	1982
Junior United States senator from Pennsylvania	Arlen Specter	Republican	1980–present	50% in 1980	1986
House of Representatives	Daniel J. Flood	Democrat	1945–1947 1951–1953 1955–31 January 1980	71% in 1976 58% in 1978	
	Raphael J. Musto	Democrat	9 April 1980–1981	27% in special election	
	James L. Nelligan	Republican	1981–1983	52% in 1980	
	Frank G. Harrison	Democrat	1983–1985	54% in 1982	
	Paul E. Kanjorski	Democrat	1985–present	59% in 1984	

Sources: United States Congress Joint Committee on Printing (2006); Barone et al. (1979); Barone and Ujifusa (1981, 1983, 1985).
*Terms of service begin 20 January for the President and 3 January for members of Congress unless noted otherwise.

Members of Congress become influential policy makers by accruing seniority in their committee assignments and developing a reputation as hardworking effective legislators among others engaged in the policy process (Herrnson, 2003; Fenno, 1973). Any district's constituents are served better in Congress by a member becoming influential in the policy process than by a series of first-term members succeeding each other.

After a quarter century of Democratic Party majorities in both chambers of Congress, elections in the early 1980s brought divided government to the legislature. The 1978 midterm elections continued Democratic control of government, with majorities in both the House of Representatives and the Senate for the 96th Congress during the final two years of Democratic President Jimmy Carter's term. The 1980 elections produced a new type of divided government, i.e., a Republican president and a Republican majority in the Senate, while the House remained in Democratic Party control for six years (1981–1987), through the end of the 99th Congress (Table 2). Within the chambers, the size of both majorities was slim in the divided government years. The political division shifted in the Senate from 58% Democratic control in 1979 to 53%, 54%, and 53% Republican control in 1981, 1983, and 1985, respectively. Democratic Party control of the Senate resumed once again in 1987, with a 55% majority. Democratic majorities in the House of Representatives ranged from 56% to 64% for the five corresponding Congresses. Divided government between the executive and legislative branches was partisan during the Nixon and Ford administrations (1969–1977), but divided government within Congress produced unprecedented levels of partisan rancor during the Reagan administration, which added new partisan complexities to policy-making efforts.

The political environment in the 1980s was an unusual one, where political actors were searching for strategies that would further their policy goals. A mundane coal fire occurred in extraordinary political conditions, resulting in political behavior that

TABLE 2. PARTY DIVISIONS AND ELECTORAL MARGINS, UNITED STATES CONGRESS AND PRESIDENTS, 1979–1989

Congress	Years	Portion House Democratic (%)	Portion Senate Democratic (%)	Presidential party (popular vote)
95th	1977–1979	67	61	Democrat (50.1%)
96th	1979–1981	64	58	
97th*	1981–1983	56	47 (GOP 53)	Republican (50.7%)
98th*	1983–1985	62	46 (GOP 54)	
99th*	1985–1987	58	47 (GOP 53)	Republican (58.8%)
100th*	1987–1989	59	55	

Sources: United States Congress Joint Committee on Printing (2006); Barone et al. (1979); Barone and Ujifusa (1981, 1983, 1985).
Key: GOP—Republican.
*Divided government.

appears odd but is understandable as a function of little seniority, divided government, and a changed perception by the public of the risk involved in living above coal fires.

Two Coal Fires, Two Congressional Responses

The origins of the two coal fires in eastern Pennsylvania were quite different, which also had political ramifications in the 1980s. The Laurel Run fire was started by a miner in 1915 at a privately owned underground mine. The fire was contained despite occasional flare-ups until the mine ceased operations in 1957. The fact that the fire was burning again became obvious at the surface in September 1962, when the fire quickly began encroaching on populated areas. Representative Flood, then a three-term incumbent with experience in the House covering 17 yr, was responsible for assisting the Luzerne County Redevelopment Authority in gaining a U.S. Department of Housing and Urban Development (HUD) grant for urban renewal in 1966. The grant provided funds to relocate 850 residents of Laurel Run impacted by the fire damage (DeKok, 2000; Nolter et al., 2007 [this volume]). Flood used his seniority in Congress to insist that federal bureaucrats visit the location and find relief for his constituents in a timely manner for a fire a few miles from his home. His constituents would not have hesitated to contact a neighbor about a commonly experienced problem. In contrast, the Centralia constituents were more reticent to seek assistance, possibly because of their responsibility in starting the fire, but also because of the political culture and history of their community.

The Centralia fire ignited in a more public manner, at an abandoned surface mine that had been used as a refuse pile for residents. The town council approved an intentional burn in 1962 to reduce the volume of trash and minimize nuisance, health, and safety risks from the insects and animals frequenting the town dump. Later, council members distanced themselves from responsibility for the decision. Although burning was a common practice, it had already been outlawed by state statute (DeKok, 2000). The unintended consequence of the action by the town officials was ignition of a coal seam. The fire spread underground on four fronts, quickly moving under populated areas of the town. Efforts to extinguish or control the fire were not successful, and the fire became a chronic condition for Centralia's residents.

The Centralia fire burned underneath the border of two local governments. Although the garbage dump was maintained by the town of Centralia, the land was located across the road in Conyingham township and was technically a responsibility of that board of supervisors. Location at the border of two local governmental jurisdictions produced a political orphan. Both the Centralia town council and the Conyingham township board of supervisors believed that the other governmental body had jurisdiction, and therefore the responsibility, for managing the fire. This political geography contributed to the lack of early government response when the problem was small. In the early 1960s, a small number of Centralia residents were relocated with funds from the state and federal government. The relocation occurred at the same time as the HUD grant at Laurel Run. Temporarily resolved, both fires burned in a chronic, controlled, and stable state until February 1981, when the two fires diverged in the political system.

On Valentine's Day, twelve-year-old Todd Dombroski was walking past his grandmother's house when he stopped to watch the smoke rising from a nearby yard. The land he was standing on suddenly gave way, and he found himself very near the fire. His sixteen-year-old cousin heard his yells for help and quickly pulled him out of harm's way (DeKok, 2000; New York Times, 1981a). Newspaper sources capitalized on the sensational nature of the story, especially since the boy was unhurt and quickly pulled out of the hole that subsided due to the fire. He was said to have been swallowed by the fire in some accounts. This incident frightened the residents of Centralia, who by that time were living with state-issued carbon monoxide monitors in their homes. This incident occurred six weeks before the second anniversary of the Three Mile Island nuclear accident near Harrisburg, Pennsylvania, which is 60 mi (96.54 km) southwest of Centralia. Everyone involved in Centralia had learned lessons from Three Mile Island. Reporters, politicians, and first responders were not going to be embarrassed by underestimating the urgency of a warning like the subsidence and smoke. Likewise, citizens were not going to be lulled into a false sense of security by statements from public officials, regardless of their basis in fact. Public opinion developed a new concern about the safety of living above a coal fire, and citizens began agitating for government to fix the problem (New York Times, 1981b).

To summarize, the physical setting of the two fires is similar. The fires occurred in a single congressional district, and the extent and duration of the fires are essentially the same. The political response to the fires was similar in 1966, but they were treated in distinctly different ways in 1984. Why?

CONGRESSIONAL AGENDA SETTING

Understanding why Congress takes up any public policy issue at a particular time is a classic problem. The initial equilibrium model was developed by Kingdon (1984, 1995). Baumgartner and Jones (1993, 2002) recently extended this model for a dynamic system in order to distinguish between routine and innovative policy making. Zahariadis (1999) reviewed the various critiques of the model, while Stone (2002) critiqued the rationality assumptions in agenda setting and offered an alternative explanation of how the paradoxical contest of conflicting goals is determined in the policy process. The initial model and the critiques of agenda setting both take a static view of the policy process. The dynamic model is more useful in understanding the congressional responses to the Eleventh Congressional District coal fires. Just as rain does not always lead to flooding, neither elections nor fires always lead to policy making.

Problems, Politics, and Policies

Kingdon (1984) described three streams flowing independently of one another, a problems stream, a political stream, and a policies stream. Policy entrepreneurs take advantage of a window of opportunity to couple the streams, which allows them to initiate congressional action to make public policy (Fig. 2). The problem stream is most obvious to all observers, both casual and professional. Policy problems abound in communities without ready access to policy makers. In this example, coal fires are well known to the science and engineering communities that study them, and to the constituents who live near them, but the fires are relatively unknown to individual members of Congress and are virtually ignored by the institution as a whole. Problems require a delivery mechanism for information to reach legislators in a manner that leads to policy making. Many problems arrive through the legislative oversight process, either by the standard route or the crisis route (McCubbins and Schwartz, 1984). Either oversight mode can initiate what Kingdon called a "focusing event," which would open a window of opportunity for a policy response to be fashioned.

The political stream operates in annual, biennial, and quadrennial cycles connected with elections. Fund-raising events, public interest campaigns by special interest groups, and the search for a public policy issue around which a candidate can produce a winning campaign strategy wax and wane with the election cycle. Politicians may use a policy problem for campaign purposes, but there is no coordination between politics and policy making. In the political stream, politicians are disconnected from policy making as they focus on electoral concerns.

The traditional separation between elections and governing in the mid-twentieth century became blurred at the end of the century. Harsh electoral rhetoric designed for the short time frame relevant in elections was transported to the governing side of election day, replacing the more typical measured and diplomatic language that had been associated with governance during the middle decades of the twentieth century. Before this change, legislators could be scrappy street fighters during elections, but when the session began, they were expected to become statesmen, which for the most part they did. During elections, policy problems are used for their ability to mobilize the base, to elicit campaign contributions, or to distinguish one party's candidate from another. The political stream has an election focus which is tangential, or perhaps merely preliminary, to policy making. The political stream, however, is critical to the policy process because it sets the boundary conditions in which policy making must occur. The election determines the majority party, the size of those majorities, and whether divided government will exist between the executive and the legislative branches, or within the bicameral legislature. Further, the political stream brings to the legislature the small number of policy issues that will affect all other policy discussions for that session of Congress.

The policy stream is the most hectic, the most studied, and possibly the least understood of the streams in Kingdon's model. Scientists, whether physical, social, or behavioral, are engaged in research to gain a better understanding of the physical and social phenomena that interest them, and for which funding can be secured. The public policy status quo influences the direction

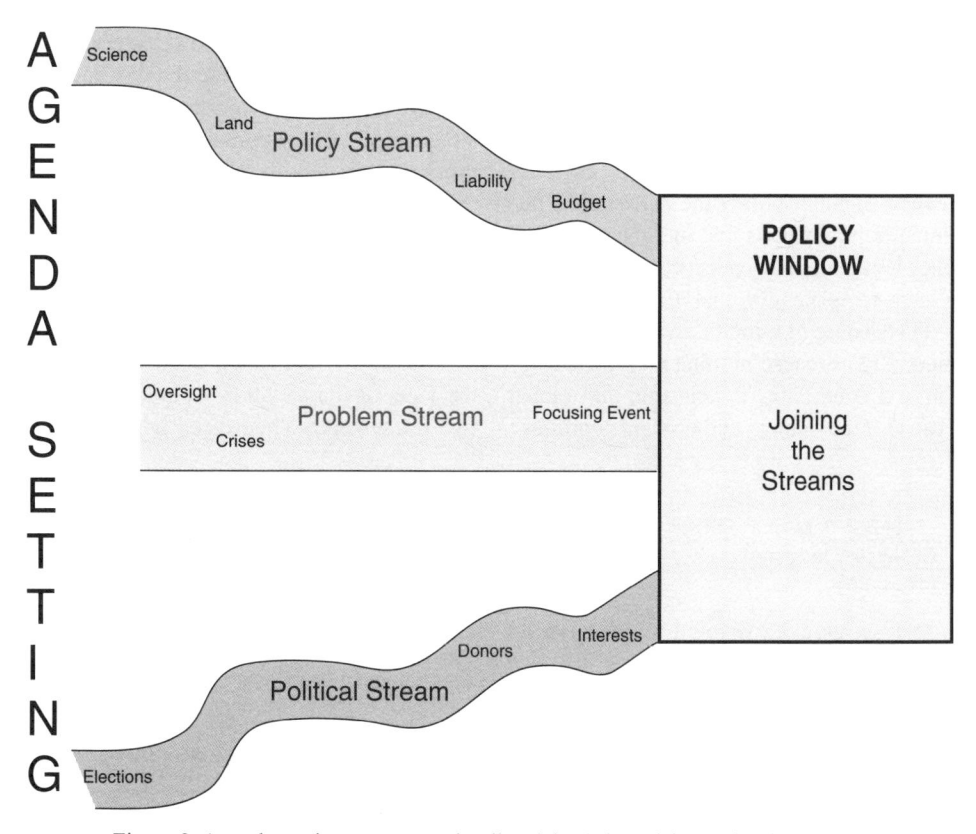

Figure 2. Agenda-setting streams and policy delta (adapted from Kingdon, 1984).

of future scientific research in part by providing funds for grants. The policy stream is frenetic with activity, without regard for coordination. Public policies are being authorized, budgeted, and implemented in an attempt to mitigate the effects of past policy problems, while at the same time, they are contributing to present and future public policy problems. Some previous policy solutions render government blind to emerging problems that do not behave like any previously known problems, resulting in the paradox of Congress taking opposing actions at the same time (Stone, 2002).

Policy Templates

The institutional context in which decisions are made is a critical component in shaping public policy. Members of Congress have institutional habits, i.e., policy frameworks or templates in which a given set of constituents, bureaucrats, and lobbyists can be mobilized quickly and efficiently to make noncontroversial adjustments to policies. A common vocabulary to understand and discuss a policy is necessary, as well as a regular place for the decision to be made (Baumgartner and Jones, 1993, p. 31). Controversial policy issues bring conflict that can break the human connections in policy frameworks but that eventually produce the means for building a new framework. Members of Congress rarely invent new solutions to policy problems until there is no reasonable alternative. Instead, they adapt old solutions to new problems. If the new policy problem reminds leaders of a recently solved problem, this solution is likely to be adapted to the novel circumstances. When the chronic Centralia coal fire flared in 1981 and presented an acute problem in the policy stream, members of Congress had four policy templates to help craft a quick solution.

Four Coal Templates

The established course of public policy, the status quo, influences the ways in which policy experts and members of Congress consider new policy issues. In the twentieth century, Congress developed four prominent templates for coal. Coal comes as a problem to Congress: (1) because of mine safety, (2) because of the aftereffects of mining, (3) because burning coal for energy produces air pollution and contributes to acid rain and global warming, and (4) because mine tailings and chat piles produce toxic chemicals that filter into the soil and groundwater through interaction with rainwater. Table 3 summarizes the key pieces of federal environmental protection legislation related to coal that were passed by both chambers of the Congress between 1969 and 1990 and became public law. Coal means a variety of things to members of Congress, depending on which policy template is used to bring a problem to their attention.

Mine safety is the oldest policy template developed for coal. Congress addressed coal mining as a health and safety issue beginning in 1952, following the Frankfurt, Illinois, cave-in, which killed 119 miners. Stronger legislation covering all underground mining, including enforcement mechanisms for fines and provisions for mitigating black lung disease, was included in the Federal Coal Mine Health and Safety Act passed in 1969 and amended in 1977.

The second template for coal is regulation of sulfur emissions in the Clean Air Act of 1970. Eastern coal producers were able to retain their markets for bituminous coal with higher SO_2 content on a par with lower-sulfur-content western-source coal by stipulating that scrubbers be used at the combustion point to reduce sulfur emissions (Ackerman and Hassler, 1981).

The third template is mitigating environmental degradation due to surface mining, covered in the Surface Mining Control and Reclamation Act of 1977 (SMCRA). The fourth template is toxic waste abatement covered in the Superfund legislation, the Comprehensive Environmental Response, Compensation, and Liability Act of 1980 (CERCLA), and the Superfund Amendments and Reauthorization Act of 1986 (SARA). Runoff problems from abandoned coal mines in the Appalachians were some of the earliest Superfund sites. While the coal industry was prosperous in Pennsylvania, coal fires were an industry problem, addressed as a regular part of the mining business. Coal fires became a public problem only after the coal industry fell on hard times in the latter twentieth century, and closures produced abandoned mines. Although fires can be chronic and a nuisance, they do not typically present urgent circumstances that compel the attention of national policy makers.

From a congressional standpoint, there is no explicit coal-fires policy. When called upon to respond, Congress reacts to the facts of the situation without a broader theoretical legislative framework. When faced with the facts in the early 1980s

TABLE 3. MAJOR FEDERAL LAWS ABOUT THE ENVIRONMENT AND COAL, 1969–1986

Coal-policy template	Federal legislation
Mine safety	Federal Coal Mine Health and Safety Act 1969, amended 1977 (PL* 91-173; PL 95-164)
Acid rain and global warming	Clean Air Act 1970, amended 1977 (PL 91-604; PL 95-95)
Postmining cleanup	Surface Mining Control and Reclamation Act 1977 (PL 95-87)
Soil and water pollution	Comprehensive Environmental Response, Compensation, and Liability Act (CERCLA or more commonly, Superfund) 1980 (PL 96-510)
	Superfund Amendments and Reauthorization Act (SARA) 1986 (PL 99-499)

Source: Adapted from Appendix 1 *in* Vig and Kraft (2006).
*Public Law.

of citizens in Centralia endangered by an underground fire, the U.S. Congress provided funding through the Office of Surface Mining to relocate the affected citizens (New York Times, 1984). The House of Representatives initially passed legislation appropriating funds for relocation in May 1982, early in the 98th Congress, without hearings. The Senate amendments passed 78–11 on 4 October 1984 as one of the last measures at the end of the session. The final House vote later the same day was overwhelmingly in favor of approval, 252–60. The Centralia relocation funding became public law 98–473 on 12 October 1984. Meanwhile, the longer-burning Laurel Run fire did not receive relocation funding in 1984, although there had been a pattern of requesting assistance (Aurand and Nolter, 2004). While the Laurel Run fire was chronic, it did not have the urgent component in 1984 that might have spurred congressional action. The coal-fire flareup in 1966 was in Representative Flood's hometown, and, more importantly, occurred when he had enough seniority to stretch an existing policy template (urban renewal) to fit his constituents' problems with structural damage to their homes.

POLICY LESSONS FROM THE ELEVENTH DISTRICT COAL FIRES

The solution Congress developed for the Centralia coal fire in 1984 was a special appropriation routed through the Office of Surface Mining to buy property and relocate the town's residents. This policy problem arose at a time in the political stream when the congressional delegation was nearly as weak as possible—a freshman representative from the minority party in the House of Representatives with one month's experience, combined with both senators in their first terms as well as being members of a new majority party that had not ruled in a quarter century. Neither Representative Nelligan, nor Senators Heinz or Specter, was in a position to build a policy consensus to address the problem in Centralia in 1982.

In 1966, no one living in a coal region was very concerned about a coal fire, which was a common occurrence, until the underground fire had encroached on populated areas and compromised the structural integrity of buildings. Events in the next decades increased public worries nationally about exposure to toxic wastes and fumes. The Three Mile Island nuclear accident occurred in March 1979. The Environmental Protection Agency plan to evacuate Love Canal, New York, was released in May 1980. By 1981, the public had been sensitized to the many health dangers associated with long-term exposure to a wide variety of common things, including coal dust, which causes Black Lung disease. By early 1982, the dioxin contamination at Times Beach, Missouri, was in the news, and an Environmental Protection Agency (EPA) buyout plan was announced in February 1983. By 1984, the policy and problem streams were coupling in Congress around the dangers of toxic waste. No evidence existed to quell public fears in Centralia that the coal fire was not producing carcinogens that would sicken and even kill them if they continued living above the fire (New York Times, 1982). Smoke that had been a nuisance in 1966 became cause enough to create a health crisis in 1984.

When the Centralia coal fire entered the problem stream and met the political and policy streams, not much indicated that this was a policy whose time had come. One of the criticisms of Kingdon's (1984) model is that the conditions to open the policy window are not well established. None of the model modifications to date has considered the effect of political energy in the system. How large the majority in Congress is, how much seniority a member has, and the subject matter of their committee specialization all contribute to political energy. The political energy in the coal-fires policy process provides a mechanism to establish boundary conditions for policy action.

None of the existing coal policy templates was used to solve the Centralia constituents' problem in the early 1980s, although not for lack of effort. Superfund could not be stretched to clean up private property that was not generating toxic waste, unlike 1966, when the urban renewal framework could easily be stretched to cover substandard housing in a rural area. Freshman Republican Nelligan could not act as a policy entrepreneur from the minority. Freshman Democrat Harrison could not act as a policy entrepreneur, even from the majority. Neither Pennsylvania senator had sufficient seniority to adapt one of the existing public policy templates for the Centralia problem as had happened in 1966. But 1984 was a presidential election year, and Pennsylvania was an important state in the Electoral College calculations for the Republicans. The one-time appropriation to buy out the residents of Centralia was made without hearings and was one of the final actions of the 98th Congress before the elections. The buyout plan was supported by both the president and Pennsylvania's governor.

The Centralia coal fire was handled in 1966 by referral by Representative Flood to an appropriate bureaucratic agency. An existing policy template was used to help citizens with substandard housing. A policy entrepreneur simply needed to make the connection to eligibility from urban to rural citizens. In 1984, the policy problem went beyond Congress and was seen by the president and the governor as a political and legal liability like Love Canal or Times Beach. The template for hazardous waste was to remove the constituents from the source of contamination and danger. So, two coal fires burning in the same congressional district in 1984 would be treated differently: Congress would appropriate funds to alleviate the public problem where constituents whose health was being endangered by toxic fumes generated by a coal fire needed to be relocated away from the fire; Congress would not take action in the case of private property being undermined by a coal fire (New York Times, 1983).

The 1984 Centralia buyout program solution did not resolve the policy problem. Twenty years later, stories about the fire run in the nation's newspapers, and there are still ten to twelve residents who refuse to leave their houses, even though there is no longer a town surrounding them. At the time, many Centralia residents were critical of the government response, both those who accepted the buyout offer and those who did not. The Centralia buyout resolved a political problem for a president facing re-election, while it left the larger policy problem of coal fires unresolved.

Kingdon's (1984) three streams move issues to the political agenda, much like physical streams move sediment. The changes in energy in a physical stream determine how much sediment is moved and where it is deposited or caught in eddies (Morisawa, 1968). The magnitude of partisan majorities in the legislature, seniority, and committee assignments can be used to provide a first approximation of political energy in Laurel Run and Centralia.

Public policy scholars understand much more about how the system works in "normal" conditions, i.e., when the system is static. Much less is understood about what might happen to the policy process either during a "flood," i.e., when election margins are unusually large resulting in veto-proof majorities, or during a "drought," when elections may produce razor thin margins, which make it difficult to accomplish much of anything legislatively. The legislative response to these two coal fires helps to illustrate the boundary conditions between congressional action and inaction.

REFERENCES CITED

Ackerman, B.A., and Hassler, W.T., 1981, Clean Coal/Dirty Air: Or How the Clean Air Act Became a Multibillion-Dollar-Bailout for High-Sulfur Coal Producers and What Should Be Done about It: New Haven, Yale University Press, 193 p.

Aurand, H., and Nolter, M.A., 2004, Centralia and Laurel Run, comparison of anthracite mine fires: Geological Society of America Abstracts with Programs, v. 36, no. 5, p. 226.

Barone, M., and Ujifusa, G., 1981, The Almanac of American Politics 1982: Washington, D.C., Barone and Company, 1258 p.

Barone, M., and Ujifusa, G., 1983, The Almanac of American Politics 1984: Washington, D.C., National Journal, 1402 p.

Barone, M., and Ujifusa, G., 1985, The Almanac of American Politics 1986: Washington, D.C., National Journal, 1593 p.

Barone, M., Ujifusa, G., and Matthews, D., 1979, The Almanac of American Politics 1980: New York, E.P. Dutton, 1055 p.

Baumgartner, F.R., and Jones, B.D., 1993, Agendas and Instability in American Politics: Chicago, University of Chicago Press, 298 p.

Baumgartner, F.R., and Jones, B.D., eds., 2002, Policy Dynamics: Chicago, University of Chicago Press, 360 p.

DeKok, D., 2000, Unseen Danger: A Tragedy of People, Government, and the Centralia Mine Fire: San Jose, iUniverse.com, Inc., 299 p.

Fenno, R.F., 1973, Congressmen in Committee: Boston, Little Brown and Company, 302 p.

Herrnson, P.S., 2003, Congressional Elections: Campaigning at Home and in Washington (fourth edition): Washington, D.C., CQ Press, 362 p.

Kingdon, J.W., 1984, Agendas, Alternatives and Public Policies: Boston, Little Brown, 240 p.

Kingdon, J.W., 1995, Agendas, Alternatives and Public Policies (second edition): New York, Harper Collins, 254 p.

McCubbins, M.D., and Schwartz, T., 1984, Congressional oversight overlooked: Police patrols versus fire alarms: American Journal of Political Science, v. 28, no. 1, p. 165–179, doi: 10.2307/2110792.

Morisawa, M., 1968, Streams: Their Dynamics and Morphology: New York, McGraw-Hill Book Company, 175 p.

New York Times, 1981a, Boy's mishap renews fears on two-decade mine fire: New York Times, 20 February 1981.

New York Times, 1981b, Fire and fear surface as Centralia mines smolder: New York Times, 10 August 1981.

New York Times, 1982, Health study involving mine fire in Pennsylvania cancelled by U.S.: New York Times, 8 March 1982.

New York Times, 1983, Chance to flee divides town atop a mine fire: New York Times, 4 December 1983.

New York Times, 1984, Town plagued by fire gets federal assistance: New York Times, 28 March 1984.

Nolter, M.A., Vice, D.H., and Aurand, H., Jr., 2007, this volume, Comparison of Pennsylvania anthracite mine fires: Centralia and Laurel Run, *in* Stracher, G.B., ed., Geology of Coal Fires: Case Studies from Around the World: Geological Society of America Reviews in Engineering Geology, v. XVIII, doi: 10.1130/2007.4118(18).

Stone, D., 2002, Policy Paradox: The Art of Political Decision Making: New York, W.W. Norton, 394 p.

United States Congress Joint Committee on Printing, 2006, Biographical Directory of the United States Congress 1774–2005: Washington, D.C., Government Printing Office, 2236 p.

United States Department of Commerce, Bureau of the Census, 1984, Congressional District Databook of the 98th Congress: 1990 Census of Housing and Population.

Vig, N.J., and Kraft, M.E., 2006, Environmental Policy: New Directions for the Twenty-First Century (sixth edition): Washington, D.C., CQ Press, 434 p.

Zahariadis, N., 1999, Ambiguity, time, and multiple streams, *in* Sabatier, P.A., ed., Theories of the Policy Process: Boulder, Colorado, Westview Press, p. 73–93.

MANUSCRIPT ACCEPTED BY THE SOCIETY 7 MARCH 2007

Index

A
Abandoned Mine Land Inventory System (AMLIS), 2
Absaroka Mountains, 170
acceleration, 33
ACIRL. *See* Australian Coal Industry Research Laboratories
activation energies, 53–54, 78
active sites, 56
additives, 51–52
adiabatic oxidation, 33
adiabatic oxidation methods, 61–64
adsorption and desorption of water moisture, 55, 57
Advanced Very High Resolution Radiometer (AVHRR), 242
aeromagnetic anomalies, 182–186
Africa, 245
agenda setting, 274–277
agglomeration, 48
aging, 37–38, 45–46
agriculture, effects on, 32
Aldridge Creek, 41, 102, 105
alunogen, 92, 94
AMLIS. *See* Abandoned Mine Land Inventory System
Anglo coal mines, 246
Angren coal deposit, 111
anorthite, 133
ANSYS software, 214
Archer till, 173
Argentina, 245
aromatic rings, 37
Arrhenius plots, 34
arsenic poisoning, 96
Asia, 110–112
asphaltenes, 98
Australia, 102, 245. *See also* Callide Coalfield; Northern Coalfield; Southland Colliery
Australian Coal Industry Research Laboratories (ACIRL), 62
AVHRR. *See* Advanced Very High Resolution Radiometer
Azerbaijan, 179

B
bacterial activity, 40
Baijigou coal mine, 224
barometric pressure, 40
basalt, 205
basket test, 66–68
Bayinshandan, 24
Beijing Remote Sensing Corporation (BRSC), 25
Benxi Formation, 24
Big Mine Run Road, 264
Biot number, 35, 69
bitumen, 36, 98
blending, stockpiles and, 46
blowing up in finite time, 35
borehole surveys, 206–207, 209
breccias, 138–139, 150. *See also* chimney structures
brown coals, 58–59
BRSC. *See* Beijing Remote Sensing Corporation
buchites, 118. *See also* paralavas
Buck Mountain coal bed, 261, 262, 263, 269
Burning Coal Draw, 163
Burning Mountain, 179, 245
buyouts, 262, 271–278

C
calcite, 146–149
calcium acetate, 52
Callide Coalfield, 63
Canada, 245. *See also* Aldridge Creek
carbonaceous shale, 40
Carbondale mine fire, 4–8
carbon dioxide
    Carbondale mine fire and, 7
    Centralia fire and, 264, 266
    coal-mine fires and, 1, 2
    Large mine fire and, 9
    low-temperature oxidation and, 37
    remote-sensing for estimation of, 240, 246–247
    temperature, oxygen concentration and, 4, 5, 6
carbon monoxide
    Carbondale mine fire and, 7
    Centralia fire and, 262, 266, 274
    as index gas, 41
    low-temperature oxidation and, 37
    subsurface temperature and, 12–13
    temperature, oxygen concentration and, 4, 5, 6
carboxyl groups, 37
carcinogens, 59
Carter, Jimmy, 273
catalysis, 55, 56–57
catenary effect, 56–57
caving characteristics, 40
cellular grout suppressants, 43
Centralia fire
    factors preventing control of, 264–266
    fire fronts of, 263–264
    geologic setting of, 262–263
    GLS processes in, 93
    Laurel Run fire versus, 261, 268–269
    overview of, 246, 261–262, 266
    policy and, 271–278
CERCLA. *See* Comprehensive Environmental Response, Compensation, and Liability Act
Chatkalsky Range, 112
Chelyabinsk Coal Basin, 146
chemical kinetics, 36–37, 63–64, 75
chimney structures, 104, 118, 160, 162
China
    coal fires in, 244–245
    detection of coal fires in, 199–200
    incidence of coal fires in, 33
    remote-sensing data for coal-mining areas in, 219–227
    statistics on coal-mine fires in, 39
    thermal-geological analysis and, 249–258
    See also Liu Huangou Coalfield; Longgu mine; Ruqigou-Gulaben coal basin; Wuda Coalfield
Clean Air Act, 276
climate, spontaneous combustion and, 25
clinker
    Centralia fire and, 264
    combustion metamorphism and, 103–104, 106–108
    gas-altered substrate (GAS) and, 93
    methods for dating, 161–162
    overview of, 118, 158–161
    of Powder River Basin, 156–157, 162–173
    See also paralavas
clinopyroxene, 133, 138, 141–142
coal dust, 43–44
coal factors, risk rating and, 20
coal fires, overview of, 200–202
Coal Mine Health and Safety Act, 276
coal mines
    control and suppression of fires in, 41–43
    hazard assessment for, 39
    origins of fires in, 39
    overview of spontaneous combustion in, 39
    before self-heating in, 40
    self-heating in, 41
coal rank, 54
coal seams, 25
coal templates, 276–277
Coaltemp program, 18
coccinite, 92
coking, 98, 105
colza oil, 48
combustion. *See* spontaneous combustion
combustion line–combustion zone, 26–27
combustion metamorphism
    characteristics of representative complexes of, 99–102
    control of distribution of fossil-fuel fires and, 103–104
    duration and age of fossil-fuel fires and, 105
    in geological history of sedimentary basins, 108–112
    in Hatrurim Basin, 133–135, 148–151
    heat transfer during, 104–105
    natural remanent magnetization and, 177, 178–179
    overview of, 97–98, 112–113
    physical conditions, general features and, 105–107
    process overview, 98
    in Rotowaro Coalfield, 117, 120–121, 130
    structure of complexes of, 104
    See also clinker
combustion plane–combustion system, 27
combustion spot–combustion center, 26, 27
compaction, 45, 48
Comprehensive Environmental Response, Compensation, and Liability Act (CERCLA), 276
condensation, gas-vent minerals and, 92–93
Congress, 262, 268, 271–278
Conrad, Joseph, 49
contact metamorphism, 137

Cook Mountain, 172
copper acetate, 52
coquimbite, 92
cover depth, 40
CPT method. *See* crossing-point temperature method
cristobalite, 117, 126, 127
critical ambient temperature, 70
crossing-point temperature (CPT) method, 64–66. *See also* transient method (TM)
cubical containers, 73
cutting, as ignition source, 39

D
Dead Sea Rift Valley, 105, 110, 133, 137, 151
deep-depth exploration, 206
degassing, self-heating and, 38
degree of oxidation, 85
density, detection and, 205
Department of Minerals and Energy (South Africa), 15
depth, estimation of, 213–217
detection of concealed coal fires, 199–210, 224–226
dewatering, 57
diagenesis, 137
differential scanning calorimetry (DSC), 38, 60
differential thermal analysis (DTA), 38, 60
digging, 42, 48
DIPOLI-3 software, 180
distillation, 98
Dombroski, Todd, 262, 274
double ratioing method, 213
downeyite, 92
drip condensates, 93
drying method, spontaneous combustion and, 57
DSC. *See* differential scanning calorimetry
DTA. *See* differential thermal analysis
dust, coal, 43–44

E
economic implications, overview of, 32–33
EHAC. (Explosion Hazard Advisory Committee). *See* Wits-Ehac Index
electrical resistivity, 205–206, 243
Emery Coalfield, 93
environmental hazards, overview of, 32–33
erosion, clinker and, 156, 158, 173
error estimates, 31
ethene, as index gas, 41
Europe, 245
"exhibiting blow up," 35
Explosion Hazard Advisory Committee (EHAC). *See* Wits-Ehac Index
explosion pipes, 133, 151–152
explosions, 33, 39, 49
extinguishing methods, 199, 202–203

F
faulting, spontaneous combustion and, 40
fayalite, 117, 124, 126–129, 144
FCC index, 64–65
feldspars, 127
Felix coal zone, 161, 163–166
Feng Chakravorty Cochrane index. *See* FCC index

filling, gallery, 42
fine-grained loose coals, 39
fireclays, 124
fire retardants, 42–43
fission-track analysis. *See* zircon fission-track (ZFT) analysis
fissures, 118, 160. *See also* gas-vent minerals
F-K parameter, 35–36
FLIR cameras. *See* forward looking infrared radiometer cameras
Flood, Daniel J., 268, 272, 274, 277
flow condensates, 93
foam/water injection, 42
focusing events, 275
Forestville, Pennsvylvania, 94–95
Fort Union Formation, 155, 156, 161
forward looking infrared radiometer (FLIR) cameras, 211, 213
fractures, 2, 39, 265
Frank-Kamenetskii (F-K) analysis
  coal dust and, 44
  critical value of F-K parameter and, 36
  infinite slab analysis and, 34–35
  overview of, 32, 34–36
  shipping of coal and, 48–49
  spontaneous combustion and, 34
  See also hot-storage test
freezing, gas-vent minerals and, 92
fulgurites, 195
fumaroles, 106
fuzzy logic techniques, 21

G
gallery filling, 42
gamma-ray logging, 206–207, 209
GAS. *See* gas-altered substrate
gas-altered substrate (GAS), 93–94
gases. *See* hydrocarbon gases; index gases; seam gases
gas-liquid–altered substrate (GLAS), 94
gas-liquid–precipitation (GLP), 94–95
gas-liquid–solidification (GLS), 93
gas reaction ± liquid-solidification (GRLS), 95
gas-vent minerals
  combustion metamorphism and, 104, 118
  gas-altered substrate and, 93–94
  gas-liquid–altered substrate and, 94
  gas-liquid–precipitation and, 94–95
  gas-liquid–solidification and, 93
  gas reaction ± liquid-solidification and, 95
  overview of, 91–92, 95
  sublimation and, 92–93
genesis-type model of spontaneous combustion, 23, 25, 29
geodetic surveys, 203–204
geoelectrical investigations, 205–206
geological exploration, 204–205
geologic setting, 25
Geo-Slope Temp/W finite-element modeling software, 214, 215
geothermal gradient, spontaneous combustion and, 40
germanium sulfide, 94–95
Gevanim Formation, 183–186
Ghareb Formation, 135, 137, 150, 178, 179

GIS (geographic information systems), 15, 20–21
GLAS. *See* gas-liquid–altered substrate
Glasser Tests, 15, 19–20
glasses, 127, 133, 146
Glauber's salt, 201
GLP. *See* gas-liquid–precipitation
GLS. *See* gas-liquid–solidification
government. *See* Congress; policy
Graham's ratio, 41
Gravel Terrace, 164, 170
greenhouse gases
  Carbondale mine fire and, 4–6, 7, 8
  gases from underground mine fires and, 2
  Large mine fire and, 6–8, 9–10
  overview of production of, 1, 13, 33
  Percy mine fire and, 8–12
  prevalence of fires and, 1–2
  remote-sensing for estimation of, 239, 246–247
  self-heating experiments and, 2–4
greigite, 179
GRLS. *See* gas reaction ± liquid-solidification
Grootegulux Mine, 20
grossite, 106
groundwater, 32, 39, 158, 173
grouting, 47–48
Gulaben coal basin. *See* Ruqigou-Gulaben coal basin
Gurim anticline, 151
Gurim Dome, 192

H
hanging-wall conditions, 40
Harrison, Frank, 272
Hat Creek fire, 112
Hatrurim Basin
  aeromagnetic anomalies in, 180
  geologic description of, 135–137
  hydrogarnet-zeolite-calcite rocks and hornfelses of, 146–149
  origin of paralavas of, 148–151
  overview of, 133–135, 151–152
  paralava mineral composition of, 142–146
  paralavas of olive unit of, 137–138, 141
  petrography of paralavas, host rocks of, 140, 141–142
  temperature and oxygen fugacity of, 148
Hatrurim Formation. *See* Mottled Zone complex
hatrurite, 106
hazard modeling, 39–40
Hazeva Formation, 135
health. *See* human health
heat, 2
heat capacity, 54
heat of reaction, 54
heat release rate, 34, 68–70
heat-release-rate method (HRRM), 68–70
heat sink mechanism, 55
heat transfer, 55, 104–105, 213
Hebron anomaly, 183–186
Heinz, H. John, III, 272, 277
Helan Mountains, 24
Helan Shan mountain range, 220
hematite, 122–126, 205
hornfelses, 135, 138–141, 146–148
hot-gas jets, 106

hot spots, 199, 202, 203–205
hot-storage test, 66–68
Housing and Urban Development (HUD), 271, 274
HRRM. *See* heat-release-rate method
Huangbaici mine, 24
HUD. *See* Housing and Urban Development
human health, 32–33, 91, 96
humidity, 55–58
Hunter Valley, 245
hydraulic sand stowing, 42
hydrocarbon gases, 133, 151, 178–179, 192, 195
hydrocarbons, 59, 201
hydrogarnet-zeolite-calcite, 146–149
hydroperoxides, 36–37
hydrothermal alteration, 137, 150
hydrous sulfates, 92
hydroxyl groups, 37
hyperkeratosis, 96

I
ignition
    causes of, 39, 76, 103, 240
    moisture and, 55
    in Powder River Basin, 158
    See also time to ignition
IMCO test, 49–50
index gases, 41, 76
India, 244. *See also* Jharia Coalfield; Raniganj coal belt
Indonesia, 245
induction logging, 207–210
inert gas injection, 42
inflexion-point temperature (IPT), 65
infrared exploration, 203–204. *See also* forward looking infrared radiometer cameras; thermal anomalies
inhibition, 38, 51–52
initial rate of heating (IRH), 40, 61–62
inorganic additives, 51–52
Institute for Cokemaking and Fuel Technology, 60
internal surface area, 56
International Organization for Standardization (ISO), 49–50
international standards, 48–49
IPT. *See* inflexion-point temperature
Iran, 99
IRH. *See* initial rate of heating
iron, 48, 118–126, 130
ISO. *See* International Organization for Standardization
isochemical processes, 91, 92–93
Israel, 99, 105
Issyk-Kul region, 245

J
Jharia Coalfield, 32–33, 41, 212–213, 244
Jhingurdah seam, 51
Jordan, 99
Junggar Basin, 110

K
Kalimantan, 245
Kamenetskii. *See* Frank-Kamenetskii (F-K) analysis

Kanjorski, Paul, 272
Kazakhstan, 100
Kenderlyk Depression, 105
Kidod Formation, 151
kinetic constants, 53–54
kinetics, 36–37, 63–64, 75
Knobloch coal zone, 155, 161, 170–172
Kunlin Mountains, 24
Kupakupa coal seams, 119
Kuraminsky Range, 112
Kuznetsk Coal Basin, 103, 108, 110

L
Lake De Smet coal zone, 161, 163, 165, 166
land-cover analysis, 219–227
Landsat-5, 43, 219–227, 229
Landsat-7, 219–227, 229, 237, 242
Large mine fire, 6–8, 9–10
LAST function. *See* linear anomaly surface transect function
Laurel Run fire
    Centralia fire versus, 261, 268–269
    factors aiding control of, 267–268
    fire front of, 267
    geologic setting of, 266–267
    overview of, 261–262, 266
    policy and, 271–278
leveling, 48
liability indexes, 40, 65
lightning, 38–39, 76, 179, 195, 240
lignite, 58–59, 98
linear anomaly surface transect (LAST) function, 211, 214–217
Little Thunder Creek, 163
Little Wolf Mountains, 164, 168–170
Liu Huangou Coalfield, 41
Llewellyn Formation, 93, 263, 264
Locust Mountain anticline, 263–264, 268
Longgu mine, 42
loose coals, 39
Love Canal, 277
low-rank coals, 58–59
low-spatial-resolution satellites (LSRS), 230
low-temperature oxidation. *See* oxidation
LSRS. *See* low-spatial-resolution satellites
lung cancer, 96
Luzerne County Redevelopment Authority, 266, 274

M
machinery, 39
maghemite, 179
magmatism, 195
magnetic anomalies
    detection of concealed coal fires and, 199, 205
    interpretation of data for Mottled Zone and, 180–182
    methodology for studying, 179–180
    processes causing, 180–182
    See also natural remanent magnetization (NRM)
magnetic logging, 210
magnetite, 117, 127–129, 205
magnetite-hematite-spinel assemblages, 122–126
Mahadevan and Ramlu index. *See* MR index

Mammoth coal bed, 263
Marcelina Formation, 179
Marcos Shale, 93
Markha River, 101
mass-transfer processes, 91, 93–95
Matuyama-Brunhes paleomagnetic reversal, 173
maximum safe temperature rise, 35
MEA-1A retardant, 42
melted-vitrified scoriaceous rocks, 118. *See also* paralavas
melting, combustion metamorphism and, 106, 134
Menuha Formation, 178
metamorphism. *See* combustion metamorphism
methane
    Carbondale mine fire and, 6, 8
    coal mine fires and, 1
    Large mine fire and, 10
    remote-sensing for estimation of, 240, 246–247
    subsurface temperature and, 12–13
method of point-source inversion (MPSI), 250, 253–258
microfractures, 39
micromagnetic profiles, 186–187, 191, 193–194
microseismicity, 208
microtremors, 199
Middelburg Colliery, 33, 39, 41
millosevichite, 94
mine environmental indexes, 40
minerals. *See* gas-vent minerals
Mishash Formation, 135, 138, 177, 178
modeling, 211, 213–217, 250–253
Moderate Resolution Imaging Spectroradiometer (MODIS), 229–237
MODIS. *See* Moderate Resolution Imaging Spectroradiometer
moisture, spontaneous combustion and, 51, 55–58, 75
Mongolia, 100. *See also* Wuda Coalfield
Montana, 166–172
Monterey Formation, 134
Mottled Zone complex (Haturim Formation), 99, 105–110, 134, 137, 177–179
Mount Carmel, 263
MPSI. *See* method of point-source inversion
MR index, 65
mud-volcano provinces, 133, 151–152
Musto, Ray, 272

N
nagelschmidtite, 135
natural coal fires, defined, 38
natural remanent magnetization (NRM), 178–179
Nelligan, James, 272, 277
new standards, 50
New Vaal Colliery, 86
New Zealand. *See* Rotowaro Coalfield
Ningxia Hui autonomous region, 24, 220, 244
nitrogen, 98
nitrogen oxides, 240, 246–247
Nizhnyaya Tunguska, 112
noncatenary effects, 57
No. 6 fuel oil, 59
Northern Coalfield, 41, 51
Norway, 246
NRM. *See* natural remanent magnetization

O

oil fields, 192–194
oil shale, 98
olive unit of Hatrurim Basin, 133, 137–138, 141, 150–152
orthoferrosilite, 117, 126–129
orthorhombic sulfur, 92
osteosclerosis, 96
oven-heating test, 66–68
ovoids of paralavas, 142–146, 147
oxidation
    chemistry of low-temperature, 36–37
    combustion metamorphism and, 98, 118
    moisture content, self-heating and, 56
    particle size and surface area and, 53
    spontaneous combustion and, 25, 28, 33, 241
    stockpiles and, 45
    See also pre-oxidation
oxygen
    fire penetration distance and, 158, 201
    greenhouse gas production and, 1, 3–5, 13
    scavenging of, burn prevention and, 85, 88–90
    spontaneous combustion and, 33, 38–39, 54–55

P

paleomagnetic orientation of clinker, 162
paralavas
    formation of, 134
    of Hatrurim Formation, 137–138, 140, 142–146, 148–150
    paleomagnetic orientation of clinker and, 162
    of Rotowaro Coalfield, 117, 124, 126–130
particle size and surface area, 52–53
Pennsylvania
    coal fires in, 246
    control and suppression of fires in, 41–42
    gas-liquid–precipitation processes in, 94–95
    policy and, 271–278
    See also Centralia fire; Laurel Run fire
Percy mine fire, 8–12
periclase, 106
peroxides, 36–37
petrographic composition, 40
plagioclases, 117, 141–142
platinum resistance thermocouples, 16, 18
policy
    Centralia and Laurel Run fires and, 261, 262, 265–266, 268
    congressional agenda setting and, 274–277
    Eleventh Congressional District of Pennsylvania and, 272–274
    lessons learned and, 277–278
    overview of, 271–272
policy stream, 275–276
political stream, 275
polynuclear aromatic hydrocarbons, 59
porcellanites, 117, 120, 124, 158–160
porosity, stockpiles and, 45
Pottsville Formation, 263
Powder River Basin
    coal bed fires in, 158
    combustion metamorphism and, 101, 103
    methods for dating clinker in, 161–162
    natural coal fires in, 38
    overview of, 156–158
    overview of clinker in, 162–173
pre-oxidation, 37–38, 45–46
problem stream, 275
progressive stages and products of coalfield fires, 23, 27–29
promoters, 52
propagation of coal fires, 23, 26–27, 29
pseudo-igneous rocks, 118. *See also* paralavas
pseudosection work, 206
pseudowollastonite, 135
pyrite, 55
pyrolysis, 98
pyrometamorphism. *See* combustion metamorphism

Q

Qinling Mountains, 24
Queensland, University of, 61–62, 76

R

R_{70} self-heating rate index, 62–63
radar, 205
radioactive methods. *See* U-Th/He dating of zircons
Ramlu index. *See* MR index
Raniganj coal belt, 244
rankings, 54, 59
Raster maps, 20–21
Ravat coal-bed fire, 103, 105, 111–112
Rayleigh number, 89
reaction rates, 53
reactive surface layers, 85–89
Reagan, Ronald, 272
Reconofax thermal scanner, 246
Red Ash coal bed, 261, 266, 269
Redstone River, 156
relocation funds, 262, 271–278
remote-sensing methods
    coal-mining areas in China and, 219–227
    detection of fires using, 43
    estimation of emissions using, 239, 246–247
    methods for, 242–243
    MODIS data and, 229–237
Renown Seam, 119
resistivity method, 205–206, 243
retrograde rocks, 141
risk, 20–21
risk index classification, 40, 62
Rochelle Hills, 155–156, 159, 163–166, 169–170
room-and-pillar mines, 2, 211, 212–213
Rosebud Creek, 156
Rosebud-Robinson zone, 161
Ross coal bed, 261, 266, 267, 269
Rotowaro Coalfield
    combustion metamorphism in, 120–121
    feldspars, silica minerals, opaques, glass of, 127
    hematite and Ti-spinel assemblages of, 126
    iron silicates of, 127
    location of, 118
    magnetite-hematite-spinel assemblages of, 122–126
    overview of, 117, 119–120
    oxidation of magnetite to hematite in, 126
    paralavas of, 117, 126–130
    slags of, 117, 122–126, 127–130
runaway ignition, 33
Ruqigou-Gulaben coal basin, 32, 219–227, 229–237
Russia, 245

S

Saharonim Formation, 183–186
salammoniac, 92, 95
sanidinite facies metamorphism, 107
SARA. *See* Superfund Amendments and Reauthorization Act
scoria, 118. *See also* paralavas
seam gases, 38
sedimentary basins, 108–112
seismo-acoustic investigations, 206
self-heating
    coal properties affecting, 54–58
    hazard modeling during, 39, 41
    overview of, 31–32, 33
    in stockpiles, 44–47
    in waste heaps, 47
    See also time to ignition
self-heating experiments, 2–4, 31
Seven Foot coal bed, 263
sewage, 42
Shangshihezi Formation, 24
Shanxi Formation, 24
shipping. *See* transportation of coal
Siberia, 101, 110–112
siderite, 124
silicate, 127
simulations of coal fires, 211, 213–217
Skidmore coal bed, 263
slags, 117, 120–130
slurry/ash injection, 42
SMCRA. *See* Surface Mining Control and Reclamation Act
sodium acetate, 52
South Africa, 246. *See also* New Vaal Colliery; Witbank Coalfield
Southland Colliery, 39
South Tisra Colliery, 213
Soviet Union, 245
Specter, Arlen, 272, 277
spinel, 122–126
spoil heaps, 47–48
spontaneous combustion
    abandoned mines and, 38–39
    adiabatic methods and, 61–64
    of coal dust, 43–44
    in coal mines, 39–43
    crossing-point temperature method and, 64–66
    effect of intrinsic coal properties on, 50–55
    effect of pre-oxidation (aging) on, 37–38
    experimental procedures for assessment of, 59–61
    Frank-Kamenetskii analysis and, 34–36
    heat-release-rate method and, 34, 68–70
    hot storage test and, 66–68
    inhibition of, 38
    low-rank coals and, 58–59
    low-temperature oxidation chemistry and, 36–37
    modeling of in Wuda Coalfield, 23–29
    moisture, humidity and, 55–58
    natural coal fires and, 38
    nomenclature overview for, 77–78
    overview of, 31–32, 33, 75–76, 241–242

parameter values for, 78
in Powder River Basin, 158
reactive surface layers for prevention of, 85–89
in stockpiles, 44–47
transient method and, 70–75
transportation and, 48–50
in waste heaps, 47–49
See also self-heating; Wits-Ehac Index
"spontaneous combustion ladder," 41
spraying, waste heap fires and, 48
stages model, 23, 27–29
stalactites, 117, 123
Stanton-Empire mine, 267
steady-state approach, 66–68
stockpiles, 33, 44–47
sub-bituminous coals, 58–59
subcritical systems, 35, 36, 49–50
sublimation, 92–93
subsidence
    Centralia fire and, 262, 264, 265, 266
    clinker hillsides and, 160
    effects on, 32
    Laurel Run fire and, 266
    oxygen supply, spontaneous combustion and, 38–39
    as problem, 240
Sudan, 179
Suggate ranking, 63
Suhaitu mine, 24, 29, 256
sulfur content, 25–26, 38, 46–47, 98
sulfuric efflorescence, 201
supercooling, 92
supercritical systems, 35, 36, 49–50
Superfund Amendments and Reauthorization Act (SARA), 276
Surface Mining Control and Reclamation Act (SMCRA), 276
surface sealing, 42, 48
Svea Nord mine, 246

T

Taimur River, 104–105
Taiyuan Formation, 24
Tajikistan, 100, 111
Taqiye Formation, 135, 137, 150, 179
Taupiri coal seams, 119
Tauranga Group, 119
Te Kuiti Group, 119
temperature, 1, 12–13, 40
temperature-programmed reaction technique (TGA), 58–59
templates, policy and, 276–277
Tengiz deposit, 134
terrain, spontaneous combustion and, 25
TG. See thermal gravimetry

TGA. See temperature-programmed reaction technique
thallium poisoning, 96
thermal anomalies (LAST) function
    depth estimation and, 211, 213
    detection of concealed coal fires and, 201, 203, 242–243
    MODIS analysis of, 229–237
    surface temperatures and, 250
    See also linear anomaly surface transect function
thermal-geological models, 249–258
thermal gravimetry (TG), 60
thermal runaway, defined, 33
thermocouples, 61, 71–74
thickness, 25, 38
Thornburgh, Richard, 272
Three Mile Island, 274, 277
threshold temperature, 241
Tien Shan, 111
time to ignition, 50
Times Beach, Missouri, 277
Ti-spinel, 126
titaniferous magnetite, 127–129
TM. See transient method
tomography, 206, 207
Tongue River Valley, 155–156, 159, 161, 164–172
topsoil, effects on, 32
total temperature rise (TTR), 40, 61–62
transient method (TM), 70–75. See also crossing-point temperature (CPT) method
transportation of coal, 48–50
trench cutting, 42
tridymite, 117, 126, 127
TTR. See total temperature rise
Tungus Coal Basin, 112

U

UBC process. See upgraded brown coal process
Ukraine, 246
Ulanbuhe Desert, 24
Ulm coal zone, 161
United Nations, 50
United States, 245
University of Queensland, 61, 62, 76
upgraded brown coal (UBC) process, 58–59
upgrading, 57–58
Utah, 93
U-Th/He dating of zircons, 155, 162–163, 166–168, 170, 243
Uzbekistan, 100, 111–112

V

vegetation, 32, 267
Venezuela, 179, 245
ventilation, 42, 45

vents. See gas-vent minerals
voidage, 45
volatile content, 51, 56
voltaite, 92

W

Waikito coal region, 119
Wasatch Formation, 156, 161
waste heaps, 47–48
water vapor production, 57
Waterburg coalfield, 21
welding as ignition source, 39
Western Middle Field, 261, 262–263
Wilkes-Barre synclinorium, 266
wind pipes, 48
wire-mesh basket test, 66–68
Witbank Coalfield, 21, 33, 39
Wits-Ehac Index, 15–21, 65
Wuda Coalfield
    economic losses in, 32
    gas-vent minerals and, 92, 94
    modeling spontaneous combustion in, 23–29
    MODIS data and, 229–237
    overview of, 245
    remote-sensing data for, 219–227
    thermal-geological analysis and, 255–258
Wuda Mining Limited Liability Company, 24
Wuhai City, 24
Wuhushan mine, 24, 256
Wyodak-Anderson coal zone, 155–156, 159, 161, 163–164
Wyoming, 163–169

X

Xiashihezi Formation, 24
Xinjiang autonomous region, 24

Y

Yagnob River, 111
Yellow River, 24, 223, 226
Yellowstone caldera, 173
Youngstown mine, 8. See also Percy mine fire
Youth (Conrad), 49

Z

Zaisan Basin, 110
Zarnista, 194
zeolite, 146–149
ZFT. See zircon fission-track analysis
zircon fission-track (ZFT) analysis, 155, 161–165, 170
Zohar Formation, 180–181, 192
zones conducive to spontaneous combustion, 23, 28–29
Zoroastrim Fire Temple, 179